INDUSTRIAL WATER POLLUTION CONTROL

THIRD EDITION

McGraw-Hill Series in Water Resources and Environmental Engineering

Bailey and Ollis: *Biochemical Engineering Fundamentals*

Bishop: *Pollution and Prevention: Fundamentals and Practice*

Bouwer: *Groundwater Hydrology*

Canter: *Environmental Impact Assessment*

Chanlett: *Environmental Protection*

Chapra: *Surface Water-Quality Modeling*

Chow, Maidment, and Mays: *Applied Hydrology*

Crites and Tchobanoglous: *Small and Decentralized Wastewater Management Systems*

Davis and Cornwell: *Introduction to Environmental Engineering*

deNevers: *Air Pollution Control Engineering*

Eckenfelder: *Industrial Water Pollution Control*

Eweis, Ergas, Chang, and Schroeder: *Bioremediation Principles*

LaGrega, Buckingham, and Evans: *Hazardous Waste Management*

Linsley, Franzini, Freyberg, and Tchobanoglous: *Water Resources and Engineering*

McGhee: *Water Supply and Sewage*

Mays and Tung: *Hydrosystems Engineering and Management*

Metcalf & Eddy, Inc.: *Wastewater Engineering: Collection and Pumping of Wastewater*

Metcalf & Eddy, Inc.: *Wastewater Engineering: Treatment, Disposal, Reuse*

Peavy, Rowe, and Tchobanoglous: *Environmental Engineering*

Sawyer, McCarty, and Parkin: Chemistry for *Environmental Engineering*

Tchobanoglous, Theisen, and Vigil: *Integrated Solid Waste Management: Engineering Principles and Management Issues*

Wentz: *Hazardous Waste Management*

Wentz: *Safety, Health, and Environmental Protection*

INDUSTRIAL WATER POLLUTION CONTROL

THIRD EDITION

W. Wesley Eckenfelder, Jr.

Senior Technical Director

Eckenfelder/Brown and Caldwell

Distinguished Professor Emeritus

Vanderbilt University

Boston   Burr Ridge, IL   Dubuque, IA   Madison, WI   New York   San Francisco   St. Louis
Bangkok   Bogotá   Caracas   Lisbon   London   Madrid
Mexico City   Milan   New Delhi   Seoul   Singapore   Sydney   Taipei   Toronto

McGraw-Hill Higher Education

A Division of The **McGraw-Hill** Companies

INDUSTRIAL WATER POLLUTION CONTROL

Copyright © 2000, 1989, 1967 by The McGraw-Hill Companies, Inc. All rights reserved. Printed in the United States of America. Except as permitted under the United States Copyright Act of 1976, no part of this publication may be reproduced or distributed in any form or by any means, or stored in a database or retrieval system, without the prior written permission of the publisher.

This book is printed on acid-free paper.

2 3 4 5 6 7 8 9 0 DOC/DOC   0 9 8 7 6 5 4 3 2 1 0

ISBN 0-07-039364-8

Publisher: *Thomas Casson*
Executive editor: *Eric M. Munson*
Editorial coordinator: *Michael Jones*
Senior marketing manager: *John T. Wannemacher*
Project manager: *Susanne Riedell*
Production supervisor: *Debra R. Benson*
Freelance design coordinator: *Pam Verros*
Photo researcher: *Judy Kausal*
Supplement coordinator: *Craig S. Leonard*
Compositor: *Lachina Publishing Services*
Typeface: *10.5/12 Times Roman*
Printer: *R. R. Donnelley & Sons Company*

Library of Congress Cataloging-in-Publication Data

Eckenfelder, W. Wesley (William Wesley), 1926–
    Industrial water pollution control / W. Wesley Eckenfelder, Jr.--
3rd ed.
        p.   cm. -- (McGraw-Hill series in water resources and
environmental engineering)
    Includes index.
    ISBN 0-07-039364-8
    1. Sewage--Purification.   2. Factory and trade waste-
-Purification.   I. Title.   II. Series.
TD745.E23   2000
628.3--dc21   99-36138

http://www.mhhe.com

ABOUT THE AUTHOR

W. WESLEY ECKENFELDER, JR., received degrees in civil engineering from Manhattan College and environmental engineering from Pennsylvania State University and New York University. In 1990 he received an honorary D.Sc. degree from Manhattan College. He was associate professor at Manhattan College, professor at the University of Texas, and is presently Emeritus Distinguished Professor of Environmental Engineering at Vanderbilt University. He was a founder of the consulting firm of Hydroscience, Inc., a partner in Weston Eckenfelder & Associates (now Roy F. Weston, Inc.) and has served as board chairman of AWARE Incorporated and Eckenfelder, Inc.

He has developed continuing education courses in water pollution control for Manhattan College, the University of Texas, Vanderbilt University, the University of Queensland, Australia, the American Institute of Chemical Engineers, and the Chemical Manufacturers Association. He has organized courses and workshops in 17 countries in Asia, Europe, Africa and Latin America.

He has served as consultant to over 150 industries, municipalities, consulting firms, and governmental agencies.

He is an author or editor of 31 books and over 200 technical papers in water pollution control including *Industrial Water Pollution Control,* McGraw-Hill, (1967 and 1989); *Principles of Water Quality Management,* CBI Publishing (1980); and *Water Pollution Control,* Jenkins Publishing Co. (1978). These books have been translated into Japanese and Chinese and one into French and Italian. Professor Eckenfelder is on numerous editorial boards including *Water Technology Letters* (U.K.) and *Hazardous Waste and Hazardous Materials.*

He is a member of numerous technical societies including the Water Pollution Control Federation (honorary member), the American Institute of Chemical Engineers, and the American Society of Civil Engineers. He is an honorary member of the International Association on Water Quality.

In 1974 he was awarded the Synthetic Organic Chemicals Manufacturers Association gold medal for excellence in environmental chemistry, in 1957 the Rudolfs Medal, and in 1981 the Thomas Camp Medal from the Water Pollution Control Federation. In 1998 he received the Lawrence Cecil Award from the American Institute of Chemical Engineers. In 1999 he received the Gordon Maskew Fair Award from the American Academy of Environmental Engineers.

He is a registered Professional Engineer in the State of Texas.

CONTENTS

PREFACE

It has been 30 years since this book was first published and 10 years since the second edition. During that period not only have regulations undergone a vast change, but conventional technologies have been further refined and new technologies have been developed to meet increasingly more stringent water quality criteria. Effluent limitation on specific priority pollutants and toxicity to aquatic organisms have rendered many of the older conventional treatment facilities obsolete. The challenge today is to meet these new requirements in a way that is both environmentally acceptable and cost-effective.

In order to address these new challenges, the present volume reviews the existing theory and addresses the application of state-of-the-art technology to the solution of today's problems in industrial water pollution control.

Of necessity, this book does not develop the detailed principles or the theory of processes applicable to specific areas of water pollution control. Rather it stresses the application of these theories to specific industrial problems. Publications and texts are referenced in the bibliography for the reader who wishes a more detailed development of the theory.

Where applicable, case histories are used to illustrate the application of technology to specific industrial applications. Problems are drawn from field experience.

This book is intended as a text for the student in courses related to industrial water pollution control and as a guide for the engineer in industry, governmental agencies, and consulting engineering firms involved in developing state-of-the-art solutions to industrial water pollution control problems.

While all specific problems cannot be answered in one text, it is hoped that this volume will provide guidance and direction to those faced with the increasingly more complex solutions of water pollution control.

Special thanks to Dr. Alan Bowers of Vanderbilt University for his contribution to the chapter on chemical oxidation.

I would like to express my thanks for many useful comments and suggestions provided by colleagues who reviewed this text during the course of its development, especially to Robert W. Okey, the University of Utah; Bruce DeVantier, Southern Illinois University at Carbondale; Peter Fox, Arizona State University; Clifford W. Randall, Virginia Polytechnic University; and Paul M. Berthouex, University of Wisconsin–Madison.

W. Wesley Eckenfelder

1

SOURCE AND CHARACTERISTICS OF INDUSTRIAL WASTEWATERS

1.1
UNDESIRABLE WASTEWATER CHARACTERISTICS

Depending on the nature of the industry and the projected uses of the waters of the receiving stream, various waste constituents may have to be removed before discharge. These may be summarized as follows:

1. Soluble organics causing depletion of dissolved oxygen. Since most receiving waters require maintenance of minimum dissolved oxygen, the quantity of soluble organics is correspondingly restricted to the capacity of the receiving waters for assimilation or by specified effluent limitations.
2. Suspended solids. Deposition of solids in quiescent stretches of a stream will impair the normal aquatic life of the stream. Sludge blankets containing organic solids will undergo progressive decomposition resulting in oxygen depletion and the production of noxious gases.
3. Priority pollutants such as phenol and other organics discharged in industrial wastes will cause tastes and odors in the water and in some cases are carcenogenic. If these contaminants are not removed before discharge, additional water treatment will be required.
4. Heavy metals, cyanide, and toxic organics. The EPA has defined a list of toxic organic and inorganic chemicals that now appear as specific limitations in most permits. The identified priority pollutants are listed in Table 1.1.
5. Color and turbidity. These present aesthetic problems even though they may not be particularly deleterious for most water uses. In some industries, such as pulp and paper, economic methods are not presently available for color removal.
6. Nitrogen and phosphorus. When effluents are discharged to lakes, ponds, and other recreational areas, the presence of nitrogen and phosphorus is particularly undesirable since it enhances eutrophication and stimulates undesirable algae growth.

1

TABLE 1.1
EPA list of organic priority pollutants

Compound name	Compound name
1. Acenaphthene[†]	Dichlorobenzidine[†]
2. Acrolein[†]	28. 3,3′-Dichlorobenzidine
3. Acrylonitrile[†]	
4. Benzene[†]	Dichloroethylenes[†] (1,1-dichloroethylene
5. Benzidine[†]	and 1,2-dichloroethylene)
6. Carbon tetrachloride[†]	29. 1,1-Dichloroethylene
(tetrachloromethane)	30. 1,2-*trans*-Dichloroethylene
	31. 2,4-Dichlorophenol[†]
Chlorinated benzenes (other than	
dichlorobenzenes)	Dichloropropane and dichloropropene[†]
7. Chlorobenzene	32. 1,2-Dichloropropane
8. 1,2,4-Trichlorobenzene	33. 1,2-Dichloropropylene (1,2-
9. Hexachlorobenzene	dichloropropene)
Chlorinated ethanes[†] (including 1,2-	34. 2,4-Dimethylphenol[†]
dichloroethane, 1,1,1-trichloroethane,	
and hexachloroethane)	Dinitrotoluene[†]
10. 1,2-Dichloroethane	35. 2.4-Dinitrotoluene
11. 1,1,1-Trichloroethane	36. 2,6-Dinitrotoluene
12. Hexachloroethane	37. 1,2-Diphenylhydrazine[†]
13. 1,1-Dichloroethane	38. Ethylbenzene[†]
14. 1,1,2-Trichloroethane	39. Fluoranthene[†]
15. 1, 1,2,2-Tetrachloroethane	
16. Chloroethane (ethyl chloride)	Haloethers[†] (other than those listed
	elsewhere)
Chloroalkyl ethers[†] (chloromethyl,	40. 4-Chlorophenyl phenyl ether
chloroethyl, and mixed ethers)	41. 4-Bromophenyl phenyl ether
17. Bis(chloromethyl) ether	42. Bis(2-chloroisopropyl) ether
18. Bis(2-chloroethyl) ether	43. Bis(2-chloroethoxy) methane
19. 2-Chloroethyl vinyl ether (mixed)	
	Halomethanes[†] (other than those listed
Chlorinated napthalene[†]	elsewhere)
20. 2-Chloronapthalene	44. Methylene chloride (dichloromethane)
	45. Methyl chloride (chloromethane)
Chlorinated phenols[†] (other than those	46. Methyl bromide (bromomethane)
listed elsewhere; includes	47. Bromoform (tribromomethane)
trichlorophenols and chlorinated cresols)	48. Dichlorobromomethane
21. 2,4,6-Trichlorophenol	49. Trichlorofluoromethane
22. *para*-Chloro-*meta*-cresol	50. Dichlorodifluoromethane
23. Chloroform (trichloromethane)[†]	51. Chlorodibromomethane
24. 2-Chlorophenol[†]	52. Hexachlorobutadiene[†]
	53. Hexachlorocyclopentadiene[†]
Dichlorobenzenes[†]	54. Isophorone[†]
25. 1,2-Dichlorobenzene	55. Naphthalene[†]
26. 1,3-Dichlorobenzene	56. Nitrobenzene[†]
27. 1,4-Dichlorobenzene	

TABLE 1.1 *(continued)*

Compound name	Compound name
Nitrophenols[†] (including 2,4-dinitrophenol and dinitrocresol)	87. Trichloroethylene[†]
	88. Vinyl chloride[†] (chloroethylene)
57. 2-Nitrophenol	
58. 4-Nitrophenol	Pesticides and metabolites
59. 2,4-Dinitrophenol[†]	89. Aldrin[†]
60. 4,6-Dinitro-*o*-cresol	90. Dieldrin[†]
	91. Chlordane[†] (technical mixture and metabolites)
Nitrosamines[†]	
61. *N*-Nitrosodimethylamine	
62. *N*-Nitrosodiphenylimine	DDT and metabolites[†]
63. *N*-Nitrosodi-*n*-propylamine	92. 4,4′-DDT
64. Pentachlorophenol[†]	93. 4,4′-DDE (*p,p*′-DDX)
65. Phenol[†]	94. 4,4′-DDD (p,p′-TDE)
Phthalate esters[†]	Endosulfan and metabolites[†]
66. Bis(2-ethylhexyl) phthalate	95. α-Endosulfan-alpha
67. Butyl benzyl phthalate	96. β-Endosulfan-beta
68. Di-*n*-butyl phthalate	97. Endosulfan sulfate
69. Di-*n*-octyl phthalate	
70. Diethyl phthalate	Endrin and metabolites[†]
71. Dimethyl phthalate	98. Endrin
	99. Endrin aldehyde
Polynuclear aromatic hydrocarbons (PAH)[†]	Heptachlor and metabolites[†]
72. Benzo(a)anthracene (1,2-benzanthracene)	100. Heptachlor
	101. Heptachlor epoxide
73. Benzo(a)pyrene (3,4-benzopyrene)	
74. 3,4-Benzofluoranthene	Hexachlorocyclohexane (all isomers)[†]
75. Benzo(k)fluoranthene (11,12-benzofluoranthene)	102. α-BHC-alpha
	103. β-BHC-beta
76. Chrysene	104. γ-BHC (lindane)-gamma
77. Acenaphthylene	105. δ-BHC-delta
78. Anthracene	
79. Benzo(ghi)perylene (1,12-benzoperylene)	Polychlorinated biphenyls (PCB)[†]
80. Fluorene	106. PCB-1242 (Arochlor 1242)
81. Phenanthrene	107. PCB-1254 (Arochlor 1254)
82. Dibenzo(a,h)anthracene (1,2,5,6-dibenzanthracene)	108. PCB-1221 (Arochlor 1221)
	109. PCB-1232 (Arochlor 1232)
83. Indeno (1,2,3-cd) pyrene (2,3-*o*-phenylenepyrene)	110. PCB-1248 (Arochlor 1248)
	111. PCB-1260 (Arochlor 1260)
84. Pyrene	112. PCB-1016 (Arochlor 1016)
85. Tetrachloroethylene[†]	113. Toxaphene[†]
86. Toluene[†]	114. 2,3,7,8-Tetrachlorodibenzo-*p*-dioxin (TCDD)[†]

[†]Specific compounds and chemical classes as listed in the consent degree.

7. Refractory substances resistant to biodegradation. These may be undesirable for certain water-quality requirements. Refractory nitrogen compounds are found in the textile industry. Some refractory organics are toxic to aquatic life.
8. Oil and floating material. These produce unsightly conditions and in most cases are restricted by regulations.
9. Volatile materials. Hydrogen sulfide and volatile organics will create air-pollution problems and are usually restricted by regulation.
10. Aquatic toxicity. Substances present in the effluent that are toxic to aquatic species and are restricted by regulation.

1.2
PARTIAL LIST OF REGULATIONS WHICH AFFECT WASTEWATER TREATMENT REQUIREMENTS WITHIN THE UNITED STATES

It is not the intent of this book to discuss federal and state regulations, but a brief summary of present regulatory requirements relative to industrial water pollution control will serve as guidance to the reader. Details of these regulations can be found in the cited Code of the Federal Register (CFR) as noted.

Air

NESHAP

• Regulates carcinogenic VOCs to mass loading and concentration limits. For example, in the case of benzene anything in excess of 10 mg/l concentration in the wastewater or 10 Mg/yr. Requires off-gas capture and treatment until such limits are achieved.

National Emission Standards for Hazardous Air Pollutants (40 CFR, part 61)

• Regulates 60 VOCs to mass loading and concentration limits. Requires off-gas capture and treatment until a required percent removal is achieved.

Occupational Safety and Health Administration Standards

• Regulates hydrogen sulfide and contaminants which pose exposure risks.

Liquid

Federal Industry Point Source Category Limits (40 CFR, part 405-471)

• Mass-based for raw material processing, e.g., pulp and paper, and concentration-based for synthetic chemicals and pharmaceuticals for conventional pollutants.
• Concentration-based limits for nonconventional pollutants (metals and priority pollutants).

Regional Initiatives (e.g., Great Lakes Initiative)

• For example, concentration-based limits for total phosphorus.

State Water Quality Standards

- Limits for pollutants based on design receiving stream low flow (i.e., 7–Q10, the average 7-day low flow every ten years) for the use classification.

Local Pretreatment Limits (USEPA, PB92-129188, December 1987)

- Those regulated under point source categories, plus those required to ensure POTW (publicly owned treatment works) effluent compliance.

1.3
SOURCES AND CHARACTERISTICS OF WASTEWATERS

The volume and strength of industrial wastewaters are usually defined in terms of units of production (e.g., gallons per ton[†] of pulp or cubic meters per tonne[‡] of pulp and pounds of BOD per ton of pulp or kilograms of BOD per tonne of pulp for a pulp-and-paper-mill waste) and the variation in characteristic by a statistical distribution. In any one plant there will be a statistical variation in wasteflow characteristic. The magnitude of this variation will depend on the diversity of products manufactured and of process operations contributing waste, and on whether the operations are batch or continuous. Good housekeeping procedures to minimize dumps and spills will reduce the statistical variation. Plots showing the variation in flow resulting from a sequence of batch processes are shown in Fig. 1.1. Variation in waste flow and characteristics within a single plant are shown in Fig. 1.2.

Wide variation in waste flow and characteristics will also appear among similar industries, e.g., the paperboard industry. This is a result of differences in housekeeping and water reuse as well as of variations in the production processes. Very few industries are identical in their sequence of process operations; as a result, an industrial waste survey is usually required to establish waste loadings and their variations. Variations for several industries are shown in Table 1.2. The variation in suspended solids and BOD (biochemical oxygen demand) discharge from 11 paperboard mills is shown in Fig. 1.3.

1.4
INDUSTRIAL WASTE SURVEY

The industrial waste survey involves a procedure designed to develop a flow-and-material balance of all processes using water and producing wastes and to establish the variation in waste characteristics from specific process operations as well as from the plant as a whole. The results of the survey should establish possibilities for water conservation and reuse and finally the variation in flow and strength to undergo wastewater treatment.

[†]ton = 2000 lb.
[‡]tonne = 1000 kg.

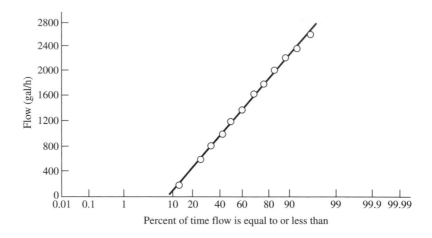

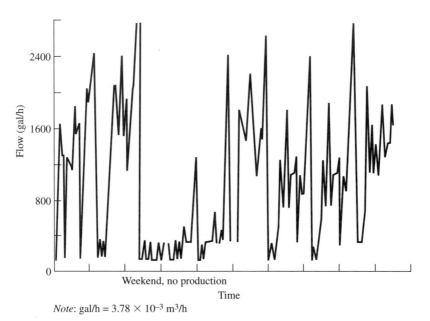

Note: gal/h = 3.78×10^{-3} m³/h

FIGURE 1.1
Variation in flow from a batch operation.

The selected method of flow measurement will usually be contingent on the physical location to be sampled. When the waste flows through a sewer, it is frequently possible to measure the velocity of flow and the depth of water in the sewer and calculate the flow from the continuity equation. Since $Q = AV$, the area in a partially filled circular sewer can be determined given the depth from Fig. 1.4. This method applies only for partially filled sewers of constant cross section. The average velocity of flow can be estimated as 0.8 of the surface velocity timed from a floating object between manholes. More accurate measurements can be obtained by the use of a current meter. In gutters or channels, either a small weir can be constructed or

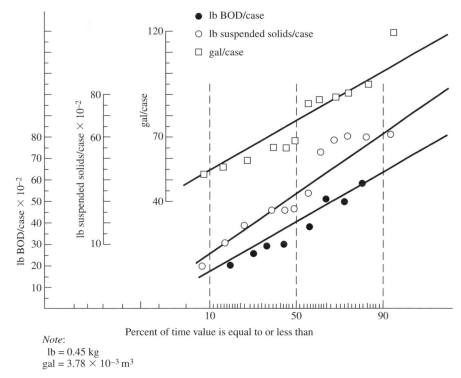

FIGURE 1.2
Daily variation in flow and characteristics: tomato waste.

Note:
lb = 0.45 kg
gal = 3.78 × 10⁻³ m³

Note:
$lb = 0.45 \text{ kg}$
$gal = 3.78 \times 10^{-3} \text{ m}^3$

TABLE 1.2
**Variation in flow and waste characteristics for
some representative industrial wastes**

Waste	Flow, gal/production unit % frequency			BOD, lb/production unit % frequency			Suspended solids, lb/production unit % frequency		
	10	50	90	10	50	90	10	50	90
Pulp and paper[†]	11,000	43,000	74,000	17.0	58.0	110.0	26.0	105.0	400.0
Paperboard[†]	7,500	11,000	27,500	10	28	46	25	48	66
Slaughterhouse[‡]	165	800	4,300	3.8	13.0	44	3.0	9.8	31.0
Brewery[§]	130	370	600	0.8	2.0	44	0.25	1.2	2.45
Tannery[¶]	4.2	9.0	13.6	575[††]	975	1400	600[††]	1900	3200

[†] Tons paper production.
[‡] 1000 lb live weight kill.
[§] bbl beer.
[¶] Pounds of hides; sulfides as S vary from 260 mg/l (10%) to 1230 mg/l (90%).
[††] As mg/l.

Note:
$gal = 3.78 \times 10^{-3} \text{ m}^3$
$lb = 0.45 \text{ kg}$
$ton = 907 \text{ kg}$
$bbl \text{ beer} = 0.164 \text{ m}^3$

7

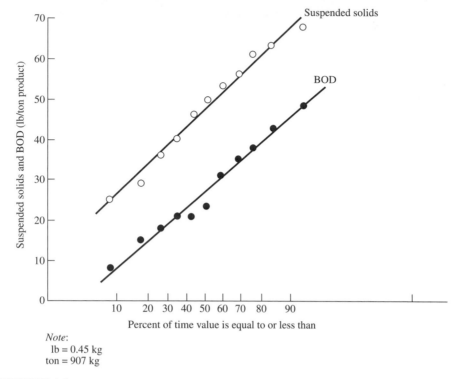

FIGURE 1.3
Variations in suspended solids and BOD from 11 paperboard mills.

the flow can be estimated as above by measuring the velocity and depth of flow in the channel. In some cases the flow can be obtained from the pumping rate and the duration of pumping of a wastestream. Total waste flow from an industrial plant can be measured by use of a weir or other suitable measuring device. In certain instances the daily waste flow can be estimated from water consumption records.

The general procedure to be followed in developing the necessary information with a minimum of effort can be summarized in four steps:

1. Develop a sewer map from consultation with the plant engineer and an inspection of the various process operations. This map should indicate possible sampling stations and a rough order of magnitude of the anticipated flow.
2. Establish sampling and analysis schedules. Continuous samples with composites weighted according to flow are the most desirable, but these either are not always possible or do not lend themselves to the physical sampling location. The period of sample composite and the frequency of sampling must be established according to the nature of the process being investigated. Some continuous processes can be sampled hourly and composited on an 8-, 12-, or even 24-h basis, but those that exhibit a high degree of fluctuation may require a 1- or 2-h composite and analysis. More frequent samples are rarely required, since most industrial waste-treatment processes have a degree of built-in equalization and storage capacity. Batch processes should be composited during the course of the batch dump.

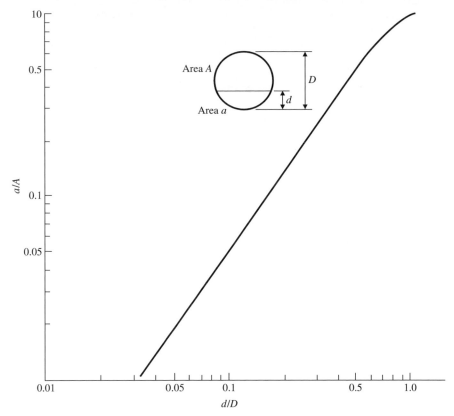

FIGURE 1.4
Determination of waste flow in partially filled sewers.

3. Develop a flow-and-material-balance diagram. After the survey data are collected and the samples analyzed, a flow-and-material balance diagram should be developed that considers all significant sources of waste discharge. How closely the summation of the individual sources checks the measured total effluent provides a check on the accuracy of the survey. A typical flow-and-material-balance diagram for a corn-processing plant is shown in Fig. 1.5.
4. Establish statistical variation in significant waste characteristics. As was previously shown, the variability of certain waste characteristics is significant for waste-treatment plant design. These data should be prepared as a probability plot showing frequency of occurrence.

The analyses to be run on the samples depend both on the characteristics of the samples and on the ultimate purpose of the analysis. For example, pH must be run on grab samples, since it is possible in some cases for compositing to result in neutralization of highly acidic and basic wastes, and this would yield highly misleading information for subsequent design. Variations in BOD loading may require 8-h or shorter composites for certain biological-treatment designs involving short detention periods, while 24-h composites will usually suffice for aerated lagoons with many days' retention under completely mixed conditions. Where constituents

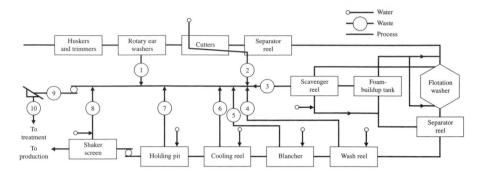

Line	1	2	3	4	5	6	7	8	9	10
from→	Washer	Cutters	Reel	Reel	Blanch	Cool	Hold	Shaker	Sewer	Screen
to→				Sewer					Screen	Treatment
Flow, g/min	21.7	27.0	10.4	18.0	4.5	24.5	16.9	2.1	125.1	121
BOD, lb/d	2,500	2,300	390	973	610	1,630	186		8,600	6,250
COD, lb/d	3,640	4,640	555	1,030	870	2,140	192		13,000	9,980
SS, lb/d	1,820	2,480	184	281	144	530	50		5,500	1,700
VSS, lb/d	1,740	2,360	95	91	92	266	38		4,700	1,900
Analysis:										
BOD, mg/l	9,830	7,112	3,130	4,600	11,300	5,630	918		5,730	6,200
COD, mg/l	14,000	14,400	4,450	4,780	16,100	7,280	950		3,670	6,030
SS, mg/l	6,950	7,660	1,460	1,300	2,670	1,830	250		3,670	1,170
VSS, mg/l	6,690	7,290	760	420	1,710	910	190		3,140	1,030

Note:
gal/min = 3.78×10^{-3} m³/min
lb/d   = 0.45 kg/d

FIGURE 1.5
Waste flow diagram and material balance at a corn plant.

such as nitrogen or phosphorus are to be measured to determine required nutrient addition for biological treatment, 24-h composites are sufficient since the biological system possesses a degree of buffer capacity. One exception is the presence of toxic discharges to a biological system. Since a one-shot dose of certain toxic materials can completely upset a biological-treatment process, continuous monitoring of such materials is required if they are known to exist. It is obvious that the presence of such materials would require separate consideration in the waste-treatment design. Other waste-treatment processes may require similar considerations in sampling schedules.

Data from industrial waste surveys are highly variable and are usually susceptible to statistical analysis. Statistical analysis of variable data provides the basis for process design. The data are reported in terms of frequency of occurrence of a particular characteristic, which is that value of the characteristic that may be expected to be equaled or not exceeded 10, 50, or 90 percent of the time. The 50 percent chance value is approximately equal to the median. Correlation in this manner will linearize variable data, as shown in Fig. 1.6. The probability of occurrence of any value, such as flow, BOD, or suspended solids, may be determined as shown in the probability plot. This can also be determined by a standard computer program.

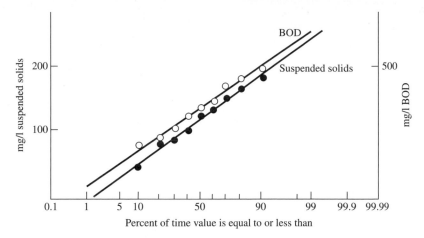

FIGURE 1.6
Probability of occurrence of BOD and suspended solids in raw waste.

The suspended solids and BOD values are each arranged in order of increasing magnitude. n is the total number of solids or BOD values and m is the assigned serial number from 1 to n. The plotting positions $m/(n + 1)$ are equivalent to the percent occurrence of the value. The actual values are then plotted against the percent occurrences on probability paper, as indicated in Fig. 1.6. A smooth curve of best fit may usually be drawn by eye or, if desirable, it may be calculated by standard statistical procedures. The probability of occurrence of any values can be obtained. The statistical calculation is shown in Example 1.1. In order to extrapolate the results of an industrial waste survey to future production, it is desirable to relate waste flow and loading to production schedules. Since some effluent-producing operations do not vary directly with production increase or decrease, the scale-up is not always linear. This is true of the cannery operation shown in Fig. 1.7, in which six process operations were independent of the number of washing and cleaning rigs in operation.

EXAMPLE 1.1. For small amounts of industrial waste survey data (i.e., less than 20 datum points), the statistical correlation procedure is as follows:

1. Arrange the data in increasing order of magnitude (first column of Table 1.3).
2. In the second column of Table 1.3, m is the assigned serial number from 1 to n where n is the total number of values.
3. The plotting position is determined by dividing the total number of samples into 100 and assigning the first value as one-half this number (third column of Table 1.3):

$$\text{Plotting position} = \frac{100}{n} + \text{previous probability}$$

For $m = 5$:

$$\text{Plotting position} = \frac{100}{9} + 38.85$$

$$= 49.95$$

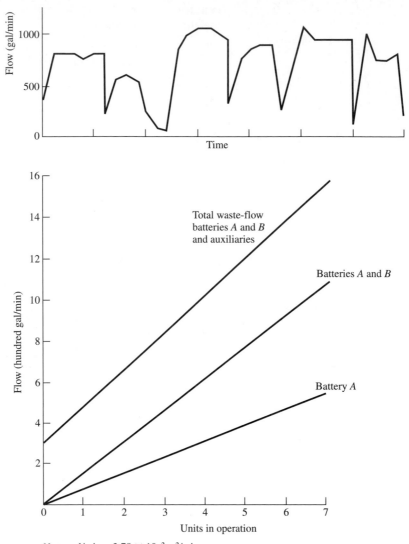

Note: gal/min = 3.78×10^{-3} m³/min

FIGURE 1.7
Variation in plant waste flow from unit operations.

4. These data are illustrated in Fig. 1.8. The standard deviation of these data (S) is calculated by

$$S = \frac{X_{84.1\%} - X_{15.9\%}}{2}$$

From Fig. 1.8:

$$S = \frac{436 - 254}{2}$$

$$= 91 \text{ mg/l}$$

TABLE 1.3
Statistical correlation of BOD data

BOD, mg/l	m	Plotting position
200	1	5.55
225	2	16.65
260	3	27.75
315	4	38.85
350	5	49.95
365	6	61.05
430	7	72.15
460	8	83.75
490	9	94.35

and mean:

$$\overline{X} = X_{50.0\%}$$

$$= 335 \text{ mg/l}$$

When large numbers of data are to be analyzed it is convenient to group the data for plotting, for example, 0 to 50, 51 to 100, 101 to 150, etc. The plotting position is determined as $m/(n + 1)$, where m is the cumulative number of points and n is the total number of observations. The statistical distribution of data serves several important functions in developing the industrial waste management program.

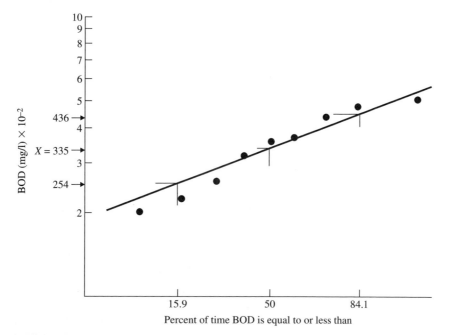

FIGURE 1.8

1.5
WASTE CHARACTERISTICS—
ESTIMATING THE ORGANIC CONTENT

Although the interpretation of most of the waste characteristics is straightforward and definitive, special consideration must be given to the organic content. The organic content of the waste can be estimated by the BOD, COD (chemical oxygen demand), TOC (total organic carbon), or TOD (total oxygen demand). Considerable caution should be exercised in interpreting these results:

1. The BOD_5 test measures the biodegradable organic carbon and, under certain conditions, the oxidizable nitrogen present in the waste. Usual practice today suppresses nitrification so that only carbonaceous oxidation is recorded as $CBOD_5$.
2. The COD test measures the total organic carbon with the exception of certain aromatics, such as benzene, which are not completely oxidized in the reaction. The COD test is an oxidation-reduction, so other reduced substances, such as sulfides, sulfites, and ferrous iron, will also be oxidized and reported as COD. NH_3^-N will not be oxidized in the COD test.
3. The TOC test measures all carbon as CO_2, and hence the inorganic carbon (CO_2, HCO_3^-, and so on) present in the wastewater must be removed prior to the analysis or corrected for in the calculation.
4. The TOD test measures organic carbon and unoxidized nitrogen and sulfur.

Remember to exercise considerable caution in interpreting the test results and in correlating the results of one test with another. Correlations between BOD and COD or TOC should usually be made of filtered samples (soluble organics) to avoid the disproportionate relationship of volatile suspended solids in the respective tests.

The BOD by definition is the quantity of oxygen required for the stabilization of the oxidizable organic matter present over 5 days of incubation at 20°C. The BOD is conventionally formulated as a first-order reaction:

$$\frac{dL}{dt} = -kL \tag{1.1}$$

which integrates to

$$L = L_o e^{-kt} \tag{1.2}$$

Since L, the amount of oxygen demand remaining at any time, is not known, Eq. (1.2) is re-expressed as

$$y = L_o(1 - e^{-kt})$$

where y is the amount of BOD exerted at time t:

$$y = L_o - L$$

or

$$y = L_o(1 - 10^{-kt}) \tag{1.3}$$

By definition, L_o is the oxygen required to stabilize the total quantity of biologically oxidizable organic matter present; if k is known, L_5 is a fixed percentage of

L_o. In order to interpret the BOD$_5$ obtained on industrial wastes, certain important factors must be considered.

It must be recognized that the oxygen consumed in the BOD test is the sum of (1) oxygen used for synthesis of new microbial cells using the organic matter present and (2) endogenous respiration of the microbial cells as shown in Fig. 1.9. (The basis for this is discussed in detail on p. 201). The rate of oxygen utilization during phase 1 is 10 to 20 times that during phase 2. In most readily degradable substrates, phase 1 is complete in 24 to 36 hours.

In wastes containing readily oxidizable substrates, e.g., sugars, there will be a high oxygen demand for the first day, as the substrate is rapidly utilized, followed by a slower endogenous rate over the subsequent days of incubation. When these data are fitted to a first-order curve over a 5-day period, a high k value will result because of the high initial slope of the curve. Conversely, a well-oxidized effluent will contain very little available substrate, and for the most part only endogenous respiration will occur over the 5-day incubation period. Since this rate of oxygen utilization is only a fraction of the rate obtained in the presence of available substrate, the resulting k rate win be correspondingly lower. Schroepfer[1] showed this by comparing the

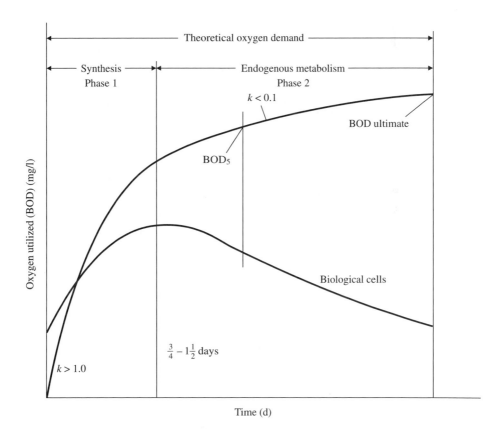

FIGURE 1.9
Reactions occurring in the BOD bottle.

k_{10} rates of a well-treated sewage effluent and raw sewage containing a large quantity of available substrate. The average rate was 0.17 per day for the sewage and 0.10 per day for the effluent. It is obvious that under these conditions a direct comparison of 5-day BODs is not valid. Typical rate constants are shown in Table 1.4.

Many industrial wastes are difficult to oxidize; they require a bacterial seed acclimated to the specific waste, or a lag period may occur which yields an erroneous interpretation of the 5-day BOD values. Stack[2] showed that the 5-day BOD of synthetic organic chemicals varied markedly depending on the acclimation of the seed used. Some typical BOD curves are shown in Fig. 1.10. Curve A is normal exertion of BOD. Curve B is representative of what might be expected from sewage which slowly acclimated to the waste. Curves C and D are characteristic of nonacclimated seed or a toxic wastewater. Acclimation of microorganisms to organics is shown in Table 1.5 and discussed on p. 203.

In the pulp-and-paper industry, both 1-day and 5-day BODs have been run. The results are shown in Fig. 1.11. Note that in the untreated wastewater, 70 percent of

TABLE 1.4
Average BOD rate constants at 20°C

Substance	k_{10}, day^{-1}
Untreated wastewater	0.15–0.28
High-rate filters and anaerobic contact	0.12–0.22
High-degree biotreatment effluent	0.06–0.10
Rivers with low pollution	0.04–0.08

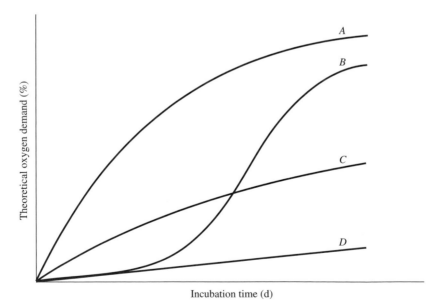

FIGURE 1.10
Characteristic BOD curves.

the 5-day BOD is exerted in 1 day, while for the effluent, only 50 percent is exerted. Referring to the discussion on p. 15, it can be interpreted that the 1-day BOD reflects the soluble organics remaining in the sample. In many cases, the 1-day BOD may provide a good control test for treatment plant performance.

TABLE 1.5
Effect of structural characteristics on bio-acclimation

1. Nontoxic aliphatic compounds containing carboxyl, ester, or hydroxyl groups readily acclimate (<4 days acclimation).

2. Toxic compounds with carbonyl groups or double bonds, 7–10 days acclimation; toxic to unacclimated acetate cultures.

3. Amino functional groups difficult to acclimate and slow degradation.

4. Seeds for dicarboxylic groups longer to acclimate compared to one for carboxylic group.

5. Position of functional group affects lag period for acclimation.

Primary butanol	4 days
Secondary butanol	14 days
Tertiary butanol	Not acclimated

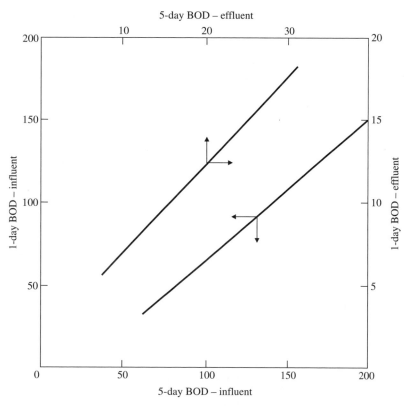

FIGURE 1.11
Relationship between 1-day and 5-day BOD for a pulp and paper-mill wastewater.

Although modifications of the BOD procedure such as the short-term test proposed by Busch[3] may eliminate some of the errors resulting from the first-order assumption and the variation in k_{10} due to substrate level, these procedures have not found broad application in industry. It is essential, therefore, that the following factors be considered in the interpretation of the BOD for an industrial waste:

1. That the seed is acclimated to the waste and all lag periods are eliminated.
2. That long-term BOD tests establish the magnitude of k_{10} on both the waste and treated effluent. In the case of acidic wastes, all samples should be neutralized before incubation.

Toxicity in a waste is usually evidenced by so-called sliding BOD values, i.e., an increasing calculated BOD with increasing dilution. If this situation exists, it is necessary to determine the dilution value below which the computed BODs are consistent.

The COD test measures the total organic content of a waste which is oxidized by dichromate in acid solution. When a silver sulfate catalyst is used, the recovery for most organic compounds is greater than 92 percent. However, some aromatics such as toluene are only partially oxidized. Since the COD will report virtually all organic compounds, many of which are either partially biodegradable or non-biodegradable, it is proportional to the BOD only for readily assimilable substances, e.g., sugars. Such a case is shown in Fig. 1.12 for a readily assimilable pharmaceutical waste. Tables 1.6 and 1.8 show the BOD and COD characteristics for a variety of industrial effluents.

Because the 5-day BOD will represent a different proportion of the total oxygen demand for raw wastes than for effluents, the BOD/COD ratio will frequently vary for effluents as compared to untreated wastes. There will be no correlation between BOD and COD when organic suspended solids in the waste are only slowly biodegradable in the BOD bottle, and, therefore, filtered or soluble samples should always be used. Pulp and fiber in a paper mill waste are an example. There will also be no correlation between BOD and COD in complex waste effluents containing refractory substances such as ABS. For this reason, treated effluents may exert virtually no BOD and yet exhibit a substantial COD.

Total organic carbon (TOC) has become a common and popular method of analysis due to its simplicity of measurement. There are presently several carbon analyzers on the market.

When considering routine plant control or investigational programs, the BOD is not a useful test because of the long incubation time. It is therefore useful to develop correlations between BOD and COD or TOC.

The theoretical oxygen demand (THOD) of a wastewater containing identified organic compounds can be calculated as the oxygen required to oxidize the organics to end products; e.g., for glucose:

$$C_6H_{12}O_6 + 6O_2 \rightarrow 6CO_2 + 6H_2O$$

$$\text{THOD} = \frac{6M_{O_2}}{M_{C_6H_{12}O_6}} = 1.07 \frac{\text{mg COD}}{\text{mg organic}}$$

For most organics, with the exception of some aromatics and nitrogen-containing compounds, the COD will equal the THOD. For readily degradable wastewaters,

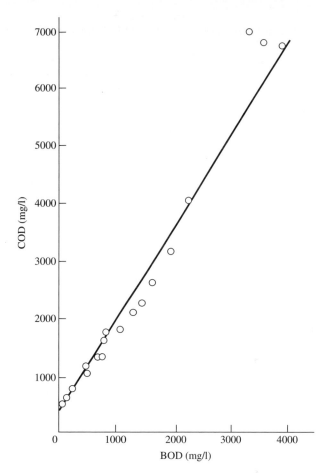

FIGURE 1.12
Relationship between BOD and COD for a pharmaceutical wastewater.

such as from a dairy, the COD will equal the $BOD_{ult}/0.92$. When the wastewater also contains nondegradable organics, the difference between the total COD and the $BOD_{ult}/0.92$ will represent the nondegradable content.

It has been found that some nondegradable organics will accumulate during biooxidation by oxidation by-products of the organics in the wastewater and by-products of endogenous metabolism. There are defined as soluble microbial products (SMP). Hence the effluent COD through biological treatment will increase over the influent nondegradable COD. This is shown in Fig. 1.13.

When the compounds are identified, the TOC can be related to COD through a carbon-oxygen balance:

$$C_6H_{12}O_6 + 6O_2 \rightarrow 6CO_2 + 6H_2O$$

$$\frac{COD}{TOC} = \frac{6M_{O_2}}{6M_C} = 2.66 \frac{mg\ COD}{mg\ organic\ carbon}$$

TABLE 1.6
Oxygen demand and organic carbon of industrial wastewaters[4]

Waste	BOD$_5$, mg/l	COD, mg/l	TOC, mg/l	BOD/TOC	COD/TOC
Chemical[†]	—	4,260	640	—	6.65
Chemical[†]	—	2,410	370	—	6.60
Chemical[†]	—	2,690	420	—	6.40
Chemical		576	122	—	4.72
Chemical	24,000	41,300	9,500	2.53	4.35
Chemical—refinery	—	580	160	—	3.62
Petrochemical	—	3,340	900	—	3.32
Chemical	850	1,900	580	1.47	3.28
Chemical	700	1,400	450	1.55	3.12
Chemical	8,000	17,500	5,800	1.38	3.02
Chemical	60,700	78,000	26,000	2.34	3.00
Chemical	62,000	143,000	48,140	1.28	2.96
Chemical	—	165,000	58,000	—	2.84
Chemical	9,700	15,000	5,500	1.76	2.72
Nylon polymer	—	23,400	8,800	—	2.70
Petrochemical	—	—	—	—	2.70
Nylon polymer	—	112,600	44,000	—	2.50
Olefin processing	—	321	133	—	2.40
Butadiene processing	—	359	156	—	2.30
Chemical	—	350,000	160,000	—	2.19
Synthetic rubber	—	192	110	—	1.75

[†] High concentration of sulfides and thiosulfates.

Depending on the organic in question, the COD/TOC ratio may vary from zero when the organic material is resistant to dichromate oxidation to 5.33 for methane. Since the organic content changes during biological oxidation, it can be expected that the COD/TOC ratio will also change. This same rationale also applies to the BOD/TOC ratio. Values of BOD and COD for a variety of organics are shown in Table 1.7. Since only biodegradable organics are removed in the activated sludge process, the COD remaining in the effluent will consist of the nondegradable organics present in the influent wastewater $[(SCOD_{nd})_i]$ and residual degradable organics (as characterized by the soluble BOD) and soluble microbial products generated in the treatment process. The SMP are not biodegradable (designated as SMP_{nd}) and, thus, exert a soluble COD (or TOC) but no BOD. Data indicate that the SMP_{nd} is 2 to 10 percent of the influent degradable COD. The actual percentage depends on the type of wastewater and the operating solids retention time (SRT) of the biological process. COD, BOD, and SMP relationships for industrial wastewaters are presented in Table 1.8, where the SMP_{nd} are assumed as 5 percent of the influent degradable SCOD.

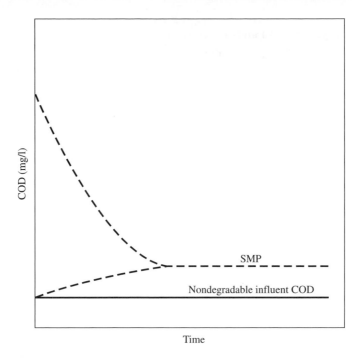

FIGURE 1.13
COD relationships during biooxidation.

The effluent total COD ($TCOD_e$) can be calculated as the sum of the degradable plus nondegradable soluble COD ($SCOD_d + SCOD_{nd}$) plus the "particulate" COD due to the effluent suspended solids (TSS_e). If the effluent solids are primarily activated sludge floc carryover, their COD can be estimated as 1.4 TSS_e. This is expressed as follows:

$$TCOD_e = (SCOD_{nd})_e + (SCOD_d)_e + 1.4\,TSS_e \qquad (1.4)$$

$$(SCOD_{nd})_e = SMP_{nd} + (SCOD_{nd})_i \qquad (1.5)$$

$$(SCOD_{nd})_i = SCOD_i - (SCOD_d)_i \qquad (1.6)$$

$$(TCOD)_e = SCOD_i - (SCOD_d)_i + SMP_{nd} + (SCOD_d)_e + 1.4\,TSS_e \qquad (1.7)$$

The degradable SCOD of the influent or effluent wastewater can be estimated from the ratio of BOD_5 to ultimate BOD (BOD_u) (designated as f_i or f_e). Assuming that $BOD_u = 0.92\ SCOD$, the degradable SCOD in the influent (i) or effluent (e) can be estimated as:

$$(SCOD_d)_{i/e} = \frac{(BOD_5)_{i/e}}{f_{i/e} \cdot 0.92} \qquad (1.8)$$

TABLE 1.7
**Evaluation of COD and BOD with respect to the theoretical
oxygen demand of selected organic compounds[3]**

Chemical group	THOD, mg/mg	Measured COD, mg/mg	$\dfrac{COD}{THOD}$, %	Measured BOD$_5$, mg/mg	$\dfrac{BOD_5}{THOD}$, %
Aliphatics					
Methanol	1.50	1.05	70	1.12	75
Ethanol	2.08	2.11	100	1.58	76
Ethylene glycol	1.26	1.21	96	0.39	29
Isopropanol	2.39	2.12	89	0.16	7
Maleic acid	0.83	0.80	96	0.64	77
Acetone	2.20	2.07	94	0.81	37
Methyl ethyl ketone	2.44	2.20	90	1.81	74
Ethyl acetate	1.82	1.54	85	1.24	68
Oxalic acid	0.18	0.18	100	0.16	89
Group average			91		56
Aromatics					
Toluene	3.13	1.41	45	0.86	28
Benzaldehyde	2.42	1.98	80	1.62	67
Benzoic acid	1.96	1.95	100	1.45	74
Hydroquinone	1.89	1.83	100	1.00	53
O-Cresol	2.52	2.38	95	1.76	70
Group average			84		58
Nitrogenous organics					
Monoethanolamine	2.49	1.27	51	0.83	34
Acrylonitrile	3.17	1.39	44	nil	0
Aniline	3.18	2.34	74	1.42	44
Group average			58		26

The effluent TCOD can then be estimated by combining Eqs. (1.4) through (1.8).

$$(TCOD)_e = SCOD_i - \left[\frac{(BOD_5)_i}{f_i \cdot 0.92}\right] + SMP_{nd} + \left[\frac{(BOD_5)_e}{f_e \cdot 0.92}\right] + 1.4\,TSS_e \quad (1.9)$$

The calculations relative to BOD, COD, and TOC are illustrated in Example 1.2.

The BOD, COD, and TOC tests are gross measures of organic content and as such do not reflect the response of the wastewater to various types of biological treatment technologies. It is therefore desirable to partition the wastewater into several categories, as shown in Fig. 1.14. The distinction between sorbable and non-sorbable BOD is important in selecting the appropriate process configuration for the control of sludge quality, as discussed on p. 254.

TABLE 1.8
COD, BOD, and SMP relationships for industrial wastewaters

Wastewater	Influent BOD, mg/L	Influent COD, mg/L	Effluent BOD, mg/L	Effluent COD, mg/L	SMP_{nd},[†] mg/L	$(COD_{nd})e$,[‡] mg/L	BOD_5/COD_{deg},[§]
Pharmaceutical	3,290	5,780	23	561	261	526	0.60
Diversified chemical	725	1,487	6	257	62	248	0.56
Cellulose	1,250	3,455	58	1,015	122	926	0.47
Tannery	1,160	4,360	54	561	190	478	0.28
Alkylamine	893	1,289	12	47	62	29	0.69
Alkyl benzene sulfonate	1,070	4,560	68	510	202	405	0.25
Viscose rayon	478	904	36	215	35	160	0.61
Polyester fibers	208	559	4	71	24	65	0.40
Protein process	3,178	5,355	5	245	256	237	0.59
Tobacco	2,420	4,270	139	546	186	332	0.59
Propylene oxide	532	1,124	49	289	42	214	0.56
Paper mill	380	686	7	75	31	64	0.58
Vegetable oil	3,474	6,302	76	332	298	215	0.55
Vegetable tannery	2,396	11,663	92	1,578	504	1,436	0.22
Hardboard	3,725	5,827	58	643	259	554	0.67
Saline organic chemical	3,171	8,597	82	3,311	264	3,185	0.56
Coke	1,618	2,291	52	434	93	354	0.79
Coal liquid	2,070	3,160	12	378	139	360	0.70
Textile dye	393	951	20	261	35	230	0.53
Kraft paper mill	308	1,153	7	575	29	564	0.50

[†] $0.05 \, (COD_{deg})_i$

[‡] $(COD_{nd})e = SCOD_e - [(BOD_5)_e/0.65]$

[§] $(COD_d)_i = COD_i - (COD_{nd})e + SMP_{nd}$

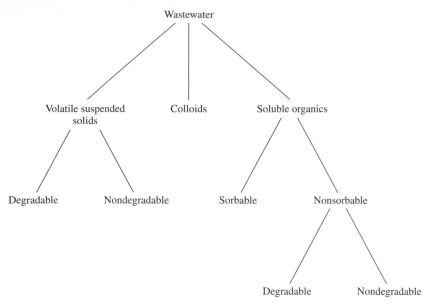

FIGURE 1.14
Partition of organic constituents of a wastewater.

EXAMPLE 1.2. A wastewater contains the following:

 150 mg/l ethylene glycol
 100 mg/l phenol
 40 mg/l sulfide (S^{2-})
 125 mg/l ethylene diamine hydrate (ethylene diamine is essentially nonbiodegradable)

(*a*) Compute the COD and TOC.
(*b*) Compute the BOD_5 if the k_{10} is 0.2/day.
(*c*) After treatment, the BOD_5 is 25 mg/l. Estimate the COD (k_{10} = 0.1/day).

Solution.

(*a*) The COD is computed:

 Ethylene glycol

$$C_2H_6O_2 + 2.5O_2 \rightarrow 2CO_2 + 3H_2O$$

$$COD = \frac{2.5(32)}{62} \times 150 \text{ mg/l} = 194 \text{ mg/l}$$

 Phenol

$$C_6H_6O + 7O_2 \rightarrow 6CO_2 + 3H_2O$$

$$COD = \frac{7(32)}{94} \times 100 \text{ mg/l} = 238 \text{ mg/l}$$

Ethylene diamine hydrate

$$C_2H_{10}N_2O + 2.5O_2 \rightarrow 2CO_2 + 2H_2O + 2NH_3$$

$$COD = \frac{2.5(32)}{78} \times 125 \text{ mg/l} = 128 \text{ mg/l}$$

Sulfide

$$S^{2-} = 2O_2 \rightarrow SO_4^{2-}$$

$$COD = \frac{2(32)}{32} \times 40 \text{ mg/l} = 80 \text{ mg/l}$$

The total COD is 640 mg/l.

The TOC is computed:

Ethylene glycol

$$\frac{24}{62} \times 150 \text{ mg/l} = 58 \text{ mg/l}$$

Phenol

$$\frac{72}{94} \times 100 \text{ mg/l} = 77 \text{ mg/l}$$

Ethylene diamine

$$\frac{24}{78} \times 125 \text{ mg/l} = 39 \text{ mg/l}$$

The total TOC is 174 mg/l.

(b) The ultimate BOD can be estimated:

$$(194 \text{ mg/l} + 238 \text{ mg/l} + 80 \text{ mg/l}) \times 0.92 = 471 \text{ mg/l}$$

Thus

$$\frac{BOD_5}{BOD_{ult}} = (1 - 10^{-(5 \times 0.2)}) = 0.9$$

BOD$_5$ is 471 mg/l $\times$ 0.9 = 424 mg/l.

(c) The BOD$_{ult}$ of the effluent is

$$\frac{25 \text{ mg/l}}{1 - 10^{-(5 \times 0.1)}} = \frac{25 \text{ mg/l}}{0.7} = 36 \text{ mg/l}$$

The COD is 36/0.92 = 39 mg/l. Therefore the COD will be 128 mg/l + 39 mg/l + residual by-products.

1.6
MEASURING EFFLUENT TOXICITY

The standard technique for determining the toxicity of a wastewater is the bioassay, which estimates a substance's effect on a living organism. The two most common types of bioassay tests are the chronic and the acute. The chronic bioassay estimates longer-term effects that influence the ability of an organism to reproduce, grow, or behave normally; the acute bioassay estimates short-term effects, including mortality.

The acute bioassay exposes a selected test organism, such as the fathead minnow or *Mysidopsis bahia* (mysid shrimp), to a known concentration of sample for a specified time (typically 48 or 96 h, but occasionally as short as 24 h). The acute toxicity of the sample is generally expressed as the concentration lethal to 50 percent of the organisms, denoted by the term LC_{50}. The chronic bioassay exposes a selected test organism to a known concentration of sample for longer periods of time than the acute bioassay, usually 7 days with daily renewal. The toxicity of the sample is currently expressed as the IC_{25} value, which represents the sample concentration that produces 25 percent inhibition to a chronic characteristic of the test species (e.g., growth weight or reproduction). NOEC is the concentration at which there was no observed effect.

The LC_{50} and IC_{25} values are determined through statistical analysis of mortality-time data or weight- or reproduction-time data, respectively. The lower the LC_{50} or IC_{25} values, the more toxic the wastewater. The bioassay data can be expressed as a concentration of a specific compound (e.g., mg/l), or in the case of whole-effluent (i.e., overall effluent) toxicity, as a dilution percentage or as toxic units. Toxic units can be calculated as 100 times the inverse of the dilution percentage expression of whole-effluent toxicity; a whole-effluent toxicity of 25% is equal to 100/25 or 4 toxic units. This results in a more logical measure where increasing values indicate increasing toxicity. The toxic unit expression of the data is applicable to acute or chronic tests using any organism. It is simply a mathematical expression.

Various organisms are used to measure toxicity. The organism and life stage (e.g., adult or juvenile) selected depends on the receiving stream salinity, the stability and nature of the expected contaminants, and the relative sensitivity of different species to the effluent. Different organisms exhibit different toxicity thresholds to the same compound (Table 1.9), and there is a relatively large variability in

TABLE 1.9
Acute toxicity of selected compounds (96-h LC_{50})

	Units	Fathead minnow	*Daphnia magna*	Rainbow trout
Organics				
Benzene	mg/l	42	35	38
1,4-Dichlorobenzene	mg/l	3.72	3.46	2.89
2,4-Dinitrophenol	mg/l	5.81	5.35	4.56
Methylene chloride	mg/l	326	249	325
Phenol	mg/l	39	33	35
2,4,6-Trichlorophenol	mg/l	5.91	5.45	4.62
Metals				
Cadmium	μg/l	38	0.29	0.04
Copper	μg/l	3.29	0.43	1.02
Nickel	μg/l	440	54	—

toxicity for a single compound and any one test species as a result of biological factors. This is shown in Fig. 1.15 for a petroleum refinery effluent. It should also be noted that variability in a plant effluent will result in highly variable effluent toxicity as shown in Fig. 1.16. In addition, results for multiple tests vary because of factors such as the species of organism, test conditions, and the number of replicates (i.e., duplicates) and organisms used, as well as the laboratory conducting the test (with greater variability observed when more than one lab is involved).

The precision of toxicity test results decreases significantly as the actual toxicity decreases. For example, in one series of bioassays conducted with *Mysidopsis bahia*, for an LC_{50} of 10 percent (10 toxic units), the 95 percent confidence level was about 7 to 15 percent, whereas for an LC_{50} of 50 percent (2 toxic units), it was 33 to 73 percent. This variation is a result of the statistical nature of the test. At high LC_{50} values, there is a low mortality rate, which, if the test is done with only a few organisms, can result in a wide range of actual LC_{50} concentrations. If, conversely, there is a high mortality, the results are more precise, since a greater percentage of the organisms was affected by the sample, thus giving a more statistically precise estimate of actual toxicity. It must be kept in mind that any such test gives only estimates of the actual toxicity and should be accorded the precision of an estimate.

Because of the wide variability in bioassay test results, numerous confirmatory data points are needed to be confident that the true extent of the toxicity problem has been estimated. No single data point should be relied upon for any conclusion. This may require long-term operation of bench, pilot-scale, or full-scale systems to determine the effectiveness of treatment.

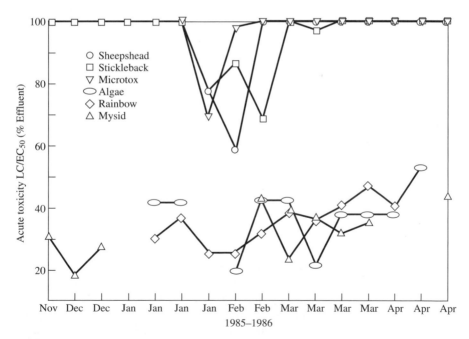

FIGURE 1.15
Acute toxicity of six species to refinery effluent.[5]

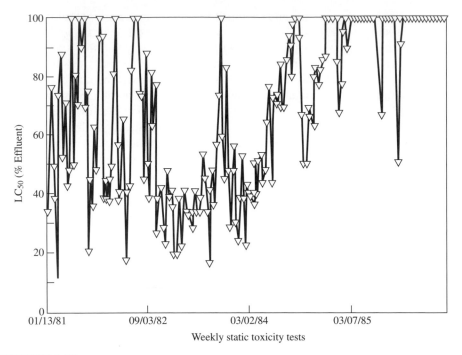

FIGURE 1.16
Acute toxicity of refinery effluent to stickleback.[5]

A thorough toxicity reduction program will involve a large number of tests on treated and untreated samples. For the initial screening stages, one should consider employing either a shortened, simplified bioassay technique (or surrogate test) such as a 24-h or 48-h version of the required test, or a rapid aquatic toxicity test such as Microtox®, IQ™, or Ceriofast™ to determine whether there is a correlation between the surrogate test and the required test. If a reasonably strong correlation exists, the surrogate test allows much quicker data turnaround, usually at a much lower cost.

Microtox (Microbics Corporation, Carlsbad, California) uses a freeze-dried marine luminescent microorganism *Vibrio fischeri* incubated at constant temperature in a high-salt, low-nutrient growth medium. Light outputs are measured to determine the effect of the wastewater on the luminescence of the organisms. Microtox has been found to correlate reasonably well in many domestic wastewater cases and in some relatively simple industrial cases. Figure 1.17 shows a correlation for a carbon-treated chemical plant effluent.

IQ (Aqua Survey, Flemington, New Jersey) uses less than 24-h-old Cladoceran species *(Daphnia magna, Daphnia pulex)* hatched from the eggs provided with the test kit. The test organisms are exposed to serial dilutions of the samples for 1 h. Then, the organisms are fed a fluorometrically tagged sugar substrate for 15 min. The intensity of the fluorescence of the control organisms is compared with the fluorescence of the test organisms to determine the effect of the wastewater on the capacity of the organisms to digest the substrate. Figure 1.18 shows a correlation for a pharmaceutical plant effluent.

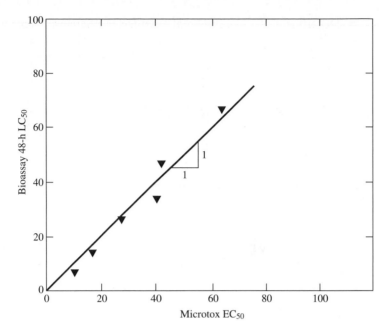

FIGURE 1.17
Bioassay/Microtox comparison—carbon-treated effluent.

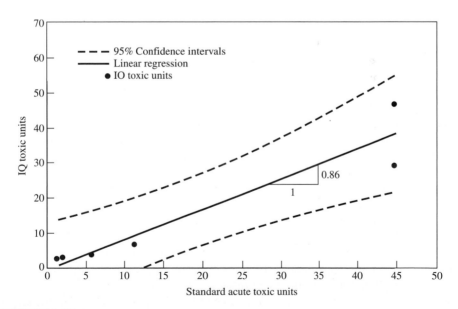

FIGURE 1.18
IQ toxic units versus standard acute toxic units for a pharmaceutical effluent.

Ceriofast (Department of Environmental Engineering Sciences, University of Florida, Gainesville, Florida) uses an in-house culture of *Ceriodaphnia dubia.* The 24- to 48-h test organisms are exposed to dilutions of the samples for 40 min. Then, the test organisms and the controls are fed for 20 min with a yeast substrate that contains a nontoxic fluorescent stain. The presence of fluorescence in the intestinal tract of the control organisms is compared with the presence or absence of fluorescence in the test organisms to determine the effect of the wastewater on the feeding ability of the organisms.

Toxicity Identification of Effluent Fractionation

Toxicity identification of effluent fractionation investigations determine, generically or specifically, the cause of effluent toxicity. Either the actual effluent is fractionated by chemical and/or physical means, or a sample that simulates the known toxic effect is synthesized. The goal is to measure the toxicity of each suspected key component in the absence of other toxic components, but with identical or equivalent nontoxic background components.

The procedure generally involves sample manipulation to eliminate toxicity associated with specific chemical groups. Results of toxicity tests on the treated samples are compared with tests on the unmanipulated effluent. Any difference found indicates that the constituent, or class of constituents, removed is likely to be responsible for the toxicity.

The specific components to be isolated, and the means for doing this, are quite varied and may require extensive investigation. Some techniques are now fairly standard and reliable, others may need to be further developed specifically for the subject wastewater. However, this evaluation can result in significant savings of time and money if properly executed.

Before beginning the fractionation, a thorough plan needs to be developed, one that must address the specific conditions encountered at the facility. Not all separation options (summarized in Fig. 1.19) are employed in every fractionation, and treatments vary from case to case.

The first step is to study the process flowsheet, as well as any long-term data on the chemical makeup of the discharge and plant production. This analysis may provide clues to the source of the toxicity. (However, these are just clues, since only the response of the organism in the actual sample is evidence of a toxic effect. In some cases, a compound whose concentration is above the reported toxic level is not harmful in the actual wastewater stream because it is unavailable to the organisms. For example, many heavy metals are toxic at very low levels in soft water but not in hard water.)

If nothing conclusive results from the process analysis (which will often be the case), a literature search should be performed to obtain information on toxicity of similar wastewaters and toxicity of the toxic compounds known to be present. These data will aid in later investigations.

The next step is to actually fractionate the discharge sample. In all cases, blanks (i.e., control samples) should be analyzed to make certain that toxicity is not

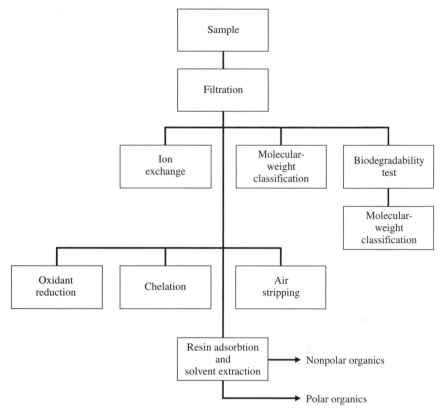

FIGURE 1.19
Various techniques can be employed to fractionate a wastewater sample.

being introduced by the fractionation or testing procedure. The following manipulations may be considered:

- *Filtration.* Filtration is generally performed first to determine whether the toxicity is related to the soluble or insoluble phase of the sample. Typically, 1-μm glass fiber filters that have been prewashed with ultrapure water are used. The insoluble phase should be resuspended in control water to be sure that filtration, not adsorption on the filter medium, removed the toxicity. A 0.45-μm filter should be used in order to determine if the colloidal fraction is responsible for the toxicity.
- *Ion exchange.* Inorganic toxicity can be studied by using cationic and anionic exchange resins to remove potentially toxic inorganic compounds or ions.
- *Molecular weight classification.* Evaluating the molecular weight distribution of the influent, and the toxicity of each molecular-weight range, can often narrow the list of suspected contaminants.
- *Biodegradability test.* Controlled biological treatment of effluent samples in the lab can result in almost compete oxidation of the biodegradable portion of

organics. Bioassay analysis can then quantify the toxicity associated with the nonbiodegradable components, as well as the reduction in toxicity attainable by biological treatment.

- *Oxidant reduction.* Residual chemical oxidants carried over from a process (e.g., chlorine and chloramines used for disinfection, or ozone and hydrogen peroxide used in sludge conditioning) can be toxic to most organisms. A simple batch reduction of these oxidants at various concentrations, using an agent such as sodium thiosulfate, will indicate the toxicity of any remaining oxidants.
- *Metal chelation.* The toxicity of the sum of all cationic metals (with the exception of mercury) can be determined by chelation of samples, using varying concentrations of ethylenediaminetetraacetic acid (EDTA) and evaluating the change in toxicity.
- *Air stripping.* Batch air stripping at acid, neutral, and basic pH can remove essentially all volatile organics. At a basic pH, ammonia is removed as well. Thus, if both volatile organics and ammonia are suspected toxicants, an alternative ammonia removal technique, such as a zeolite exchange, should be used. (Note that ammonia is toxic in the nonionized form, so ammonia toxicity is very pH dependent.)
- *Resin adsorption and solvent extraction.* Specific nonpolar organics can sometimes be identified as toxics, using a resin-adsorption/solvent-extraction process. A sample is adsorbed on a long-chain organic resin, the organics are reextracted from the resin with a solvent (e.g., methanol), and the sample's toxicity is determined via a bioassay.

Source Analysis and Sorting

In a typical wastewater collection system, multiple streams from various sources throughout the plant combine into fewer and fewer streams, ultimately forming a single discharge stream or treatment plant influent. To identify exactly where in the process the toxicity originates, one performs source analysis and sorting. This procedure starts with the influent to the treatment plant and proceeds upstream to the various points of wastestream combination, until the sources of key toxicants have been identified.

Treatment characteristics for each source are evaluated to determine whether the stream can be detoxified by the existing end-of-pipe technology at the facility, usually a biological treatment system. During this evaluation, the relative contribution of each source to effluent toxicity can be defined. (Methods for reducing or eliminating contributions through source treatment at the individual processing unit will be addressed in a later phase of the program.)

Gathering and interpreting this information for a large number of sources requires a well-planned and organized program; a procedure such as the one outlined in Fig. 1.19 can be used. Sources are first classified (before treatment) according to the following criteria:

- Bioassay toxicity, in terms of mg/L of the key chemical component
- Flow, as a percentage of the total effluent

- Concentration of the key chemical component [e.g., organic material expressed as total organic carbon (TOC)]
- Biodegradability

Relatively nonbiodegradable wastestreams are the most likely to induce toxic effects in the discharge and should be evaluated carefully. Some may require physical or chemical treatment because they are relatively nonbiodegradable. Others may have relatively high biodegradation rates but result in high residual levels after biodegradation. These will require additional testing to assess whether they remain significantly toxic following biological treatment.

Wastestreams found to be highly biodegradable have low probabilities for inducing toxic effects. However, their actual impact on effluent toxicity must be confirmed. This can be done by feeding a composite wastestream, of all (or most of) the streams to a continuous-flow biological reactor, and then determining the reactor effluent's toxicity.

It is also necessary to determine whether interactions are occurring between streams. To do so, the individual wastestream toxicities are compared with the toxicity of a biological-reactor effluent after treating a blend of the streams. If the toxicity (expressed in toxic units) of a blend of several samples is exactly additive, then no synergistic or antagonist effects are taking place. If the measured value for the composite stream is lower than the calculated value (i.e., the composite sample is nontoxic), the streams are synergistic; if higher, they are antagonistic.

In many cases the cause of effluent toxicity cannot be isolated, and a correlation must be made on total effluent COD as shown in Fig. 1.20 for a petroleum refinery effluent.

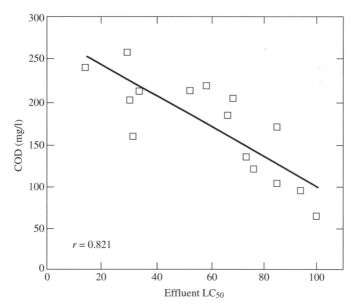

FIGURE 1.20
Toxicity/COD correlation for wastewater treatment plant effluent.

1.7
IN-PLANT WASTE CONTROL AND WATER REUSE

Waste Minimization

Before end-of-pipe wastewater treatment or modifications to existing wastewater treatment facilities to meet new effluent criteria are considered, a program of waste minimization should be initiated.

Reduction and recycling of waste are inevitably site- and plant-specific, but a number of generic approaches and techniques have been used successfully across the country to reduce many kinds of industrial wastes.

Generally, waste minimization techniques can be grouped into four major categories: inventory management and improved operations, modification of equipment, production process changes, and recycling and reuse. Such techniques can have applications across a range of industries and manufacturing processes, and can apply to hazardous as well as nonhazardous wastes.

Many of these techniques involve source reduction—the preferred option on EPA's hierarchy of waste management. Others deal with on- and off-site recycling. The best way to determine how these general approaches can fit a particular company's needs is to conduct a waste minimization assessment, as discussed below. In practice, waste minimization opportunities are limited only by the ingenuity of the generator. In the end, a company looking carefully at bottom-line returns may conclude that the most feasible strategy would be a combination of source reduction and recycling projects.

Waste minimization approaches as developed by the U.S. EPA are shown in Table 1.10. In order to implement the program, an audit needs to be made as described in Table 1.11. Case studies from three industries following rigorous source management and control are listed in Table 1.12. Pollution reduction can be directly achieved in several ways.

1. *Recirculation.* In the paperboard industry, white water from a paper machine can be put through a save-all to remove the pulp and fiber and recycled to various points in the paper making process.
2. *Segregation.* In a soap and detergent case, clean streams were separated for direct discharge. Concentrated or toxic streams may be separated for separate treatment.
3. *Disposal.* In many cases concentrated wastes can be removed in a semidry state. In the production of ketchup, after cooking and preparation of the product, residue in the kettle bottoms is usually flushed to the sewer. The total discharge BOD and suspended solids can be markedly reduced by removal of this residue in a semidry state for disposal. In breweries the secondary storage units have a sludge in the bottom of the vats which contain both BOD and suspended solids. Removal of this as a sludge rather than flushing to the sewer will reduce the organic and solids load to treatment.
4. *Reduction.* It is common practice in many industries, such as breweries and dairies, to have hoses continuously running for cleanup purposes. The use of automatic cutoffs can substantially reduce the wastewater volume. The use of drip pans to catch products, as in a dairy or ice cream manufacturing plant,

TABLE 1.10
Waste minimization approaches and techniques

Inventory management and improved operations

- Inventory and trace all raw materials
- Purchase fewer toxic and more nontoxic production materials
- Implement employee training and management feedback
- Improve material receiving, storage, and handling practices

Modification of equipment

- Install equipment that produces minimal or no waste
- Modify equipment to enhance recovery or recycling options
- Redesign equipment or production lines to produce less waste
- Improve operating efficiency of equipment
- Maintain strict preventive maintenance program

Production process changes

- Substitute nonhazardous for hazardous raw materials
- Segregate wastes by type for recovery
- Eliminate sources of leaks and spills
- Separate hazardous from nonhazardous wastes
- Redesign or reformulate end products to be less hazardous
- Optimize reactions and raw material use

Recycling and reuse

- Install closed-loop systems
- Recycle on site for reuse
- Recycle off site for reuse
- Exchange wastes

---

instead of flushing this material to the sewer considerably reduces the organic load. A similar case exists in the plating industry where a drip pan placed between the plating bath and the rinse tanks will reduce the metal dragout.

5. *Substitution.* The substitution of chemical additives of a lower pollutional effect in processing operations, e.g., substitution of surfactants for soaps in the textile industry.

A summary of cost-effective pollution control is shown in Table 1.13.

Although it is theoretically possible to completely close up many industrial process systems through water reuse, an upper limit on reuse is imposed by product quality control. For example, a closed system in a paper mill will result in a continuous buildup of dissolved organic solids. This can increase the costs of slime control, cause more downtime on the paper machines, and under some conditions cause discoloration of the paper stock. Obviously, the maximum degree of reuse is reached before these problems occur.

TABLE 1.11
Source management and control

Phase I—Preassessment

- Audit focus and preparation
- Identify unit operations and processes
- Prepare process flow diagrams

Phase II—Mass balance

- Determine raw material inputs
- Record water usage
- Assess present practice and procedures
- Quantify process outputs
- Account for emissions:
    To atmosphere
    To wastewater
    To off-site disposal
- Assemble input and output information
- Derive a preliminary mass balance
- Evaluate and refine the mass balance

Phase III—Synthesis

- Identify options
    Identify opportunities
    Target problem areas
    Confirm options
- Evaluate options
    Technical
    Environmental
    Economic
- Prepare action plan
    Waste reduction plan
    Production efficiency plan
    Training

Water requirements are also significant when reuse is considered. Water makeup for hydropulpers in a paper mill need not be treated for removal of suspended solids; on the other hand, solids must be removed from shower water on the paper machines to avoid clogging of the shower nozzles. Wash water for produce such as tomatoes does not have to be pure but usually requires chlorination to ensure freedom from microbial contamination.

By-product recovery frequently accompanies water reuse. The installation of save-alls in paper mills to recover fiber permits the reuse of treated water on cylinder showers. The treatment of plating-plant rinsewater by ion exchange yields a

TABLE 1.12
Source management and control

Case studies	Before	After
1. Chemical industry		
• Volume (m³/d)	5000	2700
• COD (t/d)	21	13
2. Hide and skin industry		
• Volume (m³/d)	2600	1800
• BOD (t/d)	3–6	2–6
• TDS (t/d)	20	10
• SS (t/d)	4–83	3–7
3. Metal preparation and finishing		
• Volume (m³/d)	450	270
• Chromium (kg/d)	50	5
• TTM (kg/d)	180	85

t = tonne = 1000 kg.

TABLE 1.13
Summary of cost-effective pollution control

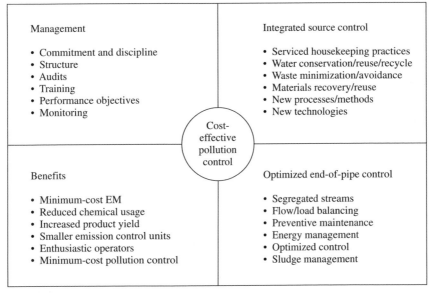

Management	Integrated source control
• Commitment and discipline	• Serviced housekeeping practices
• Structure	• Water conservation/reuse/recycle
• Audits	• Waste minimization/avoidance
• Training	• Materials recovery/reuse
• Performance objectives	• New processes/methods
• Monitoring	• New technologies

(Center: Cost-effective pollution control)

Benefits	Optimized end-of-pipe control
• Minimum-cost EM	• Segregated streams
• Reduced chemical usage	• Flow/load balancing
• Increased product yield	• Preventive maintenance
• Smaller emission control units	• Energy management
• Enthusiastic operators	• Optimized control
• Minimum-cost pollution control	• Sludge management

EM: environmental management

reusable chromic acid. Water conservation and materials recovery in nickel plating is shown in Fig. 1.21. There are many other similar instances in industry.

The best way to illustrate water reuse in industry is to cite several examples. Figure 1.22 shows simplified schematics of two water-reuse systems in a paperboard

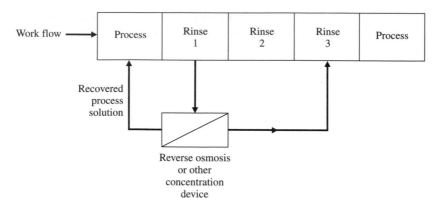

FIGURE 1.21
Water conservation and materials recovery (nickel plating).

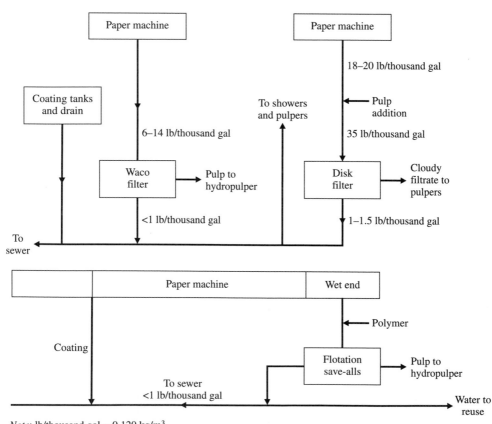

Note: lb/thousand gal = 0.120 kg/m³

FIGURE 1.22
Pulp and fiber recovery in a board mill.

mill. When a disk filter is employed, the cloudy filtrate is returned to the pulpers as makeup water and the clear filtrate is used for showers and other miscellaneous uses. The overflow is continuously discharged to the sewer. The use of a flotation save-all results in a similar operation. Waste from the clay-coating operations is usually discharged directly to the sewer.

Water conservation in brewery operations has been effected by reuse of the third rinse in kettle and vat cleaning for the next rinse, by reuse of cooling water for cleaning purposes, and by counterflow in pasteurizing and bottling operations. The total waste loading is reduced by removing the spent grains in a semidry state, by removing the yeast from the fermenters for filtration and drying, and by removing the sediment from the chill storage tanks as a slurry for separate disposal. In many cases these operations have yielded a salable by-product which has more than offset the increased costs of recovery.

Reuse at a pharmaceutical formulation plant has been effected by saving the first-rinse wastewaters from the cleaning of the formulation tank to make up the next batch of product. A significant reduction in wastewater volume and strength was achieved.

Reuse and makeup of sulfide dehairing solutions from tanning liquors have reduced both the volume and strength of wastes from this industry.

Data for water reuse in a corn-processing plant are shown in Table 1.14.

TABLE 1.14
Unit waste flows for corn-processing operation and
possible changes to reduce flow

Unit	Line	Flow, gal/min	Water use	Possible change	Estimated new flow
Rotary ear	1	21.7	Loosens and removes sand etc.	Screen 7 and reuse in washer	21.7
Cutting machines	2	27.0	Flumes kernels to flotation washer	Screen and recirculate	0.0
Scavenger reel	3	10.4	Flumes waste separated by flotation washer	Use smaller flow or remove as solid	5.2
Wash reel	4	18.0	Rewashes corn (not necessary)	Remove	8.0
Blancher	5	4.5	Overflow from makeup waste	No change	4.5
Cooling reel	6	24.5	Cool corn after blanching	Use smaller flow	10.0
Holding tank	7	16.9	Overflow from makeup water	Screen and reuse in rotary washer	0.0
Shaker screen	8	2.1	Flumes waste separated by reel	Remove as solid or reuse screen	0.0
Total		125.1			49.4

Note: gal/min = 3.78×10^{-3} m^3 /min

Spent caustic in a refinery is an alkaline waste, containing high concentrations of sulfides, mercaptans, and phenolates. Separate treatment of spent caustic may markedly reduce wastewater-treatment costs and can in some cases result in a marketable product.

In the organic chemicals industry, an example might be drawn from the Union Carbide plant at South Charleston, West Virginia. The plant is reasonably representative of what might be expected from a major multiproduct chemicals complex. A detailed study of waste load reduction by in-plant changes showed that the present plant flow of 11.1 million gal/d (42,000 m³/d) and 55,700 lb of BOD/d (25,300 kg/d) could be reduced to a flow of 8.3 million gal/d (31,400 m³/d) and 37,100 lb BOD/d (16,800 kg/d). The ways of achieving these reductions are shown in Table 1.15. Equipment revision and additions and unit shutdowns refer to those units that would be replaced. Incineration involves the option of taking the more concentrated wastewater streams and, instead of discharging to the main sewer, segregating them for incineration. Reprocessing refers to taking the tank bottoms that still have product present and reprocessing them for further product recovery.

Frequently, wastestreams can be eliminated or reduced by process modifications or improvements. A notable example of this is the use of save-rinse and spray-rinse tanks in plating lines. This measure will bring about a substantial reduction in wastewater flow and strength. In dairies, equipment modifications to collect drippings will usually substantially reduce the BOD loading to the sewer. Substitution of sizing agents in textile plants may result in a lower net pollutional load for treatment.

TABLE 1.15
General analysis of in-plant modifications

Type of modifications	Description	Total RWL reduction, %
Equipment revision and additions	Self-explanatory	25
Unit shutdown	Shutdowns due to the age of the unit or the product; these shutdowns are not a direct result of pollution considerations, but they are somewhat hastened by these considerations	10
Scrubber replacement	Replacement of scrubbers associated with amine production by burning of the off-vapors	3
Segregation, collection, and incineration	Of specific concentrated wastewater streams	35
Raw material substitutions	Self-explanatory	3
Reprocessing	Collecting tail streams from specific processes, then putting streams through an additional processing unit to recover more product and concentrate the final wastestream	
Miscellaneous small projects	A variety of modifications which individually do not represent a large reduction in RWL	21

There are many other examples from various industries. A critical evaluation of water reuse and by-product recovery possibilities should be undertaken before the development of waste-treatment needs.

Segregation of incompatible wastestreams should be considered before treatment process design criteria are developed. In some older plants, however, this may be economically unfeasible or in some cases impossible. Segregation becomes a necessity when combination of wastestreams will result in potential safety hazards. In a plating plant, for example, a mixture of acidic metal rinses with cyanide streams could produce toxic HCN.

It is not uncommon for a portion of a waste flow to contain the majority of the suspended solids loading. Only this part of the flow needs to receive treatment for solids removal. This is true, for instance, in a tannery where 90 percent of the suspended solids come from the beam house in approximately 60 percent of the flow.

Frequently, cooling water or other noncontaminated streams can be segregated and discharged directly to the receiving water.

Consideration of water reuse, by-product recovery, and waste segregation will make it possible to revise the material balance-and-flow diagram to establish the basis for process design of waste-treatment facilities. This is illustrated in Example 1.3.

Up to this point, reductions in wastewater volume and strength have been achieved by in-plant changes. The installation of wastewater treatment can be considered for the purpose of recycling rather than for disposal. This becomes particularly significant in the case of the high effluent discharge standards. From this figure a relationship may be developed for determining the daily cost for a water management system:

$$C = C_W(1 - f)Q + C_R fQ + C_D[Q(1 - f) - L]$$

where C = daily cost
  Q = production water requirements
  C_W = freshwater cost
  C_R = recycled water costs
  C_D = effluent treatment costs
  f = fraction recycled
  L = total water losses

EXAMPLE 1.3. A plant uses 10,000 gal/h (37,850 l/h) of process water with a maximum contaminant concentration of 1 lb/1000 gal (120 mg/l). The raw water supply has a contaminant concentration of 0.5 lb/1000 gal (60 mg/l) (see Fig. 1.23). Optimize a water-reuse system for this plant based on a raw water cost of 20 cents/1000 gal (5.3 cents/1000 l). The following conditions apply:

  Evaporation and product loss E = 1000 gal/h (3785 l/h)
  Contaminant addition Y = 100 lb/h (45.4 kg/h)
  Maximum discharge to receive water = 20 lb/h (9.1 kg/h)

Example calculation
Let A = 3000 gal/h (11,355 l/h). Since $A = E + B$, $B = 3000 - 1000 = 2000$ gal/h (7570 l/h) and the recycle V is 7000 gal/h (26,495 l/h).

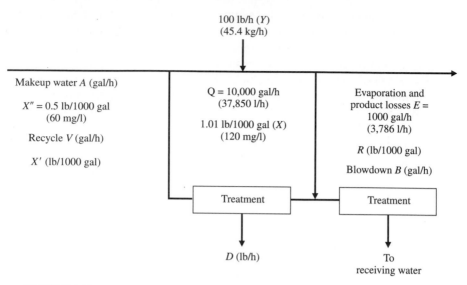

FIGURE 1.23
Water reuse and treatment balance.

By a material balance:

$$R = \frac{(A + V)X + Y}{(A + V) - E} = \frac{(10)(1.0) + 100}{10 - 1} = 12.2 \text{ lb}/1000 \text{ gal } (1464 \text{ mg}/1)$$

By a material balance:

$$QX = AX'' + VX'$$

$$X' = \frac{QX - AX''}{V} = \frac{(10)(1.0) - A(0.5)}{V} = \frac{10 - (3)(0.5)}{7}$$

$$= 1.21 \text{ lb}/1000 \text{ gal } (145 \text{ mg}/1)$$

The required efficiency of treatment for reuse is:

$$\% \text{ Removal} = \frac{R - X'}{R} \cdot 100 = \frac{12.2 - 1.21}{12.2} \cdot 100 = 90\%$$

The required treatment efficiency for discharge to the river is:

$$\frac{2(12.2) - 20}{2(12.2)} \cdot 100 = 18\%$$

The raw water cost for 3000 gal/h is (3)(24)(0.20) = \$14.40/d. The cost of effluent treatment for discharge to the river is 5 cents/1000 gallon for a total daily cost of (2)(0.05)(24), or \$2.40/d. The cost of treatment for reuse (see Fig. 1.24) is (7)(0.42)(24) = \$70.60/d. The net total cost is \$87.40/d. A similar series of calculations is made for freshwater inputs varying from 2000 to 10,000 gal/h (no reuse). The total daily water cost can then be plotted versus percent recycle as shown in Fig. 1.25.

Another study in a bleached kraft mill is shown in Fig. 1.26. Effluent 1 requires a 75 percent removal of BOD and suspended solids with no limitation on

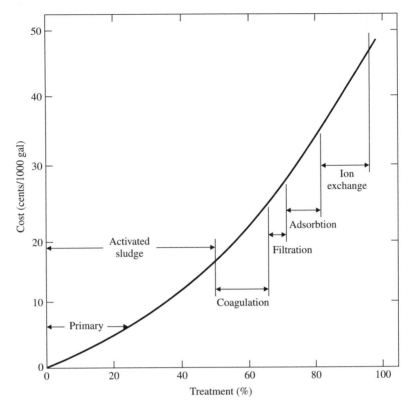

FIGURE 1.24
Relationship between total water cost and treatment (1 cent/1000 gal = 0.26 cent/1000 l).

color. Effluent 2 requires 20 mg/l BOD_5 and suspended solids without color control. Effluent 3 requires 25 mg/l BOD_5 and 30 mg/l SS with 50 units of color. It is obvious from this figure that the optimal recycle will primarily be related to the effluent quality requirement and to the cost of fresh water.

1.8
STORMWATER CONTROL

In most industrial plants, it is now necessary to contain and control pollutional discharges from stormwater. Pollutional discharges can be minimized by providing adequate diking around process areas, storage tanks, and liquid transfer points with drainage into the process sewer. Contaminated stormwater is usually collected according to the storm frequency for the area in question (e.g., a 10-year storm) in a holding basin. The collected water is then passed through the wastewater treatment plant at a controlled rate. A total storm-runoff flow and contaminant loading of a refinery petrochemical installation is shown in Fig. 1.27. An example of stormwater control is shown in Example 1.4.

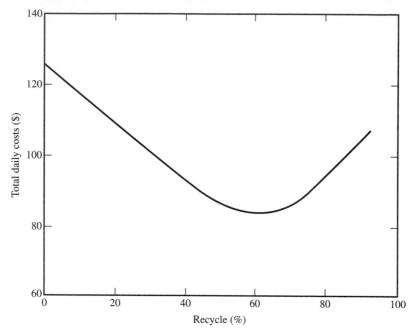

FIGURE 1.25
Relationship between total daily water cost and treated waste recycle for reuse.

EXAMPLE 1.4. If runoff is to be stored or surged prior to eventual treatment, anticipated rainfall volumes must be estimated. The probable rainfalls for periods of 1 day to 1 year are given in Table 1.16 and Fig. 1.28 for the 2-year storm.

The areas and runoff coefficients associated with a refinery complex are shown in Table 1.17. Determine:

(*a*) The cumulative runoff (million gal) as a function of time (days).
(*b*) The required minimum return rate for treatment (million gal/d or m³/d) for storage periods of 10, 30, 60, and 120 d.
(*c*) The required storage capacity (million gal or m³) for storage periods of 10, 30, 60, and 120 d.

Solution.
(*a*) Total effective areas

$$= 21.8 \times 1.0 + 115.4 \times 0.5 + 166.9 \times 0.7 + 176.7 \times 0.5 + 55.6 \times 0.9$$

$$= 334.7 \text{ acres} \quad (1.36 \times 10^6 \text{ m}^2)$$

From the table of cumulative rainfall with various time periods, the cumulative runoff as a function of time can be estimated by

$$q = i(CA)$$

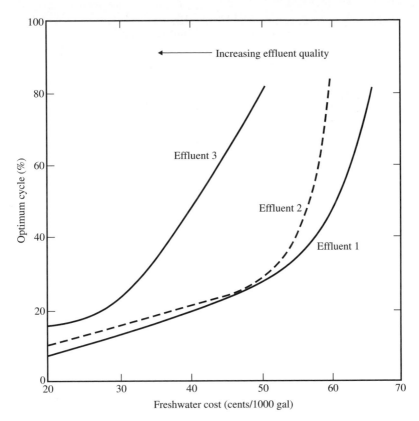

FIGURE 1.26
Optimum recycle versus freshwater cost for varied effluent quality (1 cent/1000 gal =
0.26 cent/100 l).

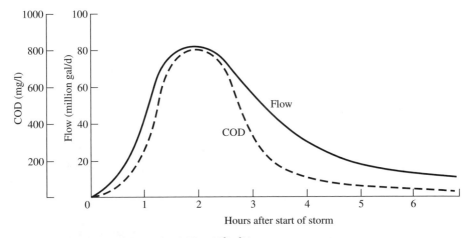

Note: million gal/d = 3.78 × 10³ m³/d

FIGURE 1.27
Total storm runoff flow and contaminant loading for a refinery installation.

TABLE 1.16

Time period, d	Cumulative rainfall, in
1	2.5
5	3.7
10	4.5
30	6.6
60	10.7
120	18.0
240	30.0
360	41.0

Note: in = 0.0254 m = 2.54 cm

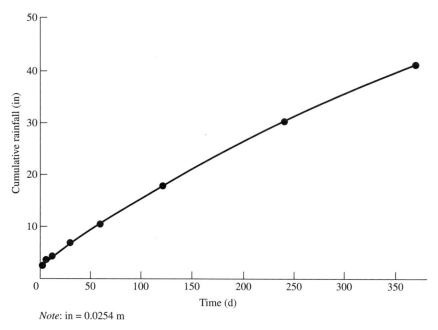

Note: in = 0.0254 m

FIGURE 1.28
Two-year recurrence level probability rainfalls.

TABLE 1.17

Type of surface	Area, acres	Runoff coefficient, C
Water surface	21.8	1.0
Dirt or gravel	115.4	0.5
Bermed tank farm	166.9	0.7
Grass	176.7	0.5
Asphalt or concrete	55.6	0.9

Note: acre = 4.05×10^3 m^2

in which CA is the effective area and i is the rainfall intensity. An example calculation for a 10-d period at 4.5 in (0.114 m) rainfall is

$$q = 4.5 \text{ in} \times 334.7 \text{ acre} \times \frac{43{,}560 \text{ ft}^2}{\text{acre}} \times \frac{0.0833 \text{ ft}}{\text{in}} \times \frac{7.48 \text{ gal}}{\text{ft}^3}$$

$$= 4.088 \times 10^7 \text{ gal or } 40.9 \text{ million gal} \qquad (1.55 \times 10^5 \text{ m}^3)$$

The results are tabulated in Table 1.18 and plotted in Fig. 1.29.

TABLE 1.18

Time, d	Runoff, million gal
1	22.7
5	33.6
10	40.9
30	60.0
60	97.2
120	163.5
240	272.5
360	372.4

Note: million gal $= 3.78 \times 10^3$ m^3

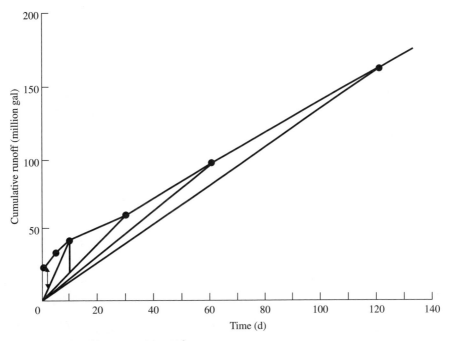

Note: million gal $= 3.78 \times 10^3$

FIGURE 1.29
Cumulative runoff curve.

(b) The minimum allowable return rate for treatment from a storage basin, assuming the stored volume will be returned at a uniform flow rate, is defined by the slope of a straight line connecting the origin and a point on the cumulative runoff curve representing the maximum practical storage period. A sample calculation for 10 d is

$$\text{Return rate} = \frac{40.9 - 0}{10 - 0} \text{ million gal/d}$$

$$= 4.1 \text{ million gal/d } (1.55 \times 10^4 \text{ m}^3/\text{d})$$

(c) The volume of runoff stored is the difference between the cumulative output and the cumulative input. This is equal to the maximum vertical distance between the cumulative runoff curve and the straight line depicting the rate of removal. An example for 10 d is

$$\text{Required volume storage} = 18 \text{ million gal} \quad (6.8 \times 10^4 \text{ m}^3)$$

The results are given in Table 1.19.

TABLE 1.19

Storage period, period d	Return rate, million gal/d	Required storage, million gal
10	4.09	18.0
30	2.00	20.9
60	1.62	24.0
120	1.36	28.0

Note: million gal/d = 3.78×10^3 m³/d
million gal = 3.78×10^3 m³

PROBLEMS

1.1. The 4-h composite COD data from a brewery effluent over a 7-d period is as follows:

980			
2800	3200	6933	3325
1380	3175	1240	6000
1250	3850	580	3100
720	2870	710	2500
8650	2600	3410	1830
7200	2743	2910	3225
2800	3600	8300	2370
2570	4066	2950	1380
1780	1550	2230	2600

Develop a statistical plot of these data in order to define the 50 and the 90 percent values.

1.2. A tomato-processing plant is shown in Fig. P1.2. The industrial waste survey data are shown in the table below. Develop a flow-and-material balance diagram for the plant sewer system. Show possible changes to reduce the flow and loading. The waste from the trimming tables is presently flumed to the sewer.

Sampling station	Process unit	Flow, gal/min	BOD, mg/l	SS, mg/1
1	Niagra washers	500	75	180
2	Rotary washers	—	90	340
3	Pasteurizer	300	30	20
4	Source cooker	10	3520	7575
5	Cook room	20	4410	8890
6	Source finisher	150	230	170
7	Trimming tables	100	450	540
8	Main outfall	1080	240	470

Note: gal/min = 3.78×10^{-3} m³/min

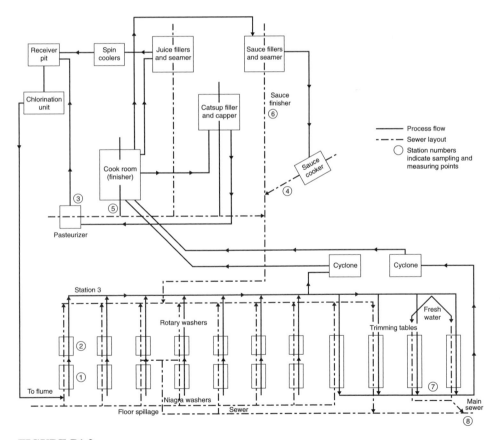

FIGURE P1.2
Flow diagram of a tomato-processing plant.

1.3. A wastewater contains the following constituents:

40 mg/l phenol
350 mg/l glucose
3 mg/l S^{2-}
50 mg/l methyl alcohol, CH_3OH
100 mg/l isophorene, $C_9H_{14}O$ (nondegradable)

(a) Compute the THOD, the COD, the TOC, and the BOD_5 assuming the k_{10} for the mixed wastewater is 0.25/d.

(b) After treatment the soluble BOD_5 is 10 mg/l with a k_{10} of 0.1/d. Compute the residual COD and TOC.

REFERENCES

1. Schroepfer, G. J.: *Advances in Water Pollution Control,* vol. 1, Pergamon Press, New York, 1964.
2. Stack, V. T.: *Proc. 8th Ind. Waste Conf.,* 1953, p. 492, Purdue University.
3. Busch, A. W.: *Proc. 15th Ind. Waste Conf.,* 1961, p. 67, Purdue University.
4. Ford, D. L.: *Proc. 23rd Ind. Waste Conf.,* 1968, p. 94, Purdue University.
5. Dorn, P. B.: *Toxicity Reduction—Evaluation and Control,* Technomic Publishing Co., 1992.

2

WASTEWATER TREATMENT PROCESSES

The schematic diagram in Fig. 2.1 illustrates an integrated system capable of treating a variety of plant wastewaters. The scheme centers on the conventional series of primary and secondary treatment processes, but also includes tertiary treatment and individual treatment of certain streams.

Primary and secondary treatment processes handle most of the nontoxic wastewaters; other waters have to be pretreated before being added to this flow. These processes are basically the same in an industrial plant as in a publicly owned treatment works (POTW).

Primary treatment prepares the wastewaters for biological treatment. Large solids are removed by screening, and grit, if present, is allowed to settle out. Equalization, in a mixing basin, levels out the hour-to-hour variations in flows and concentrations. There should be a spill pond, to retain slugs of concentrated wastes that could upset the downstream processes. Neutralization, where required, follows equalization because streams of different pH partly neutralize each other when mixed. Oils, greases, and suspended solids are removed by flotation, sedimentation, or filtration.

Secondary treatment is the biological degradation of soluble organic compounds—from input levels of 50 to 1000 mg/l BOD and even greater to effluent levels typically under 15 mg/l. This is usually done aerobically, in an open, aerated vessel or lagoon, but wastewaters may be pretreated anaerobically, in a pond or a closed vessel. After biotreatment, the microorganisms and other carried-over solids are allowed to settle. A fraction of this sludge is recycled in certain processes, but ultimately the excess sludge, along with the sedimented solids, has to be disposed of.

Many existing wastewater-treatment systems were built just for primary and secondary treatment, though a plant might also have systems for removing materials that would be toxic to microorganisms. Until recently, this was adequate, but now it is not, and so new facilities have to be designed and old facilities retrofitted to include additional capabilities to remove priority pollutants and residuals toxic to aquatic life.

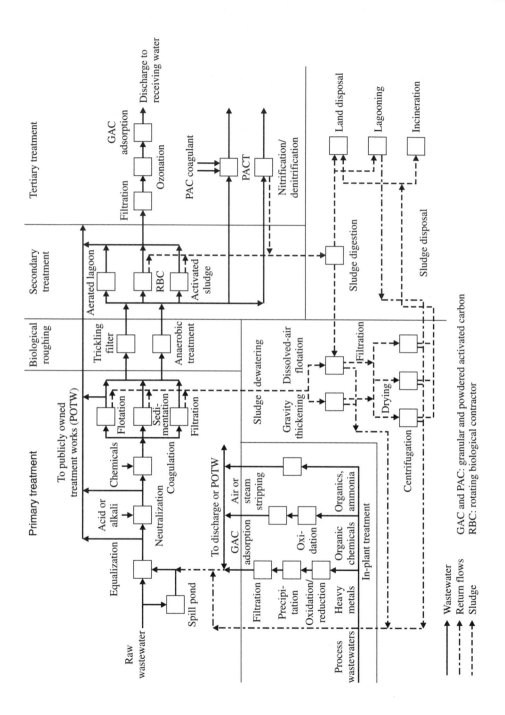

FIGURE 2.1
Alternative technologies for the treatment of industrial wastewaters.

52

Tertiary-treatment processes are added on after biological treatment in order to remove specific types of residuals. Filtration removes suspended or colloidal solids; adsorption by granular activated carbon (GAC) removes organics; and chemical oxidation also removes organics. Unfortunately, tertiary systems have to treat a large volume of wastewater, and so they are expensive. They can also be inefficient because the processes are not pollutant-specific. For example, dichlorophenol can be removed by ozonation or GAC adsorption, but those processes will remove most other organics as well, and this adds greatly to the treatment costs for removing the dichlorophenol.

In-plant treatment is necessary for streams rich in heavy metals, pesticides, and other substances that would pass through primary treatment and inhibit biological treatment. In-plant treatment also makes sense for low-volume streams rich in nondegradable materials, because it is easier and much less costly to remove a specific pollutant from a small, concentrated stream than from a large, dilute one. Processes used for in-plant treatment include precipitation, activated carbon adsorption, chemical oxidation, air or steam stripping, ion exchange, reverse osmosis, electrodialysis, and wet air oxidation.

Existing treatment systems can also be modified so as to broaden their capabilities and improve their performance; this is more widely practiced than the above options. One example is adding powdered activated carbon (PAC) to the biological-treatment process, to adsorb organics that the microorganisms cannot degrade or slowly degrade; this is marketed as the PACT process. Another example is adding coagulants at the end of the biological-treatment basin so as to remove phosphorus and residual suspended solids.

All these processes have their place in the overall wastewater-treatment scheme. The selection of a wastewater-treatment process or a combination of processes depends upon:

1. The characteristics of the wastewaters. This should consider the form of the pollutant, i.e., suspended, colloidal, or dissolved; the biodegradability; and the toxicity of the organic and inorganic components.
2. The required effluent quality. Consideration should also be given to possible future restrictions such as an effluent bioassay aquatic toxicity limitation.
3. The costs and availability of land for any given wastewater-treatment problem. One or more treatment combinations can produce the desired effluent. Only one of these alternatives, however, is the most cost-effective. A detailed cost analysis should therefore be made prior to final process design selection.

Process Selection

A preliminary analysis should be carried out to define the wastewater treatment problems as shown in Fig. 2.2.

For wastewaters containing nontoxic organics, process design criteria can be obtained from available data or from a laboratory or pilot plant program. Examples are pulp and paper mill wastewaters and food processing wastewaters. In the case of complex chemical wastewaters containing toxic and nontoxic

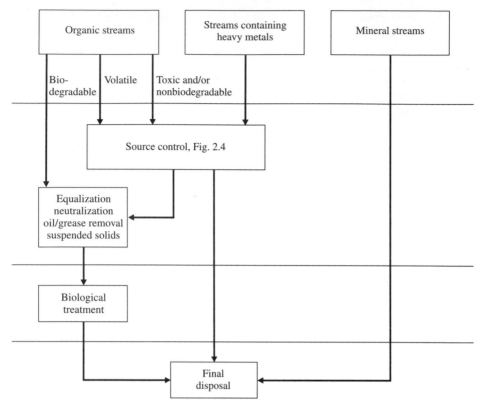

FIGURE 2.2
Conceptual approach of treatment/management program for a high (organic) strength and toxic industrial wastewater.

organics and inorganics, a more defined screening procedure is necessary to select candidate processes for treatment. A protocol has been established as shown in Fig. 2.3. If the wastewater contains heavy metals, they are removed by precipitation. Volatile organics are removed by air stripping.

Following laboratory screening, pretreatment-as-required analyses are conducted on an equalized sample as shown in Fig. 2.3. It should be noted that all significant wastewater streams in the industrial complex should be evaluated. The next step is to determine whether the wastewater is biodegradable and whether it will be toxic to the biological process at some level of concentration. The fed batch reactor (FBR) procedure is employed for this purpose. Details of this procedure are discussed on p. 268. Acclimated sludge as discussed on p. 203 should be used. If the wastewater is nonbiodegradable or toxic it should be considered for source treatment or in-plant modification. Source treatment technologies are shown in Fig. 2.4.

If the wastewater is biodegradable, it is subjected to a long-term biodegradation, usually 48 hours, in order to remove all degradable organics. It is then evaluated for aquatic toxicity and priority pollutants. If nitrification is required, a nitrification rate analysis is conducted as discussed on p. 290. If the effluent is toxic or

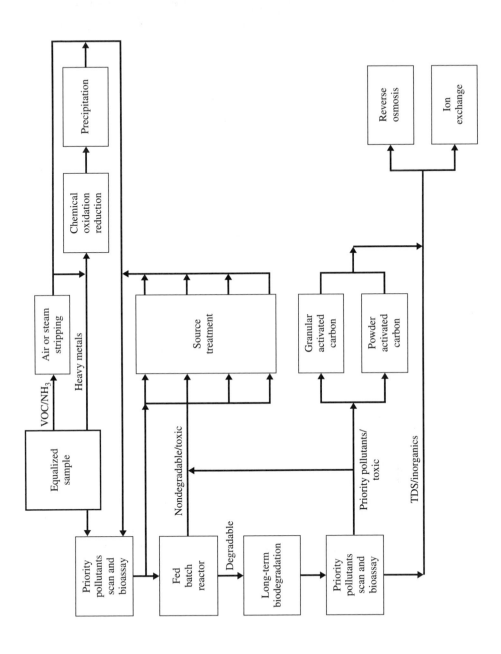

FIGURE 2.3
Screening laboratory procedures.

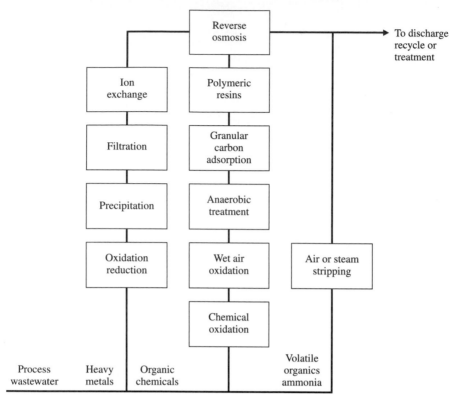

FIGURE 2.4
Applicable technologies for treatment of toxic wastewaters.

if priority pollutants have not been removed, it should be considered for source treatment or tertiary treatment using powdered or granular activated carbon. It should be noted that toxicity due to soluble microbial products (SMP) will require tertiary treatment for its removal.

When biological treatment is considered, several options exist. A screening procedure has been developed to determine the most cost-effective alternative, as shown in Fig. 2.5.

The alternatives for biological treatment are summarized in Table 2.1 and discussed in detail in Chap. 7. A screening and identification matrix for physical-chemical treatment is shown in Table 2.2. Table 2.2 should provide a guide to select applicable technologies for specific problems. A summary of physical-chemical technology applications is shown in Table 2.3. The maximum attainable effluent quality for conventional wastewater treatment processes is shown in Table 2.4.

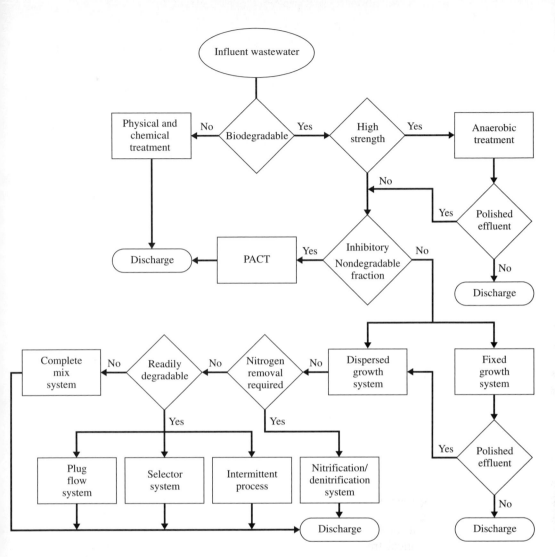

FIGURE 2.5
Simplified process selection flowsheet for biological treatment.

TABLE 2.1
Biological waste treatment

Treatment method	Mode of operation	Degree of treatment	Land requirements	Equipment	Remarks
Lagoons	Intermittent or continuous discharge; facultative or anaerobic	Intermediate	Earth dug; 10–60 days' retention		Odor control frequently required
Aerated lagoons	Completely mixed or facultative continuous basins	High in summer; less in winter	Earth basin, 8–16 ft (2.44–4.88 m) deep, 8–16 acres/(million gal/d) [8.55–17.1 m²/(m³/d)]	Pier-mounted or floating surface aerators or subsurface diffusers	Solids separation in lagoon; periodic dewatering and sludge removal
Activated sludge	Completely mixed or plug flow; sludge recycle	>90% removal of organics	Earth or concrete basin; 12–20 ft (3.66–6.10 m) deep; 75,000–350,000 ft³/(million gal/d) [0.561–2.262 m³/(m³/d)]	Diffused or mechanical aerators; clarifier for sludge separation and recycle	Excess sludge dewatered and disposed of
Trickling filter	Continuous application; may employ effluent recycle	Intermediate or high, depending on loading	225–1400 ft²/(million gal/d) [5.52–34.4 m²/10³ m³/d)]	Plastic packing 20–40 ft deep (6.10–12.19 m)	Pretreatment before POTW or activated sludge plant
RBC	Multistage continuous	Intermediate or high		Plastic disks	Solids separation required
Anaerobic	Complete mix with recycle; upflow or downflow filter, fluidized bed; upflow sludge blanket	Intermediate		Gas collection required; pretreatment before POTW or activated sludge plant	
Spray irrigation	Intermittent application of waste	Complete; water percolation into groundwater and runoff to stream	40–300 gal/(min · acre) [6.24 × 10⁻⁷ – 4.68 × 10⁻⁶ m³/(s · m²)]	Aluminum irrigation pipe and spray nozzles; movable for relocation	Solids separation required; salt content in waste limited

Note:
ft = 0.305 m
acre/(million gal · d) = 1.07 m²/(m³ · d)
ft³/(million gal · d) = 7.48 × 10⁻³ m³/(thousand m³ · d)
ft²/(million gal · d) = 2.45 × 10⁻² m²/(thousand m³ · d)
gal/(min · acre) = 1.56 × 10⁻⁸ m³/(s · m²)

58

TABLE 2.2
Screening and identification matrix for industrial wastewaters physical-chemical treatment

Stripping

Process	Parameter										Comments
	Organic compounds	Non-condensible	Temperature	Pressure	pH	O&G, mg/l	SS, mg/l	TDS, mg/l	Fe, Mn	Sol	
Air	<100 mg/l	A	DP	DP	DP	R	R	DP	R	L	$H_C > 0.005$ is recommended.
Steam	<100 mg/l to 10%	R	DP	DP	NI	R	R	DP	NI	M	Relative volatility referred to water >1.05 is recommended. Azeotrope formation is important.

Oxidation processes

Process	Parameter										Comments
	Organic compounds	Temperature °F	Pressure, psig	pH	OD, g/l	O&G, mg/l	SS, mg/l	TDS, mg/l	Fe, Mn	MW	
Wet air oxidation	A	350–650	300–3000	NI	20–200	NI	NI	DP	NI	NI	Not recommended for aromatic halogenated organics. Recommended for high COD/BOD ratio.
Supercritical water oxidation	A	750–1200	3675	NI	<10	NI	NI	L	NI	NI	
Chemical oxidation O_3	<10,000 mg/l	DP	DP	DP	DP	R	R	DP	A	NI	Catalyst and additional source of energy, e.g., ultraviolet, are important factors.
H_2O_2	A	DP	NI	DP	DP	R	R	DP	A	H	

(continued)

Adsorption and precipitation

Process	Organic compounds	Inorganic ionic species	Chemical oxidants	Parameter Temperature	pH	O&G, mg/l	SS, mg/l	TDS, mg/l	Fe, Mn	MW	Sol	Comments
Activated carbon adsorption	<10,000 mg/l	NA	NA	DP	DP	<10	<50	<10	NI	H	L	$K > 5$ mg/g, high K_{ow}, inorganics <1000 mg/l are recommended. Heavy metals may poison the carbon.
Resin adsorption	A	NA	R	DP (L)	DP	<10	<10	DP	NI	DP	M	High K_{ow} and $C_o < 0.1$ (resin capacity/3 BV) are recommended. K is a design parameter.
Chemical precipitation	NA	A	NI	DP	DP	R	NI	DP	A	NI	DP	Chelating and complexing agents interference may occur.

Membrane processes and ion exchange

Process	Organic compounds Volatile	Semi-volatile	Inorganic ionic species	Chemical oxidants	Parameter Tempera-ture	Pressure, psig	pH	O&G, mg/l	SS, mg/l	TDS, mg/l	Fe, Mn	MW, amu	Comments
Reverse osmosis	R	A	A	R	DP	<1500	DP	R	R	<10,000	R	>150	Differential osmotic pressure < 400 psi, LSI < 0, SDI < 5, and turbidity < 1 NTU are recommended.
Hyper-filtration	NA	A	NA	NI	DP	DP	DP	R	R	NI	NI	100–500	Molecular size, shape, and flexibility are important factors.

(continued)

Organic compounds

Process	Volatile	Semi-volatile	Inorganic ionic species	Chemical oxidants	Temperature	Pressure, psig	pH	O&G mg/l	SS mg/l	TDS mg/l	Fe, Mn	MW, amu	Comments
Ultrafiltration	NA	A	NA	NI	DP	10–100	DP	R	R	NI	NI	500–1,000,000	Molecular size, shape, and flexibility are important factors.
Electrodialysis/ electrodialysis reverse	R	R	A	R	DP	40–60	DP	R	R	<5000	<0.3	NI	Applied voltage is one of the design parameters. Ca < 900 mg/l is recommended.
Ion exchange resin	R	R	A	R	DP	NI	DP	R	<50 (<35)	<20,000	NI	NI	Selectivity quotient is one of the design parameters. Ion charge and volume are important factors.

A = applicable
BV = bed volume
DP = design parameter
H = high
H_C = Henry's constant
K = Freundlich isotherm coefficient
K_{ow} = octanol/water partition coefficient
M = moderate
MW = molecular weight
NA = not applicable
NTU = nephelometric turbidity units
O&G = oil and grease
R = must be removed
SDI = silt density index
Sol = solubility
SS = suspended solids
TDS = total dissolved solids

Note:
°C = $\frac{5}{9}$ (°F−32)
kPa = 6.894 psi

61

TABLE 2.3
Physical-chemical waste treatment

Treatment method	Type of waste	Mode of operation	Degree of treatment	Remarks
Ion exchange	Plating, nuclear	Continuous filtration with resin regeneration	Demineralized water recovery; product recovery	May require neutralization and solids removal from spent regenerant
Reduction and precipitation	Plating, heavy metals	Batch or continuous treatment	Complete removal of chromium and heavy metals	One day's capacity for batch treatment; 3-h retention for continuous treatment; sludge disposal or dewatering required
Coagulation	Paperboard, refinery, rubber, paint, textile	Batch or continuous treatment	Complete removal of suspended and colloidal matter	Flocculation and settling tank or sludge blanket unit; pH control required
Adsorption	Toxic or organics, refractory	Granular columns of powdered carbon	Complete removal of most organics	Powdered activated carbon (PAC) used with activated sludge process
Chemical oxidation	Toxic and refractory organics	Batch or continuous ozone or catalyzed hydrogen peroxide	Partial or complete oxidation	Partial oxidation to render organics more biodegradable

TABLE 2.4
Maximum quality attainable from waste treatment processes

Process	BOD	COD	SS	N	P	TDS
Sedimentation, % removal	10–30	—	59–90	—	—	—
Flotation, % removal[†]	10–50	—	70–95	—	—	—
Activated sludge, mg/l	< 25	‡	<20	§	§	—
Aerated lagoons, mg/l	< 50	—	>50	—	—	—
Anaerobic ponds, mg/l	> 100	—	<100	—	—	—
Deep-well disposal	Total disposal of waste					
Carbon adsorption, mg/l	< 2	< 10	< 1	—	—	—
Denitrification and nitrification, mg/l	< 10	—	—	< 5	—	—
Chemical precipitation, mg/l	—	—	< 10	—	< 1	—
Ion exchange, mg/l	—	—	< 1	¶	¶	¶

† Higher removals are attained when coagulating chemicals are used.
‡ $COD_{inf} - [BOD_{ult} \text{ (removed)}/0.9]$.
§ $N_{inf} - 0.12$ (excess biological sludge), lb.; $P_{inf} - 0.026$ (excess biological sludge), lb.
¶ Depends on resin used, molecular state, and efficiency desired.
Note: lb = 0.45 kg.

3

PRE- AND PRIMARY TREATMENT

The objective of pre- and primary treatment is to render a wastewater suitable for discharge to a POTW (publicly owned treatment works) or subsequent biological treatment. The concentrations of pollutants that make prebiological treatment desirable are summarized in Table 3.1. Common pretreatment technologies are shown in Fig. 3.1.

TABLE 3.1
Concentrations of pollutants that make prebiological treatment desirable

Pollutant or system condition	Limiting concentration	Kind of pretreatment
Suspended solids	> 125 mg/l	Sedimentation, flotation, lagooning
Oil or grease	> 35	Skimming tank or separator
Toxic ions		Precipitation or ion exchange
Pb	≤ 0.1 mg/l	
Cu + Ni + CN	≤ 1 mg/l	
Cr^{+6} + Zn	≤ 3 mg/l	
Cr^{+3}	≤ 10 mg/l	
pH	6 to 9	Neutralization
Alkalinity	0.5 lb alkalinity as $CaCO_3$/lb BOD removed	Neutralization for excessive alkalinity
Acidity	Free mineral acidity	Neutralization
Organic load variation	> 2:1	Equalization
Sulfides	> 100 mg/l	Precipitation or stripping with recovery
Ammonia	> 500 mg/l (as N)	Dilution, ion exchange, pH adjustment, and stripping
Temperature	> 38°C in reactor	Cooling

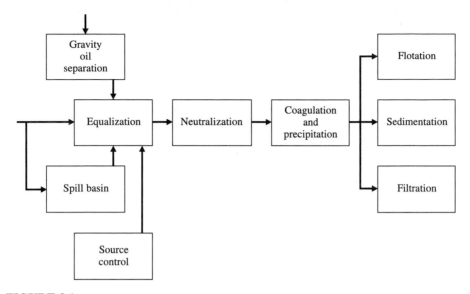

FIGURE 3.1
Pretreatment technologies.

3.1
EQUALIZATION

The objective of equalization is to minimize or control fluctuations in wastewater characteristics in order to provide optimum conditions for subsequent treatment processes. The size and type of the equalization basin varies with the quantity of waste and the variability of the wastewater stream. The basin should be of a sufficient size to adequately absorb waste fluctuations caused by variations in plant-production scheduling and to dampen the concentrated batches periodically dumped or spilled to the sewer.

The purposes of equalization for industrial treatment facilities are:

1. To provide adequate dampening of organic fluctuations in order to prevent shock loading of biological systems. The effluent concentration from a biological treatment plant will be proportional to the influent concentration. This is shown in Fig. 3.2, which represents 24-h composited samples over a period of 3 years for a petroleum refinery. As can be seen from Fig. 3.2, the effluent variability tracks the influent variability. If the wastewater is readily degradable an increase in the influent will result in a lesser increase in the effluent due to an increase in biometabolism. By contrast, if the influent contains bioinhibitors, an increased effluent concentration will result.
2. To provide adequate pH control or to minimize the chemical requirements necessary for neutralization.
3. To minimize flow surges to physical-chemical treatment systems and permit chemical feed rates compatible with feeding equipment.

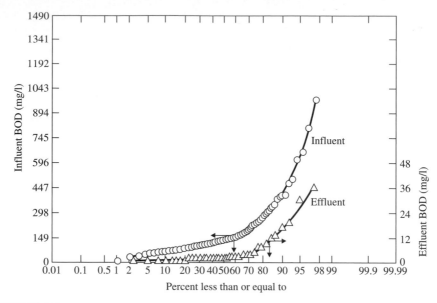

FIGURE 3.2
Variability in influent and effluent BOD.

4. To provide continuous feed to biological systems over periods when the manufacturing plant is not operating.
5. To provide capacity for controlled discharge of wastes to municipal systems in order to distribute waste loads more evenly.
6. To prevent high concentrations of toxic materials from entering the biological treatment plant.

    Mixing is usually provided to ensure adequate equalization and to prevent settleable solids from depositing in the basin. In addition, the oxidation of reduced compounds in the wastestream or the reduction of BOD by air stripping may be achieved through mixing and aeration. Methods that have been used for mixing include:

1. Distribution of inlet flow and baffling
2. Turbine mixing
3. Diffused air aeration
4. Mechanical aeration
5. Submerged mixers

The most common method is to provide submerged mixers or, in the case of readily degradable wastewater, such as that from a brewery, to use surface aerators employing a power level of approximately 15 to 20 hp/million gal (0.003 to 0.0045 kW/m^3). Air requirements for diffused air aeration are approximately 0.5 ft^3 air/gal waste (3.74 m^3/m^3). Equalization basin types are shown in Fig. 3.3.

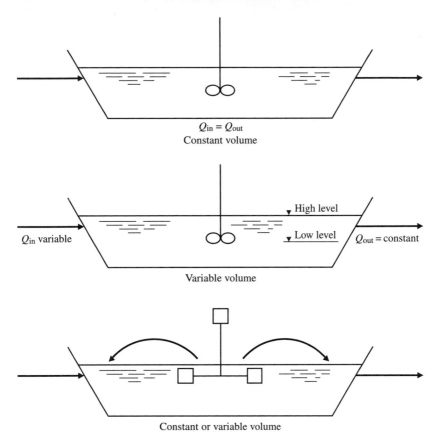

$Q_{in} = Q_{out}$
Constant volume

Q_{in} variable

High level

Low level $Q_{out} = $ constant

Variable volume

Constant or variable volume

FIGURE 3.3
Constant-volume and variable-volume equalization basins.

The equalization basin may be designed with a variable volume to provide a constant effluent flow or with a constant volume and an effluent flow that varies with the influent. The variable-volume basin is particularly applicable to the chemical treatment of wastes having a low daily volume. This type of basin may also be used for discharge of wastes to municipal sewers. It may be desirable to program the effluent pumping rate to discharge the maximum quantity of waste during periods of normally low flow to the municipal treatment facility, as shown in Fig. 3.4. Ideally, the organic loading to the treatment plant is maintained constant over a 24-h period.

Equalization basins may be designed to equalize flow, concentration, or both. For flow equalization, the cumulative flow is plotted versus time over the equalization period (for example, 24 h). The maximum volume with respect to the constant-discharge line is the equalization volume required. The required calculations are shown in Example 3.1.

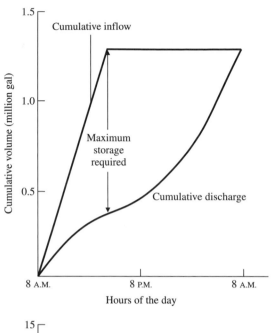

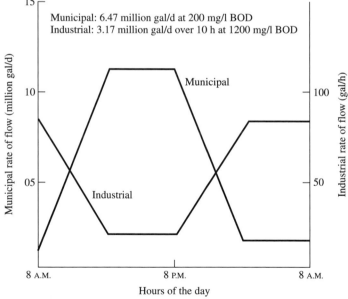

Note:
million gal = 3.78 × 10³ m³
million gal/d = 3.78 × 10³ m³/d
gal/h = 3.78 × 10⁻³ m³/h

FIGURE 3.4
Controlled discharge of an industrial wastewater to a municipal plant.

TABLE 3.2

Time	gal/min	gal	Σ gal $\times$ 10^{-3}
8	50	3,000	3.0
9	92	5,520	8.5
10	230	13,800	22.3
11	310	18,600	40.9
12	270	16,200	57.1
1	140	8,400	65.5
2	90	5,400	70.9
3	110	6,600	77.5
4	80	4,800	82.3
5	150	9,000	91.3
6	230	13,800	105.1
7	305	18,300	123.4
8	380	22,800	146.2
9	200	12,000	158.2
10	80	4,800	163.0
11	60	3,600	166.6
12	70	4,200	170.8
1	55	3,300	174.1
2	40	2,400	176.5
3	70	4,200	180.7
4	75	4,500	185.2
5	45	2,700	187.9
6	55	3,300	191.2
7	35	2,100	193.3

Note:
gal/min = 3.78 $\times$ 10^{-3} m³/min
gal = 3.78 $\times$ 10^{-3} m³

EXAMPLE 3.1. Given the data in Table 3.2, design an equalization basin for a constant outflow from the basin. The data are plotted as the summation of inflow versus time as shown in Fig. 3.5. The treatment rate is (193,300 gal/day)/(1440 min/day) = 134 gal/min (507 l/min). The required storage volume is 41,000 gal + 8000 gal residual or 49,000 gal (186 m³).

The equalization basin may be sized to restrict the discharge to a maximum concentration commensurate with the maximum permissible discharge from subsequent treatment units. For example, if the maximum effluent from an activated sludge unit is 50 mg/l BOD_5, the maximum effluent from the equalization basin may be computed and thereby provide a basis for sizing the unit.

For the case of near-constant wastewater flow and random input variations that have a normal statistical distribution of wastewater composite analyses, the required equalization retention time is[1]

$$t = \frac{\Delta t(S_i^2)}{2(S_e^2)} \tag{3.1}$$

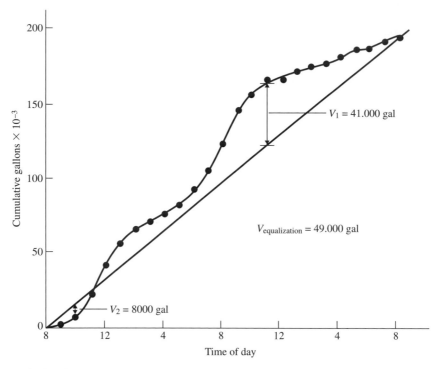

FIGURE 3.5

where t = equalization detention time, h
    Δt = time interval over which samples were composited, h
    S_i^2 = variance of the influent wastewater concentration (the square of
        the standard deviation)
    S_e^2 = variance of the effluent concentration at a specified probability
        (for example, 99 percent)

Example 3.2 illustrates this calculation.

EXAMPLE 3.2. A waste with a total flow of 5 million gal/d (0.22 m³/s or 19,000 m³/d) was characterized as shown in Fig. 3.6. Extensive data were collected every 4 h for 17 d. The average BOD was 690 mg/l and the maximum value was 1185 mg/l.

Design calculations with activated sludge systems have indicated that the effluent from the equalization basin must not exceed 896 mg/l in order to meet the effluent quality criteria of an average BOD of 15 mg/l and a maximum concentration of 25 mg/l from the activated sludge system.

Design an equalization basin to meet the desired effluent requirements. Base calculations on a 95 percent probability that the equalized effluent will be equal to or less than 896 mg/l.

Solution.

(*a*) Calculate the mean, standard deviation, and variance of the influent. These parameters may be calculated graphically from Fig. 3.6.

(*b*) From this plot obtain the 50 percent value:

$$50 \text{ percent value} \approx \bar{X} \approx 690 \text{ mg/l}$$

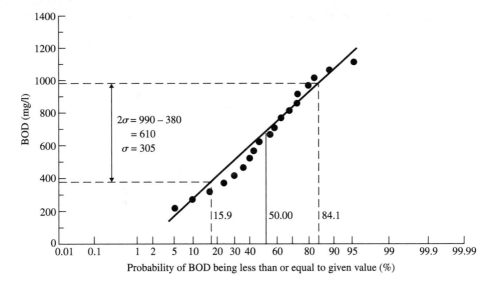

FIGURE 3.6

(c) Calculate the standard deviation, S_i, as half the difference in the values that occur at the 15.9 (50.0 minus 34.1) and 84.1 (50.0 plus 34.1) percentile levels from Fig. 3.6:

$$S_i = \frac{\text{value at } 84.1\% - \text{value at } 15.9\%}{2}$$

$$= \frac{990 - 380}{2}$$

$$= 305 \text{ mg/l}$$

(d) Calculate the variance as the square of the standard deviation:

$$S_i^2 = (305)^2$$

$$= 93{,}025 \text{ mg}^2/\text{l}^2$$

(e) Calculate the standard deviation of the effluent:

$$\bar{X} = 690 \text{ mg/l}$$

$$X_{max} = 896 \text{ mg/l}$$

The necessary condition for 95 percent of effluent BOD values to be less than 896 mg/l is

$$S_e = \frac{X_{max} - \bar{X}}{Z}$$

$$= \frac{896 - 690}{1.65}$$

$$= 125 \text{ mg/l}$$

where $Z = 1.65$ is from the normal probability tables for a 95 percent confidence level.

(*f*) Calculate the allowed effluent variance:

$$S_e^2 = (125)^2$$

$$= 15,625 \text{ mg}^2/l^2$$

(*g*) Calculate the required detention time:

$$t = \frac{\Delta t(S_i^2)}{2(S_e^2)}$$

$$= \frac{4(93,025)}{2(15,625)}$$

$$= 11.9 \text{ h}$$

$$\approx 0.5 \text{ d}$$

Where a completely mixed basin is to be used for treatment, such as in an activated sludge basin or an aerated lagoon, this volume can be considered as part of the equalization volume. For example, if the completely mixed aeration basin retention time is 8 h and the total required equalization retention time is 16 h, then the equalization basin needs to have a retention time of only 8 h.

Patterson and Menez[2] have developed a method to define equalization requirements when both the flow and the strength vary randomly. A material balance can be established for the equalization basin:

$$C_i QT + C_0 V = C_2 QT + C_2 V \tag{3.2}$$

where C_i = concentration entering the equalization basin over the sampling interval T
  T = sampling interval, i.e., 1 h
  Q = average flow rate over the sampling interval
  C_0 = concentration in the equalization basin at the start of the sampling interval
  V = volume of the equalized basin
  C_2 = concentration leaving the equalization basin at the end of the sampling interval

It is assumed that the effluent concentration is almost constant during one time interval. This is valid, assuming the time intervals are appropriately spaced.

Equation (3.2) can be rearranged to compute the effluent concentration after each time interval:

$$C_2 = \frac{C_i T + C_0 V/Q}{T + V/Q} \tag{3.2a}$$

The range of effluent concentrations can then be calculated for a range of equalization volumes V. A peaking factor PF is computed for the influent strength and flow. The effluent PF for design purposes is the ratio of the maximum concentration to the average concentration. An equalization basin design is shown in Example 3.3.

EXAMPLE 3.3. A survey of the discharge from a chemical plant showed the following results:

Time period	Mean Flow gal/min	TOC, mg/l
8 to 10 A.M.	450	920
10 to 12 noon	620	1130
12 to 2 P.M.	840	1475
2 to 4 P.M.	800	1525
4 to 6 P.M.	340	910
6 to 8 P.M.	270	512
8 to 10 P.M.	570	1210
10 to 12 midnight	1100	1520
0 to 2 A.M.	1200	1745
2 to 4 A.M.	800	820
4 to 6 A.M.	510	410
6 to 8 A.M.	570	490

Develop a plot of peaking factor versus basin volume for a variable-volume equalization system and determine the volume to yield a peaking factor of 1.2 based on mass discharge.

Solution.

Assuming the equalization basin is a completely mixed basin where no significant degradation or evaporation occurs, the differential equations that govern the constant effluent flow system for each time interval are:

$$\frac{dVi}{dt} = Q_{0i} - Q_{ei} = Q_{0i} - Q_{0\text{avg}} \tag{1}$$

$$\frac{d(V_iC_i)}{dt} = V_i\frac{dC_i}{dt} + C_i\frac{dV_i}{dt} = V_i\frac{dC_i}{dt} + C_i(Q_{0i} - Q_{0\text{avg}}) = Q_{0i}C_{0i} - Q_{ei}C_i \tag{2}$$

where
V_i = volume in the basin at time t for time interval i, gal
Q_{0i} = influent flow rate at time interval i, gpm
Q_{ei} = effluent flow rate at time interval i, gpm
$Q_{0\text{avg}}$ = daily average influent flow rate, gpm
C_{0i} = influent concentration at time interval i, mg/l
C_i = concentration in the basin and effluent concentration at time t for time interval i, mg/l

Assuming that Q_{0i}, Q_{ei}, and C_{0i} are constant during each time interval, separation of variables, integration, and rearrangement of Eqs. (1) and (2) lead to the following expressions for the volume $V_i(f)$ and the TOC concentration $C_i(f)$ in the basin at the end of each time interval, respectively.

$$V_i(f) = V_{(i-1)}(f) + (Q_{0i} - Q_{0avg})\Delta t_i \tag{3}$$

$$C_i(f) = C_{0i} - \frac{A}{(1 + B\Delta t_i)^D} \tag{4}$$

where

$$A = C_{0i} - C_{(i-1)}(f)$$

$$B = \frac{Q_{0i} - Q_{0avg}}{V_{(i-1)}(f)}$$

$$D = \frac{Q_{0i}}{Q_{0i} - Q_{0avg}}$$

$$
\begin{aligned}
V_{(i-1)}(f) &= \text{volume in the basin at the end of time interval } i-1, \text{ gal} \\
\Delta t_i &= \text{time interval } i, \text{ min} \\
C_{(i-1)}(f) &= \text{concentration in the basin and effluent concentration at the} \\
&\quad \text{end of time interval } i-1, \text{ mg/l}
\end{aligned}
$$

The TOC mass $M_{ei}(f)$ that left the equalization basin during each time interval can be calculated by

$$M_{ei}(f) = Q_{0avg} \int_0^{\Delta t_i} C_i \, dt = Q_{0avg}\left[C_{0i}\,\Delta t_i - \frac{A[[1 + B\Delta t_i]^{1-D} - 1]}{B(1 - D)} \right] \tag{5}$$

The procedure to developing the spreadsheet to calculate the masses leaving the basin at each time interval for a given volume of basin is:

1. The values for the first time interval (first row in the table below) are calculated as follows:
   - With the mean influent flow and concentration, the influent volume and mass of TOC are calculated.
   - The effluent flow rate, which is the same for all time intervals, is calculated by dividing the total influent volume by the total time.
   - For an initial basin volume (guess), the final volume is calculated by using Eq. (3).
   - For an initial concentration (guess), the final concentration is calculated by using Eq. (4).
   - The mass of TOC that left the basin is calculated by using Eq. (5).
2. The values for the other time intervals are calculated in a similar fashion but the initial volume and concentration for each time interval are the final volume and concentration of the previous time interval.
3. The volume guessed for the first time interval is modified until the maximum initial or final volume in the basin is the selected volume.
4. The concentration guessed for the first time interval is modified until it matches the final concentration for the last time interval.
5. The maximum and average masses leaving the basin are determined and the peaking factor is calculated as the ratio of the maximum to average values.

The results of the calculations for a basin volume of 310,000 gal are presented in the following table. Repetition of the calculations for different volumes allows the plotting of the peaking factor as a function of the basin volume as presented in Fig. 3.7.

Equalization basin design—Variable volume

Time period	Mean flow, gpm	Mean TOC, mg/l	Influent Volume, gal	Influent TOC, lb	Effluent Flow, gpm	Effluent TOC, lb	Basin Volume Initial gal	Basin Volume Final gal	Basin Concentration Initial mg/l	Basin Concentration Final mg/l
8 to 10	450	920	54,000	414	672.5	673	278,200	251,500	1009	993
10 to 12	620	1130	74,400	701	672.5	681	251,500	245,200	993	1028
12 to 14	840	1475	100,800	1240	672.5	745	245,200	265,300	1028	1174
14 to 16	800	1525	96,000	1221	672.5	828	265,300	280,600	1174	1278
16 to 18	340	910	40,800	310	612.5	842	280,600	240,700	1278	1225
18 to 20	270	512	32,400	138	672.5	791	240,700	192,400	1225	1125
20 to 22	570	1210	68,400	690	672.5	767	192,400	180,100	1125	1151
22 to 24	1100	1520	132,000	1673	672.5	843	180,100	231,400	1151	1327
0 to 2	1200	1745	144,000	2096	672.5	960	231,400	294,700	1327	1504
2 to 4	800	820	96,000	657	672.5	946	294,700	310,000	1504	1318
4 to 6	510	410	61,200	209	672.5	829	310,000	290,500	1318	1150
6 to 8	570	490	68,400	280	672.5	725	290,500	278,200	1150	1009

Mean = 802 lb
Maximum = 960 lb
Peaking factor 1.20

In most cases with a variable flow, a variable-volume basin will be most effective, as shown in Fig. 3.8. Load balancing for a pharmaceutical wastewater is shown in Fig. 3.9.

When occasional clumps or spills are anticipated, for example, 1 percent of the time, a spill basin with an automatic bypass activated by a monitor should be used, as shown in Fig. 3.10. The spill basin may be required for surges in organics (e.g., TOC), TDS, temperature, or specific toxic compounds.

3.2
NEUTRALIZATION

Many industrial wastes contain acidic or alkaline materials that require neutralization prior to discharge to receiving waters or prior to chemical or biological treatment. For biological treatment, a pH in the biological system should be maintained between 6.5 and 8.5 to ensure optimum biological activity. The biological process itself provides a neutralization and a buffer capacity as a result of the production of CO_2, which reacts with caustic and acidic materials. The degree of preneutralization required depends, therefore, on the ratio of BOD removed and the causticity or acidity present in the waste. These requirements are discussed in Chap. 6.

Types of Processes

Mixing Acidic and Alkaline Wastestreams

This process requires sufficient equalization capacity to effect the desired neutralization.

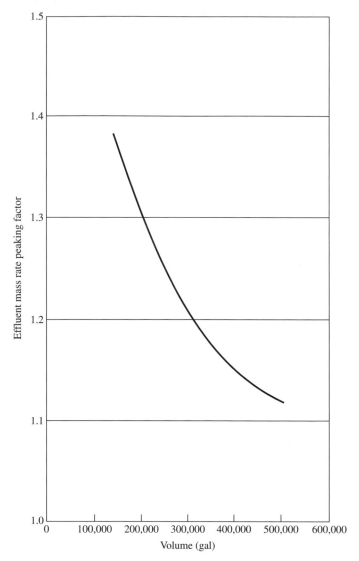

FIGURE 3.7
Variation in effluent mass rate peaking factor with basin volume for a variable-volume equalization basin.

Acid Wastes Neutralization through Limestone Beds

These may be downflow or upflow systems. The maximum hydraulic rate for downflow systems is 1 gal/(min · ft²) [4.07 × 10⁻² m³/(min · m²)] to ensure sufficient retention time. The acid concentration should be limited to 0.6 percent H_2SO_4 if H_2SO_4 is present to avoid coating of limestone with nonreactive $CaSO_4$ and excessive CO_2 evolution, which limits complete neutralization. High dilution or dolomitic limestone requires longer detention periods to effect neutralization. Hydraulic loading rates can be increased with upflow beds, because the products

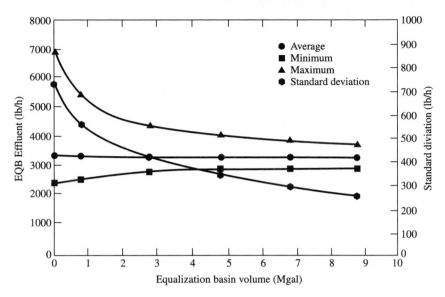

FIGURE 3.8*a*
Effect of variable-volume equalization on EQB effluent BOD loading.

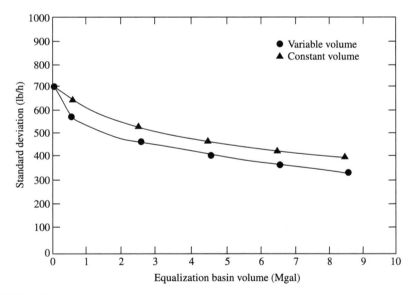

FIGURE 3.8*b*
Comparison of variable-volume and constant-volume equalization on EQB effluent BOD loading.

of reaction are swept out before precipitation. Since pH control is related to the bed depth, limestone beds are applicable only to wastewaters in which the influent acidity is relatively constant with time. A limestone bed system is shown in Fig. 3.11. The design of a limestone bed is shown in Example 3.4.

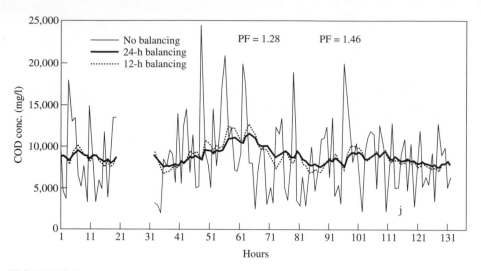

FIGURE 3.9
Load balancing analysis.

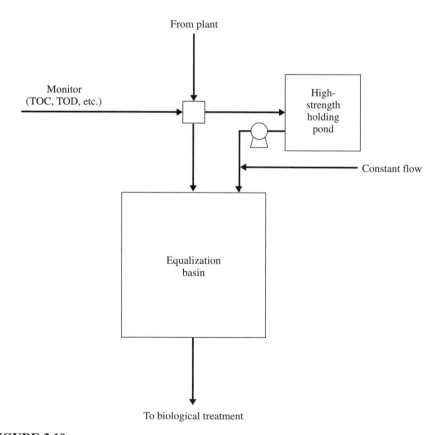

FIGURE 3.10
Use of a high-strength holding pond for spills.

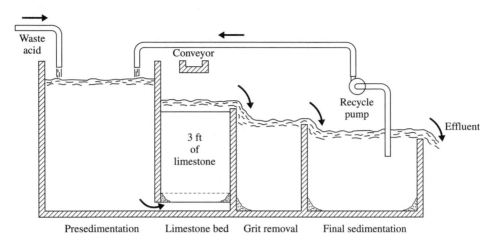

FIGURE 3.11
Simplified flow diagram of limestone neutralization.[3]

EXAMPLE 3.4. A wastewater flow of 100 gpm (0.38 m³/min) with 0.1 N H_2SO_4 requires neutralization prior to secondary treatment. This flow is to be neutralized to a pH of 7.0 using a limestone bed. Figure 3.12 presents the results of a series of laboratory pilot tests using a 1-foot- (30.5-cm-) diameter limestone bed. These data are for upflow units, with the effluent being aerated to remove residual CO_2. Assume limestone is 60 percent reactive.

Design a neutralization system specifying:
(*a*) Most economical bed depth of limestone
(*b*) Weight of acid per day to be neutralized
(*c*) Limestone requirements on an annual basis

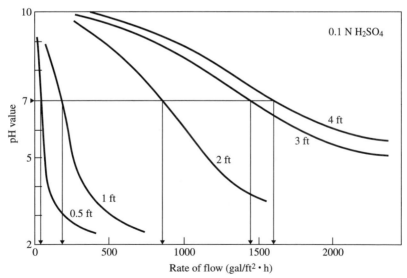

Note: ft = 30.5 cm

FIGURE 3.12

Solution.

(*a*) *Most economical bed depth.* Hydraulic loadings to get pH 7.0 with various depths of limestone bed: From the figure the allowable hydraulic loadings are estimated to be:

Depth, ft	0.5	1.0	2.0	3.0	4.0
Hydraulic loading, gal/(ft^2 · h)	42	180	850	1440	1600

Note: ft = 30.5 cm
        gal/(ft^2 · h) = 4.07 × 10^{-2} m^3/(m^2 · h)

Flow rate per unit limestone volume: The required flow rate per unit volume of bed can be calculated by:

$$Q/V = \frac{\text{hydraulic loading}}{\text{bed depth}}$$

Depth, ft	0.5	1.0	2.0	3.0	4.0
Q/V, gal/(ft^3 · h)	84	180	425	480	400

Note: gal/(ft^3 · h) = 0.134 m^3/(m^3 · h)

By plotting the flow rate per unit limestone volume against the limestone bed depth, the most economical bed depth of limestone is found to be ≈ 3 ft (0.91 m). This is the depth that gives the maximum flow per unit volume; see Fig. 3.13.

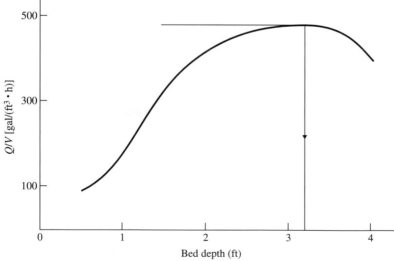

Note: ft = 30.5 cm
        gal/(ft^3 · h) = 0.134 m^3/(m^3 · h)

FIGURE 3.13

(b) *Weight of acid per day to be neutralized.* The weight of acid can be calculated by

$$\frac{100 \text{ gal}}{\text{min}} \times \frac{4900 \text{ mg H}_2\text{SO}_4}{1} \times \frac{1440 \text{ min}}{\text{day}} \times \frac{8.34 \times 10^{-6} \text{ lb}}{(\text{mg/l}) \text{ gal}}$$

$$= 5890 \text{ lb/d} \quad (2670 \text{ kg/d})$$

(c) *Annual limestone requirements.* The limestone requirements can be calculated by:

$$\frac{5890 \text{ lb}}{\text{d}} \times \frac{50 \text{ g CaCO}_3}{49 \text{ g H}_2\text{SO}_4} \times \frac{365 \text{ d}}{\text{yr}} \times \frac{1 \text{ lb limestone}}{0.60 \text{ lb CaCO}_3}$$

$$= 3,660,000 \text{ lb/yr} \quad (1,660,000 \text{ kg/yr})$$

Mixing Acid Wastes with Lime Slurries

This neutralization depends on the type of lime used. The magnesium fraction of lime is most reactive in strongly acid solutions and is useful below pH 4.2. Neutralization with lime can be defined by a basicity factor obtained by titration of a 1-g sample with an excess of HCl, boiling 15 min, followed by back titration with 0.5 N NaOH to the phenolphthalein endpoint.

In lime slaking, the reaction is accelerated by heat and physical agitation. For high reactivity, the lime reaction is complete in 10 min. Storage of lime slurry for a few hours before neutralization may be beneficial. Dolomitic quicklime (only the CaO portion) hydrates except at elevated temperature. Slaked quicklime is used as an 8 to 15 percent lime slurry. Neutralization can also be accomplished by using NaOH, Na_2CO_3, NH_4OH, or $Mg(OH)_2$.

Basic (Alkaline) Wastes

Any strong acid can be used effectively to neutralize alkaline wastes, but cost considerations usually limit the choice to sulfuric or possibly hydrochloric acid. The reaction rates are practically instantaneous, as with strong bases.

Flue gases, which may contain 14 percent CO_2, can be used for neutralization. When bubbled through the waste, the CO_2 forms carbonic acid, which then reacts with the base. The reaction rate is somewhat slower, but is sufficient if the pH need not be adjusted to below 7 to 8. Another approach is to use a spray tower in which the stack gases are passed countercurrent to the waste liquid droplets.

All of the above processes usually work better with the stepwise addition of reagents, i.e., a staged operation. Two stages, with possibly a third tank to even out any remaining fluctuations, are generally optimum.

There are a number of neutralizing agents available. Selection criteria should consider:

- Reaction rate
- Sludge production and disposal
- Safety and ease of handling for addition and storage
- Total cost including chemical feed and feed and storage equipment
- Side reactions, including dissolved salts, scale formation, and heat produced
- The effect of overdosage

The primary neutralizing agents are:

- *Basic agents—strong/weak*
- Lime in various forms—strong
- Caustic soda—strong
- Magnesium hydroxide—medium
- Sodium carbonate—weak
- Sodium bicarbonate—weak

- *Acidic agents—strong/weak*
- Sulfuric acid—strong
- Carbon dioxide—weak

The characteristics of typical neutralizing chemicals are shown in Tables 3.3 and 3.4.

System

Batch treatment is used for waste flows to 100,000 gal/d (380 m³/d). Continuous treatment employs automated pH control. Where air is used for mixing, the minimum air rate is 1 to 3 ft³/(min · ft²) [0.3 to 0.9 m³/(min · m²)] at 9 ft (2.7 m) liquid depth. If mechanical mixers are used, 0.2 to 0.4 hp/thousand gal (0.04 to 0.08 kW/m³) is required.

Control of Process

The automatic control of pH for wastestreams is one of the most troublesome, for the following reasons:

1. The relation between pH and concentration or reagent flow is highly nonlinear for strong acid–strong base neutralization, particularly when close to neutral (pH 7.0). The nature of the titration curve as shown in Fig. 3.14 favors multistaging in order to ensure close control of the pH.
2. The influent pH can vary at a rate as fast as 1 pH unit per minute.
3. The wastestream flow rates can double in a few minutes.
4. A relatively small amount of reagent must be thoroughly mixed with a large liquid volume in a short time interval.

Advantage is usually gained by the stepwise addition of chemicals (see Fig. 3.15). In reaction tank 1, the pH may be raised to 3 to 4. Reaction tank 2 raises the pH to 5 to 6 (or any other desired endpoint). If the wastestream is subject to slugs or spills, a third reaction tank may be desirable to effect complete neutralization. Okey et al.[4] have shown an advantage to combining NaOH and NaHCO₃. NaOH is employed in the first stage and NaHCO₃ in a second stage to provide a buffer and avoid pH swings due to a change in influent characteristics. Neutralization system design parameters are shown in Table 3.5. Acid waste neutralization is shown in Example 3.5.

TABLE 3.3

Summary of properties for typical neutralization chemicals

Property	Calcium carbonate ($CaCO_3$)	Calcium hydroxide ($Ca(OH)_2$)	Calcium oxide (CaO)	Hydrochloric acid (HCl)	Sodium carbonate (Na_2CO_3)	Sodium hydroxide ($NaOH$)	Sulfuric acid (H_2SO_4)
Available form	Powder, crushed (various sizes)	Powder, granules	Lump, pebble, ground	Liquid	Powder	Solid flake, ground flake, liquid	Liquid
Shipping container	Bags, barrel, bulk	Bags (50 lb),[†] bulk	Bags (80 lb), barrels, bulk	Barrels, drums, bulk	Bags (100 lb), bulk	Drum (735, 100, 450 lb)	Carboys, drums (825 lb), bulk
Bulk weight, lb/ft³	Powder 48 to 71; crushed 70 to 100	25 to 50	40 to 70	27.9%, 0.53 lb/gal[‡]; 31.45%, 9.65 lb/gal	34 to 62	Varies	106, 114
Commercial strength	—	Normally 13% $Ca(OH)_2$	75 to 99%, normally 90% CaO	27.9, 31.45, 35.2%	99.2%	98%	60° Be, 77.7%; 66° Be, 93.2%
Water solubility, lb/gal	Nearly insoluble	Nearly insoluble	Nearly insoluble	Complete	0.58 @ 32°F, 1.04 @ 50°F, 1.79 @ 68°F, 3.33 @ 86°F	3.5 @ 32°F, 4.3 @ 50°F, 9.1 @ 68°F, 9.2 @ 86°F	Complete
Feeding form	Dry slurry used in fixed beds	Dry or slurry	Dry or slurry [must be slaked to $Ca(OH)_2$]	Liquid	Dry, liquid	Solution	Liquid
Feeding type	Volumetric pump	Volumetric metering pump	Dry-volumetric, wet slurry (centrifugal pump)	Metering pump	Volumetric feeder, metering pump	Metering pump	Metering pump
Accessory equipment	Slurry tank	Slurry tank	Slurry tank, slaker	Dilution tank	Dissolving tank	Solution tank	—
Suitable handling materials	Iron, steel	Iron, steel, plastic, rubber hose	Iron, steel, plastic, rubber hose	Hastelloy A, selected plastic and rubber types	Iron, steel	Iron, steel	—
Comments	—	—	Provide means for cleaning slurry transfer pipes	—	Can cake	Dissolving solid forms generate much heat	Provide for spill cleanup and neutralization

[†] lb × 0.4536 = kg
[‡] lb/gal × 0.1198 = kg/l
[§] 0.555 (°F − 32) = °C

TABLE 3.4
Neutralization factors for common alkaline and acid reagents

Chemical	Formula	Equivalent weight	To neutralize 1 mg/l acidity or alkalinity (expressed as CaCO$_3$) requires n mg/l	Neutralization factor, assuming 100% purity for all compounds
				Basicity
Calcium carbonate	CaCO$_3$	50	1.000	1.000/0.56 = 1.786
Calcium oxide	CaO	28	0.560	0.560/0.56 = 1.000
Calcium hydroxide	Ca(OH)$_2$	37	0.740	0.740/0.56 = 1.321
Magnesium oxide	MgO	20	0.403	0.403/0.56 = 0.720
Magnesium hydroxide	Mg(OH)$_2$	29	0.583	0.583/0.56 = 1.041
Dolomitic quicklime	(CaO)$_{0.6}$(MgO)$_{0.4}$	24.8	0.497	0.497/0.56 = 0.888
Dolomitic hydrated lime	[Ca(OH)$_2$]$_{0.6}$ [Mg(OH)$_2$]$_{0.4}$	33.8	0.677	0.677/0.56 = 1.209
Sodium hydroxide	NaOH	40	0.799	0.799/0.56 = 1.427
Sodium carbonate	Na$_2$CO$_3$	53	1.059	1.059/0.56 = 1.891
Sodium bicarbonate	NaHCO$_3$	84	1.680	1.680/0.56 = 3.00
				Acidity
Sulfuric acid	H$_2$SO$_4$	49	0.980	0.980/0.56 = 1.750
Hydrochloric acid	HCl	36	0.720	0.720/0.56 = 1.285
Nitric acid	HNO$_3$	62	1.260	1.260/0.56 = 2.250
Carbonic acid	H$_2$CO$_3$	31	0.620	0.620/0.56 = 1.107

EXAMPLE 3.5. A wastewater (100 gal/min) (0.38 m^3/min) is highly acidic and requires neutralization prior to secondary treatment. This flow is to be neutralized to a pH of 7.0 by lime. The titration curve for the wastewater is shown in Fig. 3.14, from which a two-stage control neutralization system will be used with a total lime consumption of 2250 mg/l. The first stage requires 2000 mg/l and the second stage 250 mg/l.

The average lime dosage in the first stage is

$$(100 \text{ gal/min})(1440 \text{ min/d})(8.34 \text{ lb/million gal/mg/l})(2000 \text{ mg/l})$$

$$\times 10^{-6} \text{ (million gal/gal)} = 2400 \text{ lb/d} \quad (1090 \text{ kg/d})$$

The average lime dosage in the second stage is

$$100 \times 1440 \times 8.34 \times 250 \times 10^{-6} = 300 \text{ lb/d} \quad (140 \text{ kg/d})$$

The average lime dosage is 2700 lb/d (1.35 ton/d) (1230 kg/d). With this dosage and type of lime each basin should be designed with a detention time of 5 min.

$$\text{Volume} = 100 \text{ gal/min} \times 5 \text{ min} = 500 \text{ gal} (1.9 \text{ m}^3)$$

Use two tanks, 4.6 ft (1.40 m) diameter × 4.1 ft (1.25 m) deep, so that diameter and depth are approximately equal.

To maintain proper mixing in the reactor tanks, the power level required for 5 min detention time, $D/T = 0.33$, is 0.2 hp/thousand gal (40 W/m^3) (Fig. 3.16). Use 0.1-hp (75-W) mixers in each reaction tank.

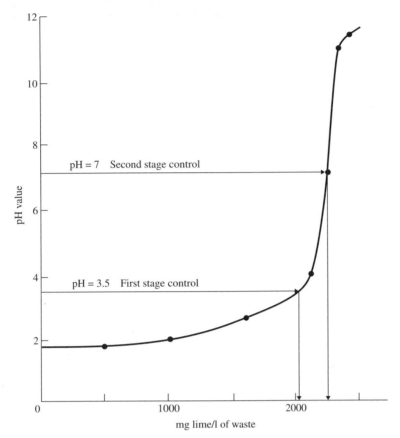

FIGURE 3.14
Lime-waste titration curve for strong acid.

Either one or two standard wall baffles, 180° apart, and one-twelfth to one-twentieth of the width of the tank diameter, located 24 in (61 cm) from the periphery of the impeller, are recommended for this operation.

3.3
SEDIMENTATION

Sedimentation is employed for the removal of suspended solids from wastewaters. The process can be considered in three basic classifications, depending on the nature of the solids present in the suspension: discrete, flocculent, and zone settling. In discrete settling, the particle maintains its individuality and does not change in size, shape, or density during the settling process. Flocculent settling occurs when the particles agglomerate during the settling period with a resulting change in size and settling rate. Zone settling involves a flocculated suspension which forms a lattice structure and settles as a mass, exhibiting a distinct interface during the settling process. Compaction of the settled sludge occurs in all sedimentation but will be considered separately under thickening.

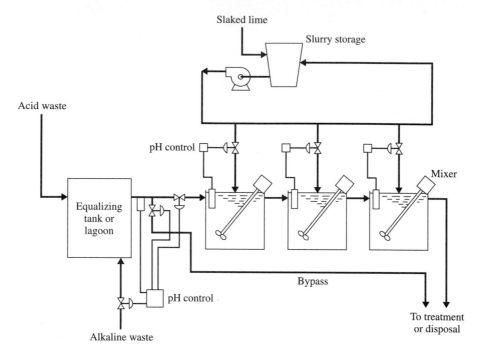

FIGURE 3.15
Multistage neutralization process.

TABLE 3.5
Neutralization system design parameters

Chemical storage tank	Liquid—use stored supply vessel
	Dry—dilute in a mix or day tank
Reaction tank:	
Size	Cubic or cylindrical with liquid depth equal to diameter
Retention time	5 to 30 min (lime—30 min)
Influent	Locate at tank top
Effluent	Locate at tank bottom
Agitator:	
Propeller type	Under 1000-gal tanks
Axial-flow type	Over 1000-gal tanks
Peripheral speeds	12 ft/s for large tanks
	25 ft/s for tanks less than 1000 gal
pH sensor	Submersible preferred to flow-through type
Metering pump or control valve	Pump delivery range limited to 10 to 1; valves have greater ranges.

The selection of neutralizing will depend on availability, chemical cost, and feeding methods.

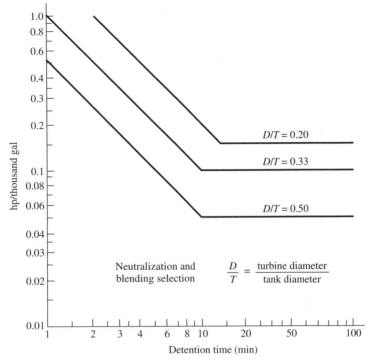

Note: hp/thousand gal = 198 W/m^3

FIGURE 3.16

Discrete Settling

A particle will settle when the impelling force of gravity exceeds the inertia and viscous forces. The terminal settling velocity of a particle is defined by the relationship

$$v = \sqrt{\frac{4g(\rho_s - \rho_l)D}{3C_d\rho_l}} \qquad (3.3)$$

where ρ_l = density of the fluid
ρ_s = density of the particle
v = terminal settling velocity of the particle
D = diameter of the particle
C_d = drag coefficient, which is related to the Reynolds number and particle shape
g = acceleration due to gravity

When the Reynolds number is small, i.e., less than 1.0 (small particles at low velocity), the viscous forces are predominant and

$$C_d = \frac{24}{N_{Re}} \qquad (3.4)$$

where

$$N_{\text{Re}} = \frac{vD\rho_l}{\mu} \tag{3.5}$$

ρ_l and μ are the density and viscosity of the liquid, respectively. Substitution of Eq. (3.4) in Eq. (3.3) yields Stokes' law:

$$v = \frac{\rho_s - \rho_l}{18\mu} gD^2 \tag{3.6}$$

As the Reynolds number increases, a transition zone occurs in which both inertia and viscous forces are effective. This occurs over a Reynolds number range of 1 to 1000, where

$$C_d = \frac{18.5}{N_{\text{Re}}^{0.6}} \tag{3.7}$$

Above a Reynolds number of 1000, viscous forces are not significant and the coefficient of drag is constant at 0.4.

The settling velocities of discrete particles, as related to diameter and specific gravity, are shown in Fig. 3.17.

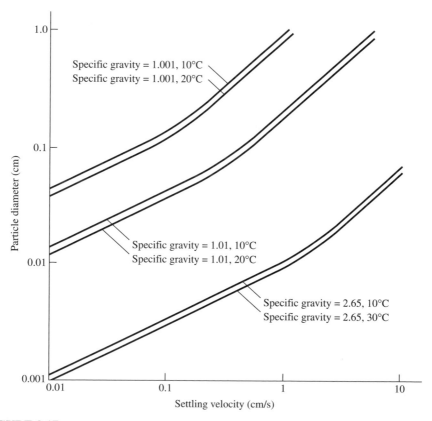

FIGURE 3.17
Settling properties of discrete particles.

Hazen[5] and Camp[6] developed relationships applicable to the removal of discrete particles in an ideal settling tank, based on the premises that the particles entering the tanks are uniformly distributed over the influent cross section and that a particle is considered removed when it hits the bottom of the tank. The settling velocity of a particle that settles through a distance equal to the effective depth of the tank in the theoretical detention period can be considered as an overflow rate:

$$v_o = \frac{Q}{A} \tag{3.8}$$

where Q = rate of flow through the tank and A = tank surface area. All particles with settling velocities greater than v_o will be completely removed, and particles with settling velocities less than v_o will be removed in the ratio v/v_o, as shown in Fig. 3.18. The removal of discrete particles is independent of tank depth and is a function only of the overflow rate.

When the suspension to be removed has a wide range of particle sizes, the total removal is defined by the relationship

$$\text{Total removal} = (1 - f_o) + \frac{1}{v_o} \int_0^{f_o} v \, df \tag{3.9}$$

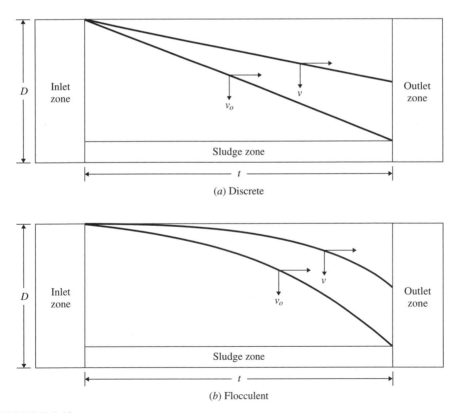

FIGURE 3.18
Ideal settling tank.

where f_o is the fraction of particles having a settling velocity equal to or less than v_o. Equation (3.9) must usually be solved by graphical integration to obtain the total removal.

The foregoing analysis is based on the performance of an ideal settling tank under quiescent conditions. In practice, however, short circuiting, turbulence, and bottom scour will affect the degree of solids removal. Dobbins[7] and Camp[6] have developed a relationship to compensate for the decrease in removals caused by turbulence (Fig. 3.17). Figure 3.19 shows the effect of turbulence in reducing the removal ratio of particles having a settling velocity v from the theoretical value of v/v_o, where v_o refers to the overflow rate. The ratio $vH/2E$ is a parameter of turbulent intensity, where H is the depth and E is a coefficient of turbulent transport. For narrow channels, Camp has shown the ratio to be equal to $122v/V$, where V is the average channel velocity.

Scour occurs when the flow-through velocity is sufficient to resuspend previously settled particles; it is defined by the relationship:

$$v_c = \sqrt{\frac{8\beta}{f} gD (S - 1)} \qquad (3.10)$$

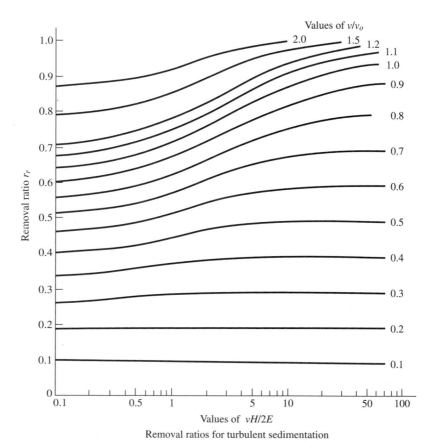

Removal ratios for turbulent sedimentation

FIGURE 3.19
Effect of turbulence on the subsidence of particles (*after Dobbins[7]*).

where v_c = velocity of scour

β = constant (0.04 for unigranular sand, 0.06 for nonuniform sticky material)

f = Weisbach-D'Arcy friction factor, 0.03 for concrete

S = particle specific gravity

g = acceleration due to gravity

Scour is usually not a problem in large settling tanks but can occur in grit chambers and narrow channels.

Flocculent Settling

Flocculent settling occurs when the settling velocity of the particle increases as it settles through the tank depth, because of coalescence with other particles. This increases the settling rate, yielding a curvilinear settling path, as shown in Fig. 3.18b. Most of the suspended solids in industrial wastes are of a flocculent nature. For discrete particles, the efficiency of removal is related only to the overflow rate, but when flocculation occurs, both overflow rate and detention time become significant.

Since a mathematical analysis is not possible in the case of flocculent suspensions, a laboratory settling analysis is required to establish the necessary parameters. The laboratory settling study can be conducted in a column of the type shown in Fig. 3.20. A minimum diameter of 5 in (12.7 cm) is recommended to minimize wall effects. Taps are located at 2-ft (0.61-m) depth intervals.

The concentration of suspended solids must be uniform at the start of the test; sparging air into the bottom of the column for a few minutes will accomplish this. It is also essential that the temperature be maintained constant throughout the test period to eliminate settling interference by thermal currents. Suspended solids are determined on samples drawn off at selected time intervals up to 120 min. The data collected from the 2- (0.61-), 4- (1.22-), and 6-ft (1.83-m) depth taps are used to develop the settling rate–time relationships.

The results obtained are expressed in terms of percent removal of suspended solids at each tap and time interval. These removals are then plotted against their respective depths and times, as shown in Fig. 3.21. Smooth curves are drawn connecting points of equal removal. The curves thus drawn represent the limiting or maximum settling path for the indicated percent; i.e., the specified percent suspended solids will have a net settling velocity equal to or greater than that shown, and would therefore be removed in an ideal settling tank of the same depth and detention time. The calculation of removal can be illustrated from the data of Fig. 3.21.

The overflow rate v_o is the effective depth, 6 ft (1.83 in), divided by the time required for a given percent to settle this distance. All particles having a settling velocity equal to or greater than v_o will be 100 percent removed. Particles with a lesser settling velocity v will be removed in the proportion v/v_o. For example, referring to Figure 3.21a, at a detention period of 60 min and a 6-ft (1.83-m) settling depth [v_o = 6 ft/h (1.83 m/h)], 50 percent of the suspended solids are completely removed; that is, 50 percent of the particles have a settling velocity equal to or greater than 6 ft/h (1.83 m/h). Particles in each additional 10 range will be removed

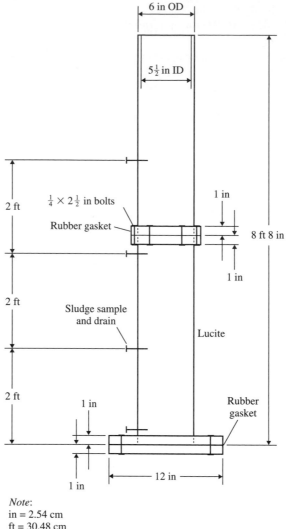

FIGURE 3.20
Laboratory settling column for the evaluation of flocculent settling.

in the proportion v/v_o or in the proportion to the average depth settled to the total depth of 6 ft (1.83 m). The average depth to which the 50 to 60 percent range has settled in Fig. 3.21*b* is 3.8 ft (1.16 m). The percent removal of this fraction is therefore 3.8 ft/6.0 ft (1.16 m/1.83 m), or 63 percent of the 10 percent. Each subsequent percent range is computed in a similar manner, and the total removal developed as shown in Table 3.6.

The total removal of suspended solids of 62.4 percent can be accomplished at an overflow rate of 6 ft/h = 1080 gal/(d · ft²) [44 m³/(d · m²)] at a retention period of 60 min. In a similar manner various percent removals and their associated overflow rates and detention periods can be computed.

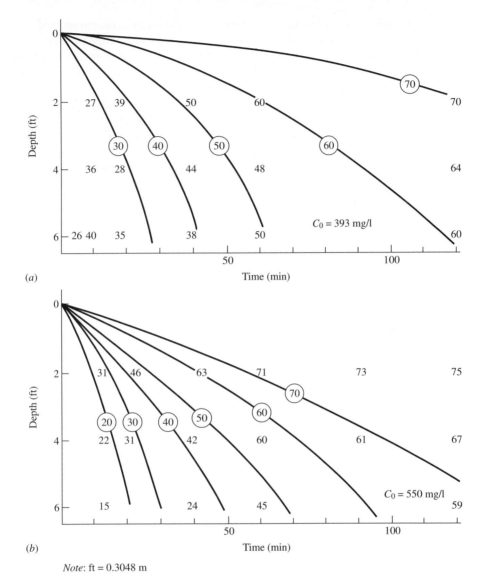

(a)

(b)

Note: ft = 0.3048 m

FIGURE 3.21
Flocculent settling relationships.

TABLE 3.6
Percent ranges for total removal of suspended solids

SS range, %	d/d_o	SS removal, %
0–50	1.0	50
50–60	0.64	6.4
60–70	0.25	2.5
70–100	0.05	1.5
Total removal		62.4

93

Since the degree of flocculation is influenced by the initial concentration of suspended solids, settling tests should be run over the anticipated range of suspended solids in the influent waste. In many wastes, a fraction of the measured suspended solids is not removed by settling, so the curves developed from the laboratory analysis will approach the removal as a limit.

Since the data obtained from the laboratory analysis represent ideal settling conditions, criteria for prototype design must account for the effects of turbulence, short circuiting, and inlet and outlet losses. The net effect of these factors is a decrease in the overflow rate and an increase in the detention time over that derived form the laboratory analysis. As a general rule, the overflow rate will be decreased by a factor of 1.25 to 1.75, and the detention period increased by a factor of 1.50 to 2.00. The development of these relationships in shown in Example 3.6.

EXAMPLE 3.6. Laboratory data were obtained on the settling of a paper mill waste (Table 3.7). Design a settling tank to produce a maximum effluent suspended solids of 150 mg/l.

Solution.

With the values given in Table 3.7, Fig. 3.21 can be drawn and Table 3.8 can be constructed by calculating the percent removal, the velocity, and the overflow rate of different times as indicated above.

By plotting the SS removal versus overflow rate and time, Figs. 3.22 and 3.23 can be drawn.

From Figs. 3.22 and 3.23:

For 393 mg/l:

$$\text{Percent removal} = \frac{393 - 150}{393} \times 100 = 62$$

$$\text{Overflow rate} = 770 \text{ gal/(d} \cdot \text{ft}^2)[31.4 \text{ m}^3/(\text{d} \cdot \text{m}^2)]$$

$$\text{Detention time} = 74 \text{ min}$$

For 550 mg/l:

$$\text{Percent removal} = \frac{550 - 150}{550} \times 100 = 73$$

$$\text{Overflow rate} = 540 \text{ gal/(d} \cdot \text{ft}^2)[22.0 \text{ m}^3/(\text{d} \cdot \text{m}^2)]$$

$$\text{Detention time} = 104 \text{ min}$$

Design:

$$\text{Overflow rate} = \frac{540}{1.5} = 360 \text{ gal/(d} \cdot \text{ft}^2)[14.7 \text{ m}^3/(\text{d} \cdot \text{m}^2)]$$

$$\text{Detention time } t = \frac{104 \text{ min}}{60 \text{ min/h}} \times 1.75 = 3 \text{ h}$$

For 1 million gal/d:

$$\text{Area } A = \frac{\text{flow}}{\text{overflow rate}} = \frac{10^6 \text{ gal/d}}{360 \text{ gal/(d} \cdot \text{ft}^2)} = 2780 \text{ ft}^2 \text{ (258 m}^2)$$

$$\text{Effective depth} = \frac{t \times \text{flow}}{A}$$

$$= \frac{3 \text{ h} \times 10^6 \text{ gal/d}}{2780 \text{ ft}^2} \times \frac{0.134 \text{ ft}^3/\text{gal}}{24 \text{ h/d}}$$

$$= 6 \text{ ft} \ (1.8 \text{ m})$$

TABLE 3.7
Settling data

Time, min	2 ft	4 ft	6 ft
Removal, %			
Time, min	2 ft	4 ft	6 ft
Initial solids, 393 mg/l			
5			26
10	27	36	40
20	39	28	35
40	50	44	38
60	60	48	50
120	70	64	60
Initial solids, 550 mg/l			
15	31	22	15
20	46	31	
40	63	42	24
60	71	60	45
90	73	61	
120	75	67	59

Note: ft = 30.48 cm

TABLE 3.8

Time, min	Velocity, ft/h	Removal of SS, %	Overflow rate, gal/(d · ft²)
$C_0 = 393$ mg/l			
27.5	13.1	41.6	2360
42.0	8.6	49.5	1550
62.0	5.8	57.3	1050
115.0	3.1	65.8	560
$C_0 = 550$ mg/l			
20	18.0	36.7	3250
27	13.3	46.8	2400
47	7.7	56.5	1400
66	5.4	62.5	980
83	3.8	70.8	690

Note:
ft/h = 0.305 m/h
gal/(d · ft²) = 4.07 × 10⁻² m³/(d · m²)

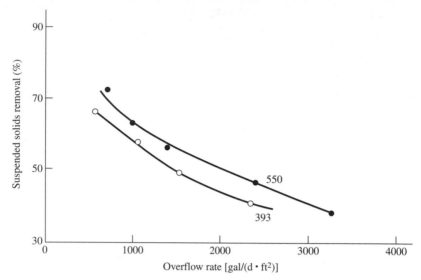

Note: gal/(d · ft²) = 4.07 × 10⁻² m³/(d · m²)

FIGURE 3.22

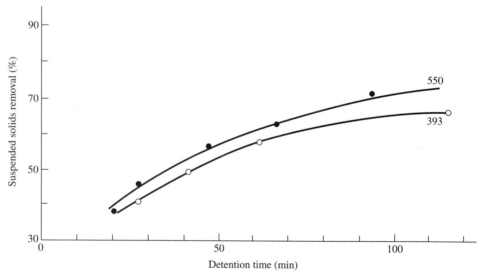

FIGURE 3.23

The sedimentation performances for various pulp and paper-mill wastes are summarized in Table 3.9.

Tanning wastewaters[9] with an initial suspended solids of 1200 mg/l showed 69 percent reduction with a retention period of 2 h. BOD removals of 86.9 percent were obtained from the settling of cornstarch wastes.[10]

The effect of the initial concentration of suspended solids on the sedimentation efficiency of a pulpwood waste is shown in Fig. 3.24.

TABLE 3.9
Settling characteristics of pulp and paper-mill wastes[8]

Type of waste	Flow, million gal/day	Raw SS, ppm	Raw BOD, ppm	Temperature, °F	Removal % SS	Removal % BOD	Detention time, h	OR, gal/ (d · ft²)
Paperboard	4.5	2,500	450	85	90	67	5.35	504
	0.75	136		85	90	50	1.15	940
	1.36	10,000	360	62	85	24	5.40	430
	2.5	1,185	395		96.1	19	5.3	525
	31	524	195	110	42	25	9.4	438
	30	850	250	95	80	25	0.5	1910
	3.3	2,000		90	85		2.6	1028
	0.25	50	100	100	80	25	4.5	39
	0.301	1,150	250	110	98	50		90
	35	4,000	200	100	90	10–15	1.5	374
Specialty	9.4	203	97	81	94	86	2.56	832
	2.2	6,215	120	120		90	1.5	157
	1.8	665	620	95	91	58	0.5	406
	50	120	85	100	80	16	18.2	477
Fine paper	6	200		65	95	90	3.9	695
	6.0	254	235	90	50	34	2.2	890
	9.9	500	364	70-100	90	35	2.4	1120
	3.5	300	250	65	95	48	6.0	372
	7.5-9.0	560	126	65	80	42	4.0	670
Miscellaneous	7	430	250	70	70	20	1.8	505
	14	1,000	330	73	65	60		911
	25	75	100		90	0.0	6.9	17
	17	100	425		95	50	5.9	846
	0.5	200	200	85	90		1.9	1590
	1.0	1,000	900	100		95	2.9	509

Note:
million gal/d = 3.75 × 10^3 m³/d
°C = $\frac{5}{9}$ (°F − 32)
gal/(d · ft²) = 4.07 × 10^{-2} m³/(d · m²)

Zone Settling

Zone settling is characterized by activated sludge and flocculated chemical suspensions when the concentration of solids exceeds approximately 500 mg/l. The floc particles adhere together and the mass settles as a blanket, forming a distinct interface between the floc and the supernatant. The settling process is distinguished by

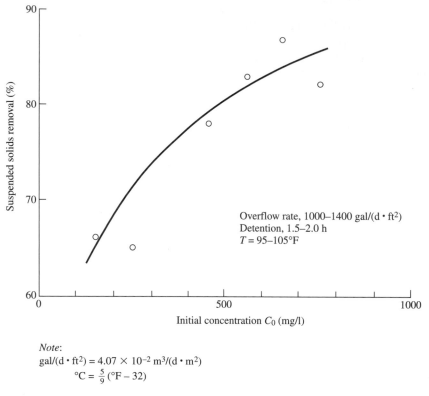

Note:

$$\text{gal/(d} \cdot \text{ft}^2) = 4.07 \times 10^{-2} \text{ m}^3/(\text{d} \cdot \text{m}^2)$$
$$°C = \tfrac{5}{9}(°F - 32)$$

FIGURE 3.24
Effect of initial suspended solids concentration on percent removal from a pulp and paper-mill wastewater.

four zones, as shown in Fig. 3.25. Initially, all the sludge is at a uniform concentration, *A* in Fig. 3.25.

During the initial settling period the sludge settles at a uniform velocity. The settling rate is a function of the initial solids concentration *A*. As settling proceeds, the collapsed solids *D* on the bottom of the settling unit build up at a constant rate. *C* is a zone of transition through which the settling velocity decreases as the result of an increasing concentration of solids. The concentration of solids in the zone settling layer remains constant until the settling interface approaches the rising layers of collapsed solids, III, and a transition zone occurs. Through the transition zone *C*, the settling velocity will decrease because of the increasing density and viscosity of the suspension surrounding the particles. When the rising layer of settled solids reaches the interface, a compression zone occurs in stage IV.

In the separation of flocculent suspensions, both clarification of the liquid overflow and thickening of the sludge underflow are involved. The overflow rate for clarification requires that the average rise velocity of the liquid overflowing the tank be less than the zone settling velocity of the suspension. The tank surface area requirements for thickening the underflow to a desired concentration level are

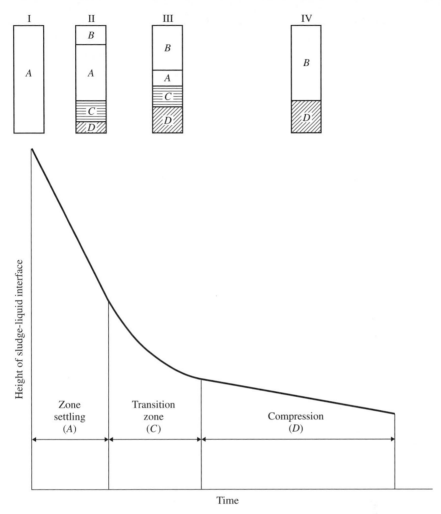

FIGURE 3.25
Settling properties of a flocculated sludge.

related to the solids loading to the unit and are usually expressed in terms of a mass loading (pounds of solids per square foot per day or kilograms per square meter per day) or a unit area (square feet per pound of solids per day or square meters per kilogram per day).

The mass loading concept for the thickening of industrial sludges is developed in Chap. 11.

Laboratory Evaluation of Zone Settling and Calculation of Solids Flux

The settling properties of flocculated sludges can be evaluated in a liter cylinder equipped with a slow-speed stirrer rotating at a speed of 4 to 5 revolutions per hour (r/h). The effect of the stirrer is to simulate the hydraulic motion and rake action in a clarifier and to break the stratification and arching action of the settling sludge. In some cases, an initial flocculation of the sludge occurs when it is added to the graduated cylinder. The settling and compaction curve is developed by plotting the height of the sludge interface versus the time of settling.

Clarifiers

Clarifiers may be either rectangular or circular. In most rectangular clarifiers, scraper flights extending the width of the tank move the settled sludge toward the inlet end of the tank at a speed of about 1 ft/min (0.3 m/min). Some designs move the sludge toward the effluent end of the tank, corresponding to the direction of flow of the density current. A typical unit is shown in Fig. 3.26.

Circular clarifiers may employ either a center-feed well or a peripheral inlet. The tank can be designed for center sludge withdrawal or vacuum withdrawal over the entire tank bottom.

Circular clarifiers are of three general types. With the center-feed type, the water is fed into a center well and the effluent is pulled off at a weir along the outside. With a peripheral-feed tank, the effluent is pulled off at the tank center. With a rim-flow clarifier, the peripheral feed and effluent drainoff are also along the clarifier rim, but this type is usually used for larger clarifiers.

The circular clarifier usually gives the optimal performance. Rectangular tanks may be desired where construction space in limited and multiple tanks are to be constructed. In addition, a series of rectangular tanks would be cheaper to construct because of the "shared wall" concept.

The reactor clarifier is another variation where the functions of chemical mixing, flocculation, and clarification are combined in the highly efficient solids contact unit. This combination achieves the highest overflow rate and the highest effluent quality of all clarifier designs.

The circular clarifier can be designed for center sludge withdrawal or vacuum withdrawal over the entire tank bottom. Center sludge withdrawal requires a min-

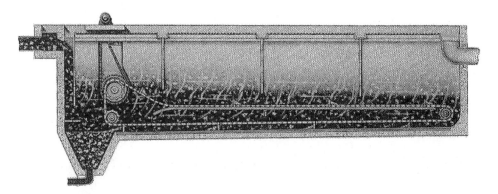

FIGURE 3.26
Rectangular clarifier.

imum bottom slope of 1 in/ft (8.3 cm/m). The flow of sludge to the center well is largely hydraulically motivated by the collection mechanism, which serves to overcome inertia and avoid sludge adherence to the tank bottom. The vacuum drawoff is particularly adaptable to secondary clarification and thickening of activated sludge. A circular clarifier is shown in Fig. 3.27.

The mechanisms can be of the plow type or the rotary-hoe type. The plow-type mechanism employs staggered plows attached to two opposing arms that move at about 10 ft/min (3 m/min). The rotary-hoe mechanism consists of a series of short scrapers suspended from a rotating supporting bridge on endless chains that make contact with the tank bottom at the periphery and move to the center of the tank.

An inlet device is designed to distribute the flow across the width and depth of the settling tank. The outlet device is likewise designed to collect the effluent uniformly at the outlet end of the tank. Well-designed inlets and outlets will reduce the short-circuiting characteristics of the tank. Increased weir length can be provided by extending the effluent channels back into the basin or by providing multiple effluent channels. In circular basins, inboard or radial weirs will ensure low take-off velocities. Relocation of weirs is sometimes necessary to minimize solids carryover induced by density currents and resulting in upwelling swells of sludge at the end of the settling tank. The installation of a plate below the effluent weir extending 18 in (45.7 cm) into the clarifier will deflect rising solids and permit them to resettle. Improved performance of secondary clarifiers has been achieved with this modification.

Tube settlers offer increased removal efficiency at higher loading rates and lower detention times. An immediate advantage is that modules of inclined tubes constructed of plastic can be installed in existing clarifiers to upgrade performance.

FIGURE 3.27
Circular clarifier.

There are two types of tube clarifiers, the slightly inclined and the steeply inclined units. The slightly inclined unit usually has the tubes inclined at a 5° angle. For the removal of discrete particles, an inclination of 5° has proven most efficient.

The steeply inclined unit is less efficient in removal of discrete particles, but can be operated continuously. When the tubes are inclined greater than 45°, the sludge is deposited and slides back out of the tube, forming a countercurrent flow. In practice most wastes are flocculent in nature, and removal efficiency is improved when the tubes are inclined to 60° to take advantage of the increased flocculation that occurs as the solids slide back out of the tube. Steeply inclined units are usually used where sedimentation units are being upgraded.

The hydraulic characteristics of a settling tank can be defined by a dispersion test in which dye or tracer is injected into the influent as a slug, and the concentration measured in the effluent as a function of time.

It is frequently possible to improve the performance of an existing settling basin by making modifications based on the results of a dispersion test. A comparison of a hydraulically overloaded center-feed tank converted to a peripheral-feed tank is shown in Fig. 3.28.

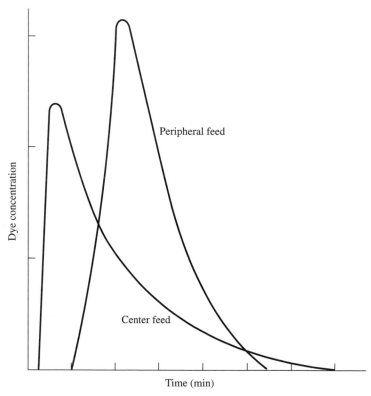

FIGURE 3.28
Dispersion characteristics of two settling tanks.

3.4
OIL SEPARATION

In an oil separator, free oil is floated to the surface of a tank and then skimmed off. The same conditions holding for the subsidence of particles apply, except that the lighter-than-water oil globules rise through the liquid. The design of gravity separators as specified by the American Petroleum Institute[11] is based on the removal of all free oil globules larger than 0.015 cm.

The Reynolds number is less than 0.5, so Stokes' law applies. A design procedure considering short circuiting and turbulence has been developed.[12]

An API oil separator is shown in Fig. 3.29. The performance of an API separator is shown in Figs. 3.30 and 3.31 and typical efficiencies shown in Table 3.10.[13] Design criteria suggest that the flow rate should not exceed 2 ft/min, the length to width ratio should be at least 5 to avoid dead spots in the separator, and the minimum depth set at 4 ft. Variability in performance reported by Rebhun and Galil[14] is shown in Fig. 3.32. The influence of initial oil concentration on separator efficiency is shown in Fig. 3.33.[13]

Plate separators include parallel plate separators and corrugated plate separators (CPS). Plate separators are designed to separate oil droplets larger than 0.006 cm. It has been found by experience that a 0.006-cm separation will generally produce a 10-mg/l free oil nonemulsified effluent. This quality can usually be met when the influent oil content is less than 1 percent. One problem with the CPS is that reduced efficiency results from high oil loadings caused by oil-droplet shear and reentrainment of the oil droplet. This is largely overcome using a cross-flow corrugated plate separator in which the separated oil rises across the direction of flow rather than against the direction of flow (plates are angled at 45° and spaced at 10 mm). The hydraulic loading varies with temperature and the specific gravity of the oil. Nominal flow rates are specified for a temperature of 20°C and a specific gravity of 0.9 for the oil. A hydraulic loading of 0.5 m³/(h · m²) of actual plate area will usually result in separation of 0.006-cm droplets. A 50 percent safety factor is usually employed for design purposes. A plate separator is shown in Fig. 3.34.

Several types of filtration devices have proved effective in removing free and emulsified oils from refinery-petrochemical wastewaters. These vary from filters with sand media to those containing special media which exhibit a specific affinity for oil. One type is an upflow unit using a graded silica medium as the filtering and coalescing section. Even the small particles and globules are separated and retained on the medium. The oil particles, which flow upward by gravity differential and fluid flow, rise through the coalescent medium and through the water phase from which the oil is separated and collected near the top of the separator. The bed is regenerated by introducing wash water at a rapid rate and evacuating the solids and remaining oil. This filtration and coalescing process is often enhanced by the use of polymer resin media. The primary application of these units are for selected in-plant streams which are dirt-free. Another application would be ballast water treatment following phase separation.

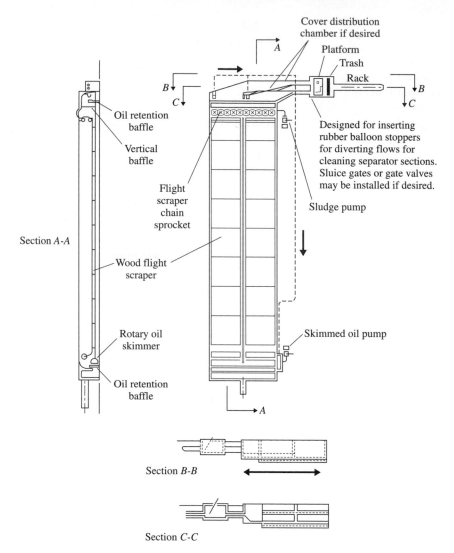

FIGURE 3.29
Example of general arrangement for API separator (*courtesy of the American Petroleum Institute*).

Emulsified oily materials require special treatment to break the emulsions so that the oily materials will be free and can be separated by gravity, coagulation, or air flotation. The breaking of emulsions is a complex art and may require laboratory or pilot-scale investigations prior to developing a final process design.

Emulsions can be broken by a variety of techniques. Quick-breaking detergents form unstable emulsions which break in 5 to 60 min to 95 to 98 percent completion. Emulsions can be broken by acidification, the addition of alum or iron salts, or the use of emulsion-breaking polymers. The disadvantage of alum or iron is the large quantities of sludge generated.

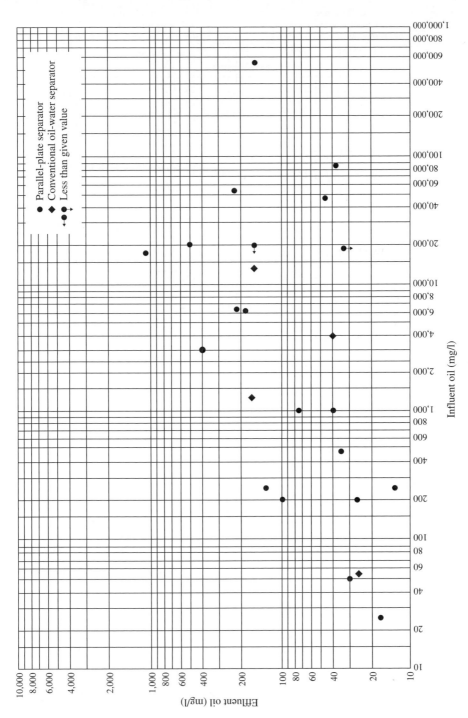

FIGURE 3.30
Correlation between influent and effluent oil levels for existing oil-water separators.

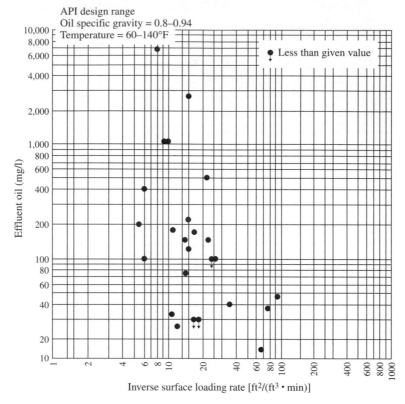

Note: The data in this figure are taken from the 1985 API Refinery Survey.

FIGURE 3.31
Performance of conventional oil-water separators.

TABLE 3.10
Typical efficiencies of oil separation units

Oil content		Oil removed, %	Type	COD removed, %	SS removed, %
Influent, mg/l	Effluent, mg/l				
300	40	87	Parallel plate	—	—
220	49	78	API	45	—
108	20	82	Circular	—	—
108	50	54	Circular	16	—
98	44	55	API	—	—
100	40	60	API	—	—
42	20	52	API	—	—
2000	746	63	API	22	33
1250	170	87	API	—	68
1400	270	81	API	—	35

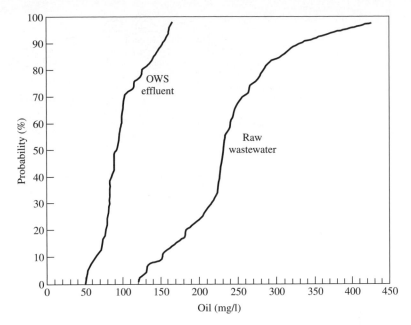

FIGURE 3.32
Oil removal by the oil-water separation (OWS) unit.

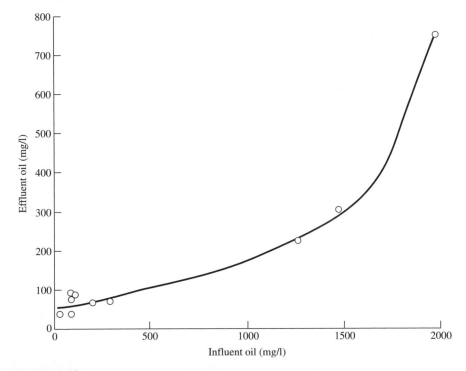

FIGURE 3.33
Effect of influent oil concentration separator efficiency.

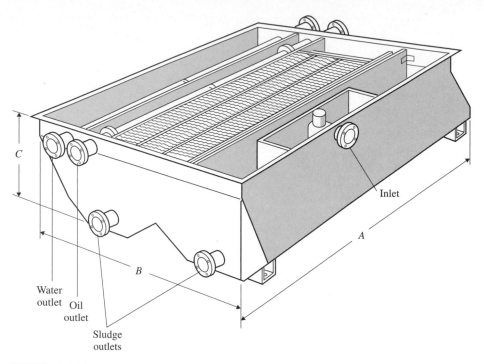

FIGURE 3.34
Cross-flow plate separator.

3.5
SOUR WATER STRIPPERS

Stripping processes are used to remove selected constituents from liquid streams. The two most prevalent pollutants found in refinery wastewaters which are susceptible to stripping are hydrogen sulfide and ammonia resulting from the destruction of essentially all the organic nitrogen and sulfur compounds during desulfurization, denitrification, and hydrotreating. The use of steam within the processes is the primary source of conveyance, as the condensation occurs simultaneously with the condensation of hydrocarbon liquids and in the presence of a hydrocarbon vapor phase which contains H_2S and NH_3.[12] Phenols also may be present in these "sour water" condensates and can be stripped from solutions, although the efficiency of removal is less than for sulfide and ammonia. Other aromatics also can be stripped from solution at various levels of efficiency. As the regulatory agencies are imposing increasingly stringent quality standards for refinery wastewater in terms of immediate oxygen demand (sulfide causative) and ammonia, the necessity of in-plant control through stripping towers may be required whether it can be justified in terms of product recovery or not.

The design criteria for sour water strippers is well documented and is outlined in detail elsewhere.[12, 15] There are various types of strippers but most involve a single tower equipped with trays or some type of packing. The feed water enters at the top of the tower and steam or stripping gas is introduced at the bottom. As H_2S

is less soluble in water than NH_3, it is more readily stripped from the solution. High temperatures (230°F or more) are required to remove NH_3, where H_2S could be stripped at 100°F if NH_3 were fixed or not present. Therefore, acidification with a mineral acid or flue gas is often used to fix the NH_3 and allow more efficient H_2S removal. Average operating characteristics of some sour water strippers are cited in Table 3.11.[15] Although acidification enhances sulfide removal, it fixes the ammonia and prevents its removal. This has led to a two-stage stripping and recovery process developed by the Chevron Research Company. The process includes a degasser-surge tank combination which allows operational flexibility. After any floating hydrocarbon is skimmed, it is routed to the first column where the H_2S is stripped and sent to a sulfur recovery plant. The water-ammonia mixture then goes to a second fractionator for ammonia stripping. The overhead ammonia, approximately 98 percent pure as it leaves the condenser, is further purified passing through a scrubber system, and is then liquefied as high-purity ammonia. The cooled water bottoms from the system are then sufficiently free of H_2S and NH_3 to satisfy most quality criteria, containing less than 5 mg/l H_2S and 50 mg/l NH_3.

3.6
FLOTATION

Flotation is used for the removal of suspended solids and oil and grease from wastewaters and for the separation and concentration of sludges. The waste flow or a portion of clarified effluent is pressurized to 50 to 70 lb/in² (345 to 483 kPa or 3.4 to 4.8 atm) in the presence of sufficient air to approach saturation. When this pressurized air-liquid mixture is released to atmospheric pressure in the flotation unit, minute air bubbles are released from solution. The sludge flocs, suspended solids, or oil globules are floated by these minute air bubbles, which attach themselves to and become enmeshed in the floc particles. The air-solids mixture rises to

TABLE 3.11
Average operating characteristics of sour water strippers

Type of stripper	Flow rate of stripping medium, SCF/gal	Removal		Temperature	
		H_2S, %	NH_3, %	Tower feed, °F	Tower bottom, °F
Steam					
With acidifying[†]	8–32	96–100	69–95	150–240	230–270
Without acidifying[‡]	4–6	97–100	0	200	230–250
Flue gas					
With steam[§]	12.7	88–98	77–90	235	235
Without steam[§]	11.9	99	8	135	140
Natural gas					
With acidifying	7.5	98	0	70–100	70–100

[†]Data from eight towers.
[§]Data from only one tower.
[‡]Data from two towers.

the surface, where it is skimmed off. The clarified liquid is removed from the bottom of the flotation unit; at this time a portion of the effluent may be recycled back to the pressure chamber. When flocculent sludges are to be clarified, pressurized recycle will usually yield a superior effluent quality since the flocs are not subjected to shearing stresses through the pumps and pressurizing system.

Air Solubility and Release

The saturation of air in water is directly proportional to pressure and inversely proportional to temperature. Pray[16] and Frohlich[17] found that the oxygen and nitrogen solubilities in water follow Henry's law over a wide pressure range. Vrablick[18] has shown that although a linear relationship exists between pressure and solubility for most industrial wastes, the slope of the curve varies, depending on the nature of the waste constituents present. The solubility of air in water at atmospheric pressure is shown in Table 3.12.

The quantity of air that will theoretically be released from solution when the pressure is reduced to 1 atm can be computed from

$$s = s_a \frac{P}{P_a} - s_a \tag{3.11}$$

where s = air released at atmospheric pressure per unit volume at 100 percent saturation, cm^3/l

      s_a = air saturation at atmospheric pressure, cm^3/l

      P = absolute pressure

      P_a = atmospheric pressure

The actual quantity of air released will depend upon turbulent mixing conditions at the point of pressure reduction and on the degree of saturation obtained in the pres-

TABLE 3.12
Air characteristics and solubilities

Temperature		Volume solubility		Weight solubility		Density	
°C	°F	ml/l	ft³/thousand gal	mg/l	lb/thousand gal	g/l	lb/ft³
0	32	28.8	3.86	37.2	0.311	1.293	0.0808
10	50	23.5	3.15	29.3	0.245	1.249	0.0779
20	68	20.1	2.70	24.3	0.203	1.206	0.0752
30	86	17.9	2.40	20.9	0.175	1.166	0.0727
40	104	16.4	2.20	18.5	0.155	1.130	0.0704
50	122	15.6	2.09	17.0	0.142	1.093	0.0682
60	140	15.0	2.01	15.9	0.133	1.061	0.0662
70	158	14.9	2.00	15.3	0.128	1.030	0.0643
80	176	15.0	2.01	15.0	0.125	1.000	0.0625
90	194	15.3	2.05	14.9	0.124	0.974	0.0607
100	212	15.9	2.13	15.0	0.125	0.949	0.0591

Values presented in absence of water vapor and at 14.7 lb/in^2 abs pressure (1 atm).

surizing system. Since the solubility in industrial wastes may be less than that in water, a correction may have to be applied to Eq. (3. 11). Retention tanks will generally yield 86 to 90 percent saturation. Equation (3.11) can be modified to account for air saturation:

$$s = s_a \left(\frac{fP}{P_a} - 1 \right) \tag{3.12}$$

where f is the fraction of saturation in the retention tank.

The performance of a flotation system depends on having sufficient air bubbles present to float substantially all of the suspended solids. An insufficient quantity of air will result in only partial flotation of the solids, and excessive air will yield no improvement. The performance of a flotation unit in terms of effluent quality and solids concentration in the float can be related to an air/solids (A/S) ratio, which is usually defined as mass of air released per mass of suspended solids in the influent waste:

$$\frac{A}{S} = \frac{s_a R}{S_a Q} \left(\frac{fP}{P_a} - 1 \right) \tag{3.13}$$

where Q = wastewater flow
        R = pressurized recycle
        S_a = influent oil and/or suspended solids

The relationship between the air/solids ratio and effluent quality is shown in Fig. 3.35. It should be noted that the shape of the curve will vary with the nature of the solids in the feed.

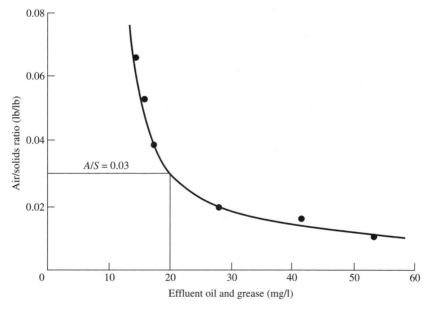

FIGURE 3.35
Effects on A/S on effluent quality.

Vrablick[18] has shown that the bubbles released after pressurization (20 to 50 lb/in^2 or 1.36 to 3.40 atm) range in size from 30 to 100 μm. The rise velocity closely follows Stokes' law. The rise velocity of a solids-air mixture has been observed to vary from 1 to 5 in/min (2.56 to 12.7 cm/min) and will increase with an increasing air/solids ratio. In the flotation of activated sludge of 0.91 percent solids and 40 lb/in^2 (276 kPa or 2.72 atm), Hurwitz and Katz[19] observed free rises of 0.3 (9), 1.2 (37), and 1.8 ft/min (55 cm/min) for recycle ratios of 100, 200, and 300 percent, respectively. The initial rise rate will vary with the character of the solids.

The primary variables for flotation design are pressure, recycle ratio, feed solids concentration, and retention period. The effluent suspended solids decrease and the concentration of solids in the float increase with increasing retention period. When the flotation process is used primarily for clarification, a detention period of 20 to 30 min is adequate for separation and concentration. Rise rates of 1.5 to 4.0 gal/(min · ft^2) [0.061 to 0.163 m^3/(min · m^2)] are commonly employed. When the process is employed for thickening, longer retention periods are necessary to permit the sludge to compact.

The principal components of a flotation system are a pressurizing pump, air-injection facilities, a retention tank, a backpressure regulating device, and a flotation unit, as shown in Fig. 3.36. The pressurizing pump creates an elevated pres-

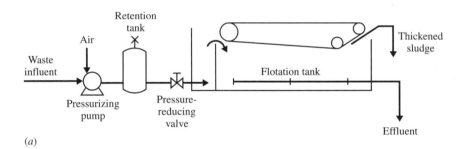

(a)

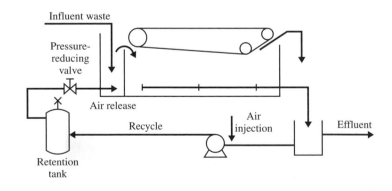

(b)

FIGURE 3.36
Schematic representation of flotation systems. (a) Flotation system without recirculation. (b) Flotation system with recirculation.

sure to increase the solubility of air. Air is usually added through an injector on the suction side of the pump or directly to the retention tank.

The air and liquid are mixed under pressure in a retention tank with a detention time of 1 to 3 min. A backpressure regulating device maintains a constant head on the pressurizing pump. Various types of values are used for this purpose. The flotation unit may be either circular or rectangular with a skimming device to remove the thickened, floated sludge, as shown in Fig. 3.37.

The induced-air flotation system, shown in Fig. 3.38, operates on the same principles as a pressurized-air DAF unit. The gas, however, is self-induced by a rotor-disperser mechanism. The rotor, the only moving part of the mechanism that is submerged in the liquid, forces the liquid through the disperser openings, thereby creating a negative pressure. It pulls the gas downward into the liquid, causing the desired gas-liquid contact. The liquid moves through a series of four cells before leaving the tank, and the float skimmings pass over the overflow weirs on each side of the unit. This type of system offers the advantages of significantly lower capital cost and smaller space requirements than the pressurized system, and current performance data indicate that these systems have the capacity to effectively remove free oil and suspended materials.

The disadvantages include higher connected power requirements than the pressurized system, performance dependent on strict hydraulic control, less chemical

FIGURE 3.37
Clarifier flotation unit.

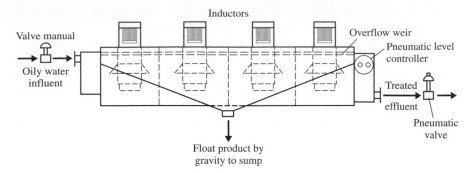

FIGURE 3.38
Induced air flotation system (*Wemco Envirotech Company*).

addition and flocculation flexibility, and relatively high volumes of float skimmings as a function of liquid throughout (3 to 7 percent of the incoming flow for induced-air systems is common compared to less than 1 percent for pressurized-air systems).

It is possible to estimate the flotation characteristics of a waste by the use of a laboratory flotation cell, as shown in Fig. 3.39. The procedure is as follows:

1. Partially fill the calibrated cylinder with waste or flocculated sludge mixture and the pressure chamber with clarified effluent or water.
2. Apply compressed air to the pressure chamber to attain the desired pressure.
3. Shake the air-liquid mixture in the pressure chamber for 1 min and allow to stand for 3 min to attain saturation. Maintain the pressure on the chamber for this period.
4. Release a volume of pressurized effluent to the cylinder and mix with the waste or sludge. The volume to be released is computed from the desired recycle ratio. The velocity of release through the inlet nozzle should be of such a magnitude as not to shear the suspended solids in the feed mixture but to maintain adequate mixing.
5. Measure the rise of the sludge interface with time. Correction must be applied to scale up the height of rise in the test cylinder to the depth of the prototype unit.
6. After a detention time of 20 min, the clarified effluent and the floated sludge are drawn off through a valve in the bottom of the cylinder.
7. Relate the effluent suspended solids to the calculated air/solids ratio as shown in Fig. 3.35. When pressurized recycle is used, the air/solids ratio is computed:

$$\frac{A}{S} = \frac{1.3 s_a R(P - 1)}{Q S_a}$$

where s_a = air saturation, cm³/l
  R = pressurized volume, l
  P = absolute pressure, atm
  Q = waste flow, l
  S_a = influent suspended solids, mg/l

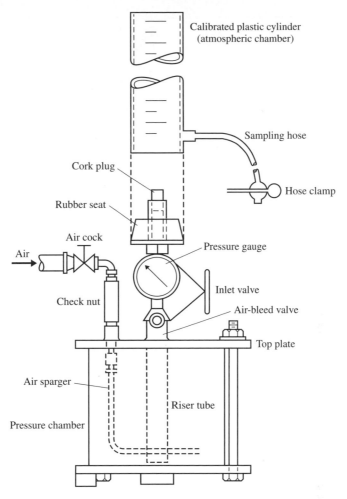

FIGURE 3.39
Laboratory flotation cell.

High-degree clarification frequently requires that flocculating chemicals be added to the influent before it is mixed with the pressurized recycle. Alum or polyelectrolytes are used as flocculating agents. Hurwitz and Katz[19] have shown that the rise rate of chemical floc will vary from 0.65 to 2.0 ft/min (20 to 61 cm/min), depending on the floc size and characteristics. Treatment efficiency with chemical addition is shown in Fig. 3.40.

Flotation design for an oily wastewater is shown in Example 3.7. Flotation units have been used for the clarification of wastewaters, for the removal of oil, and for the concentration of waste sludges. Variability in DAF performance is shown in Figs. 3.41 and 3.42. Some reported data for petroleum refinery effluents are shown in Table 3.13 and for a variety of wastewaters in Table 3.14.

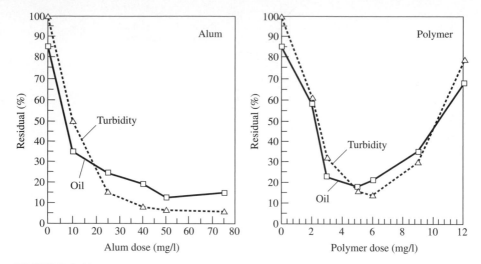

FIGURE 3.40
Treatment efficiency in chemical flocculation.[14]

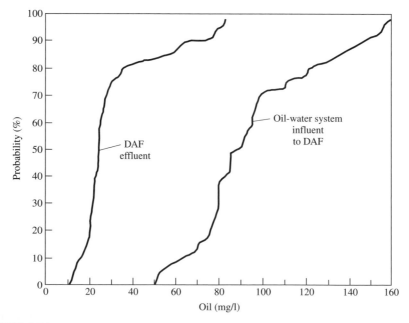

FIGURE 3.41
Oil removal by chemical flocculation and DAF.[14]

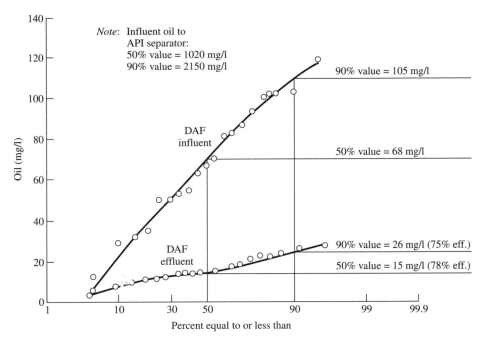

FIGURE 3.42

TABLE 3.13
DAF performance data

Influent oil, mg/l	Effluent oil, mg/l	% removal	Chemicals[†]	Configuration
1930 (90%)	128 (90%)	93	Yes	Circular
580 (50%)	68 (50%)	88	Yes	Circular
105 (90%)	26 (90%)	78	Yes	Rectangular
68 (50%)	15 (50%)	75	Yes	Rectangular
170	52	70	No	Circular
125	30	71	Yes	Circular
100	10	90	Yes	Circular
133	15	89	Yes	Circular
94	13	86	Yes	Circular
838	60	91	Yes	Rectangular
153	25	83	Yes	Rectangular
75	13	82	Yes	Rectangular
61	15	75	Yes	Rectangular
360	45	87	Yes	Rectangular
315	54	83	Yes	Rectangular

†Alum most common, 100–300 mg/l. Polyelectrolyte, 1–5 mg/l, is occasionally added.

TABLE 3.14
Air flotation treatment of oily wastewaters

| Wastewater | Coagulant, mg/l | Oil concentration, mg/l | | |
		Influent	Effluent	Removal, %
Refinery	0	125	35	72
	100 alum	100	10	90
	130 alum	580	68	88
	0	170	52	70
Oil tanker ballast water	100 alum + 1 mg/l polymer	133	15	89
Paint manufacture	150 alum + 1 mg/l polymer	1900	0	100
Aircraft maintenance	30 alum + 10 mg/l activated silica	250–700	20–50	>90
Meat packing		3830	270	93
		4360	170	96

EXAMPLE 3.7. A wastestream of 150 gal/min (0.57 m³/min) and a temperature of 103°F (39.4°C) contains significant quantities of nonemulsified oil and nonsettleable suspended solids. The concentration of oil is 120 mg/l. Reduce the oil to less than 20 mg/l. Laboratory studies showed:

Alum dose = 50 mg/l
Pressure = 60 lb/in² gage (515 kPa absolute or 4.1 relative atm)
Sludge production = 0.64 mg/mg alum
Sludge = 3 percent by weight

Calculate:

(a) The recycle rate
(b) Surface area of the flotation unit
(c) Sludge quantities generated

The air/solids ratio for effluent oil and grease of 20 mg/l is found from Figure 3.35:

$$\frac{A}{S} = 0.03 \text{ lb air released/lb solids applied}$$

At 103°F (39.4°C) the weight solubility of air is 18.6 mg/l from Table 3.12. The value of f is assumed to be 0.85.

Solution.

(a) The recycle rate is

$$R = \frac{(A/S)QS_a}{s_a(fP/P_a - 1)}$$

$$= \frac{0.03 \times 150 \times 120}{18.6(0.85 \times 515/101.3) - 1}$$

$$= 8.75 \text{ gal/min (33.1 l/min)}$$

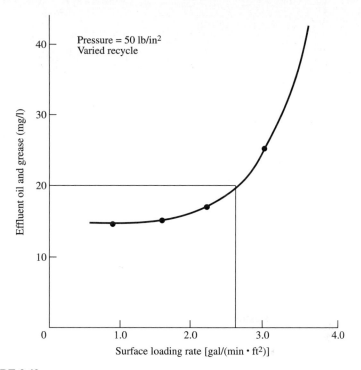

FIGURE 3.43

(*b*) The hydraulic loading for oil removal is determined from Fig. 3.43, and is 2.6 gal/(min · ft²) [0.11 m³/(min · m²)] for an effluent of 20 mg/l. The required surface area is

$$A = \frac{Q + R}{\text{loading}}$$

$$= \frac{150 + 8.75}{2.6}$$

$$= 61 \text{ ft}^2 \, (5.7 \text{ m}^2)$$

(*c*) Sludge quantities generated:

$$\text{Oil sludge} = (120 - 20) \text{ mg/l} \times 150 \text{ gal/min} \times 1440 \text{ min/d}$$

$$\times \, (\text{million gal}/10^6 \text{ gal}) \left(8.34 \, \frac{\text{lb/million gal}}{\text{mg/l}} \right)$$

$$= 180 \text{ lb/d} \, (82 \text{ kg/d})$$

$$\text{Alum sludge} = 0.64 \text{ mg sludge/mg alum} \times 50 \text{ mg/l alum}$$

$$\times\ 150\ \text{gal/min}$$

$$\times\ (1440\ \text{min/d})(\text{million gal}/10^6\ \text{gal})(8.34)$$

$$=\ 58\ \text{lb/d}\ (26\ \text{kg/d})$$

$$\text{Total sludge}\ =\ 238\ \text{lb/d}\ (108\ \text{kg/d})$$

$$\text{Total sludge volume}\ =\ 238/0.03\ \text{lb/d}\ (\text{gal}/8.34\ \text{lb})(\text{day}/1440\ \text{min})$$

$$=\ 0.66\ \text{gal/min}\ (2.5\ \text{l/min})$$

PROBLEMS

3.1. An industry is required to equalize its wastewater and discharge it such that the load in BOD/day to a POTW is constant over the 24 h. The POTW sewage flow is 6.47 million gal/d, with a BOD of 200 mg/l. The diurnal sewage variation in flow is shown in Fig. P3.1. The industrial waste flow is 3.17 million gal/d, with a BOD of 1200 mg/l, and is constant over 10 h (8 A.M. to 6 P.M.).
(*a*) Compute the volume of equalization (or holding) tank required.
(*b*) Plot the discharge curve of the industrial waste over 24 h.

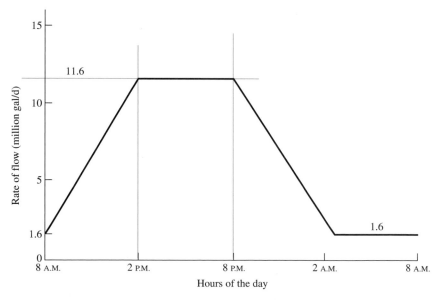

FIGURE P3.1

3.2. A survey of the discharge from a pharmaceutical plant showed the following data.

Time	Flow, gph	COD, mg/l
Day 1		
7:00 A.M.	1025	80
8:00 A.M.	600	55
9:00 A.M.	1200	48
10:00 A.M.	600	45
11:00 A.M.	720	95
12:00 P.M.	1080	66
1:00 P.M.	1200	41
2:00 P.M.	1620	39
3:00 P.M.	1200	29
4:00 P.M.	1320	138
5:00 P.M.	1020	146
6:00 P.M.	720	154
Day 2		
7:00 A.M.	960	47
8:00 A.M.	900	40
9:00 A.M.	1020	139
10:00 A.M.	900	1167
11:00 A.M.	1140	491
12:00 P.M.	1320	163
1:00 P.M.	900	90
2:00 P.M.	1320	143
3:00 P.M.	1200	88
4:00 P.M.	900	35
5:00 P.M.	1140	35
6:00 P.M.	960	47

Compute the volume of an equalization basin to yield a peaking factor of 1.2 and 1.4 based on flow through the system at constant volume and on a constant discharge rate (i.e., variable volume). For the variable-volume case the low level in the basin will be 20 percent of the daily flow.

3.3. An acidic industrial wastewater flow of 150 gal/min (0.57 m^3/min) with a peak factor of 1.2 is to be neutralized to pH 6.0 using lime. The titration curve is as follows:

pH	mg lime/l waste
1.8	0
1.9	500
2.05	1000
2.25	1500
3.5	2000
4.1	2100
5.0	2150
7.0	2200

Consider providing the necessary capacity for a maximum 2-week lime requirement for pH = 7.0 with an hourly requirement 20 percent greater.

Determine:
(a) The lime feed rate for each stage in a two-stage system.
(b) The lime storage capacity (ft³) based on the controlling requirement of either the mean monthly requirement or the maximum two-week requirement. Assume the use of pebbled quicklime (CaO), which has a bulk density of 65 lb/ft³ (1060 kg/m³).
(c) The capacity of the lime slaker and mechanism for transporting the bulk lime based on the maximum estimated requirements.
(d) The mean and maximum slaking water requirements assuming a 10 percent by weight slurry.
(e) The size of the slurry control tank based on a minimum resident time of 5 minutes.
(f) The size of the caustic storage tank based on 24-h caustic feed (NaOH) at the maximum extended usage equivalent to a hydrated lime requirement of 1000 mg/l. Assume NaOH is available with a purity of 98.9 percent and a solubility of 2.5 lb/gal (300 kg/m³).
(g) The maximum caustic (NaOH) feed rate for backup of the lime system.

3.4. A laboratory settling analysis for a pulp and paper mill effluent gave the following results for C_o = 430 mg/l, T = 29°C:

Time, min	Suspended solids removed at indicated depth, %		
	2 ft	4 ft	6 ft
10	47	27	16
20	50	34	43
30	62	48	47
45	71	52	46
60	76	65	48

Note: ft = 30.5 cm

(a) Design a settling tank to remove 70 percent of the suspended solids for 1 million gal/d flow (3785 m³/d). (Apply appropriate factors and neglect initial solids effects.)
(b) What removal will be attained if the flow is increased to 2 million gal/d (7570 m³/d)?

3.5. A wastewater has a flow of 250 gal/min (0.95 m³/min) and a temperature of 105°F (40.5°C). The concentration of oil and grease is 150 mg/l and the suspended solids 100 mg/l. An effluent concentration of 20 mg/l is required. An alum dosage of 30 mg/l is required. The required A/S is 0.04 and the surface loading rate is 2 gal/(min · ft²) [0.081 m³/(min · m²)]. The operating pressure is 65 lb/in² gage (4.4 rel. atm).

Compute:
(a) The required recycle flow
(b) The surface area of the unit
(c) The volume of sludge produced if the skimmings are 3 percent by weight

REFERENCES

1. Novotny, V., and A. J. England: *Water Res.,* vol. 8, p. 325, 1974.
2. Patterson, J. W., and J. P. Menez: *Am. Inst. Chem. Engrs. Env. Prog.,* vol. 3, p. 2, 1984.
3. Tully, T. J.: *Sewage Ind. Wastes,* vol. 30, p. 1385, 1958.
4. Okey, R. W. et al.: *Proc. 32nd Purdue Industrial Waste Conf.*, Ann Arbor Science Pub., 1977.
5. Hazen, A.: *Trans. ASCE,* vol. 53, p. 45, 1904.
6. Camp, T. R.: *Trans. ASCE,* vol. 111, p. 909, 1946.
7. Dobbins, W. E.: "Advances in Sewage Treatment Design," Sanitary Engineering Division, Met. Section, Manhattan College, May 1961.
8. Committee on Industrial Waste Practice of SED: *J. Sanit. Engrg. Div. ASCE,* December 1964.
9. Sutherland, R.: *Ind. Eng. Chem.,* p. 630, May 1947.
10. Greenfield, R. E., and G. N. Cornell: *Ind. Eng. Chem.,* p. 583, May 1947.
11. American Petroleum Institute: *Manual on Disposal of Refinery Wastes*, vol. 1, New York, 1959.
12. Azad, H. S. (editor): *Industrial Wastewater Management Handbook*, McGraw-Hill, New York, 1976.
13. Ford, D. L., private communication.
14. Galil, N., and M. Rebhum: *Water Sci. Tech.*, vol. 27, no. 7–8, p. 79, 1993.
15. Jones, H. R.: *Pollution Control in the Petroleum Industry,* Noyes Data Corp., Princeton, New Jersey, 1973.
16. Pray, H. A.: *Ind. Eng. Chem.,* vol. 44, pt. 1, p. 146, 1952.
17. Frolich, R.: *Ind. Eng. Chem.,* vol. 23, p. 548, 1931.
18. Vrablick, E. R.: *Proc. 14th Ind. Waste Conf.,* 1959, Purdue University.
19. Hurwitz, E., and W. J. Katz: "Laboratory Experiments on Dewatering Sewage Sludges by Dissolved Air Flotation," unpublished report, Chicago, 1959.

4

COAGULATION, PRECIPITATION, AND
METALS REMOVAL

4.1
COAGULATION

Coagulation is employed for the removal of waste materials in suspended or colloidal form. Colloids are particles within the size range of 1 nm (10^{-7} cm) to 0.1 nm (10^{-8} cm). These particles do not settle out on standing and cannot be removed by conventional physical treatment processes.

Colloids present in wastewater can be either hydrophobic or hydrophilic. The hydrophobic colloids (clays, etc.) possess no affinity for the liquid medium and lack stability in the presence of electrolytes. They are readily susceptible to coagulation. Hydrophilic colloids, such as proteins, exhibit a marked affinity for water. The absorbed water retards flocculation and frequently requires special treatment to achieve effective coagulation.[1]

Colloids possess electrical properties which create a repelling force and prevent agglomeration and settling. Stabilizing ions are strongly absorbed to an inner fixed layer which provides a particle charge that varies with the valence and number of adsorbed ions. Ions of an opposite charge form a diffuse outer layer which is held near the surface by electrostatic forces. The psi (ψ) potential is defined as the potential drop between the interface of the colloid and the body of solution. The zeta potential (ζ) is the potential drop between the slipping plane and the body of solution and is related to the particle charge and the thickness of the double layer. The thickness of the double layer (χ) is inversely proportional to the concentration and valence of nonspecific electrolytes, as shown in Fig. 4.1. A van der Waals attractive force is effective in close proximity to the colloidal particle.

The stability of a colloid is due to the repulsive electrostatic forces, and in the case of hydrophilic colloids, also to solvation in which an envelope of water retards coagulation.

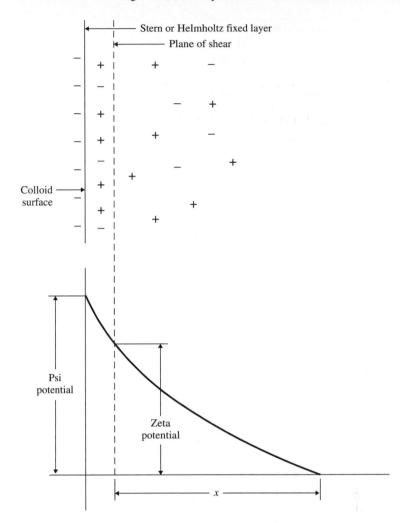

FIGURE 4.1
Electrochemical properties of a colloidal particle.

Zeta Potential

Since the stability of a colloid is primarily due to electrostatic forces, neutralization of this charge is necessary to induce flocculation and precipitation. Although
it is not possible to measure the psi potential, the zeta potential can be determined,
and hence the magnitude of the charge and resulting degree of stability can be
determined as well. The zeta potential is defined as

$$\zeta = \frac{4\pi\eta v}{\varepsilon X} = \frac{4\pi\eta EM}{\varepsilon} \tag{4.1}$$

where v = particle velocity
           ε = dielectric constant of the medium
           η = viscosity of the medium
           X = applied potential per unit length of cell
          EM = electrophoretic mobility

For practical usage in the determination of the zeta potential, Eq. (4.1) can be re-expressed:

$$\zeta \, (\text{mV}) = \frac{113,000}{\varepsilon} \, \eta(\text{poise})\text{EM}\left(\frac{\mu\text{m/s}}{\text{V/cm}}\right) \tag{4.2}$$

where EM = electrophoretic mobility, $(\mu\text{m/s})/(\text{V/cm})$.
    At 25°C, Eq. (4.2) reduces to

$$\zeta = 12.8 \, \text{EM} \tag{4.3}$$

The zeta potential is determined by measurement of the mobility of colloidal particles across a cell, as viewed through a microscope.[2, 3] Several types of apparatus are commercially available for this purpose. A recently developed Lazer Zee meter does not track individual particles, but rather adjusts the image to produce a stationary cloud of particles using a rotating prism technique. This apparatus is shown in Fig. 4.2. The computations involved in determining the zeta potential are illustrated in Example 4.1.

EXAMPLE 4.1. In a electrophoresis cell 10 cm in length, grid divisions are 160 μm at 6 $\times$ magnification. Compute the zeta potential at an impressed voltage of 35 V. The time of travel between grid divisions is 42 s and the temperature is 20°C.

Solution.

    At 20°C:

$$\eta = 0.01 \, \text{poise}$$

$$\varepsilon = 80.36$$

$$\text{EM} = \frac{v}{X} = \frac{160 \, \mu\text{m/42 s}}{35 \, \text{V/10 cm}} = 1.09\left(\frac{\mu\text{m/s}}{\text{V/cm}}\right)$$

$$\zeta(\text{mV}) = 113,000 \times \frac{\eta(\text{poise})\text{EM}(\mu\text{m/s})/(\text{V/cm})}{\varepsilon}$$

$$= 113,000 \times \frac{0.01 \times 1.09}{80.36}$$

$$= 15.3 \, \text{mV}$$

Since there will usually be a statistical variation in the mobility of individual particles, around 20 to 30 values should be averaged for any one determination. The magnitude of the zeta potential for water and waste colloids has been found to average from -16 to -22 mV with a range of -12 to -40 mV.[3] The zeta potential is unaffected by pH over a range of pH 5.5 to 9.5.

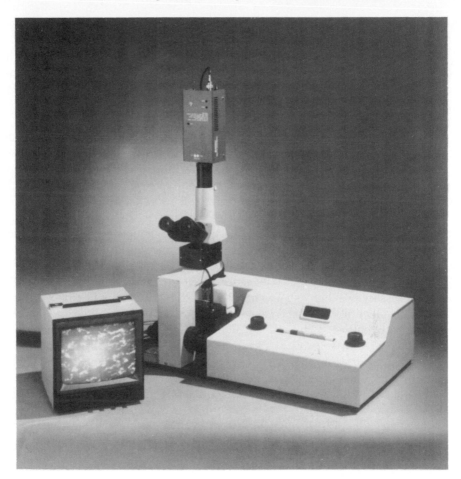

FIGURE 4.2
Lazer Zee meter for zeta potential measurement (*courtesy of Penkem Inc.*).

The zeta potential is lowered by:

1. Change in the concentration of the potential determining ions.
2. Addition of ions of opposite charge.
3. Contraction of the diffuse part of the double layer by increase in the ion concentration in solution.

Since a vast majority of colloids in industrial wastes possess a negative charge, the zeta potential is lowered and coagulation is induced by the addition of high-valence cations. The precipitating power of effectiveness of cation valence in the precipitation of arsenious oxide is

$$Na^+ : Mg^{2+} : Al^{3+} = 1 : 63 : 570$$

Optimum coagulation will occur when the zeta potential is zero; this is defined as the isoelectric point. Effective coagulation will usually occur over a zeta potential range of ± 0.5 mV.

Mechanism of Coagulation

Coagulation results from two basic mechanisms: perikinetic (or electrokinetic) coagulation, in which the zeta potential is reduced by ions or colloids of opposite charge to a level below the van der Waals attractive forces, and orthokinetic coagulation, in which the micelles aggregate and form clumps that agglomerate the colloidal particles.

The addition of high-valence cations depresses the particle charge and the effective distance of the double layer, thereby reducing the zeta potential. As the coagulant dissolves, the cations serve to neutralize the negative charge on the colloids. This occurs before visible floc formation, and rapid mixing which "coats" the colloid is effective in this phase. Microflocs are then formed which retain a positive charge in the acid range because of the adsorption of H^+. These microflocs also serve to neutralize and coat the colloidal particle. Flocculation agglomerates the colloids with a hydrous oxide floc. In this phase, surface adsorption is also active. Colloids not initially adsorbed are removed by enmeshment in the floc.

Riddick[3] has outlined a desired sequence of operation for effective coagulation. If necessary, alkalinity should first be added. (Bicarbonate has the advantage of providing alkalinity without raising the pH.) Alum or ferric salts are added next; they coat the colloid with Al^{3+} or Fe^{3+} and positively charged microflocs. Coagulant aids, such as activated silica and/or polyelectrolyte for floc buildup and zeta potential control, are added last. After addition of alkali and coagulant, a rapid mixing of 1 to 3 min is recommended, followed by flocculation, with addition of coagulant aid, for 20 to 30 min. Destabilization can also be accomplished by the addition of cationic polymers, which can bring the system to the isoelectric point without a change in pH. Although polymers are 10 to 15 times as effective as alum as a coagulant they are considerably more expensive. The mechanism of the coagulation process is shown in Fig. 4.3.

Properties of Coagulants

The most popular coagulant in waste-treatment application is aluminum sulfate, or alum [$Al_2(SO_4)_3 \cdot 18H_2O$], which can be obtained in either solid or liquid form. When alum is added to water in the presence of alkalinity, the reaction is

$$Al_2(SO_4)_3 \cdot 18H_2O + 3Ca(OH)_2 \rightarrow 3CaSO_4 + 2Al(OH)_3 + 18H_2O$$

The aluminum hydroxide is actually of the chemical form $Al_2O_3 \cdot xH_2O$ and is amphoteric in that it can act as either an acid or a base. Under acidic conditions

$$[Al^{3+}][OH^-]^3 = 1.9 \times 10^{-33}$$

At pH 4.0, 51.3 mg/l of Al^{3+} is in solution. Under alkaline conditions, the hydrous aluminum oxide dissociates:

$$Al_2O_3 + 2OH^- \rightarrow 2AlO_2^- + H_2O$$

$$[AlO_2^-][H^+] = 4 \times 10^{-13}$$

At pH 9.0, 10.8 mg/l of aluminum is in solution.

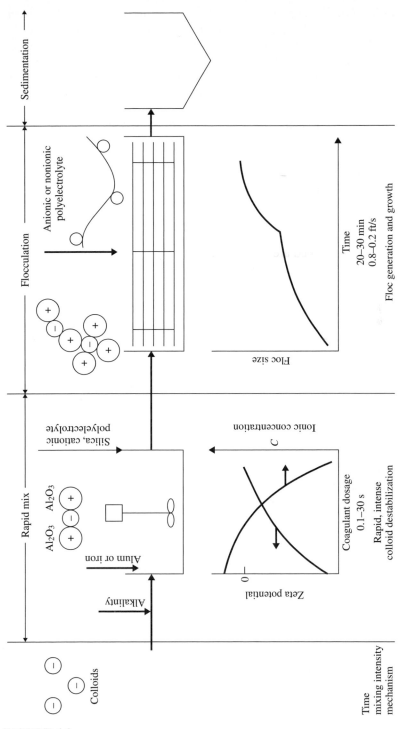

FIGURE 4.3

Mechanisms of coagulation.

129

The alum floc is least soluble at a pH of approximately 7.0. The floc charge is positive below pH 7.6 and negative above pH 8.2. Between these limits the floc charge is mixed. These relationships with respect to the zeta potential are shown in Fig. 4.4. High alum dosages used in the treatment of some industrial wastes may bring about postprecipitation of alum floc, depending on the pH of flocculation.

Ferric salts are also commonly used as coagulants but have the disadvantage of being more difficult to handle. An insoluble hydrous ferric oxide is produced over a pH range of 3.0 to 13.0:

$$Fe^{3+} + 3OH^- \rightarrow Fe(OH)_3$$

$$[Fe^{3+}][OH^-]^3 = 10^{-36}$$

The floc charge is positive in the acid range and negative in the alkaline range, with mixed charges over the pH range 6.5 to 8.0.

The presence of anions will alter the range of effective flocculation. Sulfate ion will increase the acid range but decrease the alkaline range. Chloride ion increases the range slightly on both sides.

Lime is not a true coagulant but reacts with bicarbonate alkalinity to precipitate calcium carbonate and with *ortho*-phosphate to precipitate calcium hydroxyapatite. Magnesium hydroxide precipitates at high pH levels. Good clarification usually requires the presence of some gelatinous Mg (OH)$_2$, but this makes the sludge more difficult to dewater. Lime sludge can frequently be thickened, dewatered, and calcined to convert calcium carbonate to lime for reuse.

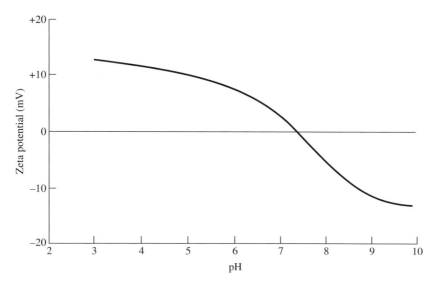

FIGURE 4.4
Zeta potential–pH plot for electrolytic aluminum hydroxide (*after Riddick*[3]).

Coagulant Aids

The addition of some chemicals will enhance coagulation by promoting the growth of large, rapid-settling flocs. Activated silica is a short-chain polymer that serves to bind together particles of microfine aluminum hydrate. At high dosages, silica will inhibit floc formation because of its electronegative properties. The usual dosage is 5 to 10 mg/l.

Polyelectrolytes are high-molecular-weight polymers which contain adsorbable groups and form bridges between particles or charged flocs. Large flocs (0.3 to 1 mm) are thus created when small dosages of polyelectrolyte (1 to 5 mg/l) are added in conjunction with alum or ferric chloride. The polyelectrolyte is substantially unaffected by pH and can serve as a coagulant itself by reducing the effective charge on a colloid. There are three types of polyelectrolytes: a cationic, which adsorbs on a negative colloid or floc particle; an anionic, which replaces the anionic groups on a colloidal particle and permits hydrogen bonding between the colloid and the polymer; and a nonionic, which adsorbs and flocculates by hydrogen bonding between the solid surfaces and the polar groups in the polymer. The general application of coagulants is shown in Table 4.1.

Laboratory Control of Coagulation

Because of the complex reactions involved, laboratory experimentation is essential to establish the optimum pH and coagulant dosage for coagulation of a wastewater. Two procedures can be followed for this purpose: the jar test, in which pH and coagulant dosage are varied to attain the optimum operating conditions; and zeta potential control as proposed by Riddick,[3] in which coagulant is added to zero zeta potential. The procedures to determine the optimum coagulant dosage using these two tests are outlined below:

1. By zeta potential measurement:[3]
   (a) Place 1000 ml of sample in a beaker.
   (b) Add the coagulant in known increments. (The optimum pH should be established either by zeta potential or by a jar-test procedure.)
   (c) Rapid-mix the sample for 3 min after each addition of coagulant; follow with a slow mix.
   (d) Determine the zeta potential after each reagent addition and plot the results as shown in Fig. 4.5. To maintain constant volume, return the sample after each determination.
   (e) If a polyelectrolyte is to be used as a coagulant aid, it should be added last.
2. By jar-test procedure:
   (a) Using 200 ml of sample on a magnetic stirrer, add coagulant in small increments at a pH of 6.0. After each addition, provide a 1-min rapid mix followed by a 3-min slow mix. Continue addition until a visible floc is formed.
   (b) Using this dosage, place 1000 ml of sample in each of six beakers.
   (c) Adjust the pH to 4.0, 5.0, 6.0, 7.0, 8.0, and 9.0 with standard alkali.

TABLE 4.1
Chemical coagulant applications

Chemical process	Dosage range, mg/l	pH	Comments
Lime	150–500	9.0–11.0	For colloid coagulation and P removal
			Wastewater with low alkalinity, and high and variable P
			Basic reactions:
			$$Ca(OH)_2 + Ca(HCO_3)_2 \rightarrow 2CaCO_3 + 2H_2O$$
			$$MgCO_3 + Ca(OH)_2 \rightarrow Mg(OH)_2 + CaCO_3$$
Alum	75–250	4.5–7.0	For colloid coagulation and P removal
			Wastewater with high alkalinity and low and stable P
			Basic reactions:
			$$Al_2(SO_4)_3 + 6H_2O \rightarrow 2Al(OH)_3 + 3H_2SO_4$$
$FeCl_3$, $FeCl_2$	35–150	4.0–7.0	For colloid coagulation and P removal
$FeSO_4 \cdot 7H_2O$	70–200	4.0–7.0	Wastewater with high alkalinity and low and stable P
			Where leaching of iron in the effluent is allowable or can be controlled
			Where economical source of waste iron is available (steel mills, etc.)
			Basic reactions:
			$$FeCl_3 + 3H_2O \rightarrow Fe(OH)_3 + 3HCl$$
Cationic polymers	2–5	No change	For colloid coagulation or to aid coagulation with a metal
			Where the buildup of an inert chemical is to be avoided
Anionic and some nonionic polymers	0.25–1.0	No change	Use as a flocculation aid to speed flocculation and settling and to toughen floc for filtration
Weighting aids and clays	3–20	No change	Used for very dilute colloidal suspensions for weighting

(d) Rapid-mix each sample for 3 min; follow this with 12-min flocculation at slow speed.

(e) Measure the effluent concentration of each settled sample.

(f) Plot the percent removal of characteristic versus pH and select the optimum pH (Fig. 4.6).

(g) Using this pH, repeat steps (b), (d), and (e), varying the coagulant dosage.

(h) Plot the percent removal versus the coagulant dosage and select the optimum dosage (Fig. 4.6).

(i) If a polyelectrolyte is used, repeat the procedure, adding polyelectrolyte toward the end of a rapid mix.

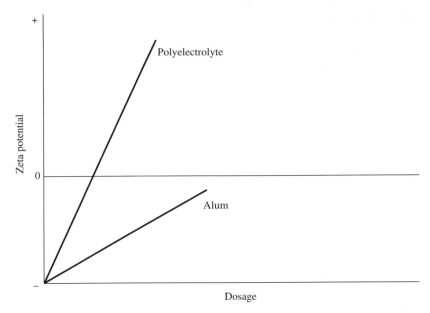

FIGURE 4.5

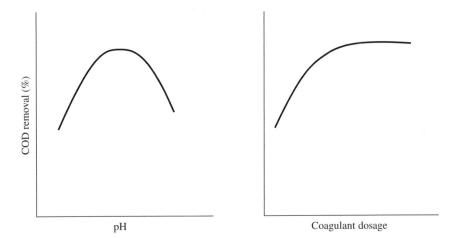

FIGURE 4.6
Characteristic plots of jar-test analysis.

Coagulation Equipment

There are two basic types of equipment adaptable to the flocculation and coagulation of industrial wastes. The conventional system uses a rapid-mix tank, followed by a flocculation tank containing longitudinal paddles which provide slow mixing. The flocculated mixture is then settled in conventional settling tanks.

A sludge-blanket unit combines mixing, flocculation, and settling in a single unit. Although colloidal destabilization might be less effective than in the conventional system, there are distinct advantages in recycling preformed floc. With lime and a few other coagulants, the time required to form a settleable floc is a function of the time necessary for calcium carbonate or other calcium precipitates to form nuclei on which other calcium materials can deposit and grow large enough to settle. It is possible to reduce both coagulant dosage and the time of floc formation by seeding the influent wastewater with previously formed nuclei or by recycling a portion of the precipitated sludge. Recycling preformed floc can frequently reduce chemical dosages, the blanket serves as a filter for improved effluent clarity, and denser sludges are frequently attainable. A sludge-blanket unit is shown in Fig. 4.7.

Coagulation of Industrial Wastes

Coagulation may be used for the clarification of industrial wastes containing colloidal and suspended solids. Paperboard wastes can be effectively coagulated with low dosages of alum. Silica or polyelectrolyte will aid in the formation of a rapid-settling floc. Typical data reported are summarized in Table 4.2.

Wastes containing emulsified oils can be clarified by coagulation.[7] An emulsion can consist of droplets of oil in water. The oil droplets are of approximately 10^{-5} cm and are stabilized by adsorbed ions. Emulsifying agents include soaps and anion-active agents. The emulsion can be broken by "salting out" with the addition of salts, such as $CaCl_2$. Flocculation will then effect charge neutralization and entrainment, resulting in clarification. An emulsion can also frequently be broken by lowering of the pH of the waste solution. An example of such a waste is that

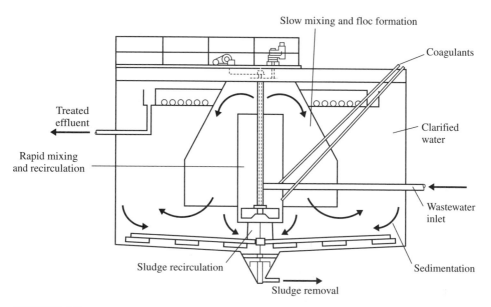

FIGURE 4.7
A reactor clarifier designed for both coagulation and settling.

TABLE 4.2
Chemical treatment of paper and paperboard wastes

Waste	Influent BOD, ppm	Influent SS, ppm	Effluent BOD, ppm	Effluent SS, ppm	Effluent pH	Coagulants Alum, ppm	Coagulants Silica, ppm	Coagulants Other, ppm	Detentions, h	Sludge, % solids	Remarks	Reference
Board		350–450		15–60		3	5		1.7	2–4		4
Board		140–420		10–40		1		10[†]	0.3	2	Flotation	4
Board		240–600		35–85					2.0	2–5	950 gal/(d · ft^2)	5
Board[‡]	127	593	68	44	6.7	10–12	10		1.3	1.76		5
Tissue	140	720	36	10–15		2	4					6
Tissue	208		33		6.6		4					6

[†] Gluc.

[‡] 15,000 gal/ton waste paper.

Note:

gal/(d · ft^2) = 4.075×10^{-2} m^3/(d · m^2)

gal/ton = $4.17 \div 10^{-3}$ m^3/t

135

produced by ball-bearing manufacture, which contains cleaning soaps and detergents, water-soluble grinding oils, cutting oils, and phosphoric acid cleaners and solvents. Treatment of this waste has been effected by the use of 800 mg/l alum, 450 mg/l H_2SO_4, and 45 mg/l polyelectrolyte. The results obtained are summarized in Table 4.3a.

The presence of anionic surface agents in a waste will increase the coagulant dosage. The polar head of the surfactant molecule enters the double layer and stabilizes the negative colloids. Industrial laundry wastes have been treated with H_2SO_4 followed by lime and alum; this has resulted in a reduction of COD of 12,000 mg/l to 1800 mg/l and a reduction of suspended solids of 1620 mg/l to 105 mg/l. Chemical dosages of 1400 mg/l H_2SO_4, 1500 mg/l lime, and 300 mg/l alum were required, yielding 25 percent by volume of settled sludge.

Laundromat wastes containing synthetic detergent have been coagulated with a cationic surfactant to neutralize the anionic detergent and the addition of a calcium salt to provide a calcium phosphate precipitate for flocculation. Typical

TABLE 4.3
Coagulation of industrial wastewaters

(*a*) Ball bearing manufacture[†]

	Analysis	
	Influent	Effluent
pH	10.3	7.1
Suspended solids, mg/l	544	40
Oil and grease, mg/l	302	28
Fe, mg/l	17.9	1.6
PO₄, mg/l	222	8.5

(*b*) Laundromat

	Influent, mg/l	Effluent, mg/l
ABS	63	0.1
BOD	243	90
COD	512	171
PO₄	267	150
CaCl₂	480	
Cationic surfactant	88	
pH	7.1	7.7

(*c*) Latex-base paint manufacture[‡]

	Influent, mg/l	Effluent, mg/1
COD	4340	178
BOD	1070	90
Total solids	2550	446

[†]800 mg/l alum, 450 mg/l H_2SO_4, 45 mg/l polyelectrolyte.
[‡]345 mg/l alum, pH 3.5–4.0.

results obtained are summarized in Table 4.3*b*. Operating the system above pH 8.5 will produce nearly complete phosphate removal.

BOD removals of 90 percent have been achieved on laundry wastes at pH 6.4 to 6.6 with coagulant dosages of 2 lb $Fe_2(SO_4)_3$/thousand gal waste[12] (0.24 kg/m^3).

Polymer waste from latex manufacture has been coagulated with 500 mg/l ferric chloride and 200 mg/l lime at pH 9.6. COD and BOD reductions of 75 and 94 percent, respectively, were achieved from initial values of 1000 mg/l and 120 mg/l. The resulting sludge was 1.2 percent solids by weight, containing 101 lb solids/thousand gal waste (12 kg/m^3) treated. Waste from the manufacture of latex base paints[8] has been coagulated with 345 mg/l alum at pH 3.0 to 4.0, yielding 20.5 lb sludge/thousand gal waste (2.5 kg/m^3) of 2.95 percent solids by weight. The treatment results are summarized in Table 4.3*c*. Wastes from paint-spray booths in automobile assembly plants have been clarified with 400 mg/l $FeSO_4$ at pH 7.0, yielding 8 percent sludge by weight.

Synthetic rubber wastes have been treated at pH 6.7 with 100 mg/l alum, yielding a sludge that was 2 percent of the original waste volume. COD was reduced from 570 to 100 mg/l, and BOD from 85 to 15 mg/l.

Vegetable-processing wastes have been coagulated[9] with lime, yielding BOD removals of 35 to 70 percent with lime dosages of approximately 0.5 lb lime/lb influent BOD (0.5 kg/kg). Results from the coagulation of textile wastewaters are shown in Table 4.4 and color removal from pulp and paper mill effluents in Table 4.5.

BOD removals of 75 to 80 percent have been obtained on wool-scouring wastes with 1 to 3 lb $CaCl_2$/lb BOD (1 to 3 kg/kg). Carbonation was used for pH control.[13]

Talinli[14] treated tannery wastewaters with 2000 mg/l lime at pH 11 and 2 mg/l nonionic polyelectrolyte, as shown in Table 4.6. The resulting sludge volume was 30 percent.

TABLE 4.4
Coagulation of textile wastewaters[10]

Plant	Coagulant	Dosage, mg/l	pH	Color[†] Influent	Color[†] Removal, %	COD Influent, mg/l	COD Removal, %
1	$Fe_2(SO_4)_3$	250	7.5–11.0	0.25	90	584	33
	Alum	300	5–9		86		39
	Lime	1200			68		30
2	$Fe_2(SO_4)_3$	500	3–4, 9–11	0.74	89	840	49
	Alum	500	8.5–10		89		40
	Lime	2000			65		40
3	$Fe_2(SO_4)_3$	250	9.5–11	1.84	95	825	38
	Alum	250	6–9		95		31
	Lime	600			78		50
4	$Fe_2(SO_4)_3$	1000	9–11	4.60	87	1570	31
	Alum	750	5–6		89		44
	Lime	2500			87		44

† Color sum of absorbances at wavelengths of 450, 550, and 650 nm.

TABLE 4.5
Color removal from pulp and paper mill effluents[11]

Plant	Coagulant	Dosage, mg/l	pH	Color Influent	Color Removal, %	COD Influent, mg/l	COD Removal, %
1	$Fe_2(SO_4)_3$	500	3.5–4.5	2250	92	776	60
	Alum	400	4.0–5.0		92		53
	Lime	1500	—		92		38
2	$Fe_2(SO_4)_3$	275	3.5–4.5	1470	91	480	53
	Alum	250	4.0–5.5		93		48
	Lime	1000	—		85		45
3	$Fe_2(SO_4)_3$	250	4.5–5.5	940	85	468	53
	Alum	250	5.0–6.5		91		44
	Lime	1000	—		85		40

TABLE 4.6
Coagulation of tannery wastewaters[14]

Parameter	Influent	Effluent	% Removal
COD	7800	2900	63
BOD	3500	1450	58
SO_4	1800	1200	33
Chromium	100	3	97

4.2
HEAVY METALS REMOVAL

There are a number of technologies available for the removal of heavy metals from a wastewater. These are summarized in Table 4.7. Chemical precipitation is most commonly employed for most of the metals. Common precipitants include OH^-, CO_3^{2-} and S^{2-}. Metals are precipitated as the hydroxide through the addition of lime or caustic to a pH of minimum solubility. However, several of these compounds are amphoteric and exhibit a point of minimum solubility. The pH of minimum solubility varies with the metal in question as shown in Fig. 4.8. Metals can also be precipitated as the sulfide (Fig. 4.8) or in some cases as the carbonate (Fig. 4.9).

In treating industrial wastewaters containing metals, it is frequently necessary to pretreat the wastewaters to remove substances that will interfere with the precipitation of the metals. Cyanide and ammonia form complexes with many metals that limit the removal that can be achieved by precipitation as shown in Fig. 4.10 for the case of ammonia. Cyanide can be removed by alkaline chlorination or other processes such as catalytic oxidation of carbon. Cyanide wastewaters containing nickel or silver are difficult to treat by alkaline chlorination because of the slow reaction rate of these metal complexes. Ferrocyanide $[Fe(CN)_6^{4-}]$ is oxidized to ferricyanide $[Fe(CN)_6^{3-}]$, which resists further oxidation. Ammonia can

TABLE 4.7
Heavy metals removal technologies

Conventional precipitation

    Hydroxide

    Sulfide

    Carbonate

    Coprecipitation

Enhanced precipitation

    Dimethyl thio carbamate

    Diethyl thio carbamate

    Trimercapto-s-triazine, trisodium salt

Other methods

    Ion exchange

    Adsorption

Recovery opportunities

    Ion exchange

    Membranes

    Electrolytic techniques

be removed by stripping, break-point chlorination, or other suitable methods prior to the removal of metals.

For many metals, such as arsenic and cadmium, coprecipitation with iron or aluminum is highly effective for removal to low residual levels. In this case the metal adsorbs to the alum or iron floc. In order to meet low effluent requirements, it may be necessary in some cases to provide filtration to remove floc carried over from the precipitation process. With precipitation and clarification alone, effluent metals concentrations may be as high as 1 to 2 mg/l. Filtration should reduce these concentrations to 0.5 mg/l or less. Carbamate salts can be employed for enhanced precipitation. Because of the chemical cost, this type of precipitation is usually employed as a polishing step following conventional precipitation. Typical results are shown in Table 4.8. Metals can be removed by adsorption on activated carbon (see p. 442), aluminum oxides, silica, clays, and synthetic material such as zeolites and resins (see Chap. 9). In the case of adsorption, higher pH favors the adsorption of cations while a lower pH favors the adsorption of anions. Complexing agents will interfere with cationic species. There will be competition from major background ions such as calcium or sodium. For chromium waste treatment, hexavalent chromium must first be reduced to the trivalent state, Cr^{3+}, and then precipitated with lime. This is referred to as the process of reduction and precipitation.

Arsenic

Arsenic and arsenical compounds are present in wastewaters from the metallurgical industry, glassware and ceramic production, tannery operation, dyestuff manufacture,

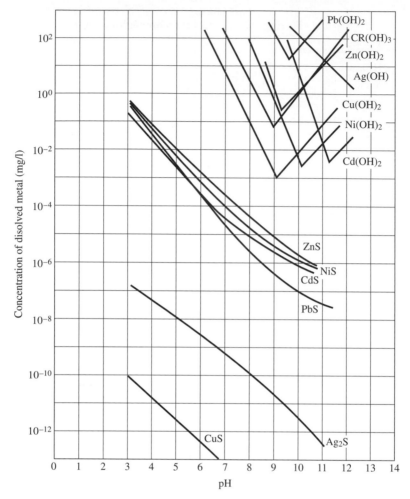

FIGURE 4.8
Heavy metals precipitation as the hydroxide and the sulfide.

pesticide manufacture, some organic and inorganic chemicals manufacture, petroleum refining, and the rare-earth industry. Arsenic is removed from wastewater by
chemical precipitation. Enhanced performance is achieved as arsenate (AsO_4^{3-},
As^{5+}) rather than arsenite (AsO_2^-, As^{3+}). Arsenite is therefore usually oxidized to
arsenate prior to precipitation. Effluent arsenic levels of 0.05 mg/l are obtainable
by precipitation of the arsenic as the sulfide by the addition of sodium or hydrogen
sulfide at pH of 6 to 7. In order to meet reported effluent levels, polishing of the
effluent by filtration would usually be required.

Arsenic present in low concentrations can also be reduced by filtration through
activated carbon. Effluent concentrations of 0.06 mg/l arsenic have been reported
from an initial concentration of 0.2 mg/l. Arsenic is removed by coprecipitation with
a ferric hydroxide floc that ties up the arsenic and removes it from solution. Effluent
concentrations of less than 0.005 mg/l have been reported from this process.

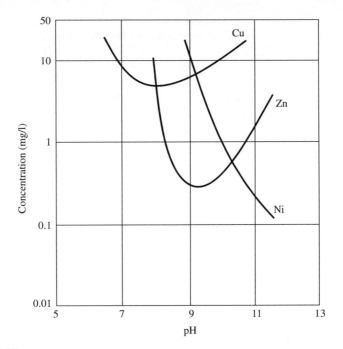

FIGURE 4.9
Residual soluble metal concentrations using Na_2CO_3.

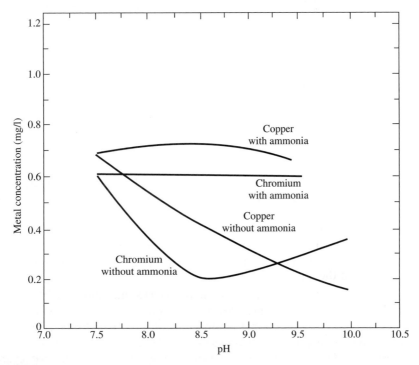

FIGURE 4.10
Comparative optimum pH values for metal removal with and without ammonia.

TABLE 4.8
Enhanced removal of soluble metals by precipitation

Metal	Input, mg/l	$Ca(OH)_2$, mg/l	$X,^†$ mg/l
Cadmium	0.4	0.2	0.04
Chromium	1.2	0.1	0.05
Copper	1.3	0.1	0.05
Nickel	3.5	0.9	0.67
Lead	7.4	0.4	0.35
Mercury	1.4	0.1	0.01
Zinc	13.5	0.2	0.09

†Enhancing chemicals:
TMT15—trimercapto triazine trisodium salt
Nalfloc Nalmet 8154 (diethyl thio carbamate)
IMP HM1 (dimethyl thio carbamate)

Barium

Barium is present in wastewaters from the paint and pigment industry; the metallurgical industry; glass, ceramic, and dye manufacturers; and the vulcanizing of rubber. Barium has also been reported in explosives manufacturing wastewater. Barium is removed from solution by precipitation as barium sulfate.

Barium sulfate is extremely insoluble, having a maximum theoretical solubility at 25°C of approximately 1.4 mg/l as barium at stoichiometric concentrations of barium and sulfate. The solubility level of barium can be reduced in the presence of excess sulfate. Coagulation of barium salts as the sulfate would be capable of reducing barium to effluent levels of 0.03 to 0.3 mg/l. Barium can also be removed from solution by ion exchange and electrodialysis, although these processes would be more expensive than chemical precipitation.

Cadmium

Cadmium is present in wastewaters from metallurgical alloying, ceramics, electroplating, photography, pigment works, textile printing, chemical industries, and lead mine drainage. Cadmium is removed from wastewaters by precipitation or ion exchange. In some cases, electrolytic and evaporative recovery processes can be employed, provided the wastewater is in a concentrated form. Cadmium forms an insoluble and highly stable hydroxide at an alkaline pH. Cadmium in solution is approximately 1 mg/l at pH 8 and 0.05 mg/l at pH 10 to 11. Coprecipitation with iron hydroxide at pH 6.5 will reduce cadmium to 0.008 mg/l; iron hydroxide at pH 8.5 reduces cadmium to 0.05 mg/l. Sulfide and lime precipitation with filtration will yield 0.002 to 0.03 mg/l at pH 8.5 to 10. Cadmium is not precipitated in the presence of complexing ions, such as cyanide. In these cases, it is necessary to pretreat the wastewater to destroy the complexing agent. In the case of cyanide, cyanide destruction is necessary prior to cadmium precipitation. A hydrogen peroxide oxidation precipitation system has been developed that simultaneously oxidizes cyanides and forms the oxide of cadmium, thereby yielding cadmium, whose recovery is feasible. Results for the hydroxide precipitation of cadmium are shown in Table 4.9.

Chromium

The reducing agents commonly used for chromium wastes are ferrous sulfate, sodium meta-bisulfite, or sulfur dioxide. Ferrous sulfate and sodium meta-bisulfite may be dry- or solution-fed; SO_2 is diffused into the system directly from gas cylinders. Since the reduction of chromium is most effective at acidic pH values, a reducing agent with acidic properties is desirable. When ferrous sulfate is used as the reducing agent, the Fe^{2+} is oxidized to Fe^{3+}; if meta-bisulfite or sulfur dioxide is used, the negative radical SO_3^{2-} is converted to SO_4^{2-}. The general reactions are

$$Cr^{6+} + Fe^{2+} \text{ or } SO_2 \text{ or } Na_2S_2O_5 + H^+ \rightarrow Cr^{3+} + Fe^{3+} \text{ or } SO_4^{2-}$$

$$Cr^3 + 3OH^- \rightarrow Cr(OH)_3\downarrow$$

Ferrous ion reacts with hexavalent chromium in an oxidation-reduction reaction, reducing the chromium to a trivalent state and oxidizing the ferrous ion to the ferric state. This reaction occurs rapidly at pH levels below 3.0. The acidic properties of ferrous sulfate are low at high dilution; acid must therefore be added for pH adjustment. The use of ferrous sulfate as a reducing agent has the disadvantage that a contaminating sludge of $Fe(OH)_3$ is formed when an alkali is added. In order to obtain a complete reaction, an excess dosage of 2.5 times the theoretical addition of ferrous sulfate must be used.

TABLE 4.9
Hydroxide precipitation treatment for cadmium

Method	Treatment pH	Initial Cd, mg/l	Final Cd, mg/l
Hydroxide precipitation	8.0	—	1.0
	9.0	—	0.54
	10.0	—	0.10
	9.3–10.6	4.0	0.20
Hydroxide precipitation plus filtration	10.0	0.34	0.054
	10.0	0.34	0.033
Hydroxide precipitation plus filtration	11.0	—	0.00075
	11.0	—	0.00070
Hydroxide precipitation plus filtration	11.5	—	0.014
Hydroxide precipitation plus filtration	—	—	0.08
Coprecipitation with ferrous hydroxide	6.0	—	0.050
Coprecipitation with ferrous hydroxide	10.0	—	0.044
Coprecipitation with alum	6.4	0.7	0.39

Reduction of chromium can also be accomplished by using either sodium meta-bisulfite or SO_2. In either case reduction occurs by reaction with the H_2SO_3 produced in the reaction. The H_2SO_3 ionizes according to the mass action:

$$\frac{(H^+)(HSO_3^-)}{(H_2SO_3)} = 1.72 \times 10^{-2}$$

Above pH 4.0, only 1 percent of sulfite is present as H_2SO_3 and the reaction is very slow. During this reaction, acid is required to neutralize the NaOH formed. The reaction is highly dependent on both pH and temperature. At pH levels below 2.0 the reaction is practically instantaneous and close to theoretical requirements.

The theoretical quantities of chemicals required for 1 ppm Cr are:

2.81 ppm $Na_2S_2O_5$ (97.5 percent)
1.52 ppm H_2SO_4
2.38 ppm lime (90 percent)
1.85 ppm SO_2

At pH levels above 3, when a basic chrome sulfate is produced, the quantities of lime required for subsequent neutralization are reduced. At pH 8.0 to 9.9, $Cr(OH)_3$ is virtually insoluble. Experimental investigations have shown that the sludge produced will compact to 1 to 2 percent by weight.

Since dissolved oxygen is usually present in wastewaters, an excess of SO_2 must be added to account for the oxidation of the SO_3^{2-} to SO_4^{2-}:

$$H_2SO_3 + \frac{1}{2}O_2 \rightarrow H_2SO_4$$

An excess dosage of 35 ppm SO_2 will usually be sufficient for reaction with the dissolved oxygen present.

The acid requirements for the reduction of Cr^{6+} depend on the acidity of the original waste, the pH of the reduction reaction, and the type of reducing agent used (for example, SO_2 produces an acid but meta-bisulfite does not). Since it is difficult, if not impossible, to predict these requirements, it is usually necessary to titrate a sample to the desired pH endpoint with standardized acid.

Many small plating plants have a total daily volume of waste of less than 30,000 gal/d (114 m³/d). The most economical system for such plants is a batch treatment in which two tanks are provided, each with a capacity of one day's flow. One tank is undergoing treatment while the other is filling. Accumulated sludge is either drawn off and hauled to disposal or dewatered on sand drying beds. A separable dry cake can be obtained after 48 hours on a sand bed. A typical batch treatment system is schematically shown in Fig. 4.11.

When the daily volume of waste exceeds 30,000 to 40,000 gal (114 to 151 m³), batch treatment is usually not feasible because of the large tankage required. Continuous treatment requires a tank for acidification and reduction, then a mixing tank, for lime addition, and a settling tank. The retention time in the reduction tank is dependent on the pH employed but should be at least 4 times the theoretical time for complete reduction. Twenty minutes will usually be adequate for flocculation. Final settling should not be designed for an overflow rate in excess of 500 gal/(d · ft²) [20 m³/(d · m²)].

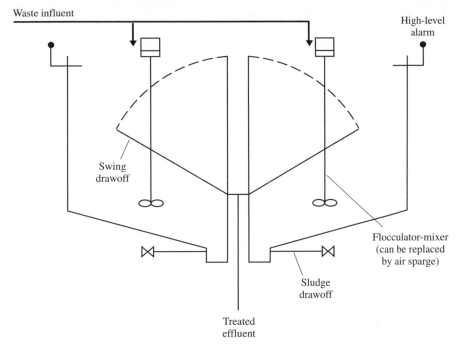

FIGURE 4.11
Batch treatment of chromium wastes.

In cases where the chrome content of the rinsewater varies markedly, equalization should be provided before the reduction tank to minimize fluctuations in the chemical feed system. The fluctuation in chrome content can be minimized by provision of a drain station before the rinse tanks.

Successful operation of a continuous chrome reduction process requires instrumentation and automatic control. pH and redox control are provided for the reduction tank. The addition of lime should be modulated by a second pH control system. A continuous chrome reduction/precipitation process is shown in Fig. 4.12. Reported results for the precipitation of chromium are shown in Table 4.10. Removal of chromium by ion exchange is shown in Table 4.11. An example of metals removal is shown in Example 4.2.

EXAMPLE 4.2. 30,000 gal/d (114 m³/d) of a waste containing 49 mg/l Cr^{6+}, 11 mg/l Cu, and 12 mg/l Zn is to be treated daily by using SO_2. Compute the chemical requirements and the daily sludge production. (Assume the waste contains 5 mg/l O_2.)

Solution.

(*a*) SO_2 requirements are as follows. For Cr^{6+}

$$1.85 \left(\frac{\text{mg } SO_2}{\text{mg } Cr^{6+}} \right) \times 49(\text{mg } Cr^{6+}/1) \times 8.34 \left(\frac{\text{lb/million gal}}{\text{mg/l}} \right)$$

$$\times \ 0.03 \text{ million gal/d} = 22.7 \text{ lb/d} \quad (10.3 \text{ kg/d})$$

and for O_2, where 1 part of O_2 requires 4 parts of SO_2:

$$4\left(\frac{mg \ SO_2}{mg \ O_2}\right) \times 5(mg \ O_2/l) \times 8.34 \times 0.03 = 5.0 \ lb/d \qquad (2.3 \ kg/d)$$

$$Total = 27.7 \ lb/d \qquad (12.6 \ kg/d)$$

(b) Lime requirements are as follows. For Cr^{3+}:

$$2.38\left(\frac{mg \ lime}{mg \ Cr^{3+}}\right) \times 49 \times 8.34 \times 0.03 = 29.2 \ lb/d \qquad (13.3 \ kg/d)$$

and for Cu and Zn (each part of Cu and Zn requiring 1.3 parts of 90 percent lime for precipitation):

$$1.3\left(\frac{mg \ lime}{mg \ Cu \ or \ Zn}\right) \times 23(mg \ Cu \ and \ Zn/l) \times 8.34 \times 0.03 = 7.5 \ lb/d \qquad (3.4 \ kg/d)$$

$$Total = 36.7 \ lb/d \qquad (16.7 \ kg/d)$$

(c) Sludge production is:

$$1.98\left(\frac{mg \ Cr(OH)_3}{mg \ Cr^{6+}}\right) \times 49 \times 8.34 \times 0.03 = 24.3 \ lb/d \ Cr(OH)_3 \qquad (11 \ kg/d)$$

$$1.53\left(\frac{mg \ sludge}{mg \ Cu \ or \ Zn}\right) \times 23 \times 8.34 \times 0.03$$

$$= 8.8 \ lb/d \ Cu(OH_2) \ and \ Zn(OH_2) \qquad (4 \ kg/d)$$

$$Total = 33.1 \ lb/d \qquad (15 \ kg/d)$$

If the sludge concentrates to 1.5 percent by weight, the volume that will require disposal each day can be calculated as follows:

$$\frac{33.1 \ lb/d}{0.015 \ lb \ solids/lb \ sludge \times 8.34 \ lb/gal} = 265 \ gal/d \qquad (1.0 \ m^3/d)$$

It should be noted that some of the copper and zinc will be soluble unless the final pH after lime addition exceeds pH 9.0.

Copper

The primary sources of copper in industrial wastewaters are metal-process pickling baths and plating baths. Copper may also be present in wastewaters from a variety of chemical manufacturing processes employing copper salts or a copper catalyst. Copper is removed from wastewaters by precipitation or recovery processes, which include ion exchange, evaporation, and electrodialysis. The value of recovered copper metal will frequently make recovery processes attractive. Ion exchange or activated carbon are feasible treatment methods for wastewaters containing copper at concentrations of less than 200 mg/l. Copper is precipitated as a relatively insoluble metal hydroxide at alkaline pH. In the presence of high sulfates, calcium

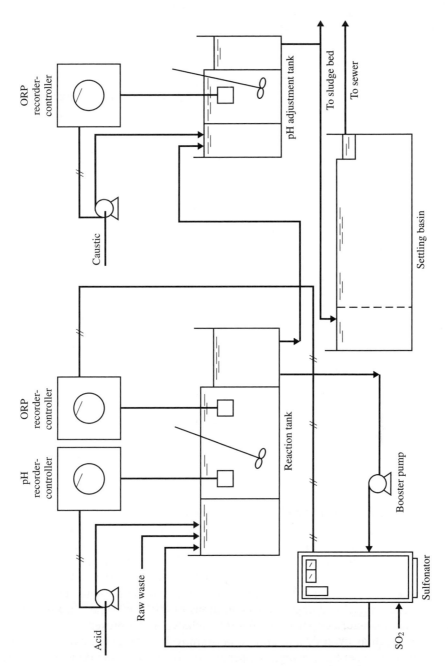

FIGURE 4.12

Continuous chrome waste treatment system (*courtesy of Fischer-Porter, Inc.*).

TABLE 4.10
Summary of trivalent chromium treatment results

| Method | pH | Chromium, mg/l | |
		Initial	Final
Precipitation	7–8	140	1.0
	7.8–8.2	16.0	0.06–0.15
	8.5	47–52	0.3–1.5
	8.8	650	18
	8.5–10.5	26.0	0.44–0.86
	8.8–10.1	—	0.6–30
	12.2	650	0.3
Precipitation with sand filtration	8.5	7400	1.3–4.6
	8.5	7400	0.3–1.3
	9.8–10.0	49.4	0.17
	9.8–10.0	49.4	0.05

TABLE 4.11
Ion exchange performance in hexavalent chromium removal[15]

| Wastewater source | Chromium, mg/l | | Resin capacity[†] |
	Influent	Effluent	
Cooling tower blowdown	17.9	1.8	5–6
	10.0	1.0	2.5–4.5
	7.4–10.3	1.0	—
	9.0	0.2	2.5
Plating rinsewater	44.8	0.025	1.7–2.0
	41.6	0.01	5.2–6.3
Pigment manufacture	1210	<0.5	—

[†]lb chromate/ft^3 resin.

sulfate will also be precipitated and will interfere with the recovery value of the copper sludge. This may dictate the use of a more expensive alkali such as NaOH to obtain a pure sludge. Cupric oxide has a minimum solubility between pH 9.0 and 10.3 with a reported solubility of 0.01 mg/l. Field practice has indicated that the maximum technically feasible treatment level for copper by chemical precipitation is 0.02 to 0.07 mg/l as soluble copper. Precipitation with sulfide at pH 8.5 will result in effluent copper concentrations of 0.01 to 0.02 mg/l. Low residual concentrations of copper are difficult to achieve in the presence of complexing agents such as cyanide and ammonia. Removal of the complexing agent by pretreatment is essential for high copper removal. Copper cyanide is effectively removed on activated carbon. A summary of copper results after hydroxide precipitation is shown in Table 4.12.

TABLE 4.12
Summary of copper results after hydroxide precipitation treatment[15]

Source (treatment)	Copper concentration, mg/l	
	Initial	Final
Metal processing (lime)	204–385	0.5
Nonferrous metal processing (lime)	—	0.2–2.3
Metal processing (lime)	—	1.4–7.8
Electroplating (caustic, soda ash + hydrazine)	6.0–15.5	0.09–0.24 (solution)
		0.30–0.45 (total)
Machine plating (lime + coagulant)	—	2.2
Metal finishing (lime)	—	0. 19 average
Brass mill (lime)	10–20	1–2
Plating (CN oxidation, Cr reduction neutralization)	11.4	2.0
Wood preserving (lime)	0.25–1.1	01–0.35
Brass mill (hydrazine + caustic)	75–124	0.25–0.85
Silver plating (CN oxidation, lime + $FeCl_3^+$ + filtration	30 (average)	0.16–0.3
Copper sulfate manufacture (lime)	433	0.14–1.25
		(0.48 average)
Integrated circuit manufacture (lime)	0.23	0.05

Fluorides

Fluorides are present in wastewaters from glass manufacturing, electroplating, steel and aluminum production, and pesticide and fertilizer manufacture. Fluoride is removed by precipitation with lime as calcium fluoride. Effluent concentrations in the order of 10 to 20 mg/l are readily obtainable. Lime precipitation at a pH above 12 has created problems with solids removal, poor settling, and cementation of filters. Enhanced removal of fluoride has been reported in the presence of magnesium. The increased removal is attributed to adsorption of the fluoride ion into the magnesium hydroxide floc, resulting in effluent fluoride concentrations of less than 1.0 mg/l. Alum coprecipitation will result in effluent levels of 0.5 to 2.0 mg/l. Low concentrations of fluoride can be removed by ion exchange. Fluoride removal through ion exchange pretreated and regenerated with aluminum salts is attributable to aluminum hydroxide precipitated in a column bed. Fluoride is removed through contact beds of activated alumina, which may be employed as a polishing unit to follow lime precipitation. Fluoride concentrations of 30 mg/l from the lime precipitation process have been reduced to approximately 2 mg/l upon passage through an activated alumina contact bed. A summary of fluoride treatment processes and levels of treatment achieved is in Table 4.13.

TABLE 4.13
Summary of fluoride treatment processes and levels of treatment achieved[15]

| Treatment process | Fluoride concentration, mg/l | | Current application |
	Initial	Final	
Lime		10	Industrial
Lime	1000–3000	20	Industrial
Lime	500–1000	20–40	Industrial
Lime	200–700	6 (16-h settling)	Industrial
Lime	45	8	Industrial
Lime	4–20	5.9 (average)	Industrial
Lime	590	80	Industrial
Lime	57.8	29.1 (average) 14–16 (best)	Industrial
Lime	93,000	0.8–8.8	Industrial (pilot-scale)
Lime	—	10.6 (clarified) 10.4 (filtered)	Industrial
Lime, two-stage	1,460	9	Industrial
Lime + calcium chloride	—	12	Industrial
Lime + alum	—	1.5	Industrial
Lime + alum	2,020	2.4	Industrial (pilot-scale)
Calcium carbonate + lime, two-stage	11,100	6	Industrial

Iron

Iron is present in a wide variety of industrial wastewaters including mining operations, ore milling, chemical industrial wastewater, dye manufacture, metal processing, textile mills, petroleum refining, and others. Iron exists in the ferric or ferrous form, depending on pH and dissolved oxygen concentration. At neutral pH and in the presence of oxygen, soluble ferrous iron is oxidized to ferric iron, which readily hydrolyzes to form the insoluble ferric hydroxide precipitate. At high pH values ferric hydroxide will solubilize through the formation of the $Fe(OH)_4^-$ complex. Ferric and ferrous iron may also be solubilized in the presence of cyanide by the formation of ferro- and ferricyanide complexes. The primary removal process for iron is conversion of the ferrous to the ferric state and precipitation of ferric hydroxide at a pH of near 7, corresponding to minimum solubility. Conversion of ferrous to ferric iron occurs rapidly upon aeration at pH 7.5. In the presence of dissolved organic matter, the iron oxidation rate is reduced. Two-stage hydroxide precipitation or sulfate precipitation will reduce iron to 0.01 mg/l.

Lead

Lead is present in wastewaters from storage-battery manufacture. Lead is generally removed from wastewaters by precipitation as the carbonate, $PbCO_3$, or the hydroxide, $Pb(OH)_2$. Lead is effectively precipitated as the carbonate by the addition of soda ash, resulting in effluent-dissolved lead concentrations of 0.01 to 0.03 mg/l at a pH of 9.0 to 9.5. Precipitation with lime at pH 11.5 resulted in effluent concentrations of 0.019 to 0.2 mg/l. Precipitation as the sulfide to 0.01 mg/l can be accomplished with sodium sulfide at a pH of 7.5 to 8.5.

Manganese

Manganese and its salts are found in wastewaters from manufacture of steel alloy, dry-cell batteries, glass and ceramics, paint and varnish, and inks and dye. Among the many forms and compounds of manganese only the manganous salts and the highly oxidized permanganate anion are appreciably soluble. The latter is a strong oxidant that is reduced under normal circumstances to insoluble manganese dioxide. Treatment technology for the removal of manganese involves conversion of the soluble manganous ion to an insoluble precipitate. Removal is effected by oxidation of the manganous ion and separation of the resulting insoluble oxides and hydroxides. Manganous ion has a low reactivity with oxygen and simple aeration is not an effective technique below pH 9. It has been reported that even at high pH levels, organic matter in solution can combine with manganese and prevent its oxidation by simple aeration. A reaction pH above 9.4 is required to achieve significant manganese reduction by precipitation. The use of chemical oxidants to convert manganous ion to insoluble manganese dioxide in conjunction with coagulation and filtration has been employed. The presence of copper ion enhances air oxidation of manganese, and chlorine dioxide rapidly oxidizes manganese to the insoluble form. Permanganate has successfully been employed in the oxidation of manganese. Ozone has been employed in conjunction with lime for the oxidation and removal of manganese. The drawback in the application of ion exchange is the nonselective removal of other ions, which increases operating costs.

Mercury

The major consumptive user of mercury in the United States is the chlor-alkali industry. Mercury is also used in the electrical and electronics industry, explosives manufacturing, the photographic industry, and the pesticide and preservative industry. Mercury is used as a catalyst in the chemical and petrochemical industry. Mercury is also found in most laboratory wastewaters. Power generation is a large source of mercury release into the environment through the combustion of fossil fuel. When scrubber devices are installed on thermal power plant stacks for sulfur dioxide removal, accumulation of mercury is possible if extensive recycle is practiced. Mercury can be removed from wastewaters by precipitation, ion exchange, and adsorption. Mercury ions can be reduced upon contact with other metals such as copper, zinc, or aluminum. In most cases mercury recovery can be achieved by distillation. For precipitation, mercury compounds must be oxidized to the mercuric ion. Table 4.14 shows effluent levels achievable by candidate technology.

Nickel

Wastewaters containing nickel originate from metal-processing industries, steel foundries, motor vehicle and aircraft industries, printing, and in some cases, the chemicals industry. In the presence of complexing agents such as cyanide, nickel may exist in a soluble complex form. The presence of nickel cyanide complexes interferes with both cyanide and nickel treatment. Nickel forms insoluble nickel hydroxide upon the addition of lime, resulting in a minimum solubility of 0.12 mg/l at pH 10 to 11. Nickel hydroxide precipitates have poor settling properties. Nickel can also be precipitated as the carbonate or the sulfate associated with recovery systems. In practice, lime addition (pH 11.5) may be expected to yield residual nickel concentrations in the order of 0.15 mg/l after sedimentation and filtration. Recovery of nickel can be accomplished by ion exchange or evaporative recovery, provided the nickel concentrations in the wastewaters are at a sufficiently high level. Results of nickel precipitation are shown in Table 4.15.

Selenium

Selenium may be present in various types of paper, fly ash, and metallic sulfide ores. The selenious ion appears to be the most common form of selenium in wastewater except for pigment and dye wastes, which contain selenide (yellow cadmium selenide). Selenium can be removed from wastewaters by precipitation as the sulfide at a pH of 6.6. Effluent levels of 0.05 mg/l are reported. Ferric hydroxide coprecipitation at pH 6.2 will reduce selenium to 0.01 to 0.05 mg/l. Alumina adsorption results in effluent levels of 0.005 to 0.02 mg/l.

TABLE 4.14
Mercury removal, effluent levels

Technology	Effluent, $\mu g/l$
Sulfide precipitation	10–20
Alum coprecipitation	1–10
Iron coprecipitation	0.5–5
Ion exchange	1–5
Carbon adsorption	
Influent	——
High	20
Moderate	2
Low	0.25

TABLE 4.15
**Comparison of lime versus lime-plus-sulfide precipitation
of nickel in electroplating wastewaters[16]**

	Wastewater		
Parameter	A	B	C
Treatment pH	8.5	8.75	9.0
Initial nickel, mg/l	119.0	99.0	3.2
Lime treatment			
Clarifier effluent	12.0	16.0	0.47
Filter effluent	9.4	12.0	0.07
Lime plus sulfide			
Clarifier effluent	11.0	7.0	0.35
Filter effluent	3.5	4.2	0.20

Silver

Soluble silver, usually in the form of silver nitrate, is found in wastewaters from the porcelain, photographic, electroplating, and ink manufacturing industries. Treatment technology for the removal of silver usually considers recovery because of the high value of the metal. Basic treatment methods include precipitation, ion exchange, reductive exchange, and electrolytic recovery. Silver is removed from wastewater by precipitation as silver chloride, which is an extremely insoluble precipitate resulting in the maximum silver concentration at 25°C of approximately 1.4 mg/l. An excess of chloride will reduce this value, but greater excess concentrations will increase the solubility of silver through the formation of soluble silver chloride complexes. Silver can be selectively precipitated as silver chloride from a mixed-metal wastestream without initial wastewater segregation or concurrent precipitation of other metals. If the treatment conditions are alkaline, resulting in precipitation of hydroxides of other metals along with the silver chloride, acid washing of the precipitated sludge will remove contaminated metal ions, leaving the insoluble silver chloride. Plating wastes contain silver in the form of silver cyanide, which interferes with the precipitation of silver as the chloride salt. Oxidation of the cyanide with chlorine releases chloride ions into solution, which in turn react to form silver chloride directly. Sulfide will precipitate silver from photographic solutions as the extremely insoluble silver sulfide. Ion exchange has been employed for the removal of soluble silver from wastewaters. Activated carbon will remove low concentrations of silver. The mechanism reported is one of reductive recovery by formation of elemental silver at the carbon surface. Reported results indicate that the carbon is capable of retaining silver to 9 percent of its weight at a pH of 2.1 and 12 percent of its weight at a pH of 5.4. Alum or iron coprecipitation will reduce silver to 0.025 mg/l and hydroxide precipitation at pH 11 to 0.02 mg/l.

Zinc

Zinc is present in wastewater streams from steelworks, rayon yarn and fiber manufacture, ground wood-pulp production, and recirculating cooling water systems employing cathodic treatment. Zinc is also present in wastewaters from the plating and metal-processing industry. Zinc can be removed by precipitation as zinc hydroxide with either lime or caustic. The disadvantage of lime addition is the concurrent precipitation of calcium sulfate in the presence of high sulfate levels in the wastewater. An effluent soluble zinc of less than 0.1 mg/l has been achieved at pH 11.0. A summary of hydroxide precipitation results is shown in Table 4.16.

TABLE 4.16
Summary of hydroxide precipitation treatment results for zinc wastewaters[15]

Industrial source	Zinc concentration, mg/l		Comments[†]
	Initial	Final	
Zinc plating	—	0.2–0.5	pH 8.7–9.3
General plating	18.4	2.0	pH 9.0
	—	0.6	Sand filtration
	55–120	1.0	pH 7.5
	46	2.9	pH 8.5
		1.9	pH 9.2
		2.8	pH 9.8
		2.9	pH 10.5
Vulcanized fiber	100–300	1.0	pH 8.5–9.5
Tableware plant	16.1	0.02–0.23	Sand filtration
Viscose rayon	26–120	0.86–1.5	—
	70	3–5	pH 5
	20	1.0	—
Metal fabrication	—	0.5–1.2	Sedimentation
		0.1–0.5	Sand filtration
Radiator manufacture		0.33–2.37	Sedimentation
		0.03–0.38	Sand filtration
Blast furnace gas scrubber water	50	0.2	pH 8.8
Zinc smelter	744	50	
	1500	2.6	
Ferroalloy waste	11.2–34	0.29–2.5	
	3–89	4.2–7.9	
Ferrous foundry	72	1.26	Sedimentation
		0.41	Sand filtration
Deep coal mine— acid water	33–7.2	0.01–10	

[†]All treatment involved precipitation plus sedimentation. Special or additional aspects of treatment are indicated under Comments. (Patterson, 1985)

Results for reverse osmosis treatment of zinc wastewaters are shown in Table 4.17. Electrolytic treatment of zinc cyanide wastewaters results are shown in Table 4.18.

TABLE 4.17
Pilot-scale results for reverse osmosis treatment of zinc wastewaters[17]

Industrial source	Zinc concentration, μg/l		
	Feed	Permeate	Percent removal
Zinc cyanide plating rinse	1,700	30	98
Steam electric power plant	300	53	82
	780	3	99
Textile mill	7,200	140	98
	5,400	6,600	−20
	460	250	46
	520	360	31
	7,200	360	95
	1,400	30	98
	4,100	180	96
	1,200	22	98
	24,000	430	98
	9,700	37	>99
Cooling tower blowdown	10,000	300	97

TABLE 4.18
Electrolytic treatment of zinc cyanide wastes

Waste	Parameter	Concentration, mg/l	
		Initial	Final
A	Zinc	352	0.7
	Cyanide	258	12.0
B	Zinc	117	0.3
	Copper	842	0.5
	Cyanide	1230	<0.1

Summary

Effluent concentrations achievable by metals removal or processes are summarized in Table 4.19. A detailed discussion of metals removal has been presented by Patterson.[15]

TABLE 4.19
Effluent levels achievable in heavy metals removal[14]

Metal	Achievable effluent concentration, mg/l	Technology
Arsenic	0.05	Sulfide ppt with filtration[†]
	0.06	Carbon adsorption
	0.005	Ferric hydroxide co-ppt
Barium	0.5	Sulfate ppt
Cadmium	0.05	Hydroxide ppt at pH 10–11
	0.05	Co-ppt with ferric hydroxide
	0.008	Sulfide precipitation
Copper	0.02–0.07	Hydroxide ppt
	0.01–0.02	Sulfide ppt
Mercury	0.01–0.02	Sulfide ppt
	0.001–0.01	Alum co-ppt
	0.0005–0.005	Ferric hydroxide co-ppt
	0.001–0.005	Ion exchange
Nickel	0.12	Hydroxide ppt at pH 10
Selenium	0.05	Sulfide ppt
Zinc	0.1	Hydroxide ppt at pH 11

[†]*Note:* ppt = precipitation

PROBLEM

4.1. A metal-finishing plant has a wastewater flow of 72,000 gal/d (273 m^3/d) with the following characteristics:

Cr^{6+} 75 mg/l
Cu   10 mg/l
Ni   8 mg/l

Design a reduction and precipitation plant for
1. Continuous flow
2. Batch flow
(*a*) Develop the ORP control points using SO_2 as the reducing agent.
(*b*) Compute the quantity of sludge and drying bed area assuming the sludge concentrates to 2 percent, is applied to the bed to a depth of 18 in (0.46 m), and is removed every 5 days at 12 percent.
(*c*) Compute the residual soluble metal if a terminal pH of 8.5 is used for the precipitation.

REFERENCES

1. Mysels, K. J.: *Introduction to Colloid Chemistry,* Interscience Publishers, New York, 1959.
2. Black, A. P., and H. L. Smith: *J. Am. Water Works Assoc.,* vol. 54, p. 371, 1962.

3. Riddick, T. M.: *Tappi*, vol. 47, pt. 1, p. 171A, 1964.
4. Palladino, A. J.: *Proc. 10th Ind. Waste Conf.,* May 1955, Purdue University.
5. Knack, M. F.: *Proc. 4th Ind. Waste Conf.,* 1949, Purdue University.
6. Leonard, A. G., and R. G. Keating: *Proc. 13th Ind. Waste Conf.,* 1946, Purdue University.
7. Bloodgood, D., and W. J. Kellenher: *Proc. 7th Ind. Waste Conf.,* 1952, Purdue University.
8. Eckenfelder, W. W., and D. J. O'Connor: *Proc. 10th Ind. Waste Conf.,* May 1955, p. 17, Purdue University.
9. Webster, R. A.: *Sewage Ind. Wastes,* vol. 25, pt. 12, p. 1432, December 1953.
10. Olthof, M. G., and W. W. Eckenfelder: *Textile Chemist and Colorist,* vol. 8, pt. 7, p. 18, 1976.
11. Olthof, M. G., and W. W. Eckenfelder: *Water Res.,* vol. 9, p. 853, 1975.
12. Southgate, B. A.: *Treatment and Disposal of Industrial Waste Waters,* His Majesty's Stationery Office, London, 1948, p. 186.
13. McCarthy, Joseph A.: *Sewage Works J.,* vol. 21, pt. 1, p. 75, January 1949.
14. Talinli, I.: *Wat. Sci. Tech.,* 29, 9, p. 175, 1994.
15. Patterson, J. W.: *Industrial Wastewater Treatment Technology,* Butterworth Publishers, Boston, 1985.
16. Robinson, A. and J. Sum: U.S. EPA 600/2-80-139, June 1980.
17. Cawley, W. (ed.): *Treatability Manual,* vol. III, U.S. EPA 600/8-80-042-C, July 1980.

5

AERATION AND MASS TRANSFER

5.1
MECHANISM OF OXYGEN TRANSFER

Aeration is used for transferring oxygen to biological-treatment processes, for stripping solvents from wastewaters, and for removing volatile gases such as H_2S and NH_3.

Aeration is a gas-liquid mass-transfer process in which interphase diffusion occurs when a driving force is created by a departure from equilibrium. In the gas phase, the driving force is a partial pressure gradient; in the liquid phase, it is a concentration gradient.

The rate of molecular diffusion of a dissolved gas in a liquid is dependent on the characteristics of the gas and the liquid, the temperature, the concentration gradient, and the cross-sectional area across which diffusion occurs. The diffusional process is defined by Fick's law:

$$N = -D_L A \frac{dc}{dy}$$
(5.1)

where
N = mass transfer per unit time
A = cross-sectional area through which diffusion occurs
dc/dy = concentration gradient perpendicular to cross-sectional area
D_L = diffusion coefficient through the liquid film

If it is assumed that equilibrium conditions exist at the interface, the mass-transfer process can be reexpressed as:

$$N = \left(-D_g A \frac{dp}{dy}\right)_1 = \left(-D_L A \frac{dc}{dy}\right)_2 = \left(-D_e A \frac{dc}{dy}\right)_3$$
(5.2)

where
D_g = diffusion coefficient through the gas film
D_e = eddy diffusion coefficient of the gas in the body of the liquid
D_L = diffusion coefficient through the liquid film

Since the systems dealt with in waste treatment involve high degrees of turbulence, the eddy diffusivity will be several orders of magnitude greater than the coefficients of molecular diffusivity, and this need not be considered as a rate-controlling step. An exception may be large aerated lagoons or aeration in flowing rivers.

Lewis and Whitman[1] developed the two-film concept which considers stagnant films at the gas and liquid interfaces through which mass transfer must occur. Equation (5.2) can be reexpressed in terms of a liquid and gas film:

$$N = K_L A(C_s - C_L) = K_g A(P_g - P) \tag{5.3}$$

where N = mass of oxygen transferred per unit area
       A = interfacial surface area
       C_s = oxygen saturation concentration
       C_L = concentration of oxygen in the liquid
       K_L = liquid-film coefficient, defined as D_L/Y_L
       K_g = gas-film coefficient, defined as D_g/Y_g
       Y_L = liquid-film thickness
       Y_g = gas-film thickness

For sparingly soluble gases, such as oxygen and CO_2, the liquid-film resistance controls the rate of mass transfer; for highly soluble gases, such as ammonia, the gas-film resistance controls the transfer rate. Most of the mass-transfer applications in waste treatment are liquid-film controlled. Increasing the fluid turbulence will decrease the film thickness and hence increase K_L. Danckwertz[2] has defined the liquid-film coefficient as the square root of the product of the diffusivity and the rate of surface renewal:

$$K_L = \sqrt{D_L r} \tag{5.4}$$

The rate of surface renewal r can be considered as the frequency with which fluid with a solute concentration C_L is replacing fluid from the interface with a concentration C_s. High degrees of fluid turbulence will increase r.

Dobbins[3] has proposed a relationship that describes the aforementioned transfer mechanisms:

$$K_L = (D_L r)^{1/2} \coth\left(\frac{r Y_L^2}{D_L}\right)^{1/2} \tag{5.5}$$

When the surface-renewal rate is zero, K_L is equal to D_L/Y_L and the transfer is controlled by molecular diffusion through the surface film. As r increases, K_L becomes equal to $\sqrt{D_L r}$ and the transfer is a function of the rate of surface renewal.

For liquid-film-controlled processes, Eq. (5.3) can be reexpressed in concentration units:

$$\frac{1}{V}N = \frac{dc}{dt} = K_L \frac{A}{V}(C_s - C_L) \tag{5.6}$$

in which V is the liquid volume and

$$K_L \frac{A}{V} = K_L a$$

$K_L a$ is an overall film coefficient and is usually used to compute the transfer rate.

The most important application of aeration in waste treatment is the transfer of oxygen to biological-treatment processes and the natural reaeration of streams and other watercourses.

The equilibrium concentration of oxygen in contact with water, C_s, is defined by Henry's law:

$$p = HC_s \qquad (5.7)$$

where p = partial pressure of oxygen in the gas phase and H = Henry's constant, which is proportional to temperature and is influenced by the presence of dissolved solids.

The solubility of oxygen in water at various temperatures is summarized in Table 5.1. As temperature and dissolved solids concentrations increase, Henry's constant also increases, thereby reducing C_s. It is therefore usually necessary in industrial wastes to measure solubility experimentally.

In aeration tanks, where air is released at an increased liquid depth, the solubility of oxygen is influenced both by the increasing partial pressure of the air

TABLE 5.1
Solubility of oxygen (mg/l) at various temperatures, elevations, and total dissolved solids levels

Temperature			Elevation, ft							TDS (sit sea level), ppm			
°F	°C	0	1000	2000	3000	4000	5000	6000	400	800	1500	2500
32.0	0	14.6	14.1	13.6	13.1	12.6	12.1	11.7	—	—	—	—
35.6	2	13.8	13.3	12.8	12.4	11.9	11.5	11.1	13.74	13.68	13.58	13.42
39.2	4	13.1	12.6	12.2	11.8	11.4	10.9	10.5	13.04	12.98	12.89	12.75
42.8	6	12.5	12.0	11.6	11.2	10.8	10.4	10.0	12.44	12.38	12.29	12.15
46.4	8	11.9	11.4	11.0	10.6	10.2	9.9	9.5	11.85	11.80	11.70	11.58
50.0	10	11.3	10.9	10.5	10.1	9.8	9.4	9.1	11.25	11.20	11.12	11.00
53.6	12	10.8	10.4	10.1	9.7	9.4	9.0	8.6	10.76	10.71	10.64	10.52
57.2	14	10.4	10.0	9.6	9.3	8.9	8.6	8.3	10.36	10.32	10.25	10.15
60.8	16	10.0	9.6	9.2	8.9	8.6	8.3	8.0	9.96	9.92	9.85	9.75
64.4	18	9.5	9.2	8.9	8.5	8.2	7.9	7.6	9.46	9.43	9.36	9.27
68.0	20	9.2	8.8	8.5	8.2	7.9	7.6	7.3	9.16	9.13	9.06	8.97
7.16	22	8.8	8.5	8.2	7.9	7.6	7.3	7.1	8.77	8.73	8.68	8.60
75.2	24	8.5	8.2	7.9	7.6	7.3	7.1	6.8	8.47	8.43	8.38	8.30
78.8	26	8.2	7.9	7.6	7.3	7.1	6.8	6.6	8.17	8.13	8.08	8.00
82.4	28	7.9	7.6	7.4	7.1	6.8	6.6	6.3	7.87	7.83	7.78	7.70
86.0	30	7.6	7.4	7.1	6.9	6.6	6.4	6.1	7.57	7.53	7.48	7.40
89.6	32	7.4	7.1	6.9	6.6	6.4	6.2	5.9	7.4	—	—	—
93.2	34	7.2	6.9	6.7	6.4	6.2	6.0	5.8	7.2	—	—	—
96.8	36	7.0	6.7	6.5	6.3	6.0	5.8	5.6	7.0	—	—	—
100.4	38	6.8	6.6	6.3	6.1	5.9	5.6	5.4	6.8	—	—	—
104.0	40	6.6	6.4	6.1	5.9	5.7	5.5	5.3	6.6	—	—	—

Note: ft = 0.3048 m.

entering the aeration tank and by the decreasing partial pressure in the air bubble as oxygen is absorbed. For these cases, a mean saturation value corresponding to the aeration tank middepth is used:

$$C_{s,m} = C_s \times \frac{1}{2}\left(\frac{P_b}{P_a} + \frac{O_t}{20.9}\right) \tag{5.8}$$

where P_a = atmospheric pressure
P_b = absolute pressure at the depth of air release
O_t = percent concentration of oxygen in the air leaving the aeration tank

Mueller et al.[4] have shown that oxygen saturation is a function not only of submergence but also of diffuser type. Coarse bubble units provide lower clean water saturation values than the fine bubble or jet diffusers. Saturation appears to be related to both bubble size and mixing pattern. It should further be noted that field data seems to suggest that the 0.25 depth level may be more accurate for fine bubble diffusers (Schmit et al.[5]). Therefore where possible oxygen saturation should be determined from field data.

To account for wastewater constituents, a factor is used:

$$\beta = \frac{C_s \text{ wastewater}}{C_s \text{ tap water}}$$

The ASCE committee on oxygen transfer[19] has recommended using a correction for TDS as shown in Table 5.1 to determine β.

The oxygen-transfer coefficient $K_L a$ is affected by the physical and chemical variables characteristic of the aeration system:

1. *Temperature.* The liquid-film coefficient will increase with increasing temperature. When air bubbles are involved, changes in liquid temperature will also affect the size of bubbles generated in the system. The effect of temperature on the coefficient is

$$K_L(T) = K_L(20°C)\theta^{T-20} \tag{5.9}$$

   For diffused aeration units, θ is usually taken as 1.02. Correlation's of Imhoff and Albrect[6] showed θ to be higher for low-turbulence diffused systems and lower for high-turbulence surface aeration systems. Landberg et al.[7] suggested θ of 1.012 for surface aeration systems. The effect of temperature on $K_L a$ is shown in Figure 5.1.
2. *Turbulent mixing.* Increasing the degree of turbulent mixing will increase the overall transfer coefficient.
3. *Liquid depth.* The effect of liquid depth H on $K_L a$ will depend in large measure on the method of aeration. For most types of bubble-diffusion systems, $K_L a$ will vary with depth according to the relationship

$$\frac{K_L a(H_1)}{K_L a(H_2)} = \left(\frac{H_1}{H_2}\right)^n \tag{5.10}$$

   The exponent n has a value near 0.7 for most systems. Wagner and Popel[8] evaluated several diffused aeration systems and showed that the oxygen-transfer efficiency increased 1.5 percent per foot of depth.

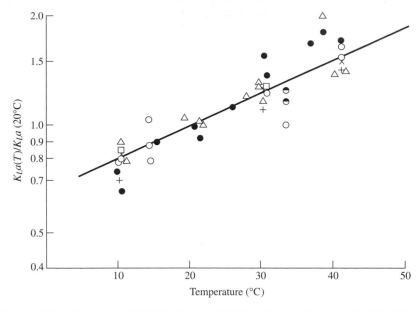

Symbol	Substance	Unit
○	Water	Spinnerette 20 holes, 0.035 mm diameter
+	Water	Spinnerette 10 holes, 0.05 mm diameter
●	Water	Aloxite stone
△	1% KCl	Aloxite stone
□	1% KCl	Spinnerette 10 holes, 0.05 mm diameter
×	50 ppm heptanoic acid	Spinnerette 20 holes, 0.035 mm diameter
◕	Water	Data of Carpani and Roxburgh[9]
△	Water	Data of Gameson and Robertson[10]

FIGURE 5.1
Relationship between the overall transfer coefficient $K_L a$ and temperature.

4. *Waste characteristics.* The presence of surface-active agents and other organics will have a profound effect on both K_L and A/V. Molecules of surface-active materials will orient themselves on the interfacial surface and create a barrier to diffusion. The excess surface concentration is related to the change in surface tension, as defined by the Gibbs equation, such that small concentrations of surface-active material will depress K_L while large concentrations will exert no further effect. The absolute effect of surfactants on K_L will also depend on the nature of the aeration surface. Less effect would be exerted at a highly turbulent liquid surface, since the short life of any interface would restrict the formation of an adsorbed film. Conversely, a greater effect would be exerted at a bubble surface because of the relatively long life of the bubble as it rises through an aeration tank. A decrease in surface tension will decrease the size of bubbles generated from an air-diffusion system. This in turn will increase A/V.

In some cases, the increase in A/V will exceed the decrease in K_L, and the transfer rate will increase over that in water. The effect of waste characteristics on $K_L a$ is defined by a coefficient α in which

$$\alpha = \frac{K_L a \text{ waste}}{K_L a \text{ water}}$$

These relationships are shown in Fig. 5.2.

Turbulence has a significant effect on α. At high turbulence levels, oxygen transfer is dependent on surface renewal and not significantly affected by diffusion through interfacial resistances. Under these conditions α may be greater than 1.0 because of the increased A/V ratio. Under low-turbulence conditions, the bulk oxygen transfer resistance is reduced but surface renewal does not yet occur, thus surfactant interfacial resistance causes a significant reduction in the oxygen transfer rate, as shown in Fig. 5.3. Increasing the TDS will increase $K_L a$ due to the generation of finer bubbles, as shown in Fig. 5.4.

In order to compare the transfer rate in water to a waste with a particular aeration device, a coefficient α has been defined as $K_L a$ (waste)/$K_L a$ (water). The coefficient α can be expected to increase or decrease and approach unity during the course of biooxidation, since the substances affecting the transfer rate are being removed in the biological process, as shown in Fig. 5.5.

In view of the effects previously mentioned, the type of aeration will have a profound effect on α. For the fine-bubble diffuser systems, values are generally lower than for coarse-bubble or surface aeration systems. Variation in mixed liquor

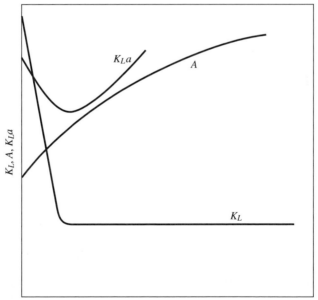

Concentration of surfactants

FIGURE 5.2
Effect of concentration of surface active agent on oxygen transfer.

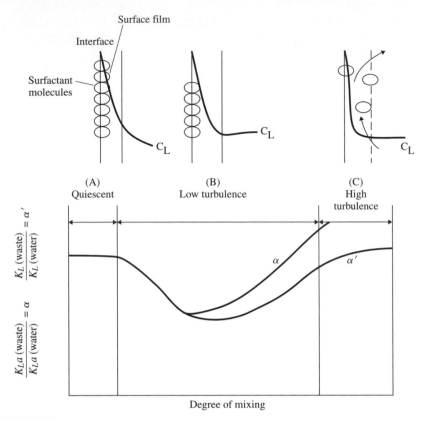

FIGURE 5.3
Effect of turbulence on oxygen transfer.

suspended solids (MLSS) from 2000 to 7000 mg/l with dome diffusers had no significant effect on $K_L a$. Reported values of α for different aeration devices are summarized in Table 5.2. In plug flow basins the α value will increase as purification occurs through the tank length. This is shown in Fig. 5.6 for two types of aeration devices.[15]

In diffused aeration systems, air bubbles are formed at an orifice from which they break off and rise through the liquid, finally bursting at the liquid surface. The velocity and shape of the air bubbles is related to a modified Reynolds number. At N_{Re} of 300 to 4000, the bubbles assume an ellipsoidal shape and rise with a rectilinear rocking motion. At N_{Re} greater than 4000, the bubbles form spherical caps. The rising velocity of the bubble is increased at high airflows because of the proximity of other bubbles and the resulting disturbances of the bubble wakes.

A general correlation for oxygen transfer from air bubbles rising through a still water column has been developed by Eckenfelder,[11] as shown in Fig. 5.7. The general relationship obtained is

$$\frac{K_L d_B}{D_L} H^{1/3} = C\left(\frac{d_B v_B \rho}{\mu}\right) \tag{5.11}$$

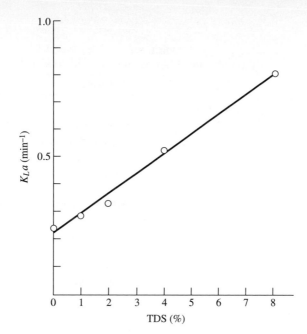

FIGURE 5.4
Effect of TDS on the oxygen transfer rate.

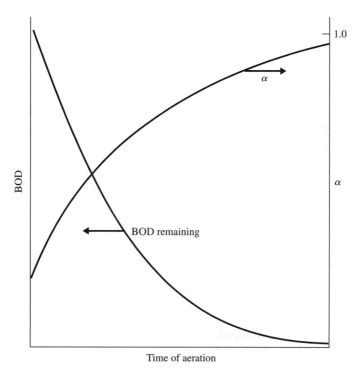

FIGURE 5.5
Change in transfer coefficient α in a biological oxidation process.

TABLE 5.2
Values of α for different aeration devices

Aeration device	Alpha factor	Wastewater
Fine bubble diffuser	0.4–0.6	Tap water containing detergent
Brush	0.8	Domestic wastewater
Coarse bubble diffuser, sparger	0.7–0.8	Domestic wastewater
Coarse bubble diffuser, wide band	0.65–0.75	Tap water with detergent
Coarse bubble diffuser, sparger	0.55	Activated sludge contact tank
Static aerator	0.60–0.95	Activated sludge treating high-strength industrial waste
Static aerator	1.0–1.1	Tap water with detergent
Surface aerators	0.6–1.2	Alpha factor tends to increase with increasing power (tap water containing detergent and small amounts of activated sludge)
Turbine aerators	0.6–1.2	Alpha factor tends to increase with increasing power; 25, 50, 190 gal tanks (tap water containing detergent)

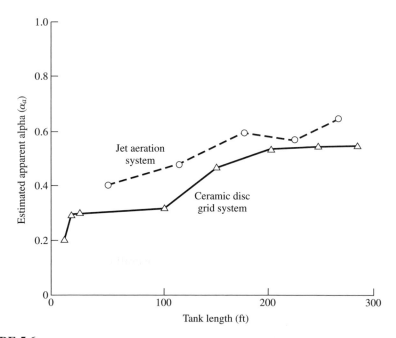

FIGURE 5.6
Estimated change in apparent alpha with tank length, Whittier Narrows, California.

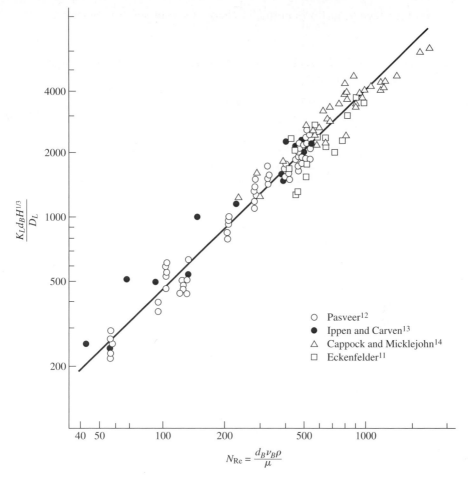

FIGURE 5.7
Correlation of bubble-aeration data.

where C = constant
   d_B = bubble diameter
   v_B = bubble velocity
   ρ = liquid density
   μ = liquid viscosity

Equation (5.11) can be expressed in terms of the overall coefficient $K_L a$ if A/V for air bubbles in an aeration tank is considered as

$$\frac{A}{V} = \frac{6G_s H}{d_B v_B V} \tag{5.12}$$

where G_s is the airflow. Equation (5.11) neglects the aeration tank liquid surface as small compared to the interfacial bubble surface.

Over the range of airflows normally encountered in aeration practice,

$$d_B \approx G_s^n \tag{5.13}$$

Equations (5.11) to (5.13) can be combined to yield a general relationship for oxygen transfer from air-diffusion systems:

$$K_L a = \frac{C'H^{2/3}G_s^{(1-n)}}{V} \tag{5.14}$$

The relationship between $K_L a$ and gas flow for several diffusion devices is shown in Fig. 5.8.

Equation (5.14) can be reexpressed in terms of the mass of oxygen transferred per diffusion unit:

$$N = C'H^{2/3}G_s^{(1-n)}(C_s - C_L) \tag{5.15}$$

where N is the mass of O_2 per hour transferred per diffuser unit.

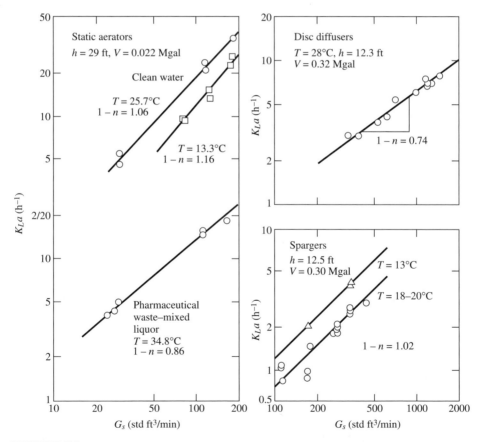

FIGURE 5.8
Effect of gas flow on $K_L a$ for various types of diffuser systems under process conditions.[22]

The oxygen transfer efficiency of a unit is defined as

$$\text{Percent transfer} = \frac{\text{wt } O_2 \text{ adsorbed/unit time}}{\text{wt } O_2 \text{ supplied/unit time}} \times 100$$

$$= \frac{K_La(C_s - C_L) \times V}{G_s \,(\text{std ft}^3/\text{min}) \times 0.232 \text{ lb } O_2/\text{lb air}} \times 100$$

$$\times \, 0.075 \text{ lbs air/std ft}^3 \qquad (5.16)$$

EXAMPLE 5.1. The following data were obtained on the oxygen transfer capacity of an air-diffusion unit in clean water.

Airflow rate	25 std ft³/min · thousand ft³)[25 1/(min · m³)]
Volume	1000 ft³ (28.3 m³)
Temperature	54°F (12°C)
Liquid depth	15 ft (4.6 m)
Average bubble diameter	0.3 cm
Average bubble velocity	32 cm/s

Time, min	C_L, mg/l
3	0.6
6	1.6
9	3.1
12	4.3
15	5.4
18	6.0
21	7.0

(a) Compute K_La and K_L.
(b) Compute the mass of O_2 per hour transferred per unit volume at 20°C and zero dissolved oxygen, and the oxygen transfer efficiency.
(c) How much oxygen will be transferred to a waste with an α of 0.82, a temperature of 32°C, and an operating dissolved oxygen of 1.5 mg/l?

Solution.

(a) At a temperature of 54°F (12°C) saturation is 10.8 mg/l. The mean saturation in the aeration tank, assuming 10 percent oxygen adsorption, is

$$C_{s,m} = C_s\left(\frac{P_b}{29.4} + \frac{O_t}{42}\right)$$

where

$$P_b = \frac{15 \text{ ft}}{2.3 \text{ ft}/(\text{lb} \cdot \text{in}^2)} + 14.7 = 21.2 \text{ lb/in}^2 \,(1.44 \text{ atm})$$

$$O_t = \frac{21(1 - 0.1)}{21(1 - 0.1) + 79} \times 100 = 19.3\% \, O_2$$

$$C_{s,m} = 10.8\left(\frac{21.2}{29.4} + \frac{19.3}{42}\right) = 10.8 \times 1.18$$

$$= 12.7 \text{ mg/l}$$

Time, min	$C_{s,m} - C_L$
3	12.1
6	11.1
9	9.6
12	8.4
15	7.3
18	6.7
21	5.7

From Eq. (5.22),

$$\log(C_{s,m} - C_L) = \log(C_{s,m} - C_O) - \frac{K_L a}{2.3}t$$

which represents a straight line on a semilog plot of log $(C_{s,m} - C_L)$ versus L. $K_L a$ can be computed from the slope of the line in Fig. 5.9.

$$K_L a = 2.3\,\frac{\log(14/9)}{10} \times 60 = 2.63/\text{h}$$

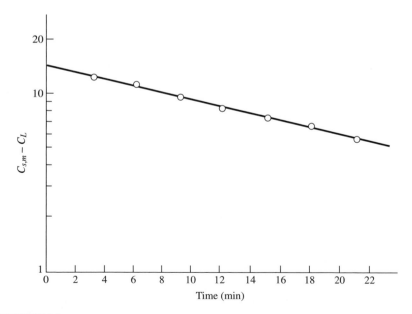

FIGURE 5.9

The interfacial area/volume ratio A/V is

$$\frac{A}{V} = \frac{6G_sH}{d_Bv_BV} = \frac{6 \times 25 \times 15 \times 60}{(0.3/30.5) \times (32/30.5) \times 3600 \times 1000} = 3.65 \ \text{ft}^2/\text{ft}^3 \ (12 \ \text{m}^2/\text{m}^3)$$

where d_B = average bubble diameter
            H = aeration liquid depth
            v_B = average bubble velocity
            V = volume

$$K_L = \frac{K_La}{A/V}$$

$$= \frac{2.63}{3.65} = 0.72 \ \text{ft/h} \ (21.96 \ \text{cm/h})$$

(b)

$$K_La_T = K_La_{20} \times 1.02^{T-20}$$

$$K_La_{20} = 1.02^8 \times 2.63 = 1.17 \times 2.63 = 3.07/\text{h}$$

$$C_{s,m(20°)} = 9.1 \times 1.18 = 10.7 \ \text{mg/l}$$

$$N_O = K_LaVC_{s,m}$$

$$= 3.07/\text{h} \times 1000 \ \text{ft}^3 \times 10.7 \ \text{mg/l} \times 7.48 \ \text{gal/ft}^3$$

$$\times 8.34 \times 10^{-6}$$

$$= 2.05 \ \text{lb} \ O_2/\text{h} \qquad (0.93 \ \text{kg} \ O_2/\text{h})$$

$$\% \ O_2 \ \text{transfer efficiency} = \frac{2.05 \ \text{lb} \ O_2/\text{h transferred}}{25 \times 60 \times 0.0746 \times 0.232 \ \text{lb} \ O_2/\text{h supplied}} \times 100$$

$$= 8.0$$

(c) The weight of oxygen transferred is determined from Eq. (5.17):

$$N = N_O \left(\frac{\beta C_{s,m} - C_L}{C_{s,m(20°)}} \right) \alpha \times 1.02^{T-20}$$

$$= 2.05 \left(\frac{0.99 \times 7.4 \times 1.18 - 1.5}{10.7} \right) \times 0.82 \times 1.02^{12}$$

$$= 1.42 \ \text{lb/h} \ (0.64 \ \text{kg/h})$$

5.2
AERATION EQUIPMENT

The aeration equipment commonly used in the industrial waste field consists of air-diffusion units, turbine aeration systems in which air is released below the rotating blades of an impeller, and surface aeration units in which oxygen transfer is accomplished by high surface turbulence and liquid sprays. Generic types of aeration equipment are shown in Fig. 5.10.

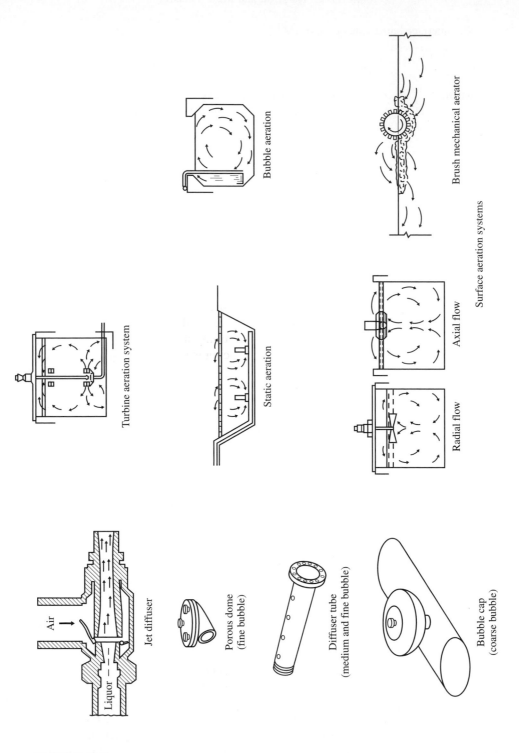

FIGURE 5.10
Aeration equipment.

A manufacturer will generally designate the oxygen transfer capability of his equipment in terms of the pounds of O_2 transferred per horsepower-hour [kg $O_2/(kW \cdot h)$] or the oxygen-transfer efficiency or pounds of O_2 transferred per hour (kg O_2/h) per diffusion unit. This is called the standard oxygen rating (SOR) in tap water, at 20°C, and zero dissolved oxygen at sea level.

The actual oxygen transferred to the wastewater (AOR) is computed from

$$N = N_O\left(\frac{\beta C_s - C_L}{9.2}\right)\alpha \times 1.02^{T-20} \tag{5.17}$$

where N = AOR, lb $O_2/hp \cdot h$ [kg $O_2/kW \cdot h$)]
     N_O = SOR, lb $O_2/hp \cdot h$ [kg $O_2/kW \cdot h$)]

Diffused Aeration Equipment

Basically, there are two types of diffused aeration equipment: units producing a small bubble from a porous medium or membrane and units using a large orifice or a hydraulic shear device to produce large air bubbles.

Porous media are either tubes or plates constructed of carborundum or other finely porous media or membranes. Tubes are placed at the sidewall of the aeration tank, perpendicular to the wall, and generate a rolling motion to maintain mixing or across the bottom of the basin. Maximum spacing is required to maintain solids in suspension; minimum spacing is required to avoid bubble coalescence. A diffused aeration system is shown in Fig. 5.11.

FIGURE 5.11
Diffused aeration system.

In order for adequate mixing to be maintained for sidewall mounts, the maximum width of the aeration tank is approximately twice its depth. This width can be doubled by placing a line of diffusion units along the centerline of the aeration tank. Figure 5.12 shows the performance of air-diffusion units in water in an aeration tank 15 ft (4.6 m) deep and 24 ft (7.3 m) wide. Fine bubble diffusers tend to clog with time, resulting in a reduced oxygen transfer efficiency as shown in Fig. 5.13. For bottom tank coverage, the tank must be drained and the diffusers cleaned.

The effect of diffuser placement on clean water oxygen transfer efficiency is shown in Table 5.3. Fine-bubble and membrane diffusers will tend to clog on the liquid side because of the precipitation of metal hydroxides and carbonates or formation of a biofilm layer. The fouling factor F is site specific and can vary from 0.2 to 0.9. F appears to increase with increasing solids retention time (SRT). The average F value for municipal wastewater treatment facilities using medium- to fine-bubble membrane diffusers is 0.6. This value will apply for most industrial

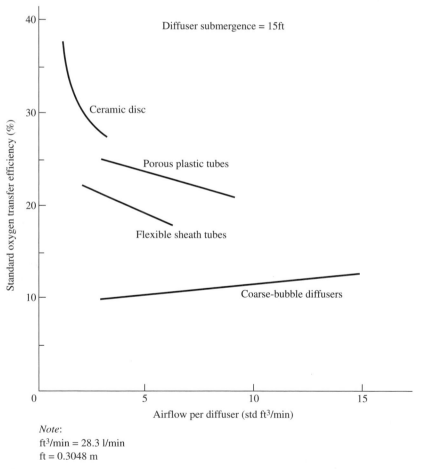

Note:
ft³/min = 28.3 l/min
ft = 0.3048 m

FIGURE 5.12
Effect of airflow rate per diffuser on oxygen transfer efficiency.

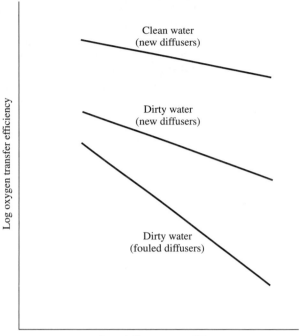

FIGURE 5.13
Change in oxygen transfer efficiency with fine-pore diffuser fouling—a hypothetical case.

TABLE 5.3
Effects of diffuser placement on clean water oxygen transfer efficiencies (OTEs) of flexible sheath tubes[15]

		Standard OTE (%) at water depth		
Placement	Airflow (std ft³/min/diffuser)	3 m	4.5 m	6 m
Floor cover (grid)	1–4	14–18	21–27	29–35
Quarter points	2–6	13–15	18–22	24–29
Mid-width	2–6	9–11	15–18	23–17
Single-spiral roll	2–6	7–11	14–18	21–28

Note: cfm × 0.47 = l/s

facilities unless other information is available. In cases where fouling is a factor, α in Eq. (5.17) should be corrected to $F\alpha$.

Large-bubble air-diffusion units will not yield the oxygen-transfer efficiency of fine-bubble diffusers since the interfacial area for transfer is considerably less. These units have the advantage, however, of not requiring air filters and of generally requiring less maintenance. These units generally operate over a wider range of airflow per unit. Performance data for coarse-bubble aeration units are shown in Fig. 5.14.

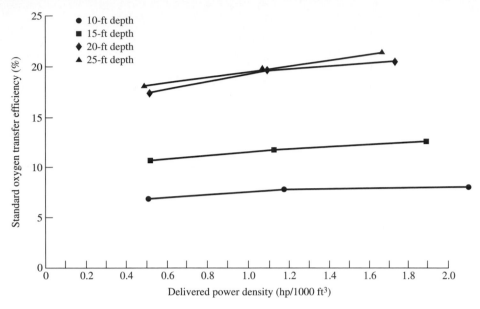

Note:
10- and 20-ft depths: 24-D-24 fixed-orifice diffusers applied in a total floor coverage configuration.
15- and 25-ft depths: 30-D-24 fixed-orifice diffusers applied in a wide-band configuration along one centerline of the tank.

FIGURE 5.14
Oxygen transfer efficiency of coarse-bubble diffusers (*courtesy of Sanitaire Corp.*).
Results are for clean water.

Static aerators consist of vertical cylindrical tubes placed at specified intervals in an aeration basin and containing fixed internal elements.

In aeration of sewage or industrial wastes, the coefficient α will usually be less than 1.0. There is some evidence that the large-bubble diffusers are less affected by the presence of surface-active materials than are the fine-bubble diffusers.

A diffused aeration design is shown in Example 5.2.

EXAMPLE 5.2. An aeration system is to operate under the following conditions:

Aeration Design

> 1000 lb O_2/h
> $T = 30°C$
> $C_L = 2$ mg/l
> $\beta = 0.95$
> $\alpha = 0.85$
> $D = 20$ ft

Use coarse bubble diffusers with a 12 percent oxygen transfer efficiency (OTE) at 15 std ft³/min per tube and 15-ft liquid depth. At 20 ft:

$$\text{OTE} = 12\left(\frac{20}{15}\right)^{0.7} = 14.7\%$$

O_2 saturation at 30°C is 7.6 mg/l. Assume 15% OTE. Then

$$O_t = \frac{21(1 - 0.15) \times 100}{79 + 21(1 - 0.15)}$$

$$= 0.18$$

$$C_{s,m} = \frac{7.6}{2}\left[\frac{14.7 + 0.433 \times 20}{14.7} + \frac{18}{21}\right]$$

$$= 9.3 \text{ mg/l}$$

$$N = 15 \text{ std ft}^3/\text{min/unit} \times 60 \times 0.232 \times 0.0746 \times 0.147$$

$$= 2.29 \text{ lb } O_2/\text{h/unit}$$

$$N = N_O\left[\frac{\beta C_{SW} - C_L}{C_s}\right]\alpha \times 1.02^{T-20}$$

$$= 2.29\left[\frac{0.95 \times 9.3 - 2.0}{9.3}\right]0.85 \times 1.02^{10}$$

$$= 1.74 \text{ lb } O_2/\text{h/unit}$$

$$\frac{1000 \text{ lb } O_2/\text{h}}{1.74} = 575 \text{ diffusers}$$

$$\text{Airflow} = 575 \times 15 = 8620 \text{ std ft}^3/\text{min}$$

$$\text{hp} = \frac{(\text{std ft}^3/\text{min})(\text{lb/in}^2)(144)}{0.7 \times 33,000} \quad (195 \text{ kW})$$

$$= \frac{8620 \times 10 \times 144}{0.7 \times 33,000} = 537$$

Turbine Aeration Equipment

Turbine aeration units disperse compressed air by the shearing and pumping action of a rotating impeller. Since the degree of mixing is independently controlled by the power input to the turbine, there are no restrictive limitations to tank geometry. Figure 5.15 shows a typical turbine aeration unit.

Air is usually fed to the turbine through a sparge ring located just beneath the impeller blades. The ratio of the turbine diameter to the equivalent tank diameter varies from 0.1 to 0.2. Most aeration applications have employed impeller tip speeds of 10 to 18 ft/s (3.1 to 5.5 m/s).

Quirk[16] showed that the oxygen transfer for a turbine aerator can be estimated.

$$O_2 \text{ transfer efficiency} = CP_d^n \tag{5.18}$$

where $P_d = \text{hp}_R/\text{hp}_c$ in which hp_R and hp_c are the horsepower (kW) of the turbine and compressor, respectively. The optimum oxygenation efficiency occurs when the power split between turbine and blower is near a 1:1 ratio.[15]

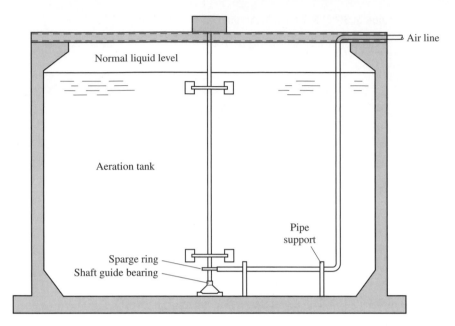

FIGURE 5.15
Typical turbine aerator installation.

The actual power drawn by the impeller will decrease as air is introduced under the impeller because of the decreased density of the aerating mixture. The transfer efficiency of turbine aeration units will vary from 1.6 to 2.9 lb O_2/hp · h [0.97 to 1.76 kg O_2/(kW · h)], depending on the turbine-blower power split.

To eliminate swirling and vortexing, baffles are normally required. In a round tank, four baffles are placed equally around the circumference of the wall. In a square tank, two baffles on opposite walls are used, while in a rectangular tank with a length/width ratio greater than 1.5 no baffles are required.

Surface-Aeration Equipment

Surface-aeration units are of two types: those employing a draft tube and those with only a surface impeller. In both types, oxygen transfer occurs through a vor-texing action and from the surface area exposure of large volumes of liquid sprayed over the surface of the aeration tank. Examples of these units are shown in Figs. 5.16 and 5.17. The transfer rate is influenced by the diameter of the impeller and its speed of rotation, and by the submergence level of the rotating element. Under optimum submergence conditions, the transfer rate per unit horsepower (W) remains relatively constant over a wide range of impeller diameters.

The quantity of oxygen transferred at the liquid surface is a function of the power level, and the overall oxygen transfer rate generally increases with increasing power level. Kormanik et al.[17] have shown a correlation between the oxygen-transfer rate

FIGURE 5.16
Low-speed surface aerator.

FIGURE 5.17
High speed surface aerator (*courtesy of Aqua Aerobic Systems Inc.*).

179

and horsepower per unit surface area, as shown in Fig. 5.18. To maintain uniform dissolved oxygen concentrations requires power levels of 6 to 10 hp/million gal (1.2 to
2.0 W/m³). The maximum area of influence for high speed and low speed surface
aerators is shown in Table 5.4. To maintain biological solids in suspension, a minimum bottom velocity of 0.4 ft/s (12 cm/s) is required for 5000 mg/l MLSS.

To prevent bottom scour, minimum depths of 6 to 8 ft (1.8 to 2.4 m) are recommended, and, to maintain solids in suspension, maximum depths of 12 ft (3.7 m)
are recommended for high-speed aerators and 16 ft (4.9 m) for low-speed aerators.

The brush aerator, which has been very popular in Europe, uses a high-speed
rotating brush that sprays liquid across the tank surface. A circular liquid motion is
induced in the aeration tank. The performance of the brush aerator is related to the
speed of rotation and submergence of the brush. The efficiencies of various aerators are shown in Table 5.5. Mixing requirements are shown in Table 5.6.

Measurement of Oxygen Transfer Efficiency

Although several procedures have been followed to estimate the transfer efficiency
of aeration devices, the non-steady-state aeration procedure has been generally
adopted as a standard. This test involves the chemical removal of dissolved oxygen
by the addition of sodium sulfite with cobalt added as a catalyst. The cobalt concentration should be 0.05 mg/l to avoid a "cobalt effect" on the magnitude of $K_L a$.
The increase in oxygen concentration is measured during aeration under specified

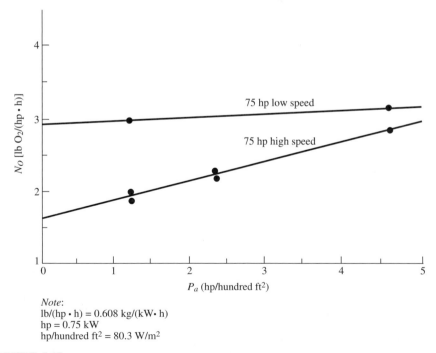

Note:
lb/(hp · h) = 0.608 kg/(kW· h)
hp = 0.75 kW
hp/hundred ft² = 80.3 W/m²

FIGURE 5.18
Comparative effect of surface area on high- and low-speed surface aerators.[17]

TABLE 5.4
Maximum area of influence for high-speed and for low-speed surface aerators [18]

	Radius of influence, ft	
Horsepower	High speed	Low speed
5	40	50
10	55	70
20	80	95
30	100	120
50	130	155
75	155	190
100	—	220
150	—	260

TABLE 5.5
Summary of aerator efficiencies

Type of aerator	Water depth, ft	OTE, %	lb O$_2$/hp · h[†]	Reference
Fine bubble				
Tubes—spiral roll	15	15–20	6.0–8.0	20
Domes—full floor coverage	15	27–31	10.8–12.4	20
Coarse bubble				
Tubes—spiral roll	15	10–13	4.0–5.2	20
Spargers—spiral roll	14.5	8.6	3.4	21
Jet aerators	15	15–24	4.4–4.8	20
Static aerators	15	10–11	4.0–4.4	20
	30	25–30	6.0–7.5	
Turbine	15	10–25	—[‡]	16
Surface aerator				
Low speed	12	—	5.9–7.5	17
High speed	12	—	3.3–5.0	17

[†] Wire horsepower must correct for overall blower efficiency.
[‡] Horsepower depends on power split.

Note:
ft = 0.3048 m
lb/(hp · h) = 0.608 kg/(kW · h)

conditions, and the overall transfer coefficient is calculated from Eq. (5.6). Dissolved oxygen can be measured by the Winkler test or by use of a dissolved oxygen probe. The suggested procedure for non-steady-state aeration is as follows:

1. Remove the dissolved oxygen in the aeration unit by adding sodium sulfite and cobalt chloride. A concentration of 0.05 mg/l of cobalt should be added, and 8 mg/l Na$_2$SO$_3$ per mg/l dissolved oxygen.
2. Thoroughly mix the tank contents. In a diffused aeration unit, aeration for 1 to 2 minutes will usually be sufficient.

TABLE 5.6
Activated sludge process mixing horsepower guide[18]

Aeration device[†]	hp/1000 ft³, activated sludge	hp/1000 ft³, aerobic digestion
High speed	1.30	2.00
Low speed	1.00	1.50
Brush ditch	0.60	1.00
Fine bubble (floor cover)	0.40	0.80
Fine bubble (gut roll)	1.00	1.50
Fine bubble (spiral roll)	0.75	1.25
Fine bubble (moving bridge)	0.60	1.00
Jet	1.00	1.50
Coarse bubble (floor cover)	0.60	1.00
Coarse bubble (gut roll)	1.00	1.50
Coarse bubble (spiral roll)	0.75	1.25
Submerged turbine	0.75	1.25

[†] Operation conditions:
hp, delivered or water horsepower
MLSS, 1500–3000 mg/l (activated sludge)
MLSS, 10,000–20,000 mg/l (aerobic digestion)

3. Start the aeration unit at the desired operating rate. Sample for dissolved oxygen at selected time intervals. (At least 5 points should be obtained before 90 percent of saturation is reached.)
4. In a large aeration tank, multiple sampling points should be selected (both longitudinally and vertically) to compensate for concentration gradients.
5. If a dissolved oxygen probe is used, it can be left in the aeration tank; the values from the probe can be recorded at appropriate time intervals.
6. Record the temperature and measure oxygen saturation. If water is being aerated, it is usually satisfactory to select the saturation value from Table 5.1. For diffused aeration units, saturation should be corrected to the tank middepth according to Eq. (5.8).
7. Compute the oxygen-transfer rate in accordance with Eq. (5.6) (see Example 5.1).

Equation (5.6) can be integrated to yield

$$C_s - C_L = (C_s - C_0)e^{-K_L a t} \qquad (5.19)$$

where C_0 = dissolved oxygen at time zero. To evaluate the constants in Eq. (5.19), a nonlinear least-squares program is recommended by ASCE[19], one that provides the best set for $K_L a$, C_0, and C_s.

It is occasionally desirable to measure the oxygen-transfer rate in the presence of activated sludge. Either a steady-state or non-steady-state procedure can be followed for this purpose.

In the aeration of activated sludge, Eq. (5.6) must be modified to account for the effect of the oxygen-utilization rate of the sludge-liquid mixture:

$$\frac{dc}{dt} = K_L a(C_s - C_L) - r_r \qquad (5.20)$$

where r_r is the oxygen-utilization rate in milligrams per liter per hour.

Under steady-state operation $dc/dt \to 0$, and K_La can be computed from the relationship

$$K_La = \frac{r_r}{C_s - C_L} \tag{5.21}$$

Aeration is continued until a constant oxygen uptake rate is obtained in order to ensure a steady state.

In the nonsteady procedure, aeration is stopped and the dissolved oxygen is allowed to approach zero by microbial respiration. Aeration is then begun and the buildup in dissolved oxygen recorded as in the non-steady-state water procedure; Fig. 5.19 shows how. K_La can be computed from the slope of a plot of dc/dt versus C_L, in accordance with a rearrangement of Eq. (5.20):

$$\frac{dc}{dt} = K_La(C_s - r_r) - K_LaC_L \tag{5.22}$$

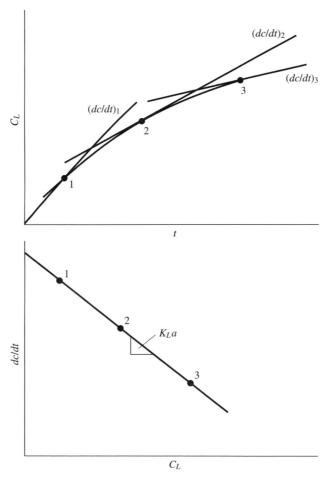

FIGURE 5.19
Non-steady-state evaluation of K_La.

since $K_La(C_s - r_r)$ will be constant for any specific operating condition. This is shown in Example 5.3.

EXAMPLE 5.3. The following data were obtained for non-steady-state aeration in an activated sludge basin:

Time, min	C_L, mg/l
0.0	0.52
0.5	0.70
1.0	0.93
2.0	1.23
3.0	1.55
4.0	1.80
5.0	2.00
10.0	2.20

Solution.

Determine the K_La.
The non-steady-state equation is:

$$\frac{dC_L}{dt} = K_La(C_s - C_L) - R_r$$

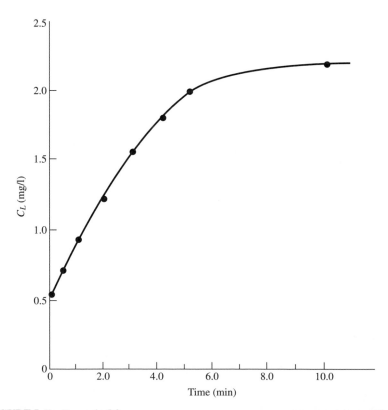

FIGURE I For Example 5.3

Rearrangement yields:

$$\frac{dC_L}{dt} = (K_L a C_s - R_r) - K_L a\, C_L$$

In a plot of dC_L/dt against C_L, the negative slope of the resulting straight line is $K_L a$, The dC_L/dt values are calculated and plotted in Table I and Figs. I and II.

TABLE I

t, min	C_L, mg/l	dC_L/dt
0	0.51	0.43
1	0.91	0.37
2	1.25	0.32
3	1.55	0.27
4	1.80	0.22
5	1.99	0.17

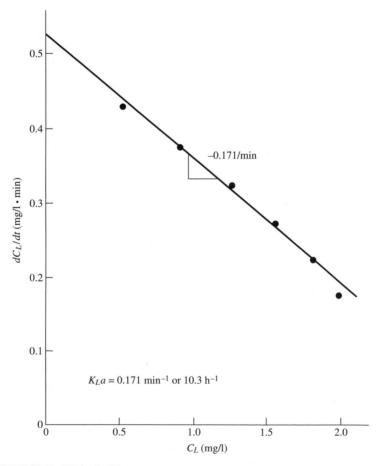

FIGURE II For Example 5.3.

Other Measuring Techniques

The off-gas analysis utilizes a hood over a portion of the aeration tank to collect the off-gases for O_2 analysis with a non-membrane-type probe. This technique is applicable for diffused aeration systems with transfer efficiencies above 5 percent.[20]

A radioactive tracer technique has been developed by Neal and Tsivoglou.[21] Their technique uses the stripping of krypton from the aeration tank to measure K_La under the exact aeration tank mixing and waste conditions. The oxygen transfer coefficient is then calculated from the known ratio of $(K_La)_{O_2}/(K_La)_{krypton}$.

5.3
AIR STRIPPING OF VOLATILE ORGANIC COMPOUNDS

The physical process of transferring volatile organic compounds (VOCs) from water into air is called *desorption* or *air stripping*. This can be accomplished by injection of water into air via spray systems, spray towers, or packed towers, or by injection of air into water through diffused or mechanical aeration systems. The most common systems in use today are packed towers or aeration systems. Aeration systems are usually used in conjunction with biological wastewater-treatment processes. Stripping efficiencies for various techniques are shown in Fig. 5.20. A typical packed tower is shown in Fig. 5.21.

Packed tower media are open-structured, chemically inert materials, usually plastic, which are selected to give high surface areas for good contacting while offering a low pressure drop through the tower. Some of the factors that affect removal of VOCs are the contact area, the solubility of the contaminant, the diffusivity of the contaminant in the air and water, and the temperature. All of these factors except diffusivity and temperature are influenced by the airflow and water-flow rates and the type of packing media. General relationships are shown in Fig. 5.22. The efficiency of transfer of contaminant from water to air depends on the mass-transfer coefficient and Henry's law constant [see Eq. (5.7)]. The mass-transfer coefficient defines the transfer of contaminant from water to air per unit volume of packing per unit time. The ability of a contaminant to be air stripped can be estimated from its Henry's law constant. A high Henry's constant indicates that the contaminant has low solubility in water and can therefore be removed by air stripping. In general, Henry's constant increases with increasing temperature and decreases with increasing solubility.

Packed Towers

Packed-tower air strippers can be designed for a wide variety of flow ranges, temperatures, and organics. One of the first steps in applying air stripping is to estimate the maximum possible removal for a given contaminant, based on its Henry's law constant, temperature, and volumetric air/water ratio. By assuming equilibrium between a volume of gas and a volume of water, the following equation can be used:

$$\frac{C_2}{C_1} = \left(1 + \frac{H_M A_w}{RT} \right)^{-1} \tag{5.23}$$

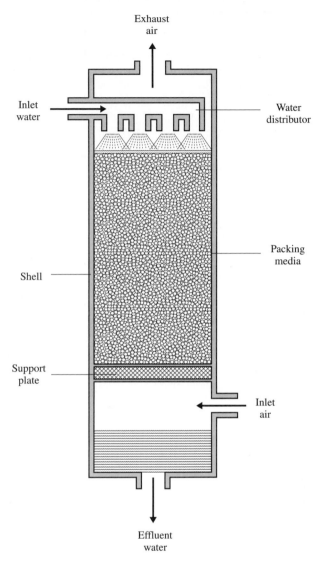

FIGURE 5.20
Typical packed tower.

where
C_2 = final concentration of organic, μg/L
C_1 = initial concentration of organic, μg/L
H_M = Henry's law constant, atm · m³/mol
A_w = volumetric air/water ratio
R = universal gas constant, 8.206×10^{-5} atm · m³/(mol · K)
T = temperature, K

While Henry's constant is not accurately defined for many organics, Table 5.7 lists estimated values for some organics that are found in wastewaters. In the scientific literature Henry's constant is often expressed as H_M (atm · m³/mol). Another form

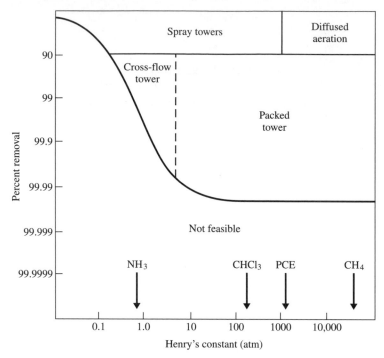

FIGURE 5.21
Stripping efficiency of various technologies.

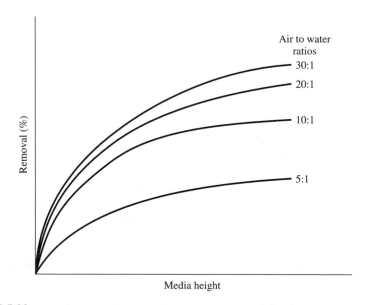

FIGURE 5.22
Illustrative relationships of air stripping removal efficiencies to media height and air to water ratios.

188

TABLE 5.7
Henry's constants for selected compounds at 20°C

Compound	Formula	Henry's constant H_M, atm · m³/mol	H_C, (µg/L)/(µg/L)
Easy to strip			
Vinyl chloride	CH_2CHCl	6.38	265
Trichloroethylene	$CCHCl_3$	0.010	0.43
1,1,1 Trichloroethane	CCH_3Cl_3	0.007	0.29
Toluene	$C_6H_5CH_3$	0.006	0.25
Benzene	C_6H_6	0.004	0.17
Chloroform	$CHCl_3$	0.003	0.12
Difficult to strip			
1,1,2-Trichloroethane	CCH_3Cl_3	7.7×10^{-4}	0.032
Bromoform	$CHBr_3$	6.3×10^{-4}	0.026
Nonstrippable			
Pentachlorophenol	$C_6(OH)Cl_5$	2.1×10^{-6}	0.000087
Dieldrin	—	1.7×10^{-8}	0.0000007

of Henry's constant is calculated as $H_c = H_M/RT$ (expressed as m³ water/m³ air). At 20°C, $H_c = 41.6H_M$, with the constant varying from 44.6 to 40.2 for a temperature range between 0 and 30°C.

The effect of temperature on the Henry's law constant is not well defined. Ashworth et al.[23] developed the following empirical relationship:

$$H_M = \exp\left(A - \frac{B}{T} \right) \qquad (5.24)$$

where A = empirical constant = 5.534 for benzene, 7.845 for trichloroethylene, 8.483 for methylene chloride

B = empirical constant = 3194 for benzene, 3702 for trichloroethylene, 4268 for methylene chloride

T = temperature, K

In general, the ability of a contaminant to be air stripped increases with increasing temperature and decreases with increasing solubility. Compounds with $H_M > 10^3$ atm · m³/mol are easy to strip, compounds with H_M between 10^{-4} and 10^3 atm · m³/mol are difficult to strip, and compounds with H_M less than 10^4 atm · m³/mol are nonstrippable.

The effect of feed temperature on the removal of soluble compounds is shown in Table 5.8.

The height of the stripping tower can be computed from the relationship

$$H = \frac{L_v}{K_L a} \left[\frac{S}{S - 1} \ln \frac{\left(\frac{C_1}{C_2}(S - 1) + 1 \right)}{S} \right] \qquad (5.25)$$

TABLE 5.8
**Influence of feed temperature on
removal of water-soluble compounds from groundwater**

Compound	Percent removal at		
	12°C	35°C	73°C
2-Propanol	10	23	70
Acetone	35	80	95
Tetrahydrofuran	50	92	>99

where L_v = volumetric liquid loading rate, ft/s
H = packing height, ft
$K_L a$ = mass-transfer coefficient
S = stripping factor = $H_c A_w$

In order to calculate the amount of packing needed to obtain the removal found from Eq. (5.25), the mass transfer coefficient $K_L a$ must be known for the VOC of interest, the type of packing media used, and the design conditions. $K_L a$ can be determined through pilot testing or calculated as follows:

$$\frac{1}{K_L a} = \left(\frac{1}{k_L a}\right) + \left(\frac{1}{H_c k_g a}\right) \tag{5.26}$$

Onda et al.[24] developed relationships to define k_L and k_g as summarized below.

$$k_L \left(\frac{\rho_L}{\mu_L g}\right)^{1/3} = 0.0051 \left(\frac{L}{a_w \mu_L}\right)^{2/3} \left(\frac{\mu_L}{\rho_L D_L}\right)^{-0.5} (a_t d_p)^{0.4} \tag{5.27}$$

$$\frac{k_G}{a_t D_g} = 5.23 \left(\frac{G}{a_t \mu_g}\right)^{0.7} \left(\frac{\mu_g}{\rho_g D_g}\right)^{1/3} (a_t d_p)^{-2} \tag{5.28}$$

$$\frac{a_w}{a_t} = 1 - \exp\left[-1.45 \left(\frac{\sigma_c}{\sigma_L}\right)^{0.75} (\mathrm{Re}_L)^{0.1} (\mathrm{Fr}_L)^{-0.05} (\mathrm{We}_L)^{-0.2}\right] \tag{5.29}$$

where k_L = liquid-phase mass-transfer coefficient
μ_L = viscosity of liquid, Pa · s
ρ_L = density of liquid, kg/m³
g = gravitational acceleration, 9.81 m/s²
L = liquid loading rate, kg/(m² · s)
a_w = wetted specific surface area, m²/m³
D_L = liquid-phase diffusivity, m²/s
a_t = total specific surface area, m²/m³
d_p = nominal packing size, m
k_g = gas-phase mass-transfer coefficient
D_g = gas-phase diffusivity, m²/s
G = gas loading rate, kg/(m² · s)
μ_g = viscosity of gas, Pa · s

ρ_g = density of gas, kg/m³
σ_c = surface tension of packing material, N/m
σ_L = surface tension of liquid, N/m
Re_L = liquid-phase Reynold's number = $L/a_t\mu_L$
Fr_L = liquid-phase Froude number = $L^2a_t/(\rho_L)^2g$
We_L = liquid-phase Weber number = $L^2/\rho_L\sigma_L a_t$

The K_La can be adjusted for different temperatures as follows:

$$K_La_{T_2} = K_La_{T_1} \times 1.024^{T_2-T_1}$$

Once the packing volume has been calculated, the tower diameter can be found from the desired pressure drop calculated in Fig. 5.23, which is a generalized correlation for predicting pressure drop in packed towers.[25] The following Eqs. (5.30) and (5.31) are the x- and y-axis values of Fig. 5.23, respectively. The packing factor is unique to the type of packing and can usually be obtained from the manufacturer.

$$\frac{L}{G}\left(\frac{\rho_g}{\rho_L}\right)^{0.5} \tag{5.30}$$

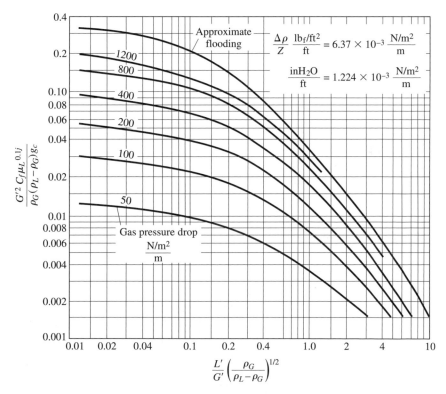

FIGURE 5.23
Generalized flooding and pressure drop curves for packed towers.

where L = liquid loading rate, lb/(ft^2· s)
          G = gas loading rate, lb/(ft^2 · s)
          ρ_g = density of gas, lb/ft^3
          ρ_L = density of liquid, lb/ft^3

$$\frac{G^2 F \mu_L{}^{0.1} J}{\rho_g (\rho_L - \rho_g) g_c} \tag{5.31}$$

where F = packing factor [97 for 16-mm ($\frac{5}{8}$-in) plastic Pall rings]
          μ = liquid viscosity, cP [kg/m · s)]
          g_c = 4.18 × 10^8 for English units (1 for SI units)
          J = 1.502 for English units (1 for SI units)

To estimate the tower size, an allowable pressure drop value is selected. Usually a pressure drop in the range of 0.25 to 0.50 in of water per foot of packed bed gives an average-size tower and a flexible operating range. Once the pressure drop is chosen, the x-axis value for Fig. 5.23 is calculated from Eq. (5.30). A vertical line is drawn from the x-axis to the intersection of the pressure drop curve corresponding to the selected value. At this intersection, the y-axis value is read. Equation (5.31) is rearranged and solved for G:

$$G = \left(\frac{(y\text{-axis value}) \rho_g (\rho_L - \rho_g) g_c}{F \mu_L{}^{0.1} J} \right)^{0.5} \tag{5.32}$$

The tower cross-sectional area is found by dividing the air mass flow rate by G, and the diameter is calculated from the area. The total packed-bed depth is calculated by dividing the volume of packing by the cross-sectional area. The total pressure drop is the pressure drop per foot of packing times the packed bed depth.

A schematic diagram of an air stripping system is shown in Fig. 5.24. The off gas containing the volatile organics must usually be treated. Applicable technologies are shown in Fig. 5.25.

EXAMPLE 5.4. Air stripper removal efficiency can be calculated from the equation

$$\frac{C_1}{C_2} = \frac{(S) \exp \left[\left(\frac{S-1}{S} \right) \left(\frac{H K_L a}{L_v} \right) \right] - 1}{S - 1}$$

where C_1 = influent concentration
          C_2 = effluent concentration
          H = packing height
          L_v = volumetric liquid loading rate = Q_L/A
          A = tower cross-sectional area = πr^2
          Q_L = water-flow rate
          S = stripping factor = $H_c A_w$
          H_c = Henry's constant
          A_w = volumetric air to water ratio = Q_G/Q_L
          Q_G = airflow rate
          $K_L a$ = mass-transfer coefficient

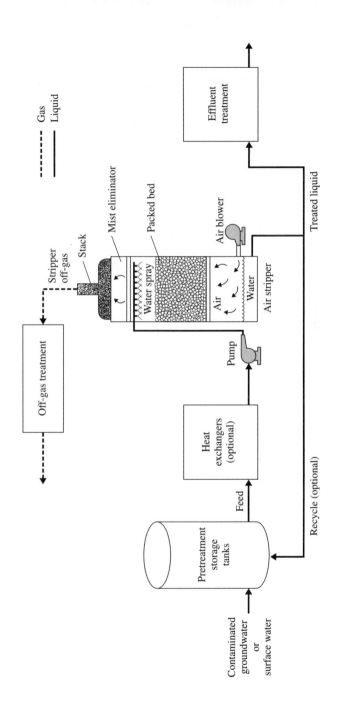

FIGURE 5.24
Schematic diagram of air stripping system.

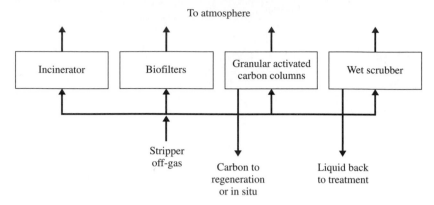

FIGURE 5.25
Process options for air stripper off-gas treatment.

If the following conditions are assumed:

$$
\begin{aligned}
H &= 40 \text{ ft} \\
L_v &= 10.03/153.9 = 0.06517 \text{ ft/s} \\
A &= \pi 7^2 = 153.9 \text{ ft}^2 \\
Q_L &= 4500 \text{ gal/min} = 10.03 \text{ ft}^3/\text{s} \\
S &= (0.2315)(19.94) = 4.616 \\
H_c &= 0.2315 \text{ for TCE at } 10°C \\
A_w &= 12{,}000/(10.03)(60) = 19.94 \\
Q_G &= 12{,}000 \text{ ft}^3/\text{min} \\
K_L a &= 0.0125 \text{ s}^{-1}
\end{aligned}
$$

then

$$
\frac{C_1}{C_2} = \frac{(4.616)\exp\left[\left(\dfrac{4.616 - 1}{4.616}\right)\left(\dfrac{(40)(0.0125)}{0.06517}\right)\right] - 1}{4.616} = 520
$$

$$
= 0.00192
$$

$$
\% \text{ Removal} = (1 - C_2/C_1) \times 100\%
$$

$$
= 99.81
$$

Steam stripping may be applied for the removal of high concentrations of volatile organics. The principal index used to estimate the steam stripping capability is the boiling point of the organic compound. A compound should exhibit a relatively low boiling point (150°C) and an acceptable Henry's law constant for effective steam stripping. A contaminated condensate is produced that may be recovered or treated further. A steam stripping system is shown in Fig. 5.26 and performance characteristics are given in Table 5.9.

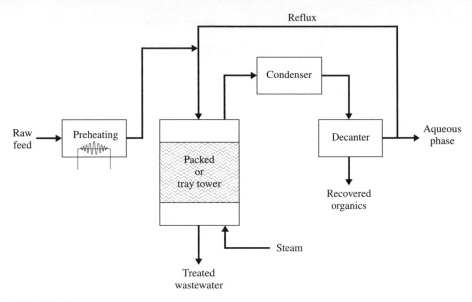

FIGURE 5.26
Typical steam stripping process.

TABLE 5.9
Full-scale industrial steam stripper performance summary

Plants using steam stripping	Stripped compound	Henry's constant, atm	Vapor pressure, mmHg @ 25°C	Concentration, ppm		Percent removal
				Influent	Effluent	
Pesticide industry						
Plant 1	Methylene chloride	177	425	<159	<0.01	99.9
Plant 2	Chloroform	188	180	70.0	<5.0	>92.6
Plant 3	Toluene	370	29	721	43.4	94.0
Organic chemicals industry						
Plant 4	Benzene	306	74	<15.4	<0.230	98.5
Plant 5	Methylene chloride	177	425	<3.02	<0.0141	99.5
	Toluene	370	29	178	<52.8	>70.3
Plant 6a	Methylene chloride	177	425	1430	<0.0153	>99.99
	Carbon tetrachloride	1280	113	<665	<0.0549	>99.99
	Chloroform	188	180	<8.81	1.15	<86.9
Plant 6b	Methylene chloride	177	425	4.73	<0.0021	>99.95
	Chloroform	188	180	<18.6	<1.9	89.8
	1,2-Dichloroethane	62	82	<36.2	<4.36	88.0
	Carbon tetrachloride	1280	113	<9.7	<0.030	99.7
	Benzene	306	74	24.1	<0.042	>99.8
	Toluene	370	29	22.3	<0.091	>99.6
Plant 7	Methylene chloride	177	425	34	<0.01	>99.97
	Chloroform	188	180	4509	<0.01	99.99
	1,2-Dichloroethane	62	82	9030	<0.01	>99.99

PROBLEMS

5.1 An air diffuser yielded a K_La of 6.5/h at 10°C. The airflow is 50 cm³/min, the bubble diameter 0.15 cm, and the bubble velocity 28 cm/s. The depth of the aeration column is 250 cm with a volume of 4000 cm³.
(a) Compute K_L.
(b) Compute K_La at 25°C.

5.2 Design an aeration system using coarse bubble diffusers with a 12 percent OTE at 15-ft liquid depth under standard conditions.

1000 lb O₂/h
$T = 30°C$
$C_L = 2$ mg/l
$\beta = 0.95$
$\alpha = 0.85$
$D = 20$ ft

5.3 The following data were collected at 29°C for non-steady state oxygen transfer in an activated sludge basin. Compute K_L and K_La at 20°C.

Time, min	C_L, mg/l
0	0.75
3	1.60
6	2.30
9	2.80
12	3.20
15	3.50
18	3.72

5.4 A pilot study was conducted to evaluate the removal of benzene by air stripping. The pilot packed-tower air stripper had a 10-in diameter and 10 ft of packing height. The tower was operated at 50°C with a flow rate of 10 gal/min and an air to water ratio of 50:1. The benzene concentration was reduced from 1400 μg/l to 200 μg/l. What is the required tower size to reduce the benzene concentration from 1600 μg/l to 10 μg/l at 50 gal/min? Assume the same packing type, liquid loading rate, air to water ratio, and temperature as the pilot study.

REFERENCES

1. Lewis, W. K., and W. G. Whitman: *Ind. Eng. Chem.,* vol. 16, p. 1215, 1924.
2. Danckwertz, P. V.: *Ind. Eng. Chem.,* vol. 43, p. 6, 1951.
3. Dobbins, W. E.: *Advances in Water Pollution Research,* vol. 2, p. 61, Pergamon Press, 1964.
4. Mueller, J. A. et al.: *Proc. 37th Ind. Waste Conf.,* May 1982, Purdue University.
5. Schmit, F. and D. Redmon: *J. Water Pollution Control Fed.,* November 1975.

6. Imhoff, K., and D. Albrecht: *Proc. 6th International Conf. on Water Pollution Research,* Jerusalem, 1972.
7. Landberg, G. et al.: *Water Research,* vol. 3, p. 445, 1969.
8. Wagner, M. R., and H. J. Popel: *Proc. Wat. Env. Fed.,* Dallas, 1996.
9. Carpani, R. E., and J. M. Roxburgh: *Can. J. Chem. Engrg,* vol. 36, p. 73, April 1958.
10. Gameson, A. H., and H. B. Robertson: *J. Appl. Chem. Engrg.,* vol. 5, p. 503, 1955.
11. Eckenfelder, W. W.: *J. Sanit. Engrg. Div. ASCE,* vol. 85, pp. 88–99, 1959.
12. Pasveer, A.: *Sewage Ind. Wastes,* vol. 27, pt. 10, p. 1130, 1955.
13. Ippen, H. T., and C. E. Carver: MIT Hydrodynamics Lab. Tech. Rep. 14, 1955.
14. Cappock, P. D., and G. T. Micklejohn: *Trans. Inst. Chem. Engrs. London,* vol. 29, p. 75, 1951.
15. *WPCF Aeration Manual of Practice,* FD-13, 1988.
16. Quirk, T. P.: personal communication.
17. Kormanik, R. et al.: *Proc. 28th Ind. Waste Conf.,* 1973, Purdue University.
18. Arthur, R. M., *Treatment Efficiency and Energy Use III, Activated Sludge Process Control,* Butterworths, Boston 1986.
19. "ASCE Standard Measurement of Oxygen Transfer in Clean Water," ASCE, New York, 1992.
20. Redman, D., and W. C. Boyle: Report to Oxygen Transfer Subcommittee of ASCE, 1981.
21. Neal, L. A., and E. C. Tsivoglou: *J. Water Pollut. Control Fed.,* vol. 46, p. 247, 1974.
22. Muller, J.: Manhattan College Summer Institute, 1996.
23. Ashworth, R. A., G. B. Howe, M. E. Mullins, and T. N. Rogers: "Air-Water Partitioning Coefficients of Organics in Dilute Aqueous Solutions," *Jour. Haz. Mat.,* vol. 18, pp. 25–36, 1988.
24. Onda, K., H. Takeuchi, and Y. Okumoto: "Mass Transfer Coefficients between Gas and Liquid Phases in Packed Columns," *J. Chem. Engrg. Japan,* vol. 1, no. 1, 1968.
25. Eckert, J. S.: *Chem. Eng. Progress,* vol. 66, March 1970.

6

PRINCIPLES OF AEROBIC
BIOLOGICAL OXIDATION

Organics are removed in a biological-treatment process by one or more mechanisms, namely sorption, stripping, or biodegradation. Table 6.1 identifies several organics and the mechanisms responsible for their removal.

6.1
ORGANICS REMOVAL MECHANISMS

Sorption

Limited sorption of nondegradable organics on biological solids occurs for a variety of organics, and this phenomenon is not a primary mechanism of organic removal in the majority of cases. An exception is Lindane, as reported by Weber and Jones,[1] who showed that while no biodegradation occurred, there was significant sorption. It is probable that other pesticides will respond in a similar manner in biological wastewater-treatment processes.

This removal mechanism is termed *partitioning* and has been related to the octanol-water partition coefficient of the organic.

$$K_{SW} = kK_{OW}{}^n \tag{6.1}$$

where K_{SW} = biosolids accumulation factor, ratio of organic sorbed and in solution, (mg/mg)(mg/l)

K_{OW} = octanol-water partition coefficient, $(mg/l)_O/(mg/l)_W$

k, n = factors that have been reported to vary from 1.38×10^{-5} to 4.3×10^{-7} (k) and from 0.58 to 1.0 $(n)^{2-5}$

In most industrial wastewaters, partitioning provides negligible SCOD removal but may be a method of bioaccumulation of certain lipid-soluble organic compounds.

The removal by adsorption can be determined through the relationship

$$\frac{C_e}{C_i} = \frac{1}{\left[1 + \dfrac{k_{SW} \cdot Xt}{\theta_c} \right]} \tag{6.1a}$$

198

TABLE 6.1
Specific removal efficiencies of priority pollutants

Compounds	Percent treatment achieved		
	Stripping	Adsorption	Biodegradation
Nitrogen compounds			
Acrylonitrile			99.9
Phenols			
Phenol			99.9
2,4-DNP			99.3
2,4-DCP			95.2
PCP		0.58	97.3
Aromatics			
1,2-DCB	21.7		78.2
1,3-DCB	—	—	—
Nitrobenzene			97.8
Benzene	2.0		97.9
Toluene	5.1	0.02	94.9
Ethylbenzene	5.2	0.19	94.6
Malogenated hydrocarbons			
Methylene chloride	8.0		91.7
1,2-DCE	99.5	0.50	
1,1,1-TCE	100.0		
1,1,2,2,-TCE	93.5		
1,2-DCP	99.9		
TCE	65.1	0.83	33.8
Chloroform	19.0	1.19	78.7
Carbon tetrachloride	33.0	1.38	64.9
Oxygenated compounds			
Acrolein			99.9
Polynuclear aromatics			
Phananthrene			98.2
Naphthalene			98.6
Phthalates			
Bis (2-ethylhexyl) phthalates			76.9
Other			
Ethyl acetate			98.8

where C_e = effluent concentration, mg/l
C_i = influent concentration, mg/l
X = mixed liquor suspended solids, mg/1
t = hydraulic detention time, d
θ_c = sludge age, d

Adsorption is not a significant factor when log K_{OW} is less than 4.

EXAMPLE 6.1. Determine the adsorption of tetrachloroethane and Lindane in the activated sludge process for the following conditions.

Tetrachloroethane $K_{OW} = 363$
Lindane $K_{OW} = 12,600$
In Eq. (6.1), $k = 3.45 \times 10^{-7}$ l/mg
$X = 3500$ mg/l
$t = 0.23$ d
$\theta_c = 6$ d

Solution.

For tetrachlorethane:

$$\frac{C_e}{C_i} = \frac{1}{\left[1 + 3.45 \times 10^{-7}\dfrac{1}{mg}(363)(3500)\dfrac{mg}{1} \times \dfrac{0.23\ d}{6\ d}\right]}$$

$$= 0.984 \quad \text{or} \quad 1.6\% \text{ adsorbed}$$

For Lindane:

$$\frac{C_e}{C_i} = \frac{1}{\left[1 + 3.45 \times 10^{-7}(12,600)(3500) \times \dfrac{0.23}{6}\right]}$$

$$= 0.633 \quad \text{or} \quad 37\% \text{ adsorbed}$$

While sorption on biomass does not seem to be a significant removal mechanism for toxic organics, sorption on suspended solids in primary treatment may be significant. The importance of this phenomenon is the fate of the organics during subsequent sludge-handling operations. In some cases, toxicity to anaerobic digestion may result or land-disposal alternatives may be restricted.

While sorption of organics on biomass is usually not significant, this is not true of heavy metals. Metals will complex with the cell wall and bioaccumulate. While low concentrations of metals in the wastewater are generally not inhibitory to the organic removal efficiency, their accumulation on the sludges can markedly affect subsequent sludge treatment and disposal operations.

Stripping

Volatile organic carbon compounds (VOCs) will air strip in biological-treatment processes, i.e., trickling filters, activated sludge, aerated lagoons. Depending on the VOC in question, both air stripping and biodegradation may occur, as shown on p. 274 and described by Kincannon and Stover.[6] Stripping of VOCs in biological treatment processes is currently receiving considerable attention in the United States, as legislation is severely limiting the permissible atmospheric emissions of VOCs.

The general relationship between the volatilization, biodegradation, and sorption of certain organic compounds is shown in Fig. 6.1 and is discussed on p. 271.

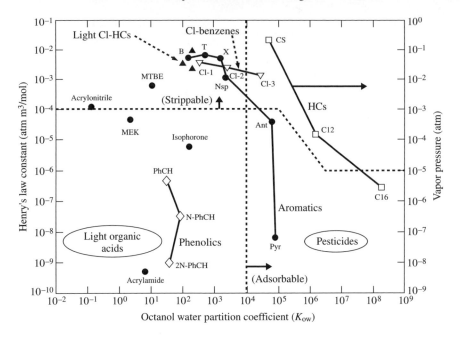

FIGURE 6.1
Categorization of organic compounds—strippability versus sorbability.[7]

Biodegradation

When organic matter is removed from solution by aerobic microorganisms, two basic phenomena occur: Oxygen is consumed by the organisms for energy, and new cell mass is synthesized. The organisms also undergo progressive autooxidation in their cellular mass. These reactions can be illustrated by the following generic equations:

$$\text{Organics} + a'O_2 + N + P \xrightarrow[K]{\text{cells}} a \text{ new cells} + CO_2 + H_2O$$

$$+ \text{ nonbiodegradable soluble residue (SMP)} \tag{6.2}$$

$$\text{Cells} + b'O_2 \xrightarrow[b]{} CO_2 + H_2O + N + P$$

$$+ \text{ nonbiodegradable cellular residue} + \text{SMP} \tag{6.3}$$

These relationships are shown in Fig. 6.2.

Of primary concern to the engineer in the design and operation of industrial waste treatment facilities are the rate at which these reactions occur, the amounts of oxygen and nutrient they require, and the quantity of biological sludge they produce.

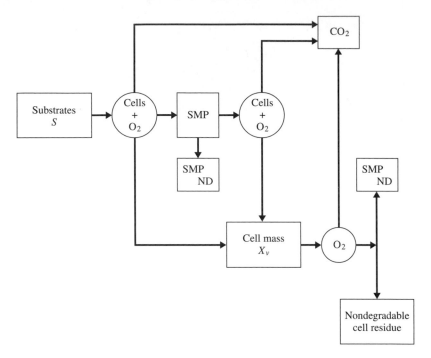

FIGURE 6.2
Mechanism of biodegradation.

In Eq. (6.2), K is a rate coefficient and is a function of the biodegradability of the organic or the mixture of organics in the wastewater. The coefficient a' is the fraction of the organics removed that is oxidized to end products for energy and the coefficient a is the fraction of organics removed that is synthesized to cell mass. The coefficient b is the fraction per day of degradable biomass oxidized, and b' is the oxygen required for this oxidation.

A small portion of the organics removed in Eq. (6.2) remains as nondegradable by-products which appear in the effluent as TOC or COD but not BOD and are defined as soluble microbial products (SMP). A portion of the cell mass generated in Eq. (6.2) remains as a nondegradable residue.

For the design or operation of a biological treatment facility, a primary objective is to balance Eqs. (6.2) and (6.3) for the wastewater in question.

Disregarding the SMP, all of the organics removed are either oxidized to end products of CO_2 and H_2O or synthesized to biomass.

Therefore,

$$a_{\text{COD}} + a'_{\text{COD}} \sim 1$$

Since the biomass is usually expressed as volatile suspended solids (VSS) and that 1.4 lbs O_2 is required to oxidize 1 lb of cells as VSS, then:

$$1.4a_{\text{VSS}} + a'_{\text{COD}} \sim 1$$

6.2
THE MECHANISMS OF ORGANIC REMOVAL BY BIOOXIDATION

The major organic-removal mechanism for most wastewaters is biooxidation in accordance with Eqs. (6.2) and (6.3).

It should be noted that, in treating industrial wastewaters, the active microbial population must be acclimated to the wastewater in question. For the more complex wastewaters, acclimation may take up to 6 weeks, as shown in Fig. 6.3 for benzidine.[8] When acclimating sludge the feed concentration of the organic in question must be less than the inhibition level, if one exists.

BOD removal from a wastewater by a biological sludge may be considered as occurring in two phases. An initial high removal of suspended, colloidal, and soluble BOD is followed by a slow progressive removal of remaining soluble BOD. Initial BOD removal is accomplished by one or more mechanisms, depending on the physical and chemical characteristics of the organic matter. These mechanisms are:

1. Removal of suspended matter by enmeshment in the biological floc. This removal is rapid and depends upon adequate mixing of the wastewater with the sludge.

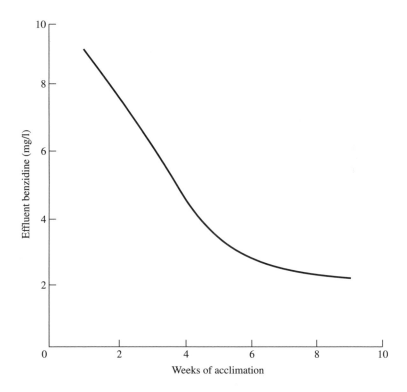

FIGURE 6.3
Acclimation for the degradation of benzidine.

2. Removal of colloidal material by physicochemical adsorption on the biological floc.
3. A biosorption of soluble organic matter by the microorganisms. There is some question as to whether this removal is the result of enzymatic complexing or is a surface phenomenon and whether the organic matter is held to the bacterial surface or is within the cell as a storage product or both. The amount of immediate removal of soluble BOD is directly proportional to the concentration of sludge present, the sludge age, and the chemical characteristics of the soluble organic matter.

The biosorption phenomenon is related to the microbial floc load in a contact time of 10 to 15 min:

$$\text{Floc load} = \frac{\text{mg BOD applied}}{\text{g VSS biological}} \tag{6.4}$$

where VSS = volatile suspended solids. The relationship between floc load and organic removal by biosorption is shown in Fig. 6.4.

The type of sludge generated markedly affects its sorptive properties. In general, biomass generated from a batch or plug flow configuration will have better sorptive properties than that generated from a complex mix configuration.

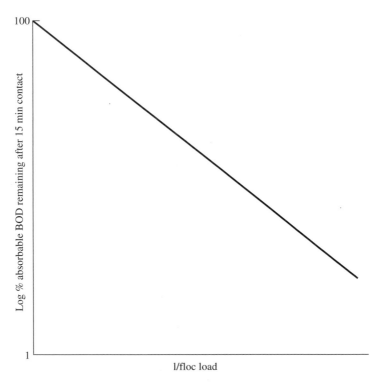

FIGURE 6.4
Biosorption relationship for soluble degradable wastewaters.

The three mechanisms begin immediately on contact of biomass with waste-water. The colloidal and suspended material must undergo sequential breakdown to smaller molecules in order that it may be made available to the cell for oxidation and synthesis. The time required for this breakdown in an acclimated system is related primarily to the characteristics of the organic matter and to the concentration of active sludge. In a complex waste mixture at high concentrations of BOD, the rate of synthesis is independent of concentration as long as all components remain and, as a result, there is a constant and maximum rate of cellular growth. With continuing aeration, the more readily removable components are depleted and the rate of growth will decrease with decreasing concentration of BOD remaining in solution. This is shown in Fig. 6.12 on p. 219.

This causes a decrease in cellular mass and cellular carbon accompanied by a corresponding decrease in cellular nitrogen, as shown in Fig. 6.5. This phenomenon has been demonstrated by Gaudy,[9] by Englebrecht and McKinney,[10] and by

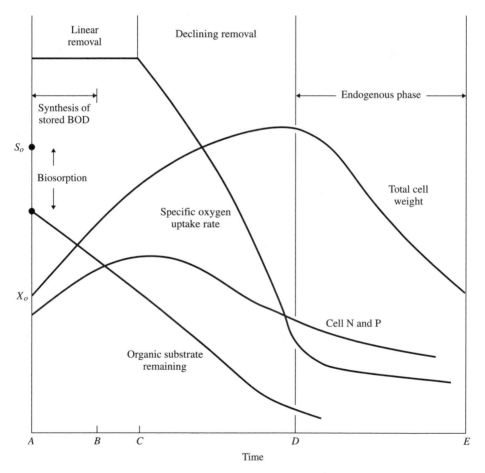

FIGURE 6.5
Reactions occurring during biooxidation.

McWhorter and Heukelekian.[11] Gaudy, in the treatment of pulp mill wastes, showed a peak in cellular carbohydrate after a 3-h aeration with a corresponding cellular protein peak after a 6-h aeration. The decrease in cellular mass after substantial exhaustion of substrate, shown by Engelbrecht and McKinney,[10] can be attributed to conversion of stored carbohydrate to cellular protoplasm.

In Fig. 6.5 declining removal occurs following biosorption. Stored carbohydrate is used by the cell over the time interval AC; this results in an increase in cellular nitrogen. The cellular nitrogen is at its peak at point C, when the stored carbohydrate is depleted. The cellular mass is shown to increase over the time interval CD (declining growth and removal). Depending on the concentration of BOD remaining at point C and the rate of removal, the sludge mass may tend to remain constant or even increase while the cellular nitrogen would remain substantially constant. Beyond point D, cell death and decay, the endogenous or autooxidation phase results in a decrease in both cell weight and cell nitrogen.

The oxygen uptake rate per unit of cell mass k_r will remain constant at a maximum rate during the log-growth phase, since substrate is not limiting the rate of synthesis.

The oxygen utilization will continue at a maximum rate until depletion of the sorbed BOD; after this it will decrease as the rate of BOD removal decreases. In wastes containing suspended and colloidal matter, the oxygen-uptake rate will also reflect the rate of solubilization and subsequent synthesis of colloidal and suspended BOD.

Sludge Yield and Oxygen Utilization

As shown in Eq. (6.3), endogenous respiration results in a degradation of cell mass. A portion of the volatile cell mass, however, is nondegradable, i.e., does not degrade in the time frame of the biological process. Quirk and Eckenfelder[12] have shown that a portion of the volatile biomass is nondegradable. As aeration proceeds the degradable portion of the biomass is oxidized, resulting in a decrease in the degradable fraction. Through kinetic and mass balances, the degradable fraction can be related to the endogenous rate coefficient and the sludge age:

$$X_d = \frac{X_d'}{1 + bX_n'\theta_c} \tag{6.5}$$

where X_d = degradable fraction of the biological VSS
          X_d' = degradable fraction of the biological VSS at generation, that is, Eq. (6.2) average of 0.8
          X_n' = nondegradable fraction of the biological VSS at generation, that is, Eq. (6.2) average of $0.2(X_d' + X_n') = 1.0$
          b = endogenous rate coefficient, d^{-1}
          θ_c = sludge age, d

This relationship is shown for a food processing wastewater in Fig. 6.6.

The degradable fraction is related to the viable or active mass. The degradable mass can be related to oxygen-uptake rates, ATP, dehydrogenase enzyme content,

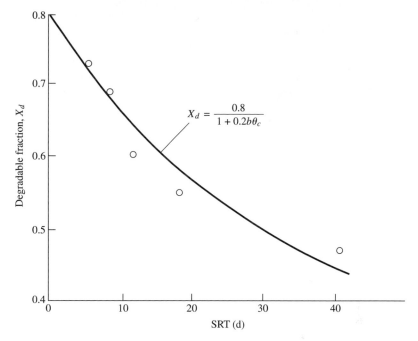

FIGURE 6.6
Relationship between degradable fraction and solids retention time (SRT) for a food processing wastewater.

or plate count measurement, as shown in Fig. 6.7. It is significant to note that, while volatile suspended solids are traditionally used as a measure of biomass, only the active mass is responsive in the process. In plant operation, some measure of active mass is necessary for process control in order to detect toxic shocks, shock loads, etc. Oxygen-uptake rate is most commonly used for this purpose.

Sludge age is defined as the average length of time the microorganisms are under aeration. In a flow-through system, i.e., no biomass recycle, sludge age is the reciprocal of the dilution rate Q/V. In order for growth to occur and to effect BOD removal, the growth rate becomes

$$\theta_c = \frac{V}{Q} \tag{6.6}$$

where θ_c is the sludge age.

In a recycle system such as an activated sludge plant, the sludge age is defined:

$$\theta_c = \frac{X_v t}{\Delta X_v} \tag{6.7}$$

where X_v = volatile suspended solids concentration, mg/l
         t = V/Q, the hydraulic detention time, d
      ΔX_v = the volatile suspended solids wasted per day in milligrams per liter, based on influent flow

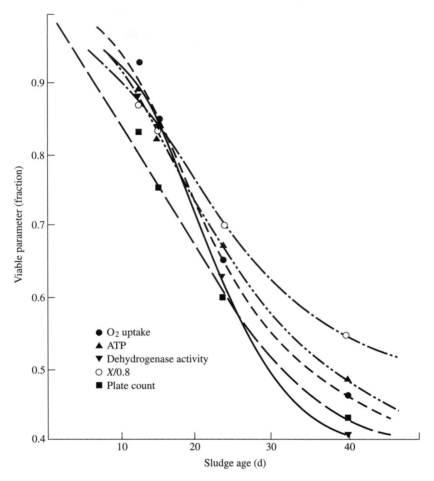

FIGURE 6.7
Viable fraction of activated sludge as related to sludge age.

Process performance can also be related to the organic loading of the process as defined by the food/microorganism ratio F/M:

$$\frac{F}{M} = \frac{S_o}{f_b X_v t} \tag{6.8}$$

in which X_v = the volatile suspended solids under aeration and f_b = the fraction which is biological.

The F/M is related to the sludge age θ_c by the relationship

$$\frac{1}{\theta_c} = a\frac{F}{M} - bX_d \tag{6.9}$$

when the effluent BOD is negligible, where a = the yield coefficient or the fraction of organic removed synthesized to biomass.

Several investigations have shown that a constant mass of biological cells is synthesized from a given weight of organic matter removed (expressed as total oxygen demand, COD). McKinney[10] has indicated that one-third of the ultimate oxygen demand (COD) of a substrate is used for energy and that two-thirds results in synthesis. Using a factor of 0.7 g VSS/g O_2 for conversion of oxygen to cellular volatile solids, 0.47 g VSS (volatile suspended solids) will be synthesized for each gram of COD removed. Variations in this value have been attributed to endogenous respiration effects. Sawyer[13] and Gellman and Heukelekian[14] have shown that for sewage and several industrial wastes, 0.5 g VSS is synthesized per gram of BOD_5 removed. Busch and Myrick[15] showed total synthesis from glucose to be 0.44 g cells/g COD. Using NO_3^- as a nitrogen source, McWhorter and Heukelekian[11] found an average synthesis of 0.315 g VSS/g COD from glucose. As shown by Pipes,[16] the reduction of nitrate to amino nitrogen requires energy such that some of the COD is consumed for this reduction, resulting in a lower yield than when ammonia is used as a nitrogen source.

When the nutrient nitrogen level is lowered below the optimum value, cell yield tends to increase as more substrate is shunted to the buildup of insoluble cell polymers.

Sludge generation from the biological oxidation of soluble substrates ($f_b = 1.0$) has been summarized by Eckenfelder[17]:

$$\Delta X_v = aS_r - bX_d X_v t \tag{6.10}$$

where S_r represents the soluble substrate removed.

In a system with sludge recycle and wastage of excess sludge, sludge age can be computed from Eq. (6.10) and is defined as

$$\theta_c = \frac{X_v t}{aS_r - bX_d X_v t} \tag{6.11}$$

In Eq. (6.11), when both θ_c and X_d are unknown, θ_c can be computed by combining Eqs. (6.5) and (6.11) for soluble substrates to yield

$$\theta_c = \frac{-(aS_r - bX_v t) + \sqrt{(aS_r - bX_v t)^2 + 4(abX_n' S_r)(X_v t)}}{2abX_n' S_r} \tag{6.12}$$

If the influent contains bioresistant VSS, such as in a pulp and paper mill wastewater, Eq. (6.10) is modified to include this contribution.

$$\Delta X_v = a[S_r + f_d f_x X_i] - bX_d f_b X_v t + (1 - f_d) f_x X_i + (1 - f_x) X_i \tag{6.13}$$

where X_i = influent VSS, mg/l
f_x = fraction of influent VSS that is degradable
f_d = fraction of degradable influent VSS degraded
f_b = fraction of mixed liquor VSS that is biomass

The degradation rate of the influent degradable VSS is a function of the solids retention time (SRT) and the specific degradation rate of the solids:

$$(1 - f_d) = e^{-k_p \theta_c} \tag{6.14}$$

where k_p = degradation rate coefficient of influent VSS, d^{-1}.

If it is assumed that 1 mg/l VSS solubilizes to generate 1 mg/l COD, then the fraction of biomass (f_b) in the overall mixed liquor can be determined as follows:

$$f_b = 1 - [(1 - f_x) + (1 - f_d)f_x]\frac{X_i}{X_v} \cdot \frac{\theta_c}{t} \tag{6.15}$$

Most pulp and fiber in pulp and paper mill wastewater are essentially non-degradable, and hence $(1 - f_d)$ is approximately unity. In food processing waste-waters, however, $(1 - f_d)$ may be less than 0.2. If the influent contains high levels of degradable VSS, the value $(1 - f_d)$ must be experimentally determined in order to accurately predict the volatile sludge production rate and true biomass yield.

The impact of influent nonvolatile suspended solids on mixed liquor charac-teristics and sludge production rate can also be significant. The quantity of inert material generated is related to the SRT, the hydraulic retention time, the fraction of influent suspended solids that is nondegradable/nonsolubilized, and the forma-tion of nonbiomass particulates in the activated sludge process.

Total sludge production can be calculated as

$$\Delta X = \frac{\Delta X_v}{f_v} \tag{6.16}$$

in which f_v = the volatile fraction of the mixed liquor suspended solids.

The impact of the total suspended solids production on the operating aeration basin MLSS concentration is a direct function of SRT or sludge age. As SRT increases, the aeration basin mixed liquor suspended solids (MLSS) will increase. Consideration must be given in secondary clarifier solids loading rate design to accommodate the inert and volatile suspended solids generation while maintaining the required SRT for substrate removal. The calculations are shown in Example 6.2.

EXAMPLE 6.2. Develop a process material balance for the following conditions.

Influent

Q = 3 million gal/d (11,360 m³/d)
S_o = 700 mg/l (soluble COD)
X_i = 40 mg/l (nondegradable)

Effluent

S_e = 150 mg/l (soluble COD)
VSS_e = 25 mg/l

Waste sludge

Q_w = 0.056 million gal/d (212 m³/d)
VSS_w = 6000 mg/l

Aeration basin

V = 0.84 mg (3180 m³)
$MLVSS$ = 2500 mg/l
b = 0.18/d
t = 0.28 d
r_r = 61 mg/l · h

where MLVSS is mixed liquor volatile suspended solids.

Solution.

$$\theta_c = \frac{X_v V}{\Delta X_v} = \frac{X_v V}{VSS_e + Q_w VSS_w}$$

$$= \frac{2500 \cdot 0.84}{25 \cdot 3 + 0.056 \cdot 6000}$$

$$= 5.1 \text{ d}$$

$$X_d = \frac{0.8}{1 + 0.2 b \theta_c}$$

$$= \frac{0.8}{1 + 0.2 \cdot 0.18 \cdot 5}$$

$$= 0.68$$

Calculation of nonbiological VSS:

$$X_{v_{NB}} = \frac{X_i \theta_c}{t}$$

$$= \frac{40 \cdot 5}{0.28} = 714 \text{ mg/l}$$

$$f_b = \frac{(2500 - 714)}{2500} = 0.71$$

Calculation of biomass yield coefficient a:

$$f_b \Delta X_v = a S_r - b X_d f_b X_v$$

$$0.71(0.056 \cdot 6000 \cdot 8.34 + 25 \cdot 3 \cdot 8.34) = a(550 \cdot 3 \cdot 8.34)$$

$$- 0.18 \cdot 0.68 \cdot 0.71 \cdot 2500 \cdot 0.84 \cdot 8.34$$

$$a = 0.29$$

Calculation of oxygen consumption coefficient a':

Oxygen uptake rate (OUR) $= 61 \text{ mg}/(\text{l} \cdot \text{h})$

$$O_2/d = 61 \cdot 0.84 \cdot 8.34 \cdot 24$$

$$= 10.256 \text{ lb } O_2/d \quad (4650 \text{ kg } O_2/d)$$

$$10{,}256 = a'(13{,}761) + 1.4 \cdot 0.18 \cdot 0.68 \cdot 0.71 \cdot 2500 \cdot 0.84 \cdot 8.34$$

$$a' = 0.59 \text{ mg } O_2/\text{mg } \Delta COD$$

Check of a and a':

$$1.4 \cdot 0.29 + 0.59 = 0.99$$

The correlations for sludge yield and endogenous rate coefficient are shown in Fig. 6.8.

Data obtained from the treatment of a soluble pharmaceutical waste are shown in Fig. 6.9. Although the synthesis coefficient a is 0.645 on a BOD_5 basis, the coefficient that results from plotting on a COD basis is 0.37.

Oxygen Utilization

It has been previously shown that the total oxygen requirements in a system are related to the oxygen consumed to supply energy for synthesis and the oxygen con-

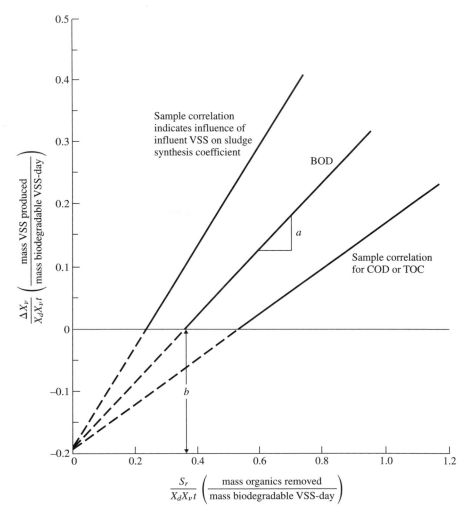

FIGURE 6.8
Determination of sludge production coefficients from Eq. (6.10).

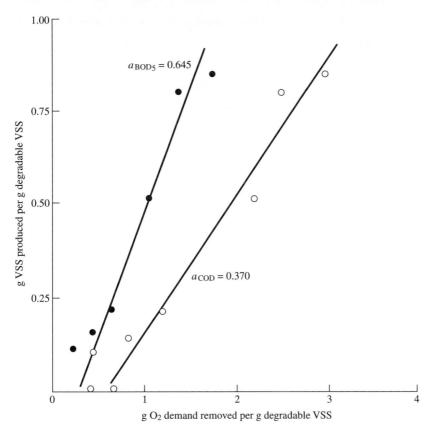

FIGURE 6.9
Cell synthesis relationship for a soluble pharmaceutical wastewater.

sumed for endogenous respiration [Eqs. (6.2) and (6.3)], in which a' is the fraction of organics consumed to oxidize to end products and b' is the autooxidation rate of the sludge. In a similar manner, the oxygen requirements for a soluble substrate can be calculated:

$$r_r = \frac{a'S_r}{t} + b'X_df_bX_v \tag{6.17}$$

in which r_r = the oxygen uptake rate, mg/(l · d).

During the log-growth phase, the specific oxygen uptake rate (SOUR) $k_r = r_r/X_v$ is constant, and hence r_r will increase with increasing synthesis of new cells. As the substrate concentration decreases, the oxygen uptake rate decreases. When the available substrate is exhausted, the oxygen uptake rate decreases to the endogenous rate, which is approximately $b'Xaf_bX_v$ in Eq. (6.18). Process material balance calculations are shown in Examples 6.2 and 6.3.

EXAMPLE 6.3. Determine the coefficients a and a' from the following data.

Soluble TOD removed across aeration = 12,000 lb/d (5450 kg/d)
Sludge reaeration (subscript e = endogenous):

$$OUR_e = 0.45 \text{ mg}/(1 \cdot \text{min})$$

$$VSS_e = 9000 \text{ mg}/1$$

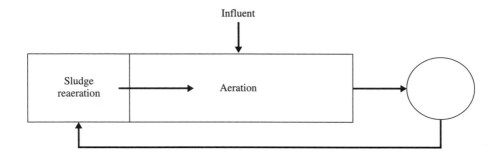

Aeration (subscript a):

$$OUR_a = 1.73 \text{ mg}/(1 \cdot \text{min})$$

$$VSS_a = 6000 \text{ mg}/1$$

$$Volume_a = 0.26 \text{ million gal } (984 \text{ m}^3)$$

Solution.

$$b'X_d = \frac{OUR_e}{VSS_e} = \frac{0.45 \text{ mg}/(1 \cdot \text{min}) \times 1440 \text{ min/d}}{9000 \text{ mg}/1} = 0.07/\text{d}$$

$$a' = \frac{8.34 V_a [OUR_a \times 1440 - (OUR_e \times 1440 \times VSS_a)/VSS_e]}{\text{soluble TOD removed}}$$

$$= \frac{8.34 \left[(\text{lb/million gal})/(\text{mg/l}) \right] \times 0.26 \text{ million gal} \times 1440 \text{ min}}{12,000 \text{ lb/d}}$$

$$\times [1.73 \text{ mg}/(1 \cdot \text{min}) - (0.45 \text{ mg}/(1 \cdot \text{min}) \times 6000 \text{ mg}/1)/900$$

$$= 0.37$$

$$a = \frac{1 - a'}{1.4} = \frac{1 - 0.37}{1.4} = 0.45$$

If the influent contains degradable VSS, Eq. (6.17) is modified to

$$r_r = a'(S_r + f_d f_x X_i) + b'X_d f_b X_v \qquad (6.18)$$

Nutrient Requirements

Several mineral elements are essential for the metabolism of organic matter by microorganisms. All but nitrogen and phosphorus are usually present in sufficient quantity in the carrier water. An exception is process wastewater generated from deionized water or high-strength industrial wastewaters. Iron and other trace nutrients may be deficient in this case. Trace nutrient requirements are shown in Table 6.2.

Sewage provides a balanced microbial diet, but many industrial wastes (cannery, pulp and paper, etc.) do not contain sufficient nitrogen and phosphorus and require their addition as a supplement.

The quantity of nitrogen required for effective BOD removal and microbial synthesis has been the subject of much research. Early work by Helmets et al.[18] indicates a nitrogen requirement of 4.3 lb N/100 lb BOD_{rem} (4.3 kg N/100 kg BOD_{rem}) and a phosphorus requirement of 0.6 lb P/100 lb BOD_{rem} (0.6 kg P/100 kg BOD_{rem}). These represent average values derived from the treatment of several nitrogen-supplemented industrial wastes. When insufficient nitrogen is present, the amount of cellular material synthesized per unit of organic matter removed increases as an accumulation of polysaccharide. At some point, nitrogen-limiting conditions restrict the rate of BOD removal. The rule-of-thumb number is BOD:N:P of 100:5:1 Nutrient-limiting conditions will also stimulate filamentous growth, as discussed on p. 255.

The nitrogen content of sludge as generated in the process has been shown to average 12.3 percent on the basis of the VSS. The nitrogen content of the sludge will decline during the endogenous phase. The nitrogen content of the nondegradable cellular mass has been shown to average 7 percent. This is shown in Fig. 6.10. The decrease in nitrogen content of an activated sludge with a soluble substrate as a function of sludge age is shown in Fig. 6.11. The phosphorus content of sludge

TABLE 6.2
Trace nutrient requirements for
biological oxidation

	mg/mg BOD
Mn	10×10^{-5}
Cu	14.6×10^{-5}
Zn	16×10^{-5}
Mo	43×10^{-5}
Se	14×10^{-10}
Mg	30×10^{-4}
Co	13×10^{-5}
Ca	62×10^{-4}
Na	5×10^{-5}
K	45×10^{-4}
Fe	12×10^{-3}
CO_3	27×10^{-4}

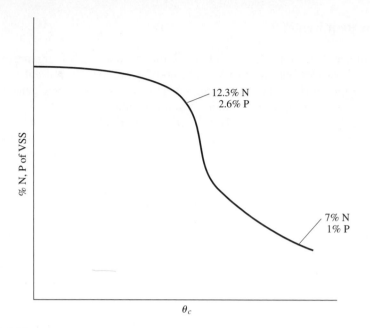

FIGURE 6.10
Nutrient requirements.

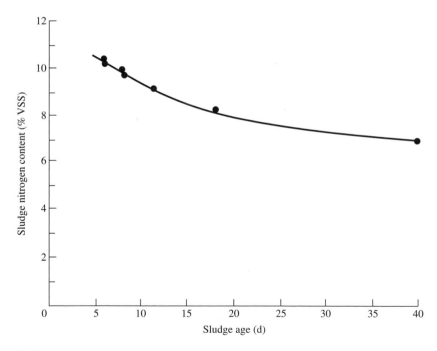

FIGURE 6.11
Nitrogen content of activated sludge as related to sludge age.

at generation has been found to average 2.6 percent with the nondegradable cellular mass having a phosphorus content of 1 percent. The nitrogen and phosphorus requirements can be calculated by considering the nitrogen and phosphorus content of the biological sludge wasted from the process:

$$N = 0.123\frac{X_d}{0.8}\Delta X_{v_b} + 0.07\frac{0.8 - X_d}{0.8}\Delta X_{v_b} \tag{6.19}$$

$$P = 0.026\frac{X_d}{0.8}\Delta X_{v_b} + 0.01\frac{0.8 - X_d}{0.8}\Delta X_{v_b} \tag{6.20}$$

Not all organic nitrogen compounds are available for synthesis. Ammonia is the most readily available form, and other nitrogen compounds must be converted to ammonia. Nitrite, nitrate, and about 75 percent of organic nitrogen compounds are also available.

Phosphorus may be fed as phosphoric acid in larger plants and ammonia as anhydrous or aqueous ammonia. In small plants, nutrients may be fed as diammonium phosphate. In many cases, in aerated lagoons treating pulp and paper mill wastewaters, nitrogen and phosphorus have not been added, but rather the retention time has been increased. Calculated rate coefficients with and without the addition of nutrients is shown in Table 6.3. Nutrient requirements are computed in Example 6.4.

EXAMPLE 6.4. An activated sludge plant treating an industrial wastewater operated under the following conditions:

Flow = 1.6 million gal/d
S_o = 560 mg/l (BOD basis)
S_e = 20 mg/l
X_v = 3000 mg/l
a = 0.55
b = 0.1/d
NH_3^-N = 5 mg/l
P = 3 mg/l
F/M = 0.4/d^{-1}
θ_c = 7 d

Compute the N and P which must be added to the process.

TABLE 6.3

Waste	K, d^{-1}	
	Without nutrients	With nutrients
Kraft paper	0.35	1.33
Board mill	0.70	3.20
Hardboard	0.34	1.66

Solution.

$$t = \frac{S_o}{X_v F/M} = \frac{560}{3000 \cdot 0.4} = 0.47 \text{ d}$$

$$X_d = \frac{0.8}{1 + (0.2 \cdot 0.1 \cdot 7)} = 0.7$$

$$\Delta X_v = aS_r - bX_d X_v t$$

$$\Delta X_v = 0.55(540) - 0.1 \cdot 0.7 \cdot 3000 \cdot 0.47$$

$$= 198 \text{ mg/l or } 2642 \text{ lb/d} \quad (1200 \text{ kg/d})$$

$$N = 0.123 \frac{0.7}{0.8} \cdot 2642 + 0.07 \cdot \frac{0.8 - 0.7}{0.8} \cdot 2642$$

$$= 284 + 23 = 307 \text{ lb/d} \quad (140 \text{ kg/d})$$

$$N_{\text{INFLUENT}} = 5 \cdot 1.6 \cdot 8.34 = 67 \text{ lb/d} \quad (30 \text{ kg/d})$$

$$N_{\text{ADDED}} = 307 - 67 - 240 \text{ lb/d}$$

$$P = 0.026 \cdot \frac{0.7}{0.8} \cdot 2642 + 0.01 \cdot \frac{0.8 - 0.7}{0.8} \cdot 2642$$

$$= 60 + 3.3$$

$$= 63.3 \text{ lb/d} \quad (29 \text{ kg/d})$$

$$P_{\text{INFLUENT}} = 3 \cdot 1.6 \cdot 8.34 = 40 \text{ lb/d} \quad (18 \text{ kg/d})$$

$$P_{\text{ADDED}} = 63.3 - 40 = 23.3 \text{ lb/d} \quad (11 \text{ kg/d})$$

Mathematical Relationships of Organic Removal

Several mathematical models have been offered to explain the mechanism of BOD removal by biological oxidation processes. All these models have shown that, at high BOD levels, the rate of BOD removal per unit mass of cells will remain constant to a limiting BOD concentration, below which the rate will become concentration-dependent and will decrease as the concentration. Wuhrmann[19] and Tischler and Eckenfelder[20] have shown that single substances are removed by a zero-order reaction to very low substrate levels. Some reactions are shown in Fig. 6.12. In a mixture of substances being removed at different rates, a constant maximum removal rate will prevail until one of the substances is completely removed. As other substances are progressively removed, the overall rate will decrease. As Gaudy, Komolrit, and Bhatla[21] have shown, sequential substrate removal will also yield a decreasing overall rate.

　　It is assumed that the volatile suspended solids concentration in the reactor is proportional to the cell mass. If volatile suspended solids are present in the influent wastewater, this assumption must be modified as discussed on p. 209.

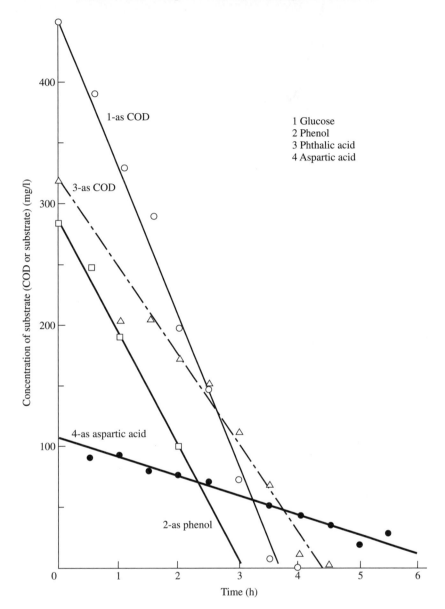

FIGURE 6.12
Zero-order removal rates for specific substrates.

In the case of a multicomponent wastewater where concurrent zero-order reactions occur, the overall removal can be formulated as shown in Fig. 6.13. The zero-order removal for three components is shown in Fig. 6.13a. When considering the total removal of all components as BOD, COD, or TOC, the overall rate will remain constant until time t_1, when component A is substantially removed. The overall rate will then decrease to reflect components B and C. At time t_2,

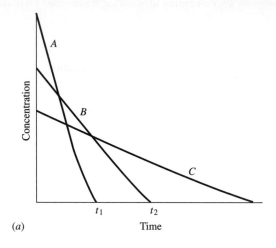

(a)

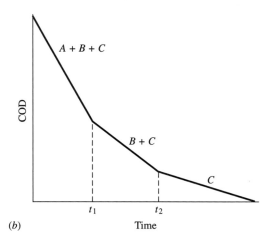

(b)

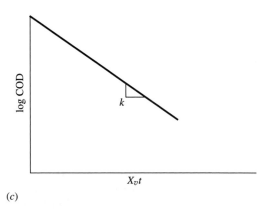

(c)

FIGURE 6.13
Schematic representation of multicomponent substrate removal.

220

component B is substantially removed and the rate will decrease to reflect only component C. For a wastewater consisting of many components, the breaks in Figure 6.13b are not apparent and a curve results. In most cases, this curve can be linearized and can be fitted to another order reaction equation [20, 22, 23] of the following form:

$$\frac{dS}{dt} = -K_n X_a \left(\frac{S}{S_o}\right)^n \tag{6.21}$$

where
$$\begin{aligned}
S &= \text{COD concentration at time } t, \text{ mg/l} \\
S_o &= \text{COD concentration at time zero, mg/l} \\
X_a &= \text{active biomass concentration, mg/l} \\
t &= \text{time, d} \\
K_n &= \text{curve fitting coefficient, d}^{-1} \\
n &= \text{function power order}
\end{aligned}$$

The active mass can be defined as

$$X_a = X_v \cdot \frac{X_d}{0.8} \cdot f_b \tag{6.22}$$

If the wastewater contains no influent VSS,

$$f_b = 1.0$$

If the influent VSS are nondegradable,

$$f_b = 1 - \frac{X_i \theta_c}{X_v t} \tag{6.23}$$

If the influent VSS are degradable, Eq. (6.15) applies.

Regardless of the actual degradation rates of the individual substrate components, the integrated form of Eq. (6.21) represents the performance of a batch or a continuous plug flow reactor (CPFR). The integrated forms of Eq. (6.21) for $n = 1$ and $n = 2$ are:

$$S_e = S_o e^{-K_1 f_b X_v t / S_o} \tag{6.24}$$

$$S_e = \frac{S_o^2}{S_o + K_2 f_b X_v t} \tag{6.25}$$

where
$$\begin{aligned}
S_e &= \text{effluent COD or BOD in a CPFR or a batch reactor, mg/l} \\
t &= \text{CPFR's hydraulic retention time or batch test reaction time, d} \\
K_1 &= \text{first-order coefficient, d}^{-1} \\
K_2 &= \text{second-order coefficient, d}^{-1}
\end{aligned}$$

Batch kinetics for a first-order approximation and a second-order approximation are shown in Figs. 6.14 and 6.15.

Plug flow kinetics following a first-order function for a pulp and paper mill wastewater are shown in Fig. 6.16. Initial biosorption in the plug flow reactor should be noted.

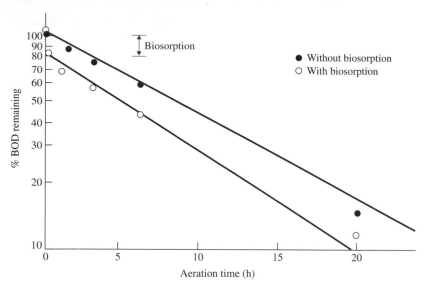

FIGURE 6.14
Batch activated sludge with and without biosorption for a pulp and paper mill wastewater.

If the individual components of the substrate are degraded at a zero-order rate, the overall rate for a CSTR will still follow a kinetic equation similar to Eq. (6.21), assuming a nonsegregated flow regime.[24, 25]

In a complete mix basin, the overall removal rate will decrease as the concentration of organics remaining in solution decreases, since the more readily degradable organics will be removed first. The kinetics of the oxidation are therefore limited by the effluent concentration and can be expressed as

$$\frac{S_o - S_e}{f_b X_v t} = K \frac{S_e}{S_o} \tag{6.26}$$

This relationship is shown in Fig. 6.17 for a soybean wastewater.

The relative biodegradability of various organic compounds is shown in Table 6.4. The rate coefficient K for various industrial wastewaters in CMAS systems is shown in Table 6.5.

In all kinetic calculations the rate coefficient K should be based on active mass in the reactor.

There are two operating conditions which influence the organic removal rate, as defined by a rate coefficient K. These are the active fraction of the biomass in the MLVSS and the aerobic portion of the biofloc. The active fraction of the MLVSS is related to the sludge age or the F/M. Increasing the F/M or decreasing the θ_c will increase the active fraction of the biomass as shown in Fig. 6.18. The second condition defines that portion of the floc which is aerobic and is related to the turbulence level or mixing intensity and bulk dissolved oxygen level in the aeration basin.

The fraction of active biomass in the mixed liquor is defined here as the biodegradable fraction X_d divided by 0.80. This is illustrated in Examples 6.5 and 6.6.

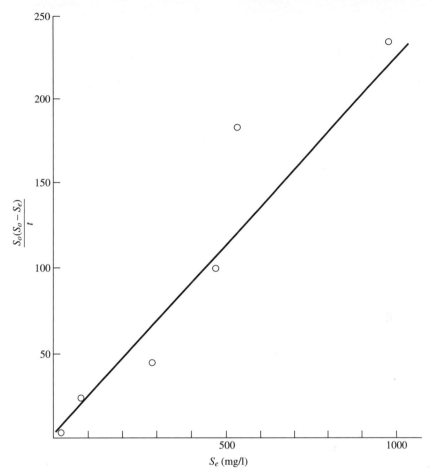

FIGURE 6.15
Batch oxidation of a chemical industry wastewater following second-order kinetics.

EXAMPLE 6.5. A complete mix activated sludge plant with an influent BOD of 800 mg/l and a K of 6 d^{-1} is operating at an SRT of 10 d. What is the effluent quality and what will the effluent quality be if the SRT is increased to 30 days? a is 0.5 and b is 0.1 d^{-1}.

Solution.

$$\frac{1}{\theta_c} = aK\frac{S_e}{S_o} - bX_d$$

At $\theta_c = 10$ d,

$$\frac{1}{10} = 0.5 \cdot 6 \cdot \frac{S_e}{S_o} - 0.1 \cdot \left(\frac{0.8}{1 + 0.2 \cdot 0.1 \cdot 10}\right)$$

$$\frac{S_e}{S_o} = 0.056$$

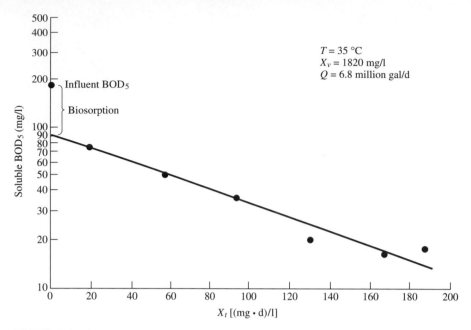

FIGURE 6.16
Plug flow BOD removal kinetics for a bleached kraft pulp and paper wastewater.

and

$$S_e = 0.056 \cdot 800 = 45 \text{ mg/l}$$

At $\theta_c = 30$ d, the K can be adjusted by Fig. 6.18:

$$K = 6 \cdot \frac{0.625}{0.83} = 4.5 \text{ d}^{-1}$$

$$\frac{1}{30} = 0.5 \cdot 4.5 \frac{S_e}{S_o} - 0.1 \cdot \left(\frac{0.8}{1 + 0.2 \cdot 0.1 \cdot 30} \right)$$

$$\frac{S_e}{S_o} = 0.0368$$

and

$$S_e = 0.0368 \cdot 800 = 30 \text{ mg/l}$$

If X_v is 2500 mg/l and $t = 0.9$ d at $\theta_c = 10$ d, the $X_v t$ at a 30-d SRT is

$$X_v t = \frac{\theta_c a S_r}{1 + \theta_c b X_d}$$

$$= \frac{30 \cdot 0.5 \cdot 770}{1 + 30 \cdot 0.1 \cdot 0.5} = 4620 \text{ (mg} \cdot \text{d)/l}$$

$$X_v = 5133 \text{ mg/l}$$

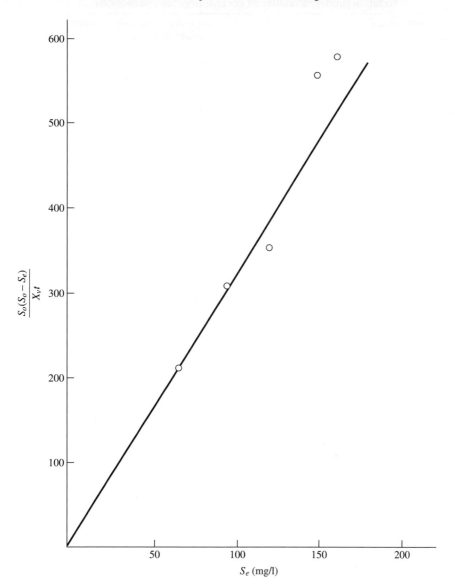

FIGURE 6.17
Complete mix kinetics for a soybean wastewater.

EXAMPLE 6.6. Determine the hydraulic retention time and the sludge age for the following condition:

S_o = 700 mg/l
S_e = 30 mg/l
K = 10/d
X_v = 3000 mg/l
a = 0.4

TABLE 6.4
Relative biodegradability of certain organic compounds

Biodegradable organic compounds[†]	Compounds generally resistant to biological degradation
Acrylic acid	Ethers
Aliphatic acids	Ethylene chlorohydrin
Aliphatic alcohols (normal, iso, secondary)	Isoprene
	Methyl vinyl ketone
Aliphatic aldehydes	Morpholine
Aliphatic esters	Oil
Alkyl benzene sulfonates with exception of porpylene-based benzaldehyde	Polymeric compounds
	Polypropylene benzene sulfonates
Aromatic amines	Selected hydrocarbons
Dichlorophenols	Aliphatics
Ethanolamines	Aromatics
Glycols	Alkyl-aryl groups
Ketones	Tertiary aliphatic alcohols
Methacrylic acid	Tertiary aliphatic sulfonates
Methyl methacrylate	Trichlorophenols
Monochlorophenols	
Nitriles	
Phenols	
Primary aliphatic amines	
Styrene	
Vinyl acetate	

[†]Some compounds can be degraded biologically only after extended periods of seed acclimation.

TABLE 6.5
Reaction rate coefficient for selected wastewaters

Wastewater source	K, d^{-1}	Temperature, °C
Vegetable tannery	1.2	20
Cellulose acetate	2.6	20
Peptone	4.0	22
Organic phosphates	5.0	21
Vinyl acetate monomer	5.3	20
Organic intermediates	5.8	8
	20.6	26
Viscose rayon and nylon	6.7	11
	8.2	19
Domestic sewage (solubles)	8.0	20
Polyester fiber	14.0	21
Formaldehyde, propanol, methanol	19.0	20
High-nitrogen organics	22.2	22
Potato processing	36.0	20

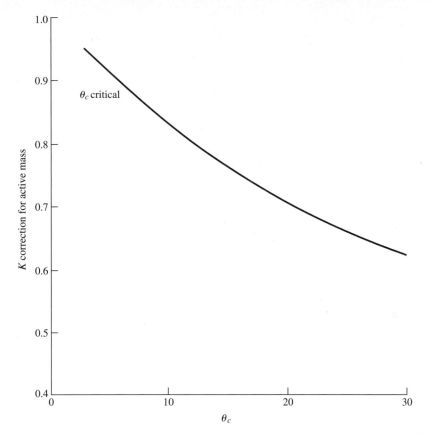

FIGURE 6.18
Relationship between K and θ_c, considering biological active mass.

What will be the required hydraulic retention time to produce the same effluent if the influent nondegradable VSS is 50 mg/l?

Solution.

Solving Eq. (6.24) for the hydraulic retention time t produces, $S_e = 30$ mg/l,

$$t = \frac{S_o S_r}{K X_v S_e}$$

$$= \frac{700 \cdot 670}{10 \cdot 3000 \cdot 30} = 0.52 \text{ d}$$

and

$$X_v t = 1560 \ (\text{mg} \cdot \text{d})/\text{l}$$

The SRT for these conditions is

$$\theta_c = \frac{X_v t}{a S_r - b X_d X_v t}$$

Assume $X_d = 0.7$. Then

$$\theta_c = \frac{1560}{0.4 \cdot 670 - 0.1 \cdot 0.7 \cdot 1560}$$

$$= \frac{1560}{159} = 9.8 \text{ d, say 10 d}$$

Check assumption for $X_d = 0.7$ of SRT = 10 d:

$$X_d = \frac{0.8}{1 + 0.2 \cdot 0.1 \cdot 10}$$

$$X_d = 0.67$$

To determine the required hydraulic retention time to produce the same effluent $S_e = 30$ mg/l if the influent nondegradable VSS = 50 mg/l, assume $t = 0.69$ d (by trial and error) and calculate the accumulation of the influent VSS in the mixed liquor ($MLVSS_i$):

$$MLVSS_i = \frac{50 \cdot 10}{0.69} = 725 \text{ mg/l}$$

The residual biomass VSS in the mixed liquor is

$$X_{vb} = 3000 - 725 = 2275 \text{ mg/l}$$

Calculate the required t:

$$t = \frac{X_{vb}t}{X_{vb}} = \frac{1560}{2275} = 0.69 \text{ d}$$

$$f_b = 2275/3000 = 0.76$$

Check sludge age:

$$\theta_c = \frac{X_v t}{(aS_r - bX_{vb}X_d t) + X_i}$$

$$= \frac{3000 \cdot 0.69}{(268 - 0.1 \cdot 2275 \cdot 0.7 \cdot 0.69) + 50}$$

$$= \frac{2070}{208} = 9.9 \text{ d}$$

The greater hydraulic retention time is required to produce the same effluent quality since the nondegradable influent VSS accumulated in the constant MLVSS concentration of 3000 mg/l. This is shown in the figure on the next page.

The size and the aerobic fraction of the biological floc are related to the operating power level in the aeration basin and the mixed liquor bulk dissolved oxygen concentration. These in turn influence both the reaction rate coefficient and the endogenous decay rate. The effect on the reaction rate coefficient has been demonstrated in treatment of wastewater from a bleached kraft pulp and paper mill. The biodegradation rate [K in Eq. (6.26)] under operation at an F/M of 0.3 d^{-1} and a conventional aeration basin power level of 200 hp/million gal averaged 4.5 d^{-1}.

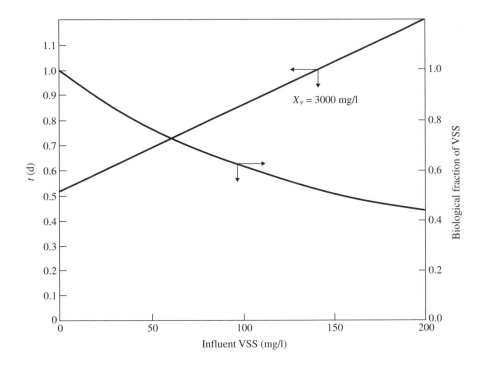

Operation at an F/M of 0.88 d^{-1} and a power level of 500 hp/million gal averaged a K of 12.5 d^{-1}. BOD removal of 92 percent was achieved under these conditions.

Two factors will influence the endogenous decay rate and the observed sludge yield. These are the degradable fraction X_d, which is a function of SRT or F/M, and the aerobic fraction of the MLVSS. Increasing the mixing intensity in the aeration basin will increase the aerobic fraction of the mixed liquor solids and the rate of endogenous respiration. This condition has been observed in the high-purity-oxygen activated sludge process in which high mixed liquor dissolved oxygen levels increased the aerobic fraction of the floc and decreased the observed sludge yield. Rickard and Gaudy[26] showed a decrease in observed sludge yield with increasing agitation at a constant F/M.

It has been observed that, at high energy levels in the aeration basin, filamentous bulking is suppressed. This was noted by Rickard and Gaudy,[26] who showed that, at a velocity gradient of 310 s^{-1}, a majority of the cells were of the filamentous type, while, at a velocity gradient of 1010/s^{-1}, filaments were reportedly absent. Zahradka[27] observed that a high power input caused the growth of sludge flocs of small and uniform size and suppressed the growth of filamentous organisms. This can be explained by the fact that, at high power levels, oxygen and BOD diffusion is not limiting with small flocs, and therefore the floc formers will dominate over the filaments. It is therefore important to consider the aeration power level and its effect on the reaction coefficient K and b in a process design or a pilot plant study.

In cases where multiple wastewaters are mixed for biological treatment, a mean reaction rate K is determined. In this case the average rate coefficient K_c can be computed:

$$\frac{1}{K_c} = \frac{\dfrac{1}{K_1}(Q_1 S_{o_1}) + \dfrac{1}{K_2}(Q_2 S_{o_2})}{Q_1 S_{o_1} + Q_2 S_{o_2}} \tag{6.27}$$

This is shown in Example 6.7. The multiple zero-order concept may also predict the performance of a combined system, where two or more different wastewater streams are treated in one unit.

EXAMPLE 6.7. Treatment of mixed industrial wastewaters. Three industrial wastewaters are to be blended and treated in an activated sludge plant. Compute the sludge age and the hydraulic retention time required to produce an effluent of 20 mg/L BOD. The characteristics of the three wastewaters are

1. $Q = 2$ million gal/d
   $S_o = 600$ mg/l
   $K = 5$ d^{-1}

2. $Q = 1$ million gal/d
   $S_o = 1200$ mg/l
   $K = 10$ d^{-1}

3. $Q = 5$ million gal/d
   $S_o = 300$ mg/l
   $K = 2$ d^{-1}

Solution.

The average influent BOD is

$$S_{o_{ave}} = \frac{2 \cdot 600 + 1 \cdot 1200 + 5 \cdot 300}{8} = 487 \text{ mg/l}$$

The average K is

$$\frac{1}{K_{ave}} = \frac{Q_1 S_{o_1}/K_1 + Q_2 S_{o_2}/K_2 + Q_3 S_{o_3}/K_3}{Q_1 S_{o_1} + Q_2 S_{o_2} + Q_3 S_{o_3}}$$

$$= \frac{0 \cdot 2 \cdot 600 + 0.1 \cdot 1 \cdot 1200 + 0.5 \cdot 5 \cdot 300}{2 \cdot 600 + 1 \cdot 1200 + 5 \cdot 300}$$

$$= 0.28$$

$$K_{ave} = 3.57 \text{ d}^{-1}$$

The mean yield coefficient a is 0.5, b is 0.1, and X_v is 3000 mg/l. For a K_{ave} of 3.57,

$$\frac{1}{\theta_c} = aK\frac{S_e}{S_o} - bX_d$$

$$= 0.5 \cdot 3.57 \cdot \frac{20}{487} - 0.1 \cdot 0.46$$

$$= 0.027$$

$$\theta_c = 37 \text{ d}$$

$$\theta_c = \frac{X_v t}{aS_r - bX_d X_v t}$$

or

$$X_v t = \frac{\theta_c a S_r}{1 + \theta_c b X_d}$$

$$= \frac{37 \cdot 0.5 \cdot 467}{1 + 37 \cdot 0.1 \cdot 0.46}$$

$$= 3200$$

$$t = \frac{3200}{3000} = 1.07 \text{ d}$$

In a two-stage activated sludge system without an intermediate clarifier, the effluent will be the same as a single-stage system as long as multiple zero-order kinetics apply and the second-stage kinetic coefficient is proportional to the squared ratio of the pretreated to raw substrate concentrations. In a two-stage system with an intermediate clarifier between the two stages, leading to the development of a biomass specifically acclimated to the substrate remaining in the first-stage effluent, the rate coefficient in the second stage will be lower than in the first stage since the more degradable compounds were removed in the first stage. The second stage K can be estimated from the relationship

$$K_2 = K_1 \left[\frac{S_1}{S_o} \right] \tag{6.28}$$

where K_2 = rate coefficient for the second stage
K_1 = rate coefficient for the first stage
S_o = influent BOD or COD to stage 1
S_1 = influent BOD or COD to stage 2

The application of Eq. (6.28) can be illustrated by using data from a two-stage activated sludge plant treating a pharmaceutical wastewater. The first stage has an influent BOD of 5825 mg/l, an effluent BOD of 540 mg/l, and a K of 3.9 d^{-1}. The influent BOD to the second stage is 540 mg/l and the K is 0.4 d^{-1}. The calculated K in stage 2, from Eq. (6.28), is

$$K_2 = K_1 \left(\frac{S_1}{S_o} \right)$$

$$= 3.9 \left(\frac{540}{5825} \right)$$

$$= 0.36 \text{ d}^{-1}$$

This is in good agreement with the measured value of 0.4 d^{-1}.

Kinetic parameters for a pulp and paper mill wastewater are shown in Table 6.6. Parameters for petroleum refinery wastewaters are shown in Table 6.7.

TABLE 6.6
Reaction rate coefficient for pulp and paper mills

Type of mill	K, d^{-1}	Temperature, °C
Oxygen bleached kraft	13.5	35
Virgin pulp and wastepaper	13.6	23
Unbleached kraft	4.5	38
Sulfite	5.0	18
Bleached sulfite	6.2	—
Bleached kraft	5.2	—
Bleached kraft	4.4	34

TABLE 6.7
Biological treatment coefficients for petroleum refinery wastewaters

Influent		Organic removal rate $K^{\dagger}$		Sludge growth coefficients				Oxygen requirement coefficients‡		Residual COD, mg/l
				BOD basis		COD basis				
BOD, mg/l	COD, mg/l	BOD, d^{-1}	COD, d^{-1}	a	bX_d	a	bX_d	a'	$b'X_d$	
244	509	4.15	2.74	—	—	—	—	0.57	0.1	106
575	981	—	7.97	—	—	0.5	0.06	0.60	0.11	53
396	782	—	5.86	—	—	0.5	0.06	0.34	0.06	100
153	428	—	2.92	0.5	0.08	0.44	0.1	0.35	0.08	22
170	600	—	5.0	—	—	0.26	0.03	0.46	0.05	100§
248	563	4.11	7.79	—	—	0.2	0.08	0.40	0.01	76
345	806	—	7.24	—	—	0.43	0.10	0.52	0.14	82
196	310	4.70	—	0.6	0.05	—	—	0.46	0.14	50
138	275	—	—	0.58	—	0.25	—	0.60	0.09	42

† At 24°C.
‡ COD basis.
§ TOD.

It is apparent that, as the organic composition of the wastewater changes the rate coefficient K in Eq. 6.26 will also change. This is not a problem for wastewaters such as those from a dairy or food processing plant since their composition remains substantially unchanged, and hence, K will remain nearly constant. Wastewaters generated from plants with multi-products and campaign production, however, will experience a constantly changing wastewater composition and, hence, a highly variable K value, as shown in Fig. 6.19.

The rate coefficient combines the effects of all removal mechanisms: biosorption, biodegradation, and volatilization, unless steps are taken to separate the effects of an individual removal mechanism. Unusually high "apparent" reaction rate coefficients may be observed when volatile organics constitute a large portion of the wastewater. Volatilization of substrate should be considered when calculated K values exceed about 30/d^{-1} at 20 to 25°C.

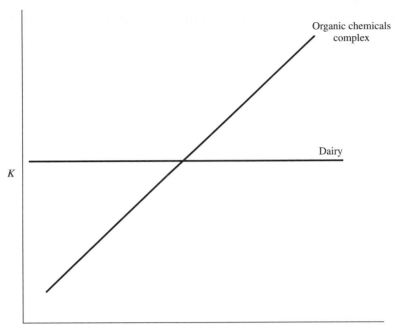

FIGURE 6.19
Variability in K as related to wastewater composition.

Industrial wastewater discharge permits typically contain two limiting condi-tions: a monthly average limit and a daily or weekly maximum limit. The treatment process must be designed and operated to reliably satisfy both of these discharge conditions.

A suggested design approach is based on a statistical distribution of the removal rate coefficient and the performance of the upstream equalization basin. For average discharge conditions, the mean K value is based on the average dis-charge limit and average influent load. These values are substituted into Eq. (6.26) to yield

$$\frac{\bar{S}_o - \bar{S}_e}{f_b X_v t} = K_{50\%} \frac{\bar{S}_e}{\bar{S}_o} \tag{6.29}$$

where $\bar{S}_o$ = average influent BOD, mg/l

$\bar{S}_e$ = soluble BOD at average permit limit, mg/l

$K_{50\%}$ = the 50 percentile value of K, d^{-1}

For the maximum permit condition, Equation (6.29) can be expressed as

$$\frac{S_{o_m} - S_{e_m}}{f_b X_v t} = K_{5\%} \frac{S_{e_m}}{S_{o_m}} \tag{6.30}$$

where S_{o_m} = maximum effluent BOD from the equalization basin, mg/l

      S_{e_m} = soluble BOD at maximum permit limit, mg/l

      $K_{5\%}$ = 5 percentile value of K, d^{-1}

The values of $X_v t$ are calculated for Eq. (6.29) and (6.30), and the larger of the two values is used for design. However, if the $X_v t$ value computed for the maximum permit condition exceeds twice the value computed for the average condition, changes in the equalization capacity or plant production schedules should be considered in order to reduce the difference. Alternatively, a less conservative value of K ($> K_{5\%}$) could be used.

Specific Organic Compounds

The kinetic removal mechanism for specific organics in an aerobic biological process has been defined by Monod:

$$\mu = \frac{\mu_m S}{K_S + S} \quad \text{and} \quad q = \frac{q_m S}{K_S + S} \tag{6.31}$$

where μ = specific growth rate, d^{-1}

     μ_m = maximum specific growth rate, d^{-1}

     S = substrate concentration, mg/l

     K_S = substrate concentration when the rate is one-half the maximum rate, mg/l

     q = specific substrate removal rate, d^{-1}

     q_m = maximum substrate removal rate, d^{-1}

The relative biodegradability of specific organics in terms of the Monod relationship is shown in Fig. 6.20.

In complete mix activated sludge (CMAS) with sludge recycle, the Monod equation can be expressed:

$$S_o - S = \frac{q_m S}{K_S + S} \cdot X_{vb} t \tag{6.32}$$

where S_o = influent substrate concentration, mg/l

     X_{vb} = biological volatile suspended solids under aeration, mg/l

     t = liquid retention time, d

Solving for S yields:

$$S = \frac{-B + (B^2 + 4S_o K_S)^{1/2}}{2} \tag{6.33}$$

in which

$$B = q_m X_{vb} t + K_S - S_o$$

The SRT in the activated sludge process for a soluble substrate can be defined as follows:

$$\theta_c = \frac{X_{vb} t}{a(S_o - S) - b X_d X_{vb} t} \tag{6.11}$$

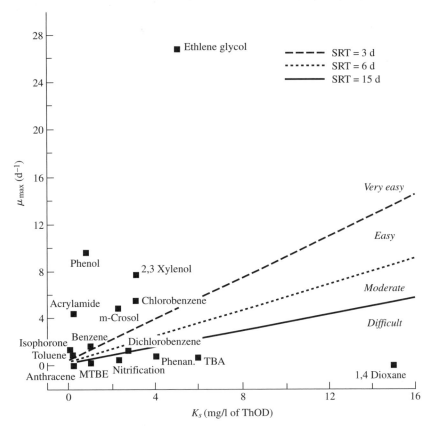

FIGURE 6.20
Relative biodegradability of specific organics at different SRT at 25°C (effluent concentration = 0.5 mg/l as COD and b = 0.11 per day).[7]

where θ_c = SRT, d
a = yield coefficient, d^{-1}
b = endogenous coefficient, d^{-1}
X_d = degradable fraction of the VSS

In complete mix activated sludge (CMAS), the effluent substrate concentration is directly related to the SRT (θ_c). Combining Eqs. (6.31) and (6.11) yields:

$$S = \frac{K_S(1 + bX_d\theta_c)}{\theta_c(q_m a - bX_d) - 1} \qquad (6.34)$$

This relationship for dichlorophenol (DCP) is shown in Fig. 6.21.

For the case of a plug flow activated sludge (PFAS) reactor with sludge recycle, the performance equation derived from the Monod relationship is:

$$\frac{1}{\theta_c} = \frac{\mu_m(S_o - S)}{(S_o - S) + CK_S} - bX_d \qquad (6.35)$$

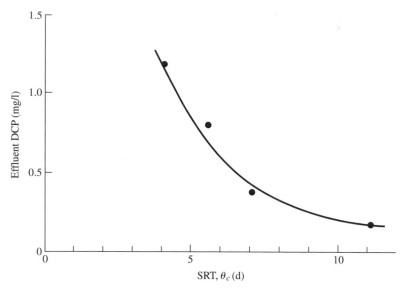

FIGURE 6.21
Effect of SRT on DCP removal.

where $C = (1 + \alpha)\ln[X(\alpha S + S_o)/(1 + \alpha)S]$
         $\alpha = R/Q$
         S_o = influent substrate concentration prior to mixing with the recycle

Effluent levels of specific priority organics, e.g., phenol, can be computed from Eqs. (6.34) or (6.35), depending on the configuration of the reactor, complete mix or plug flow.

The modified fed batch reactor (FBR) test described by Philbrook and Grady[28] is applicable to the determination to the kinetic coefficients q_m and K_S under field operating conditions. In the test, plant or pilot plant sludge at the desired SRT is placed in a 2-L reactor, and plant wastewater is added at a constant rate. In order to determine q_m, the addition rate must exceed the degradation rate. Since, in many wastewaters, the priority pollutant levels are low, the wastewater may have to be spiked to ensure a sufficient concentration of pollutant to meet the conditions of the test. It is important, however, that the concentration levels achieved in the test are below the inhibition threshold. This can be found by the shape of the concentration-time curve. The degradation rate q_m is computed as the difference in the slopes of the substrate addition rate and the residual substrate accumulation.

A second FBR test is then conducted with the addition rate of the priority pollutant equal to one-half the maximum rate determined in the first test. The steady-state concentration observed in the reactor will be K_S. FBR test data for phenol are shown in Fig. 6.22. Hoover[29] found a high variability in q_m with sludges operating under the same loading conditions with time. Based on these observations, a rou-

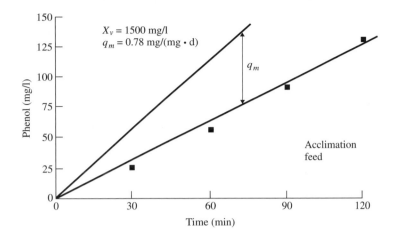

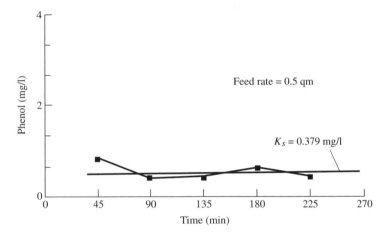

FIGURE 6.22
FBR test for the determination of q_m and K_S.

tine test program should be established at a treatment plant and values for q_m and K_S interpreted on a statistical basis.

In the pulp and paper industry attention is being focused on absorbable organic halides (AOX) and chlorinated phenolics in bleach plant effluents. Chlorophenols are amenable to aerobic mineralization, while methoxalated chlorophenols are recalcitrant to aerobic oxidation. Anaerobic pretreatment however, results in readily degradable aerobic forms. Table 6.8 summarizes the observed removal of pollutants from pulp and paper mill wastewaters using activated sludge (AS), facultative stabilization basins (FSB), and aerated stabilization basins (ASB). Substituting oxygen bleaching for conventional chlorine bleaching has a marked effect on the removal of both AOX and chlorinated phenolics as shown in Table 6.9.

TABLE 6.8
Removal performance summary for conventional pollutants and AOX

Parameter	50th and (90th) percentile values		
	AS	FSB	ASB
Conventional			
COD removal, %	54 (65)	55 (78)	57 (68)
BOD removal, %	96 (98)	96 (98)	96 (98)
NH_4-N_{effl}, mg/l	1.5 (10.1)	0.25 (5.8)	0.25 (4.5)
$(NO_2 + NO_3)$-N_{effl}, mg/l	1.4 (8.0)	4.3 (11.1)	8.0 (13.2)
VSS_{effl}, mg/l	32 (110)	62 (200)	75 (200)
AOX			
Total AOX removal, %	22 (1.7)	43 (1.3)	40 (1.3)
Filt-AOX removal, %	28 (1.5)	48 (1.2)	45 (1.2)
(Nonfilt-AOX/total AOX)$_{effl}$, %	8 (1.9)	8 (1.9)	8 (1.9)
(Nonfilt-AOX/CSS)$_{effl}$, mg/g	45 (2.8)	28 (1.9)	20 (2.6)

TABLE 6.9
**Comparison of conventional and oxygen bleaching on
activated sludge performance (kraft hardwood)[30]**

Parameter	Reduction (%) for bleaching process	
	Conventional	Oxygen
AOX	22[†]	40[†]
Chlorinated Phenols		
Phenol	39	45
Guajacols	41	79
Catacols	50	63

[†] Based on influent AOX concentrations of 136 mg/l and 57 mg/l.

EXAMPLE 6.8. Compute the SRT required in a complete mix activated sludge plant to reduce phenol from $S_o = 10$ mg/l to 15 μg/l, where:

$q_m = 1.8$ g/(g VSS · d) at 20°C
$\theta = 1.1$ (temperature coefficient)
$K_S = 100$ μg/l
$a = 0.6$
$bX_d = 0.05$ d^{-1} at 20°C = 0.033 at 10°C

Solution.

Equation (6.34) can be rearranged to yield

$$\theta_c = \frac{K_S + S}{aq_m S - bX_d(K_S + S)}$$

$$= \frac{0.1 + 0.015}{0.6 \cdot 1.8 \cdot 0.015 - 0.05(0.115)}$$

$$= 11.0 \, \text{d}$$

What is the required SRT if the temperature is reduced from 20°C to 10°C?

$$q_{m(10°)} = q_{m(20°)} \cdot 1.1^{-10}$$

$$= 1.8/2.6$$

$$= 0.69 \, \text{d}^{-1}$$

$$\theta_c = \frac{0.1 + 0.015}{0.6 \cdot 0.69 \cdot 0.015 - 0.033 \, (0.115)}$$

$$= 47.6 \, \text{d}$$

In practice in a majority of cases, a high degree of back-mixing takes place so that conditions in a plug flow tank may approach that of complete mixing. However, in order to take advantage of the kinetics of priority pollutant removal, multiple basins in series will offer distinct advantages over a single complete mix basin.

Equation (6.34) can be applied to predict the performance of multiple reactors in series as shown by the following example.

EXAMPLE 6.9. Design a single-stage and a three-stage activated sludge reactor to reduce phenol. Compute the effluent phenol level in each case.

$S_o = 10 \, \text{mg/l}$
$bX_d = 0.05$
$\theta_c = 10 \, \text{d}$
$q_m = 0.6$
$K_S = 0.2$
$a = 0.6$

Solution.

For the CMAS (one stage), $X_v t = 39.44 \, \text{mg/(l} \cdot \text{d)}$ is computed from Eqs. (6.32) and (6.33).

$$B = 0.6(39.44) + 0.2 - 10$$

$$= 13.86$$

$$S = \frac{-13.86 + [(13.86)^2 + 4 \cdot 10 \cdot 0.2]^{1/2}}{2}$$

$$= 0.14 \, \text{mg/l}$$

Hence, S for a single complete mix basin is 0.14 mg/l. We can now compute the performance of three basins in series. Each basin is considered as a complete mix basin in which the retention time is based on $(Q + R)$. The effluent from the third basin is assumed negligible. The concentration entering the first basin with a 50 percent recycle is

$$S_o = \frac{10}{1.5}$$

$$= 6.67 \text{ mg/l}$$

For the same total basin volume as in the complete mix case, $X_v t$ is reduced by a factor of 1.5:

$$X_v t = 39.44/3 \cdot 1.5$$

$$= 8.76 \text{ mg/(l} \cdot \text{d)}$$

$$B = 0.6 \cdot 8.76 + 0.2 - 6.67$$

$$= -1.21$$

$$S = \frac{+1.21 + [1.21^2 + 4 \cdot 0.2 \cdot 6.67]^{1/2}}{2}$$

$$= 1.9 \text{ mg/l}$$

In like manner, the effluent (S) from basin 2 is computed as 0.105 mg/l and from basin 3 as 0.005 mg/l.

6.3
EFFECT OF TEMPERATURE

Variations in temperature affect all biological processes. There are three temperature regimes: the mesophilic over a temperature range of 4 to 39°C, the thermophilic which peaks at a temperature of 55°C, and the psychrophilic which operates at temperatures below 4°C. For economic and geographical reasons, most aerobic biological treatment processes operate in the mesophilic range, which is shown in Fig. 6.23. In the mesophilic range, the rate of the biological reaction will increase with temperature to a maximum value at approximately 31°C for most aerobic waste systems. A temperature above 39°C will result in a decreased rate for mesophilic organisms.

Over a temperature range of 4 to 31°C

$$K_t = K_{20°} \cdot \theta^{T-20} \tag{6.36}$$

The temperature coefficient θ has been found to vary from 1.03 to 1.1 for a variety of industrial wastewaters. This relationship for a chemical wastewater is shown in Fig. 6.24 and for a bleached sulfite mill in Fig. 6.25.

It should be noted that the temperature coefficient θ for municipal wastewater has a value of 1.015, showing relatively little effect. This is because a major portion of the BOD is present as colloidal or suspended organics, which are biocoagulated with very little temperature effect. When industrial wastewaters are discharged to municipal systems, the soluble organic content is increased and hence the temperature coefficient θ is also increased.

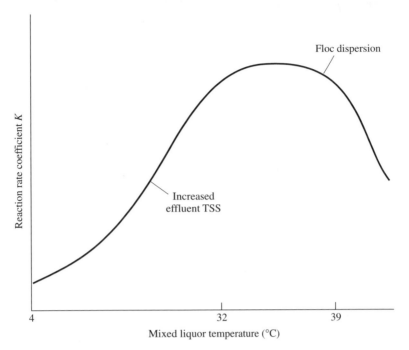

FIGURE 6.23
Effect of temperature on biological oxidation rate constant K.

Decreasing aeration basin temperature will also cause an increase in effluent suspended solids. The solids are of a dispersed nature and are nonsettleable. For example, the Union Carbide plant at South Charleston, West Virginia, had an effluent suspended solids of 42 mg/l during the summer and 104 mg/l during the winter. Removal of these suspended solids requires the addition of coagulating chemicals (see p. 371).

The endogenous coefficient b is also affected by temperature through the relationship

$$b_T = b_{20°C} \cdot 1.04^{T-20} \tag{6.37}$$

At temperatures above 96°F (35.5°C) there is a deterioration in the biological floc. Protozoa have been observed to disappear at 104°F (40°C) and a dispersed floc with filaments to dominate at 110°F (43.3°C).

In the past, hot wastewaters such as those in the pulp and paper industry were pretreated through a cooling tower so that the aeration basin temperature did not exceed 35°C. Recent air pollution regulations preclude stripping in a cooling tower without off-gas treatment. As a result, in many cases, the aeration basins must be covered and the off-gas treated. The consequence of this is a substantial temperature rise in the aeration basin due to the heat release from the exothermal biological reaction. In a recent study on a pharmaceutical wastewater, the average heat release was calculated as high as 5000 Btu/lb COD removed.

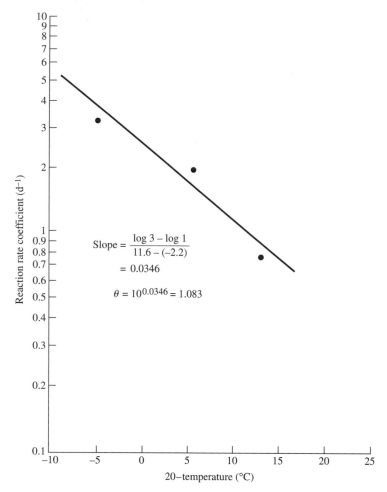

FIGURE 6.24
Effect of temperature on the reaction rate coefficient K for an agricultural chemicals
wastewater.

In one case of a pulp and paper mill effluent, the aeration basin temperature
reached 43°C. The characteristics of the sludge at 35°C and 43°C are shown in Fig.
6.26. One can see in Fig. 6.26 the absence of protozoans and the presence of filaments and dispersed floc at 43°C. The relationship between zone settling velocity
and temperature is shown in Fig. 6.27. The net effect is a flux limitation on the final
clarifier, resulting in decreased plant performance.

In a second case of an agricultural chemicals wastewater, floc dispersion
resulted at an aeration basin temperature of 36°C. In order to maintain plant performance, massive polymer doses were required. The problem was solved by
installing heat exchangers on the influent wastewater so that the influent temperature did not exceed 30°C. In a third case, treatment of a high-strength synfuels
wastewater resulted in basin temperatures of 44°C during the summer months.
Plant performance is shown in Fig. 6.28.

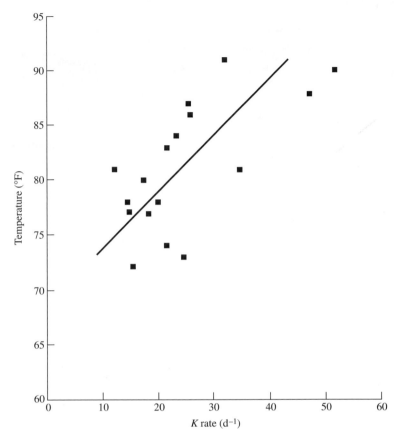

FIGURE 6.25
Effect of temperature on the reaction rate for a bleached sulfite mill wastewater.

EXAMPLE 6.10. Estimate the maximum influent temperature for a wastewater with the following characteristics.

    Influent degradable COD, S_o = 2580 mg/l
    Effluent degradable COD, S_e = 70 mg/l
    Reaction rate coefficient K = 5.0 d^{-1}
    MLVSS, X_v = 3000 mg/l
    Wastewater flow Q = 1 million gal/d

Solution.

The required aeration detention time can be calculated:

$$t = \frac{S_o(S_o - S_e)}{KX_vS_e}$$

$$= \frac{2580\,(2580 - 70)}{5 \cdot 3000 \cdot 70}$$

$$= 6.1 \text{ d}$$

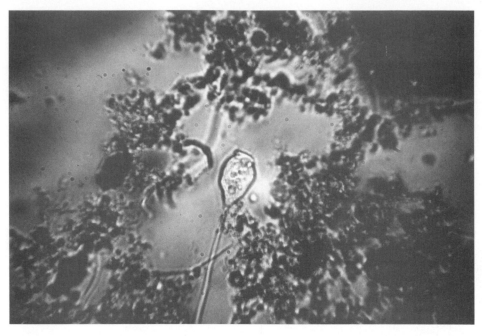

(a)

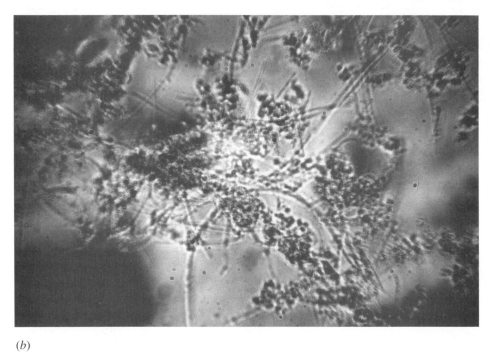

(b)

FIGURE 6.26
Sludge characteristics at (a) 35°C and (b) 43°C.

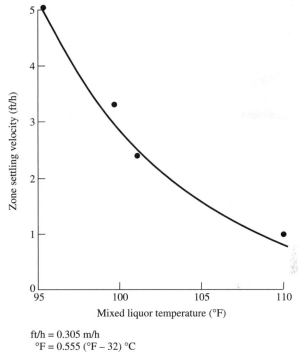

ft/h = 0.305 m/h
°F = 0.555 (°F − 32) °C

FIGURE 6.27
Effect of mixed liquor temperature on the zone settling velocity of activated sludge—pulp
and paper wastewater.

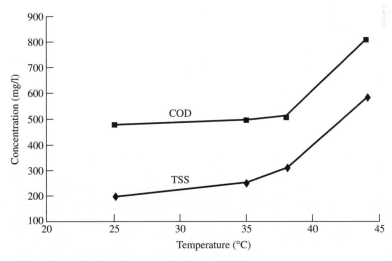

FIGURE 6.28
Effect of mixed liquor temperature on effluent quality.

The F/M is:

$$F/M = \frac{S_o}{X_v t}$$

$$= \frac{2580}{3000 \cdot 6.1}$$

$$= 0.14 \text{ d}^{-1}$$

The aeration basin volume is 6.1 million gal. The heat loss from a concrete basin can be calculated assuming a total basin surface area of 56,650 ft^2.

If we assume the basin temperature at 96°F and an average air temperature of 85°F during the summer months, the heat loss can be estimated:

$$q \text{ (Btu/h)} = UA \Delta T$$

where U = overall heat transfer coefficient, assumed as 0.35 Btu/ft$^2 \cdot$ °F $\cdot$ h
       A = exposed surface area
       ΔT = difference between the basin and air temperature
          = $0.35 \cdot 56{,}650 \cdot 11$
          = 218,102 Btu/h = 5.2×10^6 Btu/d (5.5 KJ/d)

The heat generated can be calculated by assuming that 5000 Btu is generated per pound of COD removed. COD removed is:

$$(2580 \cdot 70) \cdot 1 \cdot 8.34 = 20{,}933 \text{ lb/d } (9500) \text{ kg/d})$$

The Btu generated is:

$$20{,}933 \cdot 5000 \text{ Btu}/16 \text{ COD}_{\text{rem}} = 104.6 \times 10^6 \text{ Btu/d}$$

The net Btu increase is:

$$104.6 - 5.2 = 99.4 \times 10^6 \text{ Btu/d}$$

The temperature increase can be calculated:

$$\frac{99.4 \times 10^6 \text{ Btu/d}}{1 \text{ Btu/lb} \cdot \text{°F} \times 8.34 \times 10^6} = 11.9\text{°F} \quad (6.6\text{°C})$$

Under these conditions, the maximum allowable influent temperature is:

$$96 - 11.9 = 84.1\text{°F} \quad (29\text{°C})$$

Effect of pH

A relatively narrow effective pH range will exist for most biooxidation systems. For most processes, this covers a range of pH 5 to 9 with optimum rates occurring over the range pH 6.05 to 8.5. It is significant to note that this relates to the pH of the mixed liquor in contact with the biological growths and not the pH of the waste entering the system. The influent waste is diluted by the aeration tank contents and is neutralized by reaction with the CO_2 produced by microbial respiration. In the case of both caustic and acidic wastes, the end product is bicarbonate (HCO_3^-), which effectively buffers the aeration system at near pH 8.0. Application of the complete mixing concept is essential in order to take advantage of these reactions.

The amount of caustic that can be present in a wastestream is related to the BOD removal, which in turn will determine the CO_2 produced for reaction with the caustic. If we consider 0.9 lb or kg of CO_2 produced per lb or kg of COD removed (assuming conventional loadings) and 70 percent of this is reactive with caustic alkalinity present, then 0.63 lb or kg of caustic alkalinity (as $CaCO_3$) will be neutralized per pound or kilogram of COD removed.

Since the oxidation of organic acids results in the production of CO_2, the permissible concentration of organic acids in the wastestream is related to the reaction rate of acid degradation to CO_2.

As long as the buffering capacity of the process is maintained, the pH of the aeration tank contents should remain near pH 8.0, even under conditions of fluctuating caustic or acidic loads.

Toxicity

Toxicity in biological oxidation systems may have any of several causes:

1. An organic substance, such as phenol, which is toxic in high concentrations, but biodegradable in low concentrations.
2. Substances such as heavy metals that have a toxic threshold depending on the operating conditions.
3. Inorganic salts and ammonia, which exhibit a retardation at high concentrations.

The toxic effects of organics can be minimized by employing the complete mixing system, in which the influent is diluted by the aeration tank contents and the microorganisms are in contact only with the effluent concentration. In this way wastes with concentrations many times the toxic threshold can be successfully treated.

Heavy metals exhibit a toxicity in low concentrations to biological sludges. Acclimation of the sludge to the metal, however, will increase the toxic threshold considerably.

While an acclimated biological process is tolerant of the presence of heavy metals, the metal will concentrate in the sludge by complexing with the cell wall. Metal concentrations in the sludge as high as 4 percent have been reported. Data on heavy metal removal from petroleum refinery wastewater is shown in Table 6.10. Data on removal of metals at low concentration in a municipal plant is shown in Fig. 6.29.

TABLE 6.10
Heavy metals removal in the activated sludge process treatment of petroleum refinery wastewater

Heavy metal	Activated sludge plant	
	Influent, mg/l	Effluent, mg/l
Cr	2.2	0.9
Cu	0.5	0.1
Zn	0.7	0.4

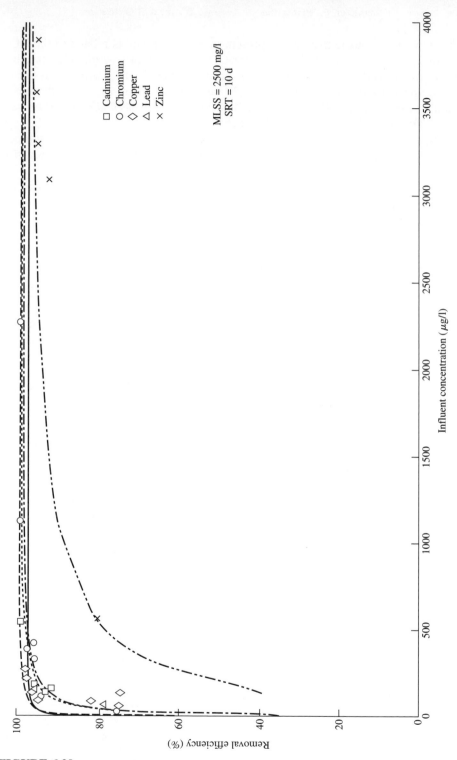

FIGURE 6.29

Heavy metals removal from municipal sewage in the activated sludge process.

This in turn creates problems relative to ultimate sludge disposal. The accumulation of copper on activated sludge with increasing sludge age is shown in Fig. 6.30.

High concentrations of inorganic salts are not toxic in the conventional sense, but rather exhibit progressive inhibition and a decrease in rate kinetics. Biological sludges, however, can be acclimated to high concentrations of salt. Processes are successfully operating with as high as 6 percent salt by weight. Frequently, high salt content will increase the effluent suspended solids. The effect of high salt concentration on suspended solids is shown in Table 6.11, which indicates that most of the biomass was nonflocculated. Monovalent ions such as Na^+ and K^+ will disperse the biological flocs while divalent ions such as Ca^{2+} and Mg^{2+} will tend to aid flocculation.[32]

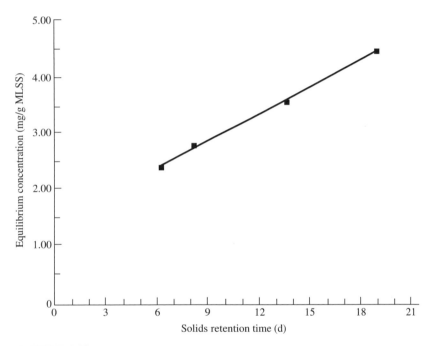

FIGURE 6.30
Copper accumulation in the activated sludge process as a function of solids retention time.

TABLE 6.11
Effect of high salt content on process performance

Influent		Effluent			
COD, mg/l	COD_f, mg/l	COD_s, mg/l	VSS, mg/l[†]	VSS, mg/l[‡]	TDS, mg/l
6437	1182	181	50	597	44,000

[†] 1.5-micron filter
[‡] 0.45-micron filter

6.4
SLUDGE QUALITY CONSIDERATIONS

One of the factors essential to the performance of the activated sludge process is effective flocculation of the sludge, with subsequent rapid settling and compaction. McKinney[12] related flocculation to the food/microorganism ratio and showed that certain organisms normally present in activated sludge deflocculate rapidly under starvation conditions. More recently, it has been shown that flocculation results from the production of a sticky polysaccharide slime layer to which organisms adhere. Flagellates are also entrapped in this slimy material. Filamentous organisms are present in most activated sludges (exceptions are found in the chemical and petrochemical industry).

Palm, Jenkins, and Parker[33] have identified three generic types of activated sludge, as shown in Fig. 6.31. Nonbulking sludge will result from plug flow or selector plant configuration or from complex organic wastewaters. Bulking sludges result from degradable wastewaters treated in a complete mix process or from oxygen or nutrient deficiencies. Pin floc usually results from low F/M (long sludge age) operation.

A number of filamentous organisms have been identified in activated sludge treating municipal and industrial wastewaters. Depending on process operating conditions, one or more of these organisms may dominate in the process, as shown in Table 6.12. Identification and control of filaments have been reported. Filament types found in various industrial wastewaters are shown in Table 6.13.[34] Proper process design and operation should not permit the filaments to overgrow the floc formers.

Filamentous overgrowth is affected by:

1. *Wastewater composition.* Wastewaters containing glucoselike saccharides (glucose, saccharin, lactose, maltose, etc.) promote filamentous growth while laundry, textile, and complex chemical wastewaters inhibit filamentous growth in a completely mixed system. In general, the more readily degradable the substrate, the more prone the system is to filamentous bulking. This can broadly be related to the reaction rate coefficient K as shown in Table 6.14.
2. *Dissolved oxygen concentration.* Oxygen must diffuse into the floc in order to be available to the organisms within the floc. The depth of oxygen penetration within the floc depends on the bulk concentration of oxygen in the surrounding liquid and on the oxygen-utilization rate of the floc. The oxygen-utilization rate is proportional to the organic loading (F/M). Hence, as the organic loading increases, the dissolved oxygen necessary to maintain a fully aerobic floc also increases. The thin filaments (1 to 4 μm) can readily obtain oxygen at concentrations less than 0.1 mg/l. The relationship between dissolved oxygen concentration to maintain a fully aerobic floc and F/M developed from the data of Palm, Jenkins, and Parker[33] is shown in Fig. 6.32. Performance data from a pulp and paper mill is shown in Fig. 6.33.

With degradable substrates at low concentrations, the filaments tend to grow. This explains why complete mix systems with low mixed liquor substrate concentrations favor filamentous growth.

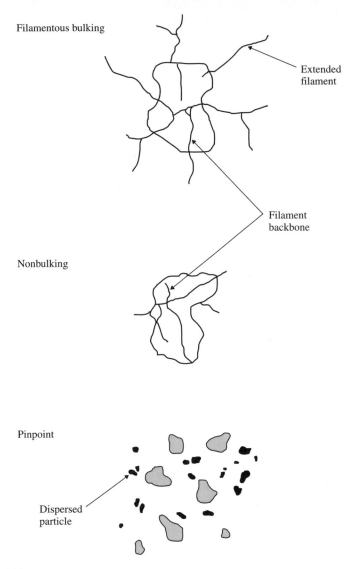

FIGURE 6.31
Activated sludge characteristics.

One of the more common causes of filamentous bulking in industrial waste-waters is inadequate nitrogen or phosphorus. There are numerous examples, particularly in the pulp and paper industry, in which severe filamentous bulking resulted from inadequate nitrogen. This is shown in Fig. 6.34. Restoration of adequate nitrogen restored a flocculated sludge within three sludge ages. Studies on a Wisconsin pulp mill indicated a minimum concentration of NH_3^-N in the effluent of 1.5 mg/l to favor zoogleal growth. Other studies indicate that higher ammonia concentrations may be required in some cases. Minimum soluble phosphorus concentrations in the effluent of 0.5 mg/l have been reported as required for optimal zoogleal growth.

TABLE 6.12
Dominant filament types as indicators of activated sludge operational problems

Suggested causative conditions	Indicative filament types
Low dissolved oxygen (DO)	Type 1701, *S. natans, H. hydrossis*
Low food/mass ratio (*F/M*)	*M. parvicella, H. hydrossis, Nocardia* spp., types 021N, 0041, 0675, 0092, 0581, 0961, and 0803
Septic wastewater sulfides	*Thiothrix* spp, *Beggiatoa* spp., and type 021N
Nutrient deficiencies	*Thiothrix* spp., *S. natans,* type 021N, and possibly *H. hydrossis* and types 0041 and 0675
Low pH	Fungi

TABLE 6.13[34]
Filament types found in industrial wastewaters

	S. natans	Type 1701	*H. hydrossis*	Type 021N	*Thiothrix* I and II	Type 1851	Type 0581	Type 0041	Type 0803	Type 0675	Type 0211	Type 0092	Type 0914	*M. parvicella*	*N. limicola*	Type 0411	*Nocardia*
Food processing and brewing	•	•	•	•	•												•
Textile						•	•	•		•		•					
Slaughterhouse and meat processing	•	•	•	•	•									•			•
Petrochemical	•	•	•	•													•
Organic chemicals				•	•							•		•			•
Pulp and paper mill	•	•	•	•	•		•	•	•	•	•	•	•	•	•	•	•

TABLE 6.14
Composite reaction rate coefficients for industrial wastewaters

Wastewater characteristics	$K_{20°C}$, d^{-1}
Readily degradable (food process, brewery)	16–30
Moderately degradable (petroleum, pulp and paper)	8–15
Poorly degradable (chemical, textile)	2–6

Therefore deficiency in substrates, such as the macro- or micronutrient concentration, residual soluble BOD, and/or dissolved oxygen concentration in the biological floc, can promote filamentous growth and sludge bulking. To illustrate these effects, consider the transfer of dissolved oxygen to the hypothetical biological floc particles, as illustrated in Fig. 6.35. Oxygen must diffuse from the bulk liquid through the floc in order to be available to the organisms within the interior of the floc particle. As it diffuses, it is consumed by the organisms within the floc.

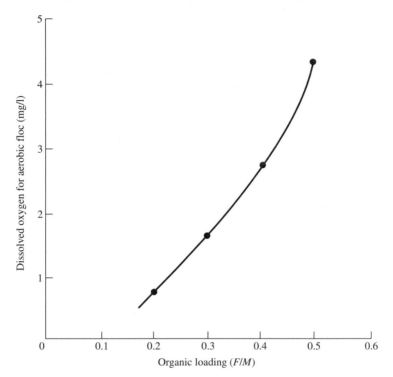

FIGURE 6.32
Relationship between dissolved oxygen and *F/M* for an aerobic floc.[33]

If there is an adequate residual of dissolved oxygen (and nutrients and organics), the rate of growth of the floc-forming organisms will exceed that of the filaments, and a flocculant well-settling sludge will result. If there is a deficiency in any of these substrates, however, the filaments, having a high surface area to volume ratio, will have a "feeding" advantage over the floc formers and will proliferate because of their higher growth rate under the adverse conditions.

Considering Case 1 in Fig. 6.35 at an *F/M* of 0.1 d^{-1}, the oxygen utilization rate is low, and even with a bulk liquid dissolved oxygen concentration of 1.0 mg/l, oxygen will fully penetrate the floc. Under these conditions, the floc formers will outgrow the filaments. In Case 2, the *F/M* is increased to 0.4 d^{-1}, causing a corresponding increase in oxygen uptake rate. If the bulk mixed liquor dissolved oxygen is maintained at 1.0 mg/l, the available oxygen will be rapidly consumed at the periphery of the floc, thus depriving a large interior portion of the floc particle of oxygen. Since the filaments have a competitive growth advantage at low dissolved oxygen levels, they will be favored and will outgrow the floc formers.

Filamentous Bulking Control

The bulk mixed liquor SBOD concentration must be sufficient to provide a driving force to penetrate the biological floc. In a complete mix basin, the concentration of

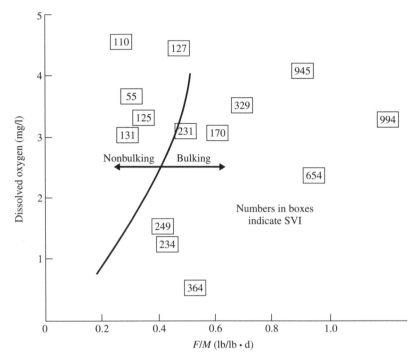

FIGURE 6.33
Relationship between *F/M* and dissolved oxygen relative to sludge bulking for a pulp and paper mill wastewater.

soluble BOD in the mixed liquor is essentially equal to the effluent concentration and is, therefore, low (< 10 mg/l) for readily degradable wastewaters. As a result, substrate penetration of the floc is not achieved, and filaments dominate the interior floc population. In order to shift the population in favor of the floc formers, sufficient driving force must be developed to penetrate the floc and favor their growth. This can be achieved by a batch or plug flow operating configuration in which a high substrate gradient (driving force) exists. Maximum growth of floc formers occurs at the influent end of the plug flow basin or in the initial period of each feed cycle of a batch activated sludge process.

Biological Selectors

A biological selector may be used for filament control instead of a plug flow or batch treatment process. In the selector, a significant portion of the soluble substrate removal occurs by biosorption. Under these conditions, the substrate gradient is high and promotes the growth of floc formers over the filaments since they have a high "sorption" capacity, whereas the filamentous organisms do not. When the wastewater is discharged from the selector to the downstream CMAS, the sol-

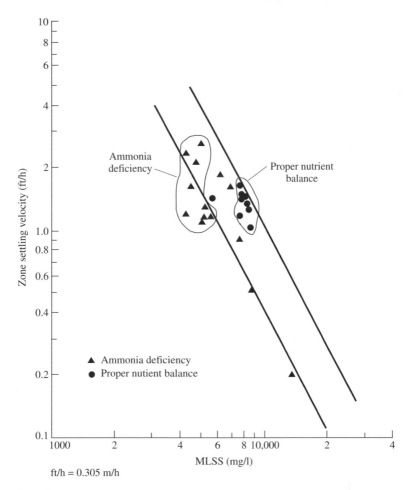

FIGURE 6.34
Effect of ammonia deficiency on zone settling velocity.

uble substrate concentration is relatively low and is available for utilization by both the floc formers and the filaments. The filaments do not predominate in the mixed liquor, however, since the principal mass of substrate removed in the selector has been initially directed to storage and subsequent growth of floc-forming biomass.

Results from parallel treatment studies of a grapefruit processing wastewater are shown in Table 6.15. Reactors 1 and 2 used an aerobic selector followed by a completely mixed aeration basin, whereas reactor 3 used a plug flow regime aeration basin. In all three cases, the SVI was below 100 ml/g, with effluent BOD concentrations ranging from 4 to 18 mg/l. The plug flow reactor, however, produced a sludge that was more readily dewatered and had superior thickening properties. A parallel complete mix activated sludge process (without selector) that was operated at the same organic loading rate as these systems suffered severe bulking problems and was shut down because it was inoperable.

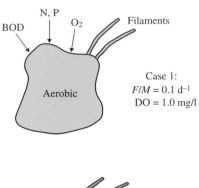

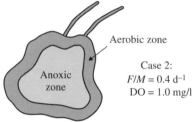

FIGURE 6.35
A mechanism of sludge bulking.

Design of Aerobic Selectors

Several methods have been proposed for process design of the aerobic biological selector. Each of these is based on the selector *F/M* or the floc loading relationship for the wastewater-sludge mixture, as defined by Eq. (6.4). The design objective is to provide sufficient biomass-wastewater contact time to remove a significant portion of the influent degradable substrate. If 60 to 75 percent of the influent degradable substrate is sorbed in the selector, then the subsequent metabolism and growth of the floc formers is usually adequate to establish a well-settling sludge. If less degradable substrate is removed because of an excessive floc load, then higher concentrations "leak" into the activated sludge reactor and support filamentous growth.

   The results of multiple batch floc load tests on a readily degradable pulp and paper mill wastewater are shown in Fig. 6.36. These data are correlated according to the relationship

$$\frac{S}{S_o} = e^{-KFL^{-1}}$$

as shown in Fig. 6.37. These data were used to select a floc loading of 100 to 150 mg COD/g VSS for operation of the aerobic selector of a bench-scale selector CMAS system.

   Equation (6.4) can be rearranged to include mixed liquor recycle

$$FL = \frac{S_o}{rX_R + r_RX_v} \tag{6.38}$$

TABLE 6.15
Treatment of a grapefruit processing wastewater

| | Aerobic selector[†] with CMAS | | Plug flow |
	Reactor 1	Reactor 2	activated sludge
Operating characteristics:			
Influent BOD, mg/l	2543	3309	3309
Effluent BOD, mg/l	18	4	6
Influent COD, mg/l	4768	4460	4460
Effluent COD, mg/l	221	139	135
MLVSS, mg/l	3431	5975	5333
SRT, d	7.2	13.2	13.5
Temperature, °C	22	22	22
SVI, ml/g	71	67	69
F/M, d^{-1}	0.32	0.24	0.20

Sludge characteristics:

| | Limiting flux to achieve a 1.5% underflow | |
	Solids flux without polymer, lb/ft$^2 \cdot$ d	Solids flux with polymer, lb/ft$^2 \cdot$ d
Selector with CMAS	5.5	47
Plug flow	24.5	62

Specific Resistance:

	Polymer dosage, lb/ton	Specific resistance, s^2/g
Selector with CMAS	3.2	190×10^6
Plug flow	2.4	104×10^6

[†] Floc load = 120 mg COD/g VSS.

where S_o = the degradable COD, mg/l
   X_R = recycle VSS, mg/l
   X_v = mixed liquor VSS, mg/l
   r = sludge recycle ratio, R/Q
   r_R = internal recycle ratio, $Q_{R/Q}$

The aerobic selector design is shown in Example 6.11.

EXAMPLE 6.11. Aerobic selector design for industrial wastewaters. Biosorption correlations are shown in Fig. 6.36. The design basis is 65 percent removal of sorbable COD in the selector. For the recycle paper case, design an aerobic selector for an influent COD of 2000 mg/l and an MLVSS of 4000 mg/l. The floc load is 150 mg COD/g VSS. The recycle sludge concentration is 8000 mg/l VSS.

Solution.

$$r = \frac{X_v}{X_r - X_v} = \frac{4000}{8000 - 4000} = 1.0$$

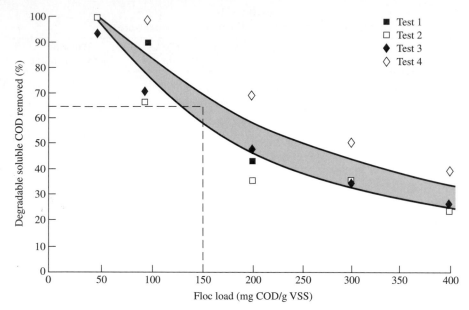

FIGURE 6.36
Floc load test results for a pulp and paper mill wastewater.

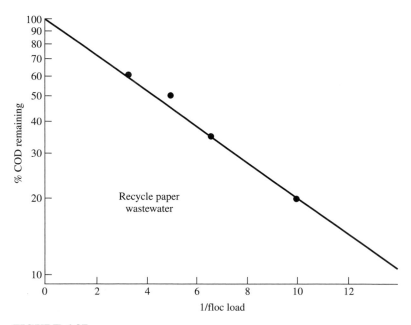

FIGURE 6.37
COD removal—floc load relationship.

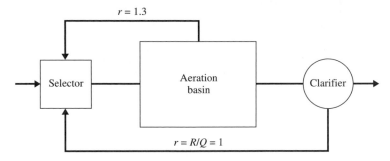

FIGURE 6.38
Selector flow sheet.

Rearranging Eq. (6.38) gives

$$r_R = \frac{S_o - \mathrm{FL}_r X_R}{\mathrm{FL} X_v}$$

$$= \frac{2000 - 0.15 \cdot 1 \cdot 8000}{0.15 \cdot 4000}$$

$$= 1.3$$

If the return sludge recycle rate is 100 percent, the internal recycle will be 130 percent. This is shown in Fig. 6.38. The selector detention time based on Q will be 0.825 h.

The sorption phenomenon provides the basis for selector design such that sufficient organics are sorbed for subsequent utilization and growth by the floc-forming organisms in preference to the growth of filamentous organisms.[35] After sorption is complete, a minimum aeration period must be provided to oxidize the sorbed organics. This is illustrated by plug flow data for a grapefruit processing wastewater, as shown in Fig. 6.39. When these date are replotted, the minimum required downstream aeration time can be determined as shown in Fig. 6.40. Metabolism of the sorbed substrate was completed at a substrate removal velocity of approximately 0.8 d^{-1}. The required minimum aeration time can then be calculated as

$$0.8 = \frac{S_r}{X_v t}$$

$$t = \frac{3600 \text{ mg/l}}{3618 \text{ mg/l} \cdot 0.8/\mathrm{d}}$$

$$= 1.24 \text{ d}$$

Many filamentous organisms are aerobic and can be destroyed by prolonged periods of anaerobiosis. Most of the bacteria, on the other hand, are facultative and can exist for extended periods without oxygen. Although available data are somewhat contradictory, it would appear that anaerobic or anoxic conditions maintained within the process will restrict the growth of these filaments. Marten and Daigger[36] recommend an anoxic selector F/M of 0.8 to 1.2 lb BOD/lb MLSS per

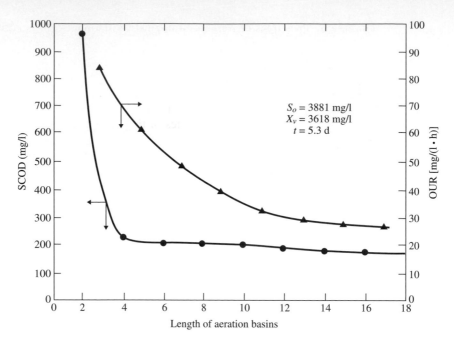

FIGURE 6.39
Change in SCOD and OUR in a plug plow aeration basin.

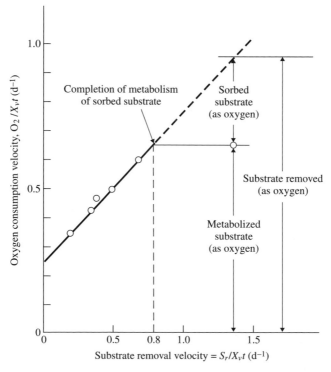

FIGURE 6.40
Stabilization time required for sorbed substrate.

day at temperatures greater than 18°C and 0.7 to 1.0 lb BOD/lb MLSS per day at temperatures less than 18°C.

The use of an anoxic zone was studied at an organic chemicals plant. Three systems were run in parallel: a conventional air system, an oxygen system, and an air system with an anoxic zone. The settling properties of the respective sludges indicated that the oxygen system yielded the best settling sludge, while the air system with insufficient dissolved oxygen demonstrated severe filamentous bulking. The anoxic-aerobic system was almost devoid of filaments, but exhibited poorer flocculation and settling than the oxygen system. These results are shown in Table 6.16.

Viscous bulking occurs when the activated sludge contains an excessive amount of extracellular biopolymers which impart a slimy, jellylike consistency to the sludge and the sludge becomes highly water retentive.[37] This hydrous sludge exhibits low settling and compaction velocities. This phenomenon has been reported to be caused by a lack of nutrients/micronutrients or the presence of a toxic compound. Low dissolved oxygen has also been shown to contribute to viscous bulking.

Chlorine or hydrogen peroxide can be added to the return sludge or the aeration basin to reduce bulking.[38] Hydrogen peroxide is selective for some filament types. Hydrogen peroxide dosages are in the order of 20 to 50 mg/l. Chlorine dosages may vary from 9 to 10 lb Cl_2/(d · 1000 lb MLSS [9 to 10 kg Cl_2/(d · 1000 kg MLSS)] in cases of severe bulking and 1 to 2 lb Cl_2/(d · 1000 lb MLSS) [1 to 2 kg Cl_2/(d · 1000 kg MLSS)] in cases of moderate bulking. Because of the growth rate of the filaments, if the hydraulic retention time exceeds 8 h, chlorination must be applied directly to the aeration basin. Since the filaments exhibit a high negative zeta potential, they can be flocculated by the addition of cationic polyelectrolytes. This treatment is expensive and the filaments are not destroyed in the process.

Nocardia foams cause significant problems in plant operation. *Nocardia* actinomycetes are slower-growing organisms compared to floc-forming organisms. The growth of *Nocardia* can be controlled by reducing the SRT to less than 3 d to create washout conditions. These short SRTs are usually not feasible if nitrification is

TABLE 6.16
Sludge characteristics under various operating conditions

System	ZSV, ft/hr	SVI, ml/g	MLSS, mg/l	F/M,[†] g/g · d	Temperature, °C	Effluent TSS, mg/l
Low DO	0.6	222	3500	0.44	33	120
Anoxic	2.0	116	3850	0.44	34	50
	2.0	129	3600	0.35	40	140
	1.4	228	2800	0.44	45	130
High DO	3.6	100	3450	0.31	35	40
	3.5	96	3350	0.61	39	160
	5.5	90	3200	0.55	45	200
Low loading	3.2	133	2700	0.28	24	20
	4.2	133	2200	0.25	35	275

[†] BOD basis.

to be employed. Aerobic selectors or anoxic zones have been successful at controlling *Nocardia* foaming when operated at low SRTs. These methods are less effective at controlling *M. parvicella*. Successful control of biological foaming problems by applying powdered hypochlorite or a hypochlorite spray directly on the foam has been reported.

6.5
SOLUBLE MICROBIAL PRODUCT FORMATION

Soluble microbial products (SMP) are generated in the activated sludge process through the biodegradation of organics and through endogenous degradation of the biomass. The SMP are oxidation by-products that are nondegradable. Pitter and Chudoba[39] have indicated that, depending on cultivation conditions, the nonbiodegradable waste products can amount to 2 to 10 percent of the COD removed. The COD, BOD, and SMP_{nd} relationships for several industrial wastewaters are summarized in Table 1.8. Data for biodegradation of a peptone-glucose mixture and a synthetic fiber wastewater are shown in Fig. 6.41. They indicate that approximately 0.20 mg of nondegradable TOC (TOC_{nd}) was produced per mg of influent TOC for the synthetic fiber wastewater. The peptone-glucose–containing waste-

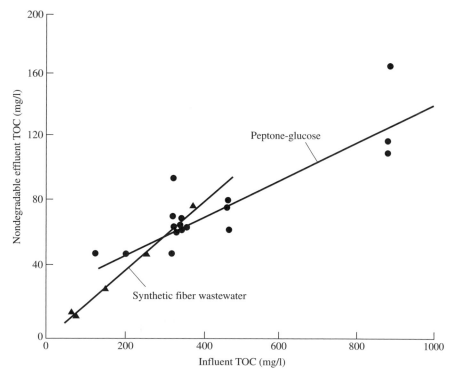

FIGURE 6.41
Nondegradable TOC as related to influent TOC.

water produced approximately 0.12 mg TOC_{nd} per mg influent TOC. In both cases, the ratios were constant over the range of influent loading conditions. This indicates that there was a constant metabolic by-product or a portion of the original substrates that was nondegradable.

Many of the metabolic by-products are of high molecular weight. The molecular weight distribution (expressed as the TOC and COD fractions) of the influent and biologically treated effluent from a plastic additives wastewater and a biologically treated effluent from a glucose-based synthetic wastewater are presented in Table 6.17. Pitter and Chudoba[39] have indicated that approximately 75 percent of the SMP_{nd} from treatment of a wastewater containing only phenol had a molecular weight above 1000. It has been further determined that some of these high-molecular-weight fractions are toxic to some aquatic organisms. Results showing the aquatic toxicity of several wastewaters from the plastics industry before and after biological oxidation are shown in Fig. 6.42. They indicate that biological treatment reduced the TOC and toxicity in most cases. In two wastewaters, however, the TOC was reduced by 32 and 78 percent, but the treated effluent was more toxic than the influent wastewater, making the oxidation by-products suspect toxicants. There is reason to believe that the high-molecular-weight SMP_{nd} strongly adsorb on activated carbon. This characteristic makes granular-activated carbon (GAC) or powdered-activated carbon (PAC) an excellent candidate process for toxicity reduction when toxicity is caused by SMP_{nd}. Inhibition to nitrification by SMP_{nd} is shown in Fig. 6.43 from the data of Chudoba.[40]

6.6
BIOINHIBITION OF THE ACTIVATED SLUDGE PROCESS

Many organics will exhibit a threshold concentration at which they inhibit the heterotropic and/or nitrifying organisms in the activated sludge process. Inhibition has been defined by the Haldane equation (or its modifications) using Monod kinetics:

$$\mu = \frac{\mu_m S}{S + K_S + S^2/K_I} - b \qquad (6.39)$$

in which K_I is the Haldane inhibition coefficient.

An example of bioinhibition by a plastics additives wastewater is shown in Fig. 6.44. The extent of bioinhibition is expressed as the ratio of the SOUR at the

TABLE 6.17
Molecular weight distribution of a biological effluent

| Molecular weight | Plastic additives wastewater | | Glucose[37] |
	Influent TOC, %	Bioeffluent TOC, %	wastewater COD, %
> 10,000	—	11.5	45
500–10,000	—	14.5	16
< 500	100	74.0	39

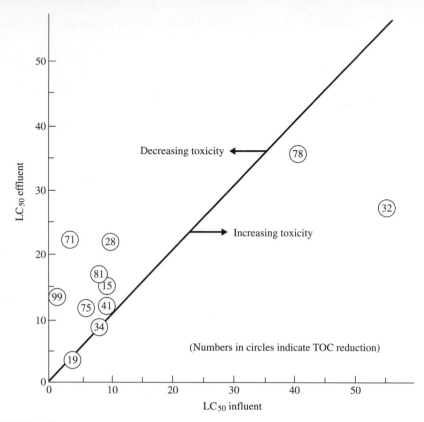

FIGURE 6.42
Effect of SMP on effluent toxicity for wastewaters from the plastics and dyestuffs industry.

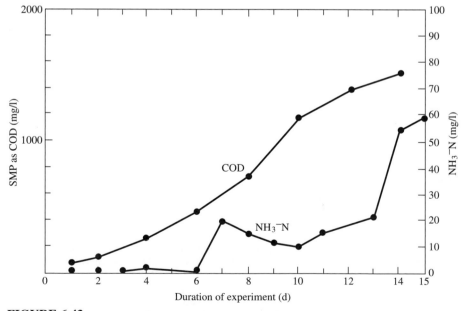

FIGURE 6.43
Relationship between SMP as COD on ammonia buildup.[40]

264

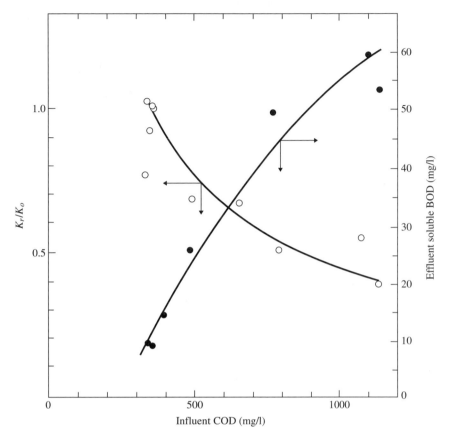

FIGURE 6.44
Activated sludge inhibition from a plastics additives wastewater.

selected influent loading concentration (K_r) to the SOUR at the no-observed-effect loading (K_o). As the concentration of influent COD (and inhibitory agent) increases, the SOUR decreases, resulting in higher effluent SBOD concentrations. In this case, the inhibition was removed by pretreating the wastewater with hydrogen peroxide (H_2O_2), thereby effecting detoxification and enhanced biodegradability. These effects are shown in Fig. 6.45. Addition of powdered activated carbon to the mixed liquor to adsorb the toxicant was also shown to reduce the inhibition.

Several acidic, aromatic, and lipid-soluble organic compounds have been demonstrated to "uncouple" oxidative phosphorylation. The result of this uncoupling effect is uncontrolled respiration and oxidation of primary substrates and intracellular metabolites. At low concentrations, uncoupling is evidenced by highly elevated oxygen utilization rates but no effect on cell growth or substrate removal. At higher concentrations, inhibition and toxicity are demonstrated by dramatic reductions in both the oxygen utilization rate and cell growth.

Volskay and Grady[41] and Watkin[42] have shown that inhibition can be competitive (the inhibitor affects the base substrate utilization), noncompetitive (the inhibitor

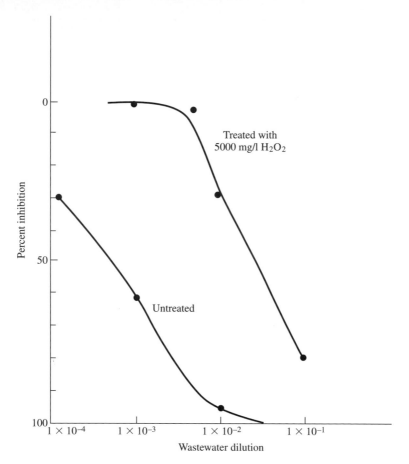

FIGURE 6.45
Detoxification of a plastics additives wastewater.

rate is influenced), or mixed, in which both rates are influenced. The effects of substrate and inhibitor concentrations on the respiration rate of a microbial culture expressed as a fraction of the rate in the absence of the inhibitor have been shown by Volskay and Grady.[41]

While the relationships described above define the mechanism of inhibition, they are of limited use in evaluating industrial wastewaters. In most cases, the inhibitor itself is not defined, variable sludge and substrate composition will influence inhibition, and interactions will frequently exist between inhibitors. The inhibition constant K_I is highly dependent on the specific enzyme system involved, which, in turn, is dependent on the history and population dynamics of the sludge. In some cases, the inhibition constant may be dependent on the particular metabolic pathways that are present in any given microbial population. For example, Watkin and Eckenfelder[43] showed a variation in K_I of 6.5 to 40.4 for different sludges and operating conditions treating 2,4-dichlorophenol and glucose. Volskay and Grady[38] showed a variation of 2.6 to 25 mg/l in the concentration of pentachlorophenol, which would cause 50 percent inhibition of oxygen utilization rates.

It is apparent, therefore, that each wastewater must be independently evaluated for its bioinhibition effects. Several protocols have been developed for this purpose. These are the fed-batch reactor (FBR) of Philbrook and Grady[28] and Watkin and Eckenfelder,[43] the OECD method 209 of Volskay and Grady,[41] and the glucose inhibition test of Larson and Schaeffer.[44] Depending on the particular wastewater, one or more of these test protocols will be applicable. Each of these methods is discussed below.

OECD Method 209

The OECD method 209 involves measurement of activated sludge oxygen uptake rate from a synthetic substrate to which the test compound has been added at various concentrations. The oxygen uptake rate is measured immediately after addition of the test compound and after 30 min of aeration. The EC_{50} value is determined as the concentration of test compound at which the oxygen uptake rate (at 30 min) is 50 percent of the uninhibited oxygen uptake rate. The OECD method uses 3,5-dichlorophenol as a reference toxicant to ensure that the test is working properly and that the biomass has the appropriate sensitivity. The reference EC_{50} value should be between 5 and 30 mg/l for the test to be valid.

Volskay and Grady[41] employed a modified OECD method to determine the toxicity of selected organic compounds. Since many of the compounds were volatile, the test was modified by using more dilute cell and substrate concentrations and by conducting the test in vessels that were sealed by the insertion of a polyfluoroethylene plug. This protocol is recommended for wastewaters containing high concentrations of volatile organics.

Fed Batch Reactor (FBR)

Fed batch reactors have been used to determine nitrification kinetics and removal kinetics of specific pollutants in activated sludge. The essential characteristics of the FBR procedure are that:

1. Substrate is continuously introduced at a sufficiently high concentration and low flow rate so that the reactor volume is not significantly changed during the test.
2. The feed rate exceeds the maximum substrate utilization rate.
3. The test duration is short and therefore allows simple modeling of biological solids growth.
4. Acclimated activated sludges are used.

A schematic diagram of the FBR is shown in Fig. 6.46. Two liters of mixed liquor are placed in the reactor, and a sample is taken for determination of oxygen utilization rate (OUR) and mixed liquor volatile and total suspended solids prior to the start of the feed flow. The feed is introduced at a flow rate of 100 ml/h, and aliquots of the reactor contents are withdrawn every 20 min for the duration of the 3-h test. The OUR is determined in situ every 30 min during the test. Suspended solids determinations are made every hour during the test.

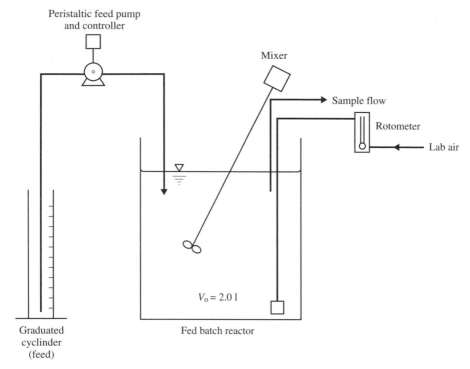

FIGURE 6.46
Fed batch reactor (FBR) configuration.

As discussed under the OECD protocol, the oxygen uptake rate will decrease in the presence of inhibition. As long as there is no inhibition, the OUR will remain constant at a maximum rate. The same limitations apply to the FBR protocol as to the OECD method.

The theoretical responses in a fed batch reactor to both inhibitory and noninhibitory substrates are depicted in Fig. 6.47. In the case where substrate is added at a sufficiently high mass and low volumetric flow rate, the maximum substrate utilization rate will be exceeded, and the change in reactor volume will be insignificant. If the FBR volume change is negligible and the mass feed rate exceeds the maximum substrate utilization rate, then a substrate concentration buildup will result in the reactor with time. Noninhibitory substrate response results in a linear residual substrate buildup in the reactor with time. The maximum specific substrate utilization rate q_{max} is calculated as the difference in slopes between the substrate feed rate and the residual substrate buildup rate divided by the biomass concentration. In the case of inhibition, substrate utilization would rapidly decrease, resulting in an upward deflection of the residual substrate concentration curve as shown in Fig. 6.47. As inhibition progresses and acute biotoxicity occurs, the trace of the residual substrate concentration should become parallel to the substrate feed rate. The inhibition constant K_I can be approximated by identifying the inhibitor concentration at the midpoint of the curvilinear portion of the substrate response.

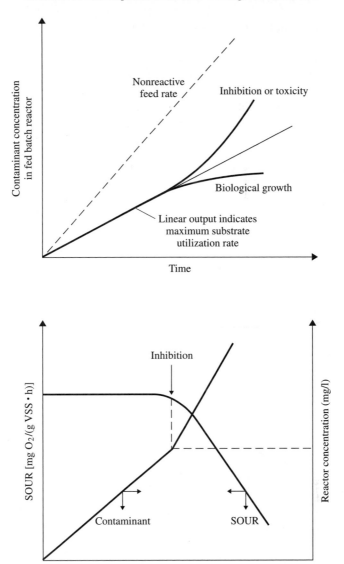

FIGURE 6.47
Theoretical fed batch reactor output with influent substrate mass flow rate greater than $q_{max} \cdot X_v$ and inhibition effects.

Glucose Inhibition Test

Larson and Schaeffer[44] developed a rapid toxicity test based on the inhibition of glucose uptake by activated sludge in the presence of toxicants. The test was subsequently modified for application to a variety of industrial wastewaters. The procedure is as follows:

1. Place 10 ml of sample into a centrifuge tube.
2. Add 10 ml of the stock glucose.
3. Add 10 ml of activated sludge to the centrifuge tube and aerate at a low rate.
4. After 60 min, add two drops of HCl and transfer tubes to the centrifuge.
5. Measure glucose concentration.
6. Sludge control—substitute 10 ml of deionized water for the sample in step 1 and perform steps 2 through 5 as before.
7. Glucose control—place 30 ml of deionized water in a centrifuge tube. Add 1 ml of stock glucose solution. Do not add sludge or aerate. Add two drops of HCl and measure glucose uptake.

The percent inhibition is calculated as follows:

$$\text{Percent inhibition} = \left[\frac{C - C_B}{C_o - C_B} \right] 100$$

where C = final glucose concentration in sample solution
           C_B = final glucose concentration in sludge control sample
           C_o = initial glucose concentration (glucose control)

Inhibition effects using the glucose test are presented in Fig. 6.48 for an organic chemicals wastewater.

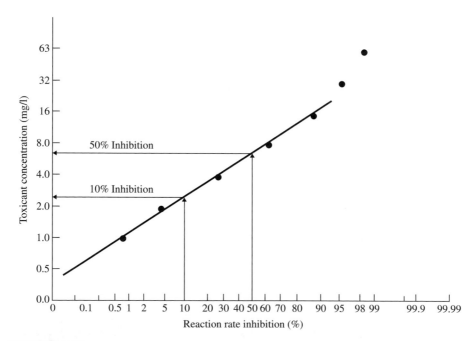

FIGURE 6.48
Determination of bioinhibition effects from the glucose inhibition test.

6.7
STRIPPING OF VOLATILE ORGANICS

Volatile organic compounds (VOCs) will be stripped from solution by oxygenation in the activated sludge process. Depending on the particular VOC, both air stripping and biodegradation may occur.

The fraction of the VOC that is stripped depends upon several factors. Compound specific factors include the Henry's law constant, the compound's biodegradation rate, and in some cases, the initial concentration of the compound and other substrates. Operational and facility design factors that influence stripping are the method of oxygenation and the power level in the aeration basin. The concentration of nonvolatile organics in the wastewater will affect the composition and mass of the sludge, the operating SRT, and hence, the fraction of VOC that is biodegraded and stripped. Biodegradation and stripping removal data for a range of VOCs are presented in Table 6.18. As a general rule, as more halogen atoms are added to the organic compound, the rate of biodegradation decreases, and the amount of compound stripped increases. This is shown in Fig. 6.49 for several chlorinated compounds in the benzene series. It should be noted that a greater SRT will usually result in less stripping since a lower power level will be employed and

TABLE 6.18
Fate of selected VOCs in the activated sludge process

Compound	Influent concentration, mg/l	SRT, d	Amount stripped, %
Toluene	100	3	12–16
	0.1	3	17
	40	3	15
	40	12	5
	0.1	6	22
Nitrobenzene	0.1	6	<1
Benzene	153	6	15
	0.1	6	16
Chlorobenzene	0.1	6	20
1,2-Dichlorobenzene	0.1	6	59
1,2-Dichlorobenzene	83	6	24
1,2,4-Trichlorobenzene	0.1	6	90
o-Xylene	0.1	6	25
1,2-Dichloroethane	150	3	92–96
1,2-Dichloropropane	180	6	5
Methyl ethyl ketone	55	7	3
	430	7	10
1, 1, 1-Trichloroethane	141	6	76

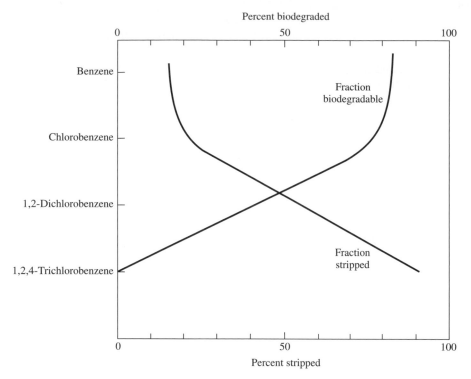

FIGURE 6.49
Relationship between biodegradation and stripping for the chlorinated benzene series.[1]

the biomass concentration will be higher. The effects of power level and the type of aeration device on the stripping of benzene are shown in Fig. 6.50, which was developed assuming an influent COD of 250 mg/l and an influent benzene concentration of 10 mg/l.

In a diffused air oxygenation system, equilibrium between the gas and liquid phases is quickly reached after the bubble is formed. Under these conditions, the amount of VOC that is stripped depends primarily on the gas to liquid ratio. Since, under most conditions, the volume of gas is relatively small, stripping is minimal. In a mechanical surface aeration system, however, the volume of gas in contact with the liquid is nearly infinite (i.e., the atmosphere), and hence, a greater quantity of VOC will be stripped.

The fraction emitted to the air of any volatile component entering an activated sludge aeration basin can be derived from the mass balance Eq. (6.39) neglecting the effects of adsorption on the biological floc:

$$Q_o C_{o,i} = Q_o C_{L,i} + r_i + r_{vi} \tag{6.40}$$

where Q_o = flow rate, l/s
          $C_{o,i}$ = influent concentration of VOC i, g/l
          $C_{L,i}$ = liquid effluent concentration of component i, g/l
          r_i = rate of biodegradation of component i, g/s
          r_{vi} = rate of volatilization of component i, g/s

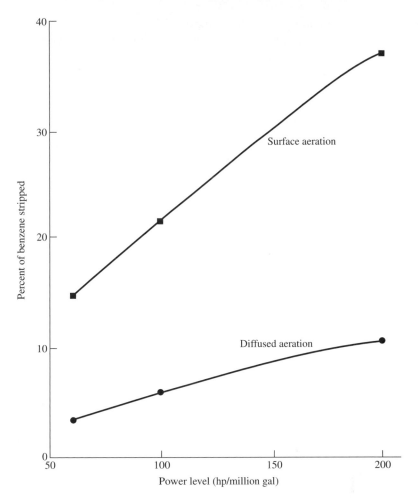

FIGURE 6.50
Stripping of benzene as related to power level and type of aerator.

For a complete mix system, the fraction of the influent VOC load that is emitted to the air (f_{air}) is given by

$$f_{air} = \frac{r_{vi}}{Q_o C_{L,i} + r_i + r_{vi}} \tag{6.41}$$

When individual compounds, pure or in a mixture, are considered, it is appropriate to use a Monod kinetic model to describe their biodegradation. The concentrations of priority pollutants normally encountered in a complete mix reactor are in the $\mu g/l$ range in order to comply with current regulations. Under these conditions, the Monod model reduces to a first-order rate expression, and the biodegradation rate of compound i (expressed as COD) is given by

$$r_i = W_i q_m X_{vb} V C_{L,i} / K_S \tag{6.42}$$

where W_i = weighting factor for compound i
          q_m = biodegradation rate constant for compound i, g COD/(g VSS · s)
          X_v = volatile solids concentration, g/l
          V = volume of reactor, l
          $C_{L,i}$ = concentration of compound i in the reactor, g/l
          K_S = Monod's half saturation constant for component i, g/l

The biomass is considered to be a mixed culture with each organic substrate being the growth-limiting factor for a specific microbial population. Under these circumstances, the fraction of biomass dedicated to a specific substrate would be proportional to the biomass yield associated with this substrate. Hence, the weighting factor would be

$$W_i = a_i C_{B,i} \left[\sum_{i=1}^{n-1} a_i C_{B,i} \right]^{-1} \tag{6.43}$$

where $C_{B,i}$ = concentration of component i that is biodegraded, g/l, and a_i = yield coefficient of component i, g VSS/g COD.

Stripping of VOC is computed by using the method of Roberts et al.[45] This model estimates stripping with surface or diffused aeration on the premise that the transfer rates for volatile solutes are proportional to one another. Since oxygen satisfies the volatility criteria and because a large mass-transfer database for it is available, oxygen has been selected as the reference compound. The overall mass-transfer coefficient for compound i is then proportional to that for oxygen at the operating conditions.

$$(K_L a)_i = \psi (K_L a)_{O_2} \tag{6.44}$$

It has been shown that the proportionality constant (ψ) depends on the liquid-phase diffusivity ratio, D_i/D_{O_2}, and is approximately constant over a wide range of temperature and mixing conditions. Values of ψ are virtually identical in clean water and wastewater, indicating that the transfer rates of dissolved oxygen and organic solutes are inhibited to the same degree as wastewater constituents.

The overall gas transfer coefficient $(K_L a)_{O_2}$ is related to the standard oxygenation rate:

$$(K_L a)_{O_2} = \frac{SOR \cdot P}{C_s \cdot V} \tag{6.45}$$

where C_s = oxygen solubility in clean water, g/l
          P = aeration basin power level, hp
          V = aeration basin volume, l
          SOR = standard oxygenation rate, g O_2/hp · h

For surface aeration, the stripping of VOCs can be computed from Eq. (6.45):

$$r_{vi} = \Psi (K_L a)_{O_2} \cdot C_{L,i} V \tag{6.46}$$

and the fraction of VOC emitted to the air becomes

$$f_{air} = \frac{\Psi K_L a_{O_2} V}{Q_o + (W_i q_m X_v V / K_s) + \Psi K_L a_{O_2} V} \tag{6.47}$$

For diffused air, assuming the exit air is in equilibrium with the liquid, the fraction of VOC remaining in the liquid phase is

$$r_{vi} = Q_{air} H_c C_{L,i} \tag{6.48}$$

in which $(H_c)_i$ = Henry's constant for compound i. The fraction of VOC emitted using diffused aeration at an airflow rate of Q_{air} is

$$f_{air} = \frac{Q_{air} H_c}{Q_o + (W_i q_m X_{vb} V / K_s) + Q_{air} H_c} \tag{6.49}$$

Example 6.12 illustrates the calculation of VOC emissions.

EXAMPLE 6.12. Determine the fraction of benzene emitted to the air under the following loading conditions for mechanical aeration and diffused aeration. The mechanical aeration system uses 60 hp/million gal and has a $(K_L a)_{O_2} = 1.52/h$, and the diffused aeration system has an airflow rate of 2.16 m³/s.

Solution.

$V = 3846 \text{ m}^3 \ (136{,}000 \text{ ft}^3)$
$Q_o = 0.178 \text{ m}^3/\text{s} \ (2820 \text{ gal/min})$
$C_o = 10 \text{ mg/l}$
$X_{vb} = 3000 \text{ mg/l}$
$S_o = 250 \text{ mg COD/l}$

$$r_i = \frac{W_i q_{mi} X_v V C_{L,i}}{K_{si}}$$

$$W_i = \frac{a_i f_{bio,i} C_{o,i}}{a(S_o - S)} \ \frac{3.08 \text{ mg COD}}{\text{mg benzene}}$$

Assuming $a_i = a$ and $f_{bio,i} = 0.87$,

$$W_i = \frac{0.87 \times 10 \times 3.08}{(250 - 20)} = 0.117$$

then for mechanical aeration,

$$f_{air,i} = \frac{0.6 \times (1.52/3600) \times 3846}{0.178 + \dfrac{0.117 \times 5.78 \times 10^{-6} \times 3000 \times 3846}{1} + 0.974}$$

$$f_{air,i} = \frac{0.974}{0.178 + 7.803 \times 0.974} = \frac{0.974}{8.955} = 0.109 = 11\%$$

Check $f_{bio,i}$:

$$f_{bio,i} = \frac{7.803}{8.955} = 0.871 = 87\% \qquad \text{Checks}$$

For diffused aeration, assume 93 percent biodegradation.

$$W_i = \frac{0.93 \times 10 \times 3.08}{(250 - 20)} = 0.125$$

$$f_{air,i} = \frac{2.16 \times 0.225}{0.178 + \dfrac{0.125 \times 5.78 \times 10^{-6} \times 3000 \times 3846}{1} + 0.486}$$

$$f_{air,i} = \frac{0.486}{0.178 + 8.336 + 0.486} = \frac{0.486}{9.0} = 0.054 = 5.4\%$$

$$f_{bio,i} = \frac{8.336}{9.0} = 0.926 = 93\% \qquad \text{Checks}$$

$S_e = 20$ mg COD/L
$\psi = 0.6$
$q_m = 5.78 \times 10^{-6}$/s
$K_S = 1.0$ mg/l
$H_c = 0.225$

Treatment of VOC Emissions

If the off-gas from the aeration basin contains high concentrations of VOC or ammonia, air quality standards may require off-gas collection and treatment. Several treatment technologies are available, including thermal and catalytic incineration, carbon adsorption, macroreticular resin adsorption, and biological degradation. The removal performance of several of these technologies is shown in Fig. 6.51 as a function of the gas phase VOC concentration. In most cases of activated sludge treatment, the gas phase VOC concentration will be low (< 50 ppmv), and incineration would be required to provide the highest removal efficiency.

Biological treatment of the off-gas is a less costly alternative to incineration and can be done with static beds of compost material such as peat or corn silage. Granular activated carbon can be added to the bed material to adsorb poorly degradable VOCs. The beds are lightly seeded with biomass and maintained at proper pH, temperature, and moisture content to support biological activity. The off-gas is passed through the bed at surface flow rates of 1 to 10 ft³/(min · ft²)(0.31 to 3.1 m³/min · m²) to provide 2 to 12 min of bed contact time. The biomass concentration depends on the type of bed material and organic loading rates.

6.8
NITRIFICATION AND DENITRIFICATION

Nitrification

Nitrogen changes in a biological-treatment process are shown in Fig. 6.52. Nitrification is the biological oxidation of ammonia to nitrate with nitrite formation as

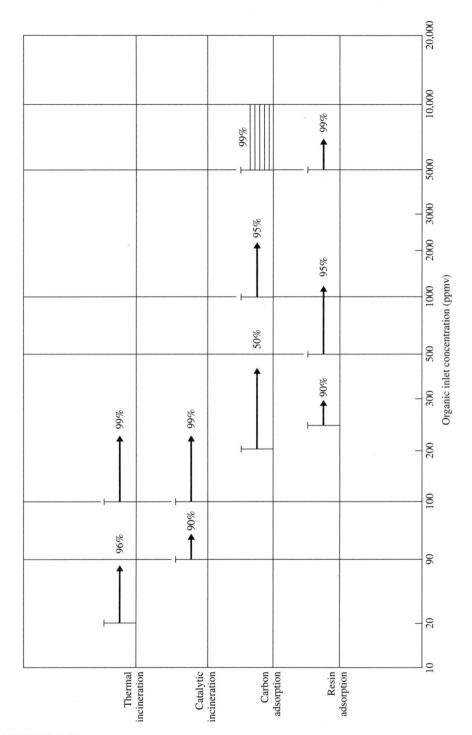

FIGURE 6.51
Performance capabilities of VOC control devices.

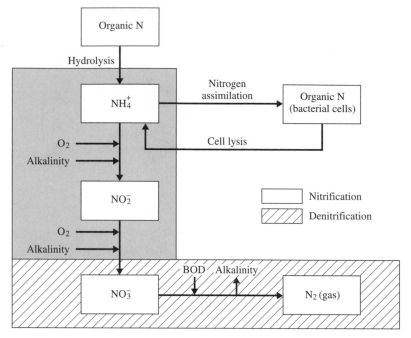

FIGURE 6.52
Nitrogen transformations.

an intermediate step. The microorganisms involved are the autotrophic species *Nitrosomonas* and *Nitrobacter,* which carry out the reaction in two steps:

$$2NH_4{}^+ + 3O_2 \rightarrow 2NO_2{}^- + 4H^+ + 2H_2O \qquad (Nitrosomonas)$$

$$2NO_2{}^- + O_2 \rightarrow 2NO_3{}^- \qquad (Nitrobacter)$$

The cell yield for *Nitrosomonas* has been reported as 0.05 to 0.29 mg VSS/mg $NH_3{}^-N$ and for *Nitrobacter* 0.02 to 0.08 mg VSS/mg $NH_3{}^-N$. A value of 0.15 mg VSS/mg $NH_3{}^-N$ is usually used for design purposes. It is generally accepted that the biochemical reaction rate of *Nitrobacter* is faster than the reaction rate of *Nitrosomonas* and hence there is no accumulation of nitrite in the process and the reaction rate of *Nitrosomonas* will control the overall reaction. Poduska[46] reviewed the results on the effect of micronutrients on the growth of nitrifying bacteria in pure culture as shown in Table 6.19.

Nitrification Kinetics

In order to maintain a population of nitrifying organisms in a mixed culture of activated sludge, the minimum aerobic sludge age $(\theta_c)_{min}$ must exceed the reciprocal of the nitrifiers' net specific growth rate:

$$(\theta_c)_{min} \geq \frac{1}{\mu_{N_T} - b_{N_T}} \qquad (6.50)$$

TABLE 6.19
Stimulatory concentration of micronutrients for nitrifying bacteria[46]

Compound	Concentration, μg/l
Calcium	0.5
Copper	0.005–0.03
Iron	7.0
Magnesium	12.5–0.03
Molybdenum	0.001–1.0
Nickel	0.1
Phosphorus	310.0
Zinc	1.0

where μ_{N_T} = nitrifiers' specific growth rate, d^{-1}, and b_{N_T} = nitrifiers' endogenous decay rate, g ΔVSS$_N$/(g VSS$_N \cdot$ d). The nitrifiers' specific growth rate is related to the specific nitrification rate

$$\mu_{N_T} = a_N q_N \tag{6.51}$$

in which q_N = specific nitrification rate, d^{-1}, and a_N = sludge yield coefficient for nitrifiers.

The specific nitrification rate in an activated sludge system depends on the concentrations of ammonia nitrogen in the effluent and dissolved oxygen as well as the pH. The effects of dissolved oxygen and effluent ammonia are defined as follows:

$$q_N = q_{N_M} \cdot \frac{NH_3^-N}{K_N + NH_3^-N} \cdot \frac{DO}{K_O + DO} \tag{6.52}$$

where K_N and K_O are the half saturation coefficients for nitrogen and oxygen, respectively. A typical value for K_N is 0.4; K_O may vary from 0 to 1.0.

The influence of mixed liquor dissolved oxygen on the nitrification rate has been somewhat controversial, partly because the bulk liquid concentration is not the same as the concentration within the floc where the oxygen is being consumed. Increased bulk liquid dissolved oxygen concentrations will increase the penetration of oxygen into the floc, thereby increasing the rate of nitrification. At a decreased SRT, the oxygen utilization rate due to carbon oxidation increases, thereby decreasing the penetration of oxygen. Conversely, at a high SRT, the low oxygen utilization rate permits higher oxygen levels within the floc, and consequently, higher nitrification rates occur. Therefore, to maintain the maximum nitrification rate, the bulk mixed liquor dissolved oxygen concentration must be increased as the SRT is decreased. This is reflected in the coefficient K_O.

The effect of pH on nitrification rate is shown in Fig. 6.53.

In treating industrial wastewaters that may inhibit nitrification, the maximum specific nitrification rate must be experimentally determined. The temperature dependence on the specific nitrification rate is given by

$$q_{N(T)} = q_{N(20°C)} \cdot 1.09^{T-20} \tag{6.53}$$

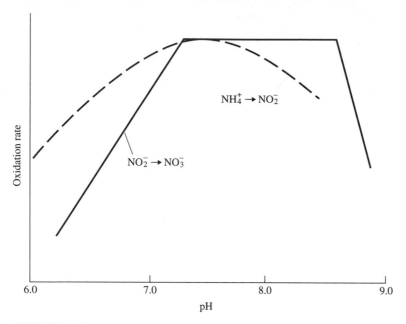

FIGURE 6.53
Effect of pH on ammonia oxidation.[47]

The endogenous decay coefficient b_N has a temperature coefficient of 1.04:

$$b_N = b_{N(20°C)} \cdot 1.04^{T-20} \tag{6.54}$$

The nitrogen to be oxidized can be computed from

$$N_{ox} = TKN - SON - N_{syn} - (NH_3{}^-N)_e$$

in which SON = nondegradable organic nitrogen and $N_{syn} = 0.08aS_r$.
    The fraction of nitrifiers can be computed from

$$f_N = \frac{0.15N_{ox}}{aS_r + 0.15N_{ox}} \tag{6.55}$$

The overall nitrification rate is

$$R_N = q_N f_N X_{vb} \tag{6.56}$$

in which R_N is the overall nitrification rate in mg/(l · d).
    The required nitrification detention time is

$$t_N = \frac{N_{ox}}{R_N}$$

In order to determine the SRT it is necessary to compute the waste activated sludge (WAS):

$$\Delta X_{vb} = (aS_r + 0.15N_{ox}) - bX_d X_{vb} t_N \tag{6.57}$$

and the SRT is

$$\theta_c = \frac{X_{vb} t_N}{\Delta X_{vb}}$$

The required oxygen is

$$O_2 = 4.33N_{ox}$$

and the required alkalinity as $CaCO_3$ is

$$ALK = 7.15N_{ox}$$

Nitrification of High-Strength Wastewaters

Wastewaters containing high ammonia concentration and negligible BOD can be treated by biological nitrification. For example, wastewater from a fertilizer manufacturing complex was treated by the activated sludge process. The $NH_4{-}N$ content of the influent wastewater varied from 339 to 420 mg/l and inorganic suspended solids varied from 313 to 598 mg/l. The TDS was 6300 mg/l. Because of the high inert suspended solids, the mixed liquor was only 20 percent volatile with a sludge volume index (SVI) of 30 to 40 ml/g. A small fragile floc was generated, which provided an effluent TSS of 55 mg/l. Alkalinity was supplied to the system in the form of sodium bicarbonate. The relationship between nitrification rate and mixed liquor temperature is shown in Fig. 6.54. The temperature correction coefficient, θ, was 1.13. This value is significantly higher than for typical domestic wastewater indicating that the nitrification rate was more sensitive to the operating

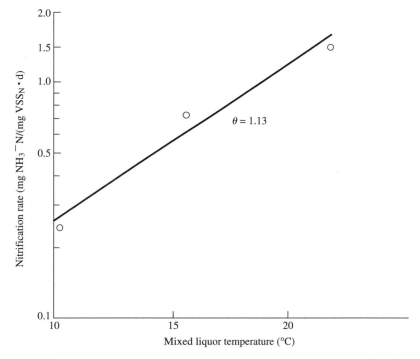

FIGURE 6.54
Relationship between nitrification rate and temperature for a fertilizer wastewater.

mixed liquor temperature. The alkalinity requirements showed considerable variation (Fig. 6.55), which was attributed to the presence of alkalinity in the influent
suspended solids.

Inhibition of Nitrification

In treating industrial wastewaters, nitrification is frequently inhibited, or in some
instances prevented, by the presence of toxic organic or inorganic compounds. This
is shown in Fig. 6.56, which shows the nitrification results for treatment of an
organic chemicals wastewater. These data show that a minimum aerobic SRT of 25
d was required to obtain complete nitrification at 22 to 24°C. The minimum SRT
required for complete nitrification of municipal wastewater at these temperatures
is approximately 4 d. An SRT of 55 to 60 d was required for complete nitrification
of this wastewater at 10°C versus approximately 12 d for municipal wastewater. It
was also shown that, at a mixed liquor temperature of 10°C, the nitrifiers were less
tolerant to variations in influent composition and temperature than were the het-
erotrophic organisms responsible for BOD removal and denitrification. Similar
results were obtained for a wastewater from a coke plant in which the nitrification

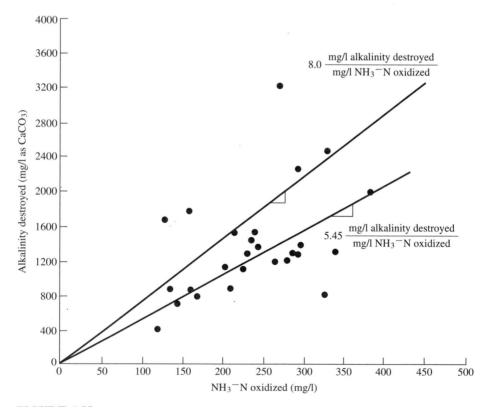

FIGURE 6.55
Alkalinity utilization in the treatment of a fertilizer wastewater.

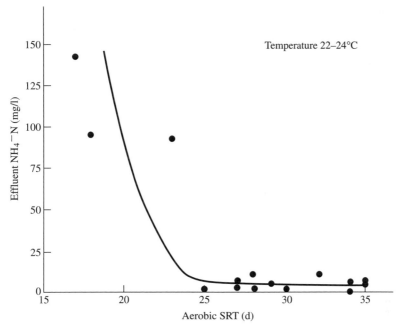

FIGURE 6.56
Nitrification relative to the aerobic SRT for an organic chemicals wastewater.[53]

rate was approximately one order of magnitude less than that for municipal wastewater, as shown in Fig. 6.57.

Blum and Speece[48] have characterized the toxicity of a variety of organic compounds to nitrification, as summarized in Table 6.20.

The effect of salt levels on nitrification were discussed by Henning and Kayser.[49] They found that fluoride concentrations of 100 mg/l reduced the nitrification rate by 80%. Sulfate had no effect at concentrations up to 50 g/l. Chlorides, however, showed significant inhibition, as shown in Fig. 6.58. They showed that nitrification rates were reduced up to 60 percent at NO_2^-N concentrations of several hundred milligrams per liter at pH 8.0.

In cases where nitrification is significantly reduced or totally inhibited, the application of powdered activated carbon (PAC) to adsorb the toxic agents may enhance nitrification. However, in some cases, excessive quantities of PAC are required to achieve single-stage nitrification. In some cases a second-stage nitrification step can be successfully employed after a first-stage biological process for removal of carbonaceous material and reduction of toxicity.

Metals have been found to be toxic to growing *Nitrosomonas* culture (Skinner and Walker[50]) with complete inhibition for the following metals and concentrations: nickel 0.25 mg/l; Cr, 0.25 mg/l; and Cu, 0.1 to 0.5 mg/l.

Cyanide toxicity to nitrifiers is shown in Fig. 6.59.[51, 52]

Un-ionized ammonia (NH_3) inhibits both *Nitrosomonas* and *Nitrobacter*, as shown in Fig. 6.60.[53] Since the un-ionized fraction increases with pH, a high pH combined with a high total ammonia concentration will severely inhibit or prevent

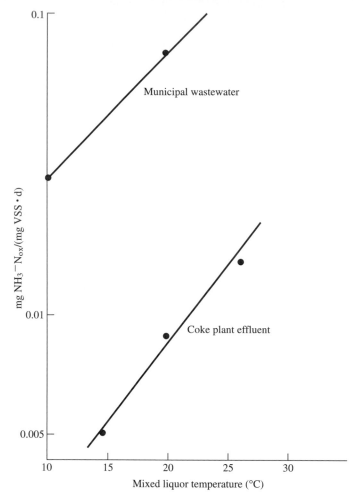

FIGURE 6.57
Relationship between nitrification rate and temperature for municipal wastewater and a coke plant effluent.

complete biological nitrification. Since *Nitrosomonas* is less sensitive to ammonia toxicity than *Nitrobacter*, the nitrification process may only be partially complete and result in accumulation of nitrite ion (NO_2^{2-}). This can have severe consequences since NO_2^{2-} is strongly toxic to many aquatic organisms whereas NO_3^-N is not. Ammonia toxicity to activated sludge biomass is rarely a problem in treating municipal wastewaters, since the concentration of total ammonia is low and the mixed liquor pH is near neutral. Industrial wastewaters with high ammonia levels and the potential for high pH excursions, however, may cause biotoxicity and loss of the nitrification process. Under these conditions, it is necessary to control the mixed liquor pH to avoid biotoxicity due to an ammonia spill or shock load. In extreme cases, two stages operated at different pH values may be required to separate the *Nitrosomonas* and *Nitrobacter* and allow complete nitrification.

TABLE 6.20
Biodegradability and biotoxicity data[48]

Compound	Bio-degradability	Rate of biodegradation, mg COD/ gVSS · h	EC$_{50}$, mg/l	
			Nitrosomonas	Heterotrophs
Cyclohexane	A	—	97	29
Octane	A	—	45	—
Decane	C	—	—	—
Dodecane	D	—	—	—
Methylene chloride	D	—	1.2	320
Chloroform	D	—	0.48	640
Carbon tetrachloride	—	—	51	130
1,1-Dichloroethane	—	—	0.91	620
1,2-Dichloroethane	—	—	29	470
1,1,1-Trichloroethane	—	—	8.5	450
1,1,2-Trichloroethane	—	—	1.9	240
1,1,1,2-Tetrachloroethane	—	—	8.7	230
1,1,2,2-Tetrachloroethane	—	—	1.4	130
Pentachloroethane	—	—	7.9	150
Hexachloroethane	—	—	32	—
1-Chloropropane	D	—	120	700
2-Chloropropane	—	—	110	440
1,2-Dichloropropane	—	—	43	—
1,3-Dichloropropane	C	—	4.8	210
1,2,3-Trichloropropane	—	—	30	290
1-Chlorobutane	D	—	120	230
1-Chloropentane	D	—	99	68
1,5-Dichloropentane	—	—	13	—
1-Chlorohexane	D	—	85	83
1-Chloroctane	—	—	420	52
1-Chlorodecane	D	—	—	40
1,2-Dichloroethylene	D	—	—	—
Trans-1,2-dichloroethylene	—	—	80	1700
Trichloroethylene	A	—	0.81	130
Tetrachloroethylene	—	—	110	1900
1,3-Dichloropropene	—	—	0.67	120
5-Chloro-1-pentyne	—	—	0.59	86
Methanol	A	26	880	20,000
Ethanol	A	32	3900	24,000
1-Propanol	A	71	980	9600
1-Butanol	A	84	—	3900
1-Pentanol	A	—	520	—
1-Hexanol	A	—	—	—
1-Octanol	A	—	67	200
1-Decanol	B	—	—	—
1-Dodecanol	B	—	140	210

TABLE 6.20 *(continued)*

Compound	Bio-degradability	Rate of biodegradation, mg COD/ gVSS · h	EC$_{50}$, mg/l	
			Nitrosomonas	Heterotrophs
2,2,2-Trichloroethanol	—	—	2.0	—
3-Chloro-1,2 propanediol	D	—	—	—
Ethylether	C	—	—	17,000
Isopropylether	D	—	610	—
Acetone	B	—	1200	16,000
2-Butanone	—	—	790	11,000
4-Methyl-2-pentanone	—	—	1100	—
Ethyl-acrylate	—	—	47	—
Butyl-acrylate	—	—	38	470
2-Chloropropionic-acid	A	24	0.04	0.18
Trichloroacetic-acid	D	0	—	—
Diethanolamine	A	16	—	—
Acetronitrile	A	—	73	7500
Acrylonitrile	A	—	6.0	52
Benzene	A	—	13	520
Toluene	A	—	84	110
Xylene	A	—	100	1000
Ethylbenzene	B	—	96	130
Chlorobenzene	D	—	0.71	310
1,2-Dichlorobenzene	—	—	47	910
1,3-Dichlorobenzene	D	—	93	720
1,4-Dichlorobenzene	D	—	86	330
1, 2,3-Trichlorobenzene	—	—	96	—
1,2,4-Trichlorobenzene	D	—	210	7700
1, 3,5-Trichlorobenzene	—	—	96	—
1, 2,3,4-Tetrachlorobenzene	—	—	20	—
1,2,4,5-Tetrachlorobenzene	D	—	9	—
Hexachlorobenzene	D	—	4	350
Benzyl alcohol	A	—	390	2100
4-Chloroanisole	—	—	—	902
2-Furaldehyde	B	37	—	—
Benzonitrile	B	—	32	470
m-Tolunitrile	—	—	0.88	290
Nitrobenzene	A	14	0.92	370
2,6-Dinitrotoluene	—	—	183	—
1-Nitronaphthalene	—	—	—	380
Naphthalene	A	—	29	670
Phenanthrene	C	—	—	—
Benzidine	D	—	—	—
Pyridine	A	—	—	—
Quinoline	A	8.5	—	—
Phenol	A	80	21	1100

TABLE 6.20 *(continued)*

Compound	Bio-degradability	Rate of biodegradation, mg COD/ gVSS · h	EC$_{50}$, mg/l	
			Nitrosomonas	Heterotrophs
m-Cresol	A	—	0.78	440
p-Cresol	A	—	27	260
2,4-Dimethylphenol	—	28.2	—	—
3-Ethylphenol	—	—	—	144
4-Ethylphenol	—	—	14	—
2-Chlorophenol	—	—	2.7	360
3-Chlorophenol	—	—	0.20	160
4-Chlorophenol	A	39.8	0.73	98
2,3-Dichlorophenol	—	—	0.42	210
2,4-Dichlorophenol	—	10.5	0.79	—
2,5-Dichlorophenol	—	—	0.61	180
2,6-Dichlorophenol	—	—	8.1	410
3,5-Dichlorophenol	—	—	3.0	—
2,3,4-Trichlorophenol	—	—	52	7.8
2,3,5-Trichlorophenol	—	—	3.9	—
2,3,6-Trichlorophenol	—	—	0.42	14
2,4,5-Trichlorophenol	—	—	3.9	23
2,4,6-Trichlorophenol	—	—	7.9	—
2,3,5,6-Tetrachlorophenol	—	—	1.3	1.5
Pentachlorophenol	—	—	6.0	—
2-Bromophenol	—	—	0.35	—
4-Bromophenol	B	—	0.83	120
2,4,6-Tribromophenol	—	—	7.7	—
Pentabromophenol	—	—	0.27	—
Resorcinol	A	57.5	7.8	—
Hydroquinone	B	54.2	—	—
2-Aminophenol	—	21.1	0.27	0.04
4-Aminophenol	—	16.7	0.07	—
2-Nitrophenol	—	14.0	11	11
3-Nitrophenol	—	17.5	—	—
4-Nitrophenol	A	16.0	2.6	160
2,4-Dinitrophenol	—	6.0	—	—

All *Nitrosomonas* and aerobic heterotroph data corrected for pKa (ionization) and H (gas/liquid partitioning).

$$A = \frac{BOD}{TOD} > 50\%; \text{ readily biodegradable} \qquad C = \frac{BOD}{TOD} < 10\%\text{–}25\%; \text{ refractory}$$

$$B = \frac{BOD}{TOD} \, 25\text{–}25\%; \text{ moderately biodegradable} \qquad D = \frac{BOD}{TOD} < 10\%; \text{ nondegradable}$$

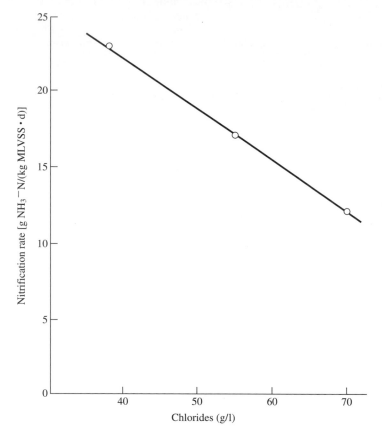

FIGURE 6.58
Nitrification kinetics at different chloride concentrations.

Unfortunately, many industrial wastewaters contain these and other compounds, which alone or together exert a greater but undetermined inhibition effect on the nitrification process. It is essential, therefore, to determine the specific nitrification rate q_N and the $(\theta_c)_{min}$ required to achieve nitrification under actual operating conditions. The value of q_N can be determined using either a batch activated sludge (BAS) test or the semicontinuous fed batch reactor method.

Batch Activated Sludge Nitrification

In the BAS test procedure, a wastewater sample is aerated in the presence of an actively nitrifying sludge. The sludge can be obtained from either a municipal activated sludge plant with negligible industrial load or developed separately from a commercially available nitrifier culture. Regardless of the sludge source, it must be possible to quantify the mass of nitrifiers in the bulk MLVSS concentration since q_N is expressed per unit of nitrifier mass (VSS_N). The initial $NH_3\text{-}N$ concentration in the BAS test should be between 20 and 50 mg/l to eliminate substrate-induced

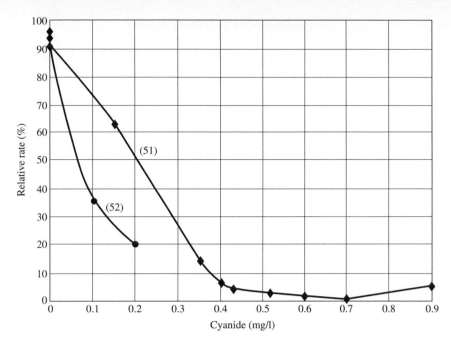

FIGURE 6.59
Relative rate of nitrification as function of cyanide.[51, 52]

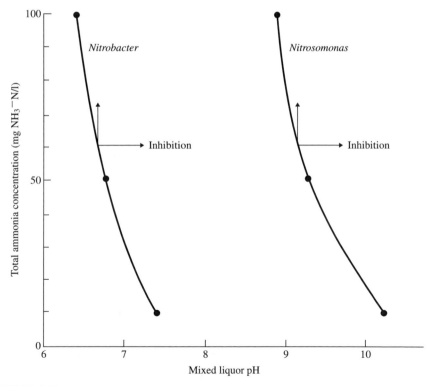

FIGURE 6.60
Ammonia inhibition in the activated sludge process.[53]

toxicity. If the wastewater contains organic nitrogen, then TKN should also be measured to account for biohydrolysis during the test. Finally, the alkalinity should be adjusted with sodium bicarbonate (NaHCO$_3$) to provide 7.15 mg alkalinity as CaCO$_3$ mg TKN plus a 50 mg/l residual. A control test should be run using the same initial NH$_3^-$N concentration. The wastewater and control sample should be vigorously aerated and aliquots withdrawn over time for analysis.

The results of a BAS nitrification test conducted at $T = 21°C$ are shown in Fig. 6.61 for a mixed liquor from a municipal wastewater treatment plant. On the basis of the plant's historical operating data and Eq. (6.54), the fraction of nitrifiers (f_N) in the MLVSS was 0.0245 mg VSS$_N$/mg VSS. Initially, the wastewater had negligible NO$_3^-$N and organic nitrogen and an NH$_3^-$N concentration of 48 mg/l. After 24 h of aeration, 38 mg/l NO$_3^-$N was produced.

The overall specific NH$_3^-$N oxidation rate was

$$\frac{38 \text{ mg}/(1 \cdot d)}{1200 \text{ mg VSS/l}} = 0.032 \text{ mg NH}_3^-\text{N}/(\text{mg MLVSS} \cdot d)$$

The nitrifier specific nitrification rate was

$$q_N = 0.032/0.0245 = 1.3 \text{ mg N}/(\text{mg VSS}_N \cdot d)$$

The nitrifier specific growth rate for $a_N = 0.15$ mg/mg was 0.195 d^{-1}. Neglecting temperature effects, and with $b_N = 0.05$d^{-1} the $(\theta_c)_{min}$ is

$$(\theta_c)_{min} = \frac{1}{0.195 - 0.05} = 6.9 \text{ d}$$

The $(\theta_c)_{min}$ for municipal wastewater at $T = 21°C$ is approximately 4 d, indicating a two-fold inhibition effect on nitrification by the industrial wastewater. The θ_c design would be determined using $(\theta_c)_{min}$ and an appropriate safety factor.

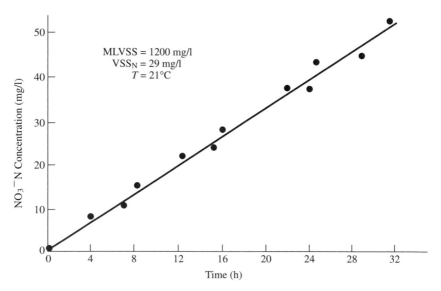

FIGURE 6.61
Results of batch nitrification test.

It should be noted that the BAS nitrification test used an existing sludge with a known f_N and VSS_N concentration to determine q_N. The actual mixed liquor established for treatment of this wastewater will have a different f_N value depending on its NH_3^-N and BOD concentrations. This nitrifier fraction and the measured q_N should be used to determine the hydraulic retention time required in the aeration basin.

Fed Batch Reactor Nitrification Test

The FBR procedure described on p. 268 can also be used to determine the nitrification rate. The requirements for sludge characterization and alkalinity addition are the same as for the BAS nitrification test. The wastewater is added to the reactor at a constant rate, and aliquots are withdrawn over time for analysis. Nitrate (NO_3^-N) and nitrite (NO_2^-N) production is the preferred method of expressing the test results and determining q_N, since it eliminates adjustments needed to account for biohydrolysis and cell synthesis. If NH_4^-N removal is used, however, then TKN and COD (or BOD) must also be measured to complete the nitrogen balance.

The results of FBR nitrification tests with and without PAC addition (200 mg/l) are presented in Fig. 6.62. The mixed liquor had a $VSS_N = 500$ mg/l, and the wastewater sample initially had significant concentrations of BOD, organic nitrogen, and NO_3^-N. The nitrification rates were calculated as the difference between the slopes of the two linear traces. They indicate that addition of 200 mg/l of PAC increased the nitrification rate from 0.6 to 1.45 mg $NO_3^-N/(mg\ VSS_N \cdot d)$. The $(\theta_c)_{min}$ can be calculated as in the BAS nitrification test procedure.

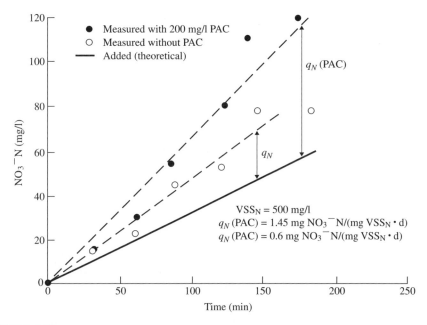

FIGURE 6.62
Determination of nitrification rate using FBR procedure with and without PAC addition.

Denitrification

Some industrial wastewaters such as those from fertilizer, explosive/propellant man-ufacture, and the synthetic fibers industry contain high concentrations of nitrates, while others generate nitrates by nitrification. Since biological denitrification gener-ates one hydroxyl ion while nitrification generates two hydrogen ions, it may be advantageous to couple the nitrification and denitrification processes to provide "internal" buffering capacity. While many organics inhibit biological nitrification, this is not generally true for denitrification. Sutton et al.[54] showed that denitrification rates for an organic chemicals plant wastewater were comparable to those observed using nitrified municipal wastewater; however, biological nitrification of the organic chemicals wastewater was severely inhibited. Denitrification uses BOD as a carbon source for synthesis and energy and nitrate as an oxygen source.

$$NO_3^- + BOD \rightarrow N_2 + CO_2 + H_2O + OH^- + \text{new cells}$$

The denitrification process consumes approximately 3.7 g COD per g NO_3^-N reduced and produces 0.45 g VSS and 3.57 g alkalinity per g NO_3^-N reduced. This amounts to one-half the alkalinity that is consumed during nitrification. Some of this alkalinity, however, is lost by reaction with the CO_2 generated by microbial respiration.

Orhon et al.[55] compared the sludge yield under aerobic conditions to anoxic conditions as shown in Table 6.21. McClintock et al.[56] showed that the biomass yield under anoxic conditions was 54 percent that under aerobic conditions and that the endogenous coefficient was 51 percent that under aerobic conditions. Comparative sludge yield will be a function of SRT.

The rate of denitrification q_{DN} is zero order to NO_3^-N concentrations of approximately 1.0 mg N/l and is determined by

$$q_{DN} = \frac{(NO_3^-N)_o - (NO_3^-N)_e}{X_{vb}t} \tag{6.58}$$

where q_{DN} = denitrification rate, g $NO_3^-N/(g \text{ VSS} \cdot d)$, and X_{vb} = nonnitrifier bio-mass under aeration, mg VSS/l. The denitrification rate is adjusted for mixed liquor temperature and bulk dissolved oxygen (DO) concentration by Eq. (6.59).

$$q_{DN_T} = q_{DN(20°)} C \cdot \theta_{DN}^{T-20} (1 - DO) \tag{6.59}$$

TABLE 6.21
Sludge yield coefficient (based on COD)[55]

Wastewater	Aerobic	Anoxic, g cell COD/g COD
Domestic sewage	0.63	0.50
Meat processing	0.64	0.51
Dairy	0.65	0.52
Confectionery	0.72	0.61

The temperature coefficient θ_{DN} varies from 1.07 to 1.20.

While oxygen inhibits denitrifying facultative bacteria depending on plant operating conditions, flocs may contain anoxic zones where denitrification will occur even when the liquid contains dissolved oxygen.

OH and Silverstein[57] showed that the IAWQ relationship was more applicable at high dissolved oxygen levels

$$q_{DN} = q_{DN\,(max)}\left(\frac{1}{1 + DO/k}\right) \tag{6.59a}$$

A suggested value for k is 0.38 mg/l. It is assumed that k will be a function of floc size and hence power level in the aeration basin.

Field experience has shown that as much as 10 to 25 percent denitrification can occur under aerobic conditions in aeration basins.

Results of denitrification of an organic chemicals plant wastewater are shown in Fig. 6.63. The temperature effect on denitrification using continuous flow and batch reactors is shown in Fig. 6.64.

The denitrification rate will depend on the biodegradability of the organics in the wastewater and the concentration of active biomass under aeration similar to

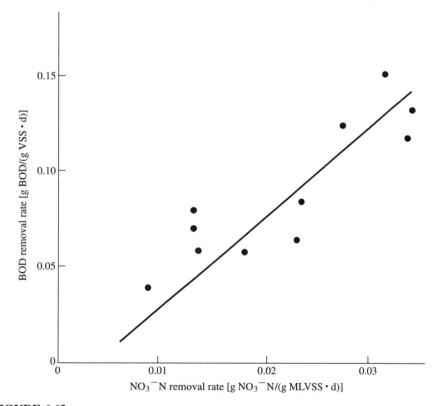

FIGURE 6.63

Relationship between nitrate reduction and BOD removal for an organic chemicals wastewater.

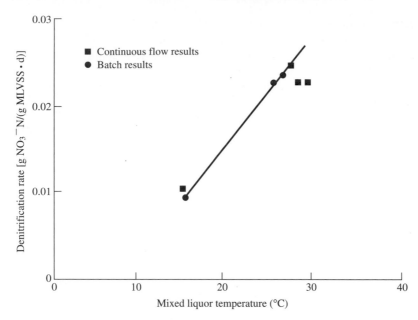

FIGURE 6.64

Relationship between denitrification rate and temperature for an organic chemicals wastewater.

the aerobic process. This, in turn, is related to both the SRT or F/M and the presence of inert solids in the sludge. As the F/M increases, the concentration of active biomass and the rate of denitrification increase. Although denitrification can occur under endogenous conditions (low F/M), using internal biomass reserves, it is very slow and requires long hydraulic retention times. Since the rate of denitrification is affected by both wastewater characteristics and process design parameters, it is usually necessary to determine the rate by experimental means. A batch denitrification test should be conducted in which sludge and wastewater are mixed under anoxic conditions [oxidation reduction potential (ORP) = − 100 mV], and the residual NO_3^-N concentration is determined with time. Depending on the organic composition of the wastewater, one of several removal rate relationships may be obtained. For a complex wastewater in a plug flow system, a pseudo first-order relationship may exist. In a complete mix system the rate will be proportional to the fraction of organics remaining in solution. Lie and Welander[58] have shown that the denitrification rate can be related to the ORP as shown in Fig. 6.65.

The denitrification rate for a wastewater can also be estimated from the oxygen uptake rate. In this case, the wastewater–anoxic sludge mixture is aerated and the SOUR determined over time. Correlation of R_{DN} and SOUR indicated that 1.0 mg NO_3^-N is equivalent to approximately 3.0 mg O_2, which is in good agreement with the theoretical value of 2.86 mg NO_3^-N/mg O_2.

In cases where a carbon source is not available in the wastewater, methanol has been used as a carbon source. Various industrial effluents can also be used as a carbon source. Baumann and Krauth[59] have summarized various carbon sources as shown in Table 6.22.

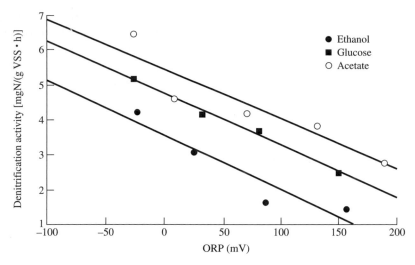

FIGURE 6.65
Denitrification activity versus ORP for activated sludge from Sjölunda pilot plant with addition of different carbon sources.

TABLE 6.22
List of possible industrial wastes or waste by-products for denitrification[59]

Industry	BOD_5, mg/l	COD, mg/l	DN rate, mg NO_3^--N/(g MLSS · h)
Chemical industry			
Deicing agent (airfield)	65,000	118,000	1.98–3.06
Glue production I	148,500	282,400	0.96–1.26
Glue production II	1,080,000	1,340,000	1.14–2.12
Pharmaceutical industry I	136,000	188,100	4.08
Pharmaceutical industry II	163,000	320,000	1.14–1.53
Photographic industry	126,000	686,000	1.59–1.70
Food industry			
Alcohol production	3,780	7,300	
Fusel oil	1,320,000	1,780,000	2.79–3.18
Milk processing industry	4,880	7,440	
Plant and vegetable processing	20,650	26,050	4.29
Slaughterhouse	183,000	246,000	1.44
Wine industry	173,100	211,100	5.40
Yeast industry	26,900	28,770	2.79–3.18
Common substrates			
Acetic acid		1,056,000	3.35
Endogenous			0.26–0.65

Temperature: 13–16°C

Denitrification in final clarifiers causes floating sludge and increased effluent suspended solids. The nitrogen gas rate production depends on the carbon source available for denitrification, the SRT, temperature and sludge concentration.[60] Henze et al.[60] estimated that 6 to 8 and 8 to 10 mg/l NO_3^--N needs to be denitrified in the sludge blanket to cause sludge flotation at 10 and 20°C, respectively. Most denitrification results from endogenous respiration and the utilization of adsorbed slowly degradable organics. This in turn is related to the active mass, which is a function of SRT. A relationship between effluent suspended solids and effluent nitrate concentration is shown in Fig. 6.66. The ATV[61] limits the thickening time in secondary clarifiers with nitrification to 1.0 to 1.5 h.

Nitrification and Denitrification Systems

A number of alternative treatment systems are available to achieve nitrification and denitrification, in which some form of aerobic-anoxic sequencing is provided. The systems differ in whether they utilize a single sludge or two sludges in separate nitrification and denitrification reactors, The single-sludge system uses one basin and clarifier and the raw wastewater or endogenous reserves as the carbon and energy sources for denitrification.

The two-sludge system uses two basins with separate clarifiers to isolate the sludges. A supplemental carbon source such as methanol (CH_3OH) is provided to the second stage for the carbon and energy source. The simplest configuration of the single-sludge system enables carbonaceous oxidation, nitrification, and denitrification to occur in a single basin by positioning the return sludge and aeration

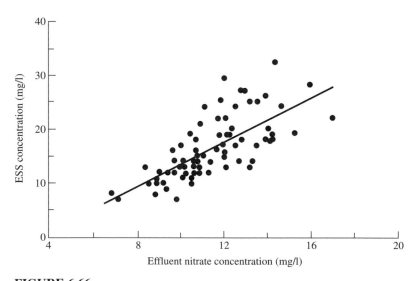

FIGURE 6.66
Effluent suspended solids concentration as a function of effluent nitrate concentration at a large nitrification/denitrification plant.[54]

equipment to maintain defined aerobic-anoxic zones in different basin sections. An alternative single-sludge system utilizes a single basin for both aeration and sedimentation by providing intermittent aeration and nonaeration cycles to yield aerobic and anoxic phases of sufficient duration to permit nitrate reduction. Two process flow configurations for single-sludge nitrification and denitrification are shown in Fig. 6.67. In the oxidation ditch (Fig. 6.67a), an aerobic zone exists in the vicinity of the aerator. As the mixed liquor passes away from the aerator, the dissolved oxygen is depleted. Anoxic conditions then exist, and denitrification occurs. This sequence is repeated around the ditch at each aerator installation.

In the internal recycle single-sludge process (Fig. 6.67b), nitrification occurs under aerobic conditions in the second basin. The second basin may be a separate tank (without intermediate clarification) or a single tank with internal baffles to isolate the aerobic and anoxic zones without short-circuiting. Each of these zones can be plug flow or CMAS. An internal recycle flow $Q_{R,\text{in}}$ is

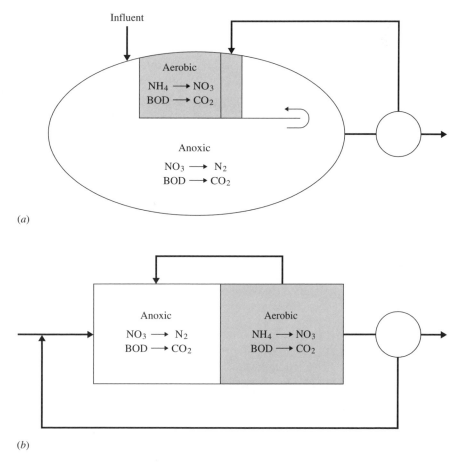

FIGURE 6.67
Alternative single-stage nitrification-denitrification systems.

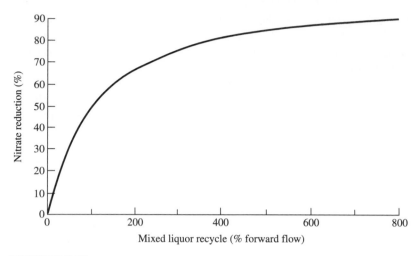

FIGURE 6.68
Theoretical nitrate removal using internal mixed liquor recycle in single-sludge system.

employed from the end of the aerobic basin to the inlet of the anoxic basin in which denitrification occurs. The total recycle ratio can be calculated by Eq. (6.60) and is shown in Fig. 6.68.

$$R = \frac{1}{(1-f)} - 1 \tag{6.60}$$

where R = total return sludge rate, percent = $(Q_R + Q_{R,in})/Q_o$
       f = overall denitrification efficiency

Nitrification and denitrification design is illustrated by Examples 6.13 and 6.14.

EXAMPLE 6.13.
(a) Design a nitrification system to produce an effluent NH_3^-N of 1 mg/l at 20°C. The influent TKN is 40 mg/l and the BOD_5 is 210 mg/l.
(b) What must the HRT and SRT be increased to if the temperature is 15°C?
(c) What will the maximum NH_3^-N concentration effluent be if the influent NH_3^-N increases to 60 mg/l at 20°C?

Conditions:

$D_o = 2.0$ mg/l
$q_{N(max)} = 1.3$ mg $NH_3^-N(mg\ VSS_N \cdot d)$
$a_N = 0.15$
$a_H = 0.6$
SON = 1 mg/l
$SBOD_e = 10$ mg/l
$X_v = 3000$ mg/l
$b_N = 0.05$ d^{-1}
$b_H = 0.1$ d^{-1}

Solution.

(a)

$$q_N = q_{N(\text{max})} \frac{(NH_3{}^-N)_e}{0.4 + (NH_3{}^-N)_e} \cdot \frac{DO}{(0.2 + DO)}$$

$$= 1.3 \frac{1}{(0.4 + 1)} \cdot \frac{2}{(0.2 + 2)} = 0.84 \text{ d}^{-1}$$

$$N_{\text{ox}} = TKN - (NH_3{}^-N)_e - 0.08 \cdot a_H S_r - SON$$

$$= 40 - 1 - 0.08 \cdot 0.6 \cdot 200 - 1 = 28.4 \text{ mg/l}$$

The nitrifier fraction is

$$f_N = \frac{0.15 \times 28.4}{0.6 \times 200 + 0.15 \times 28.4} = 0.034$$

The nitrification rate is

$$r_N = 0.84 \times 0.034 \times 3000 = 86 \text{ mg/}(l \cdot d)$$

The required hydraulic detention time is

$$t_N = \frac{28.4}{86} = 0.33 \text{ d}$$

The sludge production is

$$\Delta X_{vb} = (a_H S_r + 0.15 N_{\text{ox}}) - [b_H(1 - f_N) + b_N f_N]X_d X_{vb} t_N$$

$$= (0.6 \times 200 + 0.15 \cdot 28.4) - [0.1(1 - 0.034)$$

$$+ 0.05 - 0.034]0.62 \times 3000 \times 0.33$$

$$= 124 - 58 = 66 \text{ mg/l}$$

$$\theta_c = \frac{3000 \times 0.33}{66} = 15 \text{ d}$$

(b) The specific nitrification rate at 15°C is

$$q_{N(15°)} = q_{N(20°)}1.09^{15-20} = 0.84 \times 1.09^{-5} = 0.55 \text{ d}^{-1}$$

$$r_N = 0.55 \times 3000 \cdot 0.034 = 56 \text{ mg/}(l \cdot d)$$

$$t_N = \frac{28.4}{56} = 0.51 \text{ d}$$

The sludge production at 15°C is

$$\Delta X_{vb} = 124 - [0.082(1 - 0.034) + 0.041 \cdot 0.034]0.56 \cdot 3000 \cdot 0.511$$

$$= 124 - 69 = 55 \text{ mg/l}$$

$$\theta_c = \frac{3000 \times 0.51}{55} = 28 \text{ d}$$

(c) Assume q_N will increase to $q_{N(max)}$. Then

$$r_N = 1.3 \times 0.034 \times 3000 = 133 \text{ mg}/(1 \cdot d)$$

$$N_{ox} = r_N t_N = 133 \times 0.33 = 44 \text{ mg/l}$$

$$(NH_3^-N)_e = 60 - 1 - 9.6 - 44 = 5.4 \text{ mg/l}$$

EXAMPLE 6.14. Using the results from Example 6.14 and a denitrification rate of 0.1 mg NO_3^-N/(mg VSS $\cdot$ d), design a system to achieve 75 percent overall denitrification ($f = 0.75$), assuming complete denitrification in the anoxic zone.

Solution.

$$\frac{t_{DN}}{t_N} = \frac{q_N f_N f}{q_{DN}(1 - f_N)} = \frac{0.84 \times 0.034 \times 0.75}{0.1(1 - 0.034)} = 0.222$$

$$t_{DN} = 0.222 \times t_N = 0.33 \times 0.222 = 0.073 \text{ d}$$

The total detention time is

$$t = t_N + t_{DN} = 0.33 + 0.073 = 0.40 \text{ d}$$

The nitrogen to be denitrified is

$$N_{DN} = t_{DN} q_{DN} X_{vb}(1 - f_N) = 0.073 \times 0.1 \times 3000 = (1 - 0.034)1 = 21 \text{ mg/l}$$

The BOD removed during denitrification ($S_{r,D}$) assuming 3 mg BOD/mg N is

$$S_{r,DN} = 21 \times 3 = 63 \text{ mg/l}$$

The aerobic removal of BOD is

$$S_{r,DN} = 210 - 10 - 63 = 137 \text{ mg/l}$$

The sludge production becomes

$$\Delta X_{vb} = a_D S_{r,DN} + a_H S_r + a_N N_{ox} - [b_H(1 - f_N)t_N$$
$$+ b_{H_D}(1 - f_N)t_{DN} + b_N f_N t_N]X_d X_{vb}$$

Assuming a_D and b_D as 75 percent of a_H and b_H, respectively,

$$a_D = 0.75 \times 0.6 = 0.45$$

$$b_{H_D} = 0.75 \times 0.1 = 0.75 \text{ d}^{-1}$$

$$\Delta X_{vb} = 0.45 \times 63 + 0.6 \times 137 + 0.15 \times 28.4 - [0.1(0.966)0.33$$
$$+ 0.075(0.966)0.073 + 0.05(0.034)033]0.55 \times 3000$$
$$= 115 - [0.0377]3000 \times 0.55 = 115 - 113 \cdot 0.055 = 53 \text{ mg/l}$$

$$\theta_c = \frac{3000 \times 0.4}{53} = 23 \text{ d}$$

The total required recycle is

$$R = \frac{1}{(1 - f)} - 1 = \frac{1}{(1 - 0.75)} - 1 = 3 \text{ or } 300\%$$

If $Q_R = 50\%$, then $Q_{R,in} = 250\%$.

6.9
LABORATORY AND PILOT PLANT PROCEDURES FOR DEVELOPMENT OF PROCESS DESIGN CRITERIA

Wastewater Characterization

Wastewater characterization should be based on the equalized wastewater. Depending on the nature of the wastewater and the discharge permit requirements, the following parameters should be evaluated:

- BOD and/or COD or TOC
- Total and volatile suspended solids
- Oil and grease
- Volatile organics
- Priority pollutants
- Toxicity (bioassay)
- Nitrogen forms (TKN, NH_3, NO_2^-, NO_3^-)
- Phosphorus forms ($o\text{-}PO_4$, total P)

For wastewaters that do not contain aquatic toxicity, the following stepwise procedure is applicable to developing the necessary process design data:

1. Adjust the BOD:N:P ratio to 100:5:1, neglecting the wastewater organic nitrogen. Although organic nitrogen may be hydrolyzed to ammonia in the activated sludge process, it is initially neglected in order to ensure adequate nutrients in the experimental phase. The availability of the organic nitrogen will be reevaluated in the final process design.
2. Evaluate the wastewater's potential to promote filamentous bulking. In many cases, if the wastewater is readily degradable ($K > 6 \text{ d}^{-1}$), filamentous bulking can be expected. If in doubt, evaluate this by operating a complete mix reactor at $F/M \approx 0.4 \text{ d}^{-1}$ for 5 to 8 d to establish the proliferation of filaments.
3. Develop an acclimated mixed liquor. Determine the bioinhibition potential using the FBR procedure. If there is bioinhibition, adjust the initial feed rate of the wastewater to less than 50 percent of the inhibition threshold concentration. Operate the reactor at an F/M of 0.3 d^{-1}. As acclimation proceeds, gradually increase the feed rate until the full waste strength is being treated. For a wastewater with a low bulking potential, use a complete mix reactor. For a wastewater with a high bulking potential, acclimate the mixed liquor in a batch reactor, a sequencing batch reactor, or a biological selector.

Calculations for acclimation of a sludge using a batch fill-and-draw reactor are presented in Example 6.15.

EXAMPLE 6.15. Determine the operating conditions for a fill-and-draw acclimation procedure using a 20-l reactor volume and the following wastewater characteristics.

BOD = 2500 mg/l
TKN = 12 mg/l
NH_3^-N = 2 mg/l
$o\text{-}PO_4^-P$ = 3 mg/l

Operate the reactor using an $F/M = 0.3$ d^{-1}, with MLVSS $= 3000$ mg/l.

Solution.

$$NH_3 \text{ required} = [2500/(100/5)] - 2$$
$$= 123 \text{ mg N/l}$$

$$o\text{-PO}_4^-\text{P required} = [2500/(100/1)] - 3$$
$$= 22 \text{ mg P/l}$$

$$t = \frac{S_o}{X_v(F/M)} = \frac{2500}{(3000 \cdot 0.3)}$$
$$= 2.8 \text{ d}$$

Assuming that the wastewater is not bioinhibitory (based on FBR test results), the volume added per feeding cycle is

$$V = \frac{20}{2.8} = 7.14 \text{ l/d}$$

If the FBR results indicate that the wastewater is inhibitory, then the feed should be diluted with a weak, but readily degradable, substrate to provide an $S_o = 2500$ mg/l. The wastewater contribution to the blended feed volume should be gradually increased (about 10 percent per feed cycle) while maintaining $S_o = 2500$ mg/l. The SOUR and SVI should be measured daily and the effluent SCOD concentration measured at the end of each feed cycle.

Reactor Operation

At least three reactors should be operated in parallel over an SRT range of 3 to 12 d for a readily degradable wastewater and 10 to 40 d for a less degradable wastewater. Since floc size is related to power level, which in turn affects the reaction coefficient K, the power level in the pilot units should approximate field conditions. The SRT should be maintained by daily wasting of an appropriate mixed liquor volume (i.e., for a 10-d SRT, one-tenth of the reactor volume is wasted daily). The waste sludge mass is computed as the VSS in the wasted reactor volume, plus the VSS in the reactor effluent. The sampling and analytical schedule for the reactors is summarized in Table 6.23. If the wastewater is sorbable and readily degradable, a sequencing batch reactor (SBR) or a selector should be used to generate a nonfilamentous sludge (Fig. 6.69). For a more refractory wastewater, a complete mix reactor as shown in Fig. 6.70 should be used.

At the end of the treatability study, the degradable fraction X_d and the endogenous decay coefficient b are determined. Sludge from each reactor is washed and aerated, and the concentration of VSS is measured every 2 to 3 d until there is no further reduction in VSS. The degradable fraction and the endogenous decay coefficient can be calculated as shown in Fig. 6.71.

Figure 6.72 presents the methods for graphical analysis of the treatability reactor data. The rate coefficient K is determined by plotting $S_o(S_o - S_e)/X_v t$ versus S_e. If the influent wastewater has a variable organic composition, K will not be con-

TABLE 6.23
Biological treatment process design study sampling and analysis schedule

Analysis	Frequency
BOD	3/week
COD or TOC	Daily
O_2 uptake rate	Daily
MLVSS	Daily
Dissolved oxygen	Daily
pH	Daily
Temperature	Daily
Nitrogen[†]	2/week
Phosphorus	2/week
Bioassay[‡]	Weekly
Specific pollutants	Weekly

[†] The case of a nitrogen-deficient wastewater. If nitrification is desired, TKN, NH_3^-N, and NO_3^-N should be run 3 times per week.
[‡] The test species will depend on permit conditions.

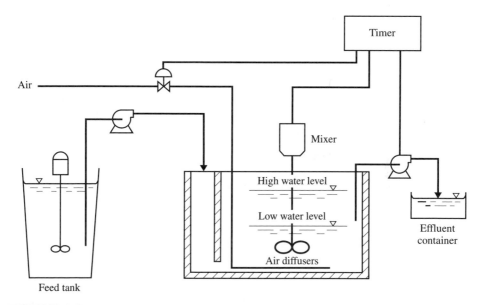

FIGURE 6.69
Pilot plant activated sludge system (SBR modification).

stant. The oxygen coefficient a' is determined as the slope of the plot of O_2/X_dX_vt versus S_r/X_dX_vt, and the endogenous respiration coefficient b' is the intercept. The sludge yield coefficient a is determined from a plot of $\Delta X_v/X_dX_vt$ versus S_r/X_dX_vt. It should be noted that ΔX_v consists of the sludge wasted per day, plus that in the reactor effluent. The design criteria for the final clarifier are determined from zone settling velocity measurements and batch flux analysis of the sludge.

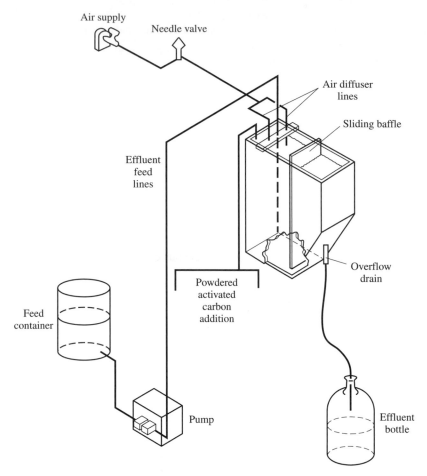

FIGURE 6.70
Biological reactor with alternative PAC addition.

Volatile Organic Carbon

The present emphasis on volatile emissions from wastewater treatment plants requires that stripping be considered in activated sludge process design where volatiles are present in the influent wastewater. There are several factors to be considered in the experimental design:

- Both the power level in the aeration basin and the type of aerator (i.e., diffused or mechanical) significantly influence stripping.
- The maximum expected concentration of each particular volatile should be employed. The degradation rate of specific volatiles will be related to both the composite wastewater composition and the process operating conditions, i.e., the SRT. Therefore, these variables must be fixed prior to volatile stripping and degradation studies.

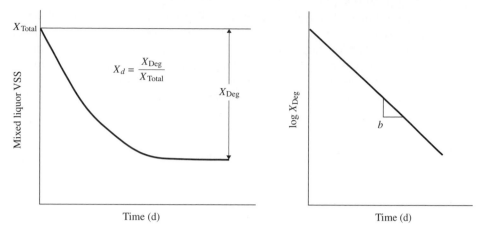

FIGURE 6.71
Determination of the degradable fraction and the endogenous coefficient.

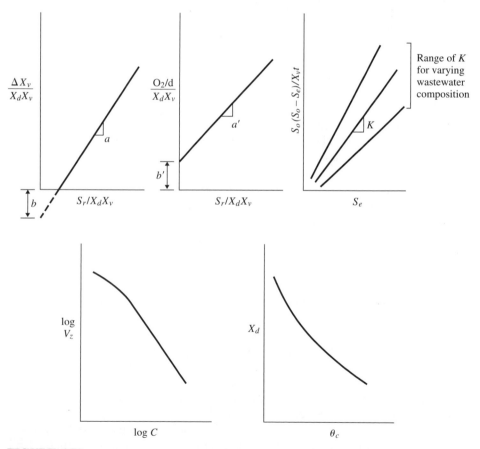

FIGURE 6.72
Parameter correlation plots.

- It has been shown that off-gas capture and recirculation will significantly enhance biodegradation of VOCs. Therefore, this process modification should be included in the pilot studies if VOC emission control is required for the plant.
- Covered aeration basins may result in a significant temperature rise due to the exothermic reaction with high-strength wastewaters. Basin temperatures in excess of approximately 38°C may result in floc dispersion. In these cases, it is necessary to monitor the reactor temperature and make appropriate temperature adjustments as required.

Once the degradation rate is defined, the calculations to determine the fraction of the volatile organics stripped can be made as described in on page 273.

Reduction of Aquatic Toxicity

Aquatic toxicity is now regulated in virtually all industrial effluent discharge permits. Toxicity can be generated by inorganics and organics that are biodegradable, nonbiodegradable, or generated within the process (SMP). One of the first steps in a treatability study, therefore, is to determine the applicability of biological treatment to the wastewater.

Two cases will be considered. In the first, there is a known toxic compound that is biodegradable. A series of reactors at varying SRT are run in order to determine the required SRT to reduce the aquatic toxicity and meet permit requirements. Am example of this using nonylphenol from a surfactant manufacturing wastewater is shown in Fig. 6.73. In the second, more common case, toxicity is caused by an unknown mixture of organics or is generated or enhanced by the production of SMP during biooxidation. In this case toxicity may be correlated to effluent COD as shown in Fig. 6.74 for a petroleum refinery wastewater. The testing protocol illustrated in Fig. 6.46 has been developed to define treatment options in this case. The biodegradability of the wastewater is determined by an FBR test. It is important that the biological sludge used in the test is fully acclimated to the wastewater. If the wastewater proves to be nondegradable and toxic, pretreatment must be considered. The primary objective of pretreatment would be detoxification and enhanced biodegradability. If the wastewater is degradable as defined by the FBR procedure, then a long-term oxidation test (e.g., 48 h aeration) is conducted to remove all degradable components. This effluent is then subjected to a bioassay. If this effluent is still toxic, then alternative pretreatment methods or tertiary treatment should be considered. Since it is not possible to distinguish between wastewater component toxicity and SMP toxicity, the wastewater should be pretreated for detoxification. A subsequent long-term oxidation test should then determine the source of the toxicity.

If the toxicity is caused by SMP, tertiary treatment with carbon should be evaluated. It is shown in Chap. 11 that, although laboratory adsorption isotherms will establish the applicability of carbon, they cannot be directly used for process design. A series of reactors should be run at various carbon dosages using the SRT required to remove degradable organics and/or priority pollutants. Each reactor

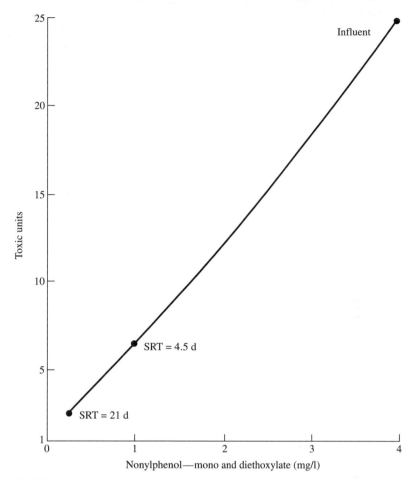

FIGURE 6.73
Effects of SRT on toxicity reduction for nonylphenolics.

should be preloaded with carbon that has been brought into equilibrium with the reactor effluent. The equilibrium carbon concentration in the reactor mixed liquor is calculated from Eq. 8.12. If the PAC-treated wastewater still exhibits aquatic toxicity, then at-source treatment or elimination of the specific toxic wastewater streams should be considered.

PROBLEMS

6.1. The following data were developed from a pilot plant study at 20°C:
   (a) Determine the K for this wastewater.
   (b) If $\theta = 1.06$, what is the K at 10°C?

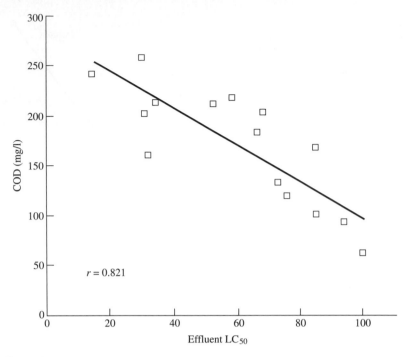

FIGURE 6.74
Toxicity/COD correlation for wastewater treatment plant effluent.

F/M, d^{-1}	X_v, mg/l	S_o, mg/l	S_e, mg/l
0.3	2000	1640	26
0.48	1980	1640	44
0.72	2000	1640	70
0.96	2056	1640	100
1.18	2050	1640	167
2.68	2100	1640	333

6.2. A complete mix activated sludge plant treating an organic chemicals wastewater yielded the following performance data:

$\text{TOD}_{\text{influent}} = 830$ mg/l
$\text{TOD}_{\text{effluent}} = 50$ mg/l
Oxygen uptake rate (OUR) = 0.49 mg/(l · min)
MLVSS = 3200 mg/l
$Q = 3.2$ million gal/d (12,100 m³/d)
$t = 0.8$ d

A sample of mixed liquor was aerated in the laboratory to determine the endogenous oxygen uptake rate where

OUR = 0.3 mg/(l · min)
MLVSS = 4700 mg/l

Calculate the endogenous coefficient b' and the coefficients a and a'.

6.3. Determine the operating F/M, MLVSS, and f_b for an activated sludge process under the following operating conditions:

$a = 0.45$
$b = 0.1$ at 20°C
$\theta_c = 10$ d
$X_i = 200$ mg/l
$f_x = 0$
$S_o = 1000$ mg/l
$S_e = 20$ mg/l
$t = 0.9$ d

6.4. Using the data from Example 6.6, compute the detention time required if the influent nondegradable VSS is 100 mg/l. If we wish to maintain the same detention time of 0.52 d, what must the MLVSS be increased to in order to maintain the same effluent quality?

6.5. The average operating and performance conditions for an activated sludge process are illustrated below.

Influent

$SCOD_o = 1018$ mg/l
$X_i = 51$ mg/l
$f_x = 0$
$Q_o = 5.19$ million gal/d

Effluent

$SCOD_e = 248$ mg/l
$VSS_e = 34$ mg/l

Waste Sludge

$Q_w = 0.08$ million gal/d
$VSS_w = 6080$ mg/l

Aeration Basin

$V = 1$ million gal/d
$MLVSS = 2295$ mg/l
$b = 0.18$ d^{-1}
$t = 0.19$ d
$OUR = 146$ mg/(l-h)

Develop a process material balance.

6.6. An untreated industrial wastewater has a BOD of 935 mg/l, negligible organic nitrogen, NH_3^-N of 8 mg/l, and a flow rate of 1.5 million gal/d. The wastewater treatment plant operates at an SRT of 15 d with MLVSS of 3000 mg/l. The effluent soluble BOD is 15 mg/l. The sludge yield coefficient is 0.6, b is 0.1 d^{-1}, and the hydraulic retention time is 1.4 d. Compute the nitrogen that must be added.

6.7. A complete mix activated sludge plant is to be designed for the following conditions:

$Q = 3.5$ million gal/d (1.32×10^4 m^3/d)
$S_o = 650$ mg/l
S (soluble) $= 20$ mg/l
$X_v = 3000$ mg/l

$a = 0.50$
$a' = 0.52$
$b = 0.1/d$ at 20°C
$\theta = 1.065$
$K = 6.0/d$ at 20°C
$b' = 0.14/d$

Compute:

   (a) The aeration volume.
   (b) The F/M.
   (c) The sludge yield.
   (d) The oxygen requirements.
   (e) The nutrient requirements.
   (f) The effluent quality at 10°C.

6.8. Develop a relationship between the NH_3^-N/BOD ratio and the aerobic volume percentage for a nitrification-denitrification channel. Consider a BOD of 200 mg/l and a range of NH_3^-N of 10 to 70 mg/l. Compute the oxygen requirements for each case. Use the following parameters:

R_N = half the maximum rate
K_{DN} = 0.06 mg NO_3-N/(mg VSS · d) at 20°C
X_v = 3000 mg/l
$a' = 0.55$
$K_{BOD,\ aerobic}$ = 8.0/day at 20°C
BOD/NO_3^-N = 3.0
$a = 0.5$
$b = 0.1\ d^{-1}$
$X_d = 0.45$
S_e = 10 mg/l

REFERENCES

1. Weber, W. J., and B. E. Jones: EPA NTIS PB86-182425/AS, 1983.
2. Matter-Mutter et al.: *Prop. Water Tech.,* vol. 12, pp. 299–313, 1980.
3. Namkung, J., and Rittman: *J. WPCF,* vol. 59, no. 7, p. 670, 1987.
4. Branghman and Pariss: *Critical Review in Microbiology,* vol. 8, p. 205, 1981.
5. Dobbs, Wang, and Govind: *Environ. Sci. & Tech.,* vol. 23, no. 9, p. 1092, 1989.
6. Kincannon, D. F., and E. L. Stover: EPA Report CR-806843-01-02, 1982.
7. Grady C. P. L. et al.: *Water Research,* vol. 30, p. 742, 1996.
8. Tabak, H. H., and E. F. Barth: *J. WPCF,* vol. 50, p. 552, 1978.
9. Gaudy, A. F.: *J. WPCF,* vol. 34, pt. 2, p. 124, February 1962.
10. Englebrecht, R. S., and R. E. McKinney: *Sewage Ind. Wastes,* Vol. .29, pt. 12(l), p. 350, December 1957.
11. McWhorter, T. R., and H. Heukeleklan: *Advances in Water Pollution Research,* vol. 2, Pergamon, New York, 1964.
12. Quirk, T., and W. W. Eckenfelder: *J. WPCF,* vol. 58, pt. 9, p. 932, 1986.
13. Sawyer, C. N.: *Biological Treatment of Sewage and Industrial Wastes,* vol. 1, Reinhold, New York, 1956.

14. Gellman, I., and H. Heukelekian: *Sewage Ind. Wastes,* vol. 25, pt. 10(l), p. 196, 1953.
15. Busch, A. W., and N. Myrick: *Proc. 15th Ind. Waste Conf.,* 1960, Purdue University.
16. Pipes, W.: *Proc. 18th Ind. Waste Conf.,* 1963, Purdue University.
17. Eckenfelder, W. W.: *Principles of Water Quality Management,* CBI, Boston, 1980.
18. Helmers, E. N, J. P. Frame, A. F. Greenbert, and C. N. Sawyer: *Sewage Ind. Wastes,* vol. 23, pt. 7, p. 834, 1951.
19. Wuhrmann, K.: *Biological Treatment of Sewage and Industrial Wastes,* vol. 1, Reinhold, New York, 1956.
20. Tischler, L. F., and W. W. Eckenfelder: *Advances in Water Pollution Research,* vol. 2, Pergamon, Oxford, England, 1969.
21. Gaudy, A. F., K. Komoirit, and M. N. Bhatla: *J. WPCF,* vol. 35, pt. 7, p. 903, July 1963.
22. Grau, P.: *Water Res.,* vol. 9, p. 637, 1975.
23. Adams, C. E., W. W. Eckenfelder, and J. Hovious: *Water Res.,* vol. 9, p. 37, 1975.
24. Van Niekerk et al.: *Wat. Sci. Tech.,* vol. 19, p. 505, 1987.
25. Argaman, Y.: *Water Research,* vol. 25, p. 1583, 1991.
26. Rickard, M. D., and Gaudy, A. F.: *J. WPCF,* vol. 49, R129, 1968.
27. Zahradka., V., *Advances in Water Pollution Research,* vol. 2, Water Pollution Control Federation, Washington, D.C., 1967.
28. Philbrook, D. M., and Grady, P. L.: *Proc. 40th Industrial Waste Conf.,* Purdue University, 1985.
29. Hoover, P.: M. S. Dissertation, Vanderbilt University, 1989.
30. Nevalainen, J. et al.: *Wat. Sci. Tech.,* vol. 24, no. 3–4, 1991.
31. Hall, E. R., and Randall, W. G.: *Wat. Sci. Tech.,* vol. 26, no. 1–12, 1992.
32. Higgins, M. J., and J. T. Novak: *Wat. Env. Res.,* vol. 69, no. 2, 1997.
33. Palm, J. C., D. Jenkins, and P. S. Parker: *J. WPCF,* vol. 52, pt. 2, p. 484, 1980.
34. Richard, M. G.: *WEF Ind. Waste Tech. Conf.,* New Orleans, 1997.
35. Eckenfelder, W. W., and J. Musterman: *Activated Sludge Treatment of Industrial Wastewaters,* Technomic Publishing, 1995.
36. Marten, W., and G. Daigger: *Water Env. Research,* vol. 69, no. 7, p. 1272, 1997.
37. Wanner, J.: *Activated Sludge Bulking and Foaming Control,* Technomic Publishing, 1994.
38. Jenkins, D., M. Richard, and G. Daigger: *Manual on the Causes and Control of Activated Sludge Bulking and Foaming Water,* Research Committee, Pretoria, S.A., 1984.
39. Pitter, J., and J. Chudoba: *Biodegradability of Organic Substances in the Aquatic Environment,* CRC Press, Boca Raton, Fla., 1990.
40. Chudoba, J.: *Water Res.,* vol. 19, no. 2, p. 197, 1985.
41. Volskay, V. T., and P. L. Grady: *J. WPCF,* vol. 60, no. 10, p. 1850, 1988.
42. Watkin, A.: Ph.D. dissertation, Vanderbilt University, 1986.
43. Watkin, A., and W. W. Eckenfelder: *Water Sci. Tech.,* vol. 21, p. 593, 1988.
44. Larson, R. J., and S. L. Schaeffer: *Water Res.,* vol. 16, p. 675, 1982.
45. Roberts, P. V. et al.: *J. WPCF,* vol. 56, no. 2, p. 157, 1984.
46. Poduska, R. A.: Ph.D. thesis, Clemson University, 1973.
47. Wong Chong, G. M., and R. C. Loehr: *Water Res.,* vol. 9, p. 1099, 1975.
48. Blum, J. W., and R.A. Speece: *Database of Chemical Toxicity to Bacteria and Its Use in Interspecies Comparisons and Correlations,* Vanderbilt University, 1990.
49. Henning, A., and R. Kayser, *42nd Purdue Ind. Waste Conf.,* p. 893, 1986.
50. Skinner, F. A., and N. Walker: *Arch. Mikrobiol.* pp. 38–339, 1961.
51. Sadick, T. E. et al.: *Proc. WEF,* vol. 1, Dallas, 1996.
52. Zacharias, B., and R. Kayser: *50th Purdue Ind. Waste Conf. Proc.,* Ann Arbor Press, 1995.

53. Anthoisen, A. C.: *J. WPCF,* vol. 48, p. 835, 1976.
54. Sutton et al.: First Workshop, Canadian-German Cooperation, Wastewater Technology Center, Burlington, Ontario, Canada, 1979.
55. Orhon, S. et al.: *Water Sci. Tech.,* vol. 34, no. 5, p. 67, 1996.
56. McClintock, S. A. et al.: *J. WPCF,* vol. 60, no. 3, 1988.
57. OH, J., and J. Silverstein: *Water Res.,* vol. 33, no. 8, p. 1925, 1999.
58. Lie, J., and R. Welander: *Water Sci. Tech.,* vol. 30, no. 6, p. 91, 1994.
59. Baumann, P., and Kh. Krauth: *Proc 2nd Specialized Conference on Pretreatment of Industrial Wastewaters,* Athens, Greece, 1996.
60. Henze H. et al.: *Water Res.,* vol. 27, no. 2, p. 231, 1993.
61. "Secondary Settling Tanks," scientific and technical report no. 6, IAWQ, 1998.

7

BIOLOGICAL WASTEWATER
TREATMENT PROCESSES

7.1
LAGOONS AND STABILIZATION BASINS

Stabilization basins are a common method of organic wastewater treatment where sufficient land area is available and where groundwater pollution from toxic organics or heavy metals is not a problem.

Stabilization basins can be divided into two classifications: the impounding and adsorption lagoon and the flow-through lagoon. In the impounding-adsorption lagoon, either there is no overflow or there is intermittent discharge during periods of high stream flow. The volumetric capacity of the basin is equal to the total waste flow less losses by evaporation and percolation. If there is intermittent discharge, the required capacity is related to the stream flow characteristics. In view of the large area requirements, impounding lagoons are usually limited to industries discharging low daily volumes of waste or to seasonal operations such as the canning industry.

The flow-through lagoon can be classified into three categories based on the dominant types of biological activity.

Type I. Facultative Ponds

Facultative ponds are divided by loading and thermal stratification into an aerobic surface and an anaerobic bottom. The aerobic surface layer will have a diurnal variation, increasing in oxygen content during the daylight hours due to algal photosynthesis and decreasing during the night, as shown in Fig. 7.1. Sludge deposited on the bottom will undergo anaerobic decomposition, producing methane and other gases. Odors will be produced if an aerobic layer is not maintained. Depths will vary from 3 to 6 ft (0.9 to 1.8 m).

313

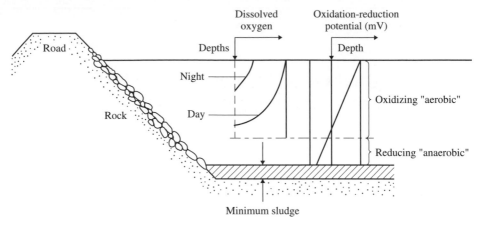

FIGURE 7.1
Waste stabilization pond—facultative type.[1]

Because oxygen generation by photosynthesis depends on light penetration, highly colored wastewaters such as textile and pulp and paper cannot be treated by this technology.

Type II. Anaerobic Ponds

Anaerobic ponds are loaded to such an extent that anaerobic conditions exist throughout the liquid volume. The biological process is the same as that occurring in anaerobic digestion tanks, i.e., primarily organic acid formation followed by methane fermentation. The depth of an anaerobic pond is selected to give a minimum surface area/volume ratio and thereby provide maximum heat retention during cold weather.

Type III. Aerated Lagoons

These range from a few days' to 2 weeks' detention, depending on the BOD removal efficiency desired. Oxygen is supplied by diffused or mechanical aeration systems, which also cause sufficient mixing to induce a significant amount of surface aeration. Depths from 6 to 15 ft (1.8 to 4.6 m) are common.

Lagoon Applications

For some industrial waste applications, aerobic ponds have been used after anaerobic ponds to provide high-degree treatment. Stabilization basins are also used to polish effluents from biological-treatment systems such as trickling filters and activated sludge.

In aerobic ponds, the amount of oxygen produced by photosynthesis can be estimated from

$$O_2 = CfS \tag{7.1}$$

where O_2 = oxygen production, lb/(acre · d) or kg/(m² · d)
        C = 0.25 if O_2 is in lb/(acre · d) or 2.8×10^{-5} if O_2 is in kg/(m² · d)
        f = light conversion efficiency, %
        S = light intensity, cal/(cm² · d)

If the light conversion efficiency is estimated as 4 percent, then $O_2 = S$. S is a function of latitude and the month of the year and may be expected to vary from 100 to 300 calories/(cm² · d) during winter and summer for a latitude 30°. This, in turn, would imply maximum loadings of 100 to 300 lb BOD_u/(acre · d) [0.011 to 0.034 kg/(m² · d)] in order to maintain any aerobic activity in the pond.

The depth of oxygen penetration in a facultative pond has been estimated as a function of surface loadings, as shown in Fig. 7.2. It should be noted that the data for Fig. 7.2 was developed from oxidation ponds treating domestic wastewater in

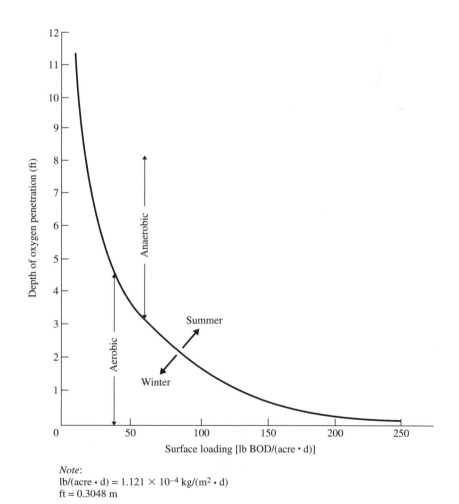

Note:
lb/(acre · d) = 1.121×10^{-4} kg/(m² · d)
ft = 0.3048 m

FIGURE 7.2
Depth of oxygen penetration in facultative ponds.[2]

California. Appropriate adjustments to this curve would have to be made for other types of wastewater being treated in other climatic conditions.

The typical green algae in waste stabilization basins are *Chlamydomonas, Chlorella,* and *Euglena.* Common blue-green algae are *Oscillatoria, Phormidium, Anacystic,* and *Anabaena.* Algae types in a pond will vary seasonally.

In the treatment of highly colored or turbid wastewaters such as from kraft pulp and paper mills, light penetration will be minimal and oxygen input will generate primarily from surface reaeration. Gellman and Berger[3] estimated oxygen input from reaeration at 45 lb O_2/(acre · d) [0.005 kg/(m^2 · d)]. The performance of stabilization basins in the pulp and paper industry from their data is shown in Fig. 7.3.

Several concepts have been employed for the design of facultative and anaerobic ponds. Functional equations for anaerobic and facultative ponds have been developed as follows:

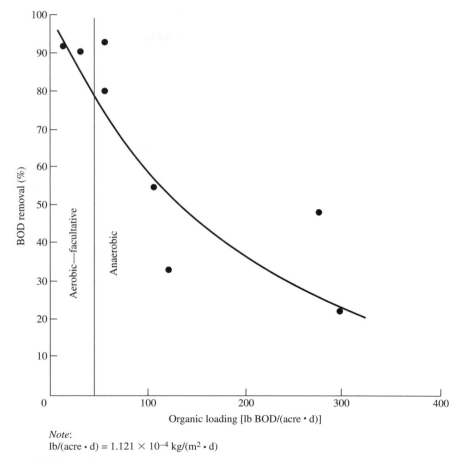

Note:
lb/(acre · d) = 1.121 × 10⁻⁴ kg/(m^2 · d)

FIGURE 7.3
Waste stabilization pond performance in the pulp and paper industry.

For a single well-mixed pond:

$$\frac{S}{S_o} = \frac{1}{1 + kt}$$ (7.2)

For multiple ponds:

$$\frac{S}{S_o} = \frac{1}{(1 + k_1 t_1)(1 + k_2 t_2) \cdots (1 + k_n t_n)}$$ (7.2a)

For an infinite number of ponds or for a plug flow pond:

$$\frac{S}{S_o} = e^{-k_n t_n}$$ (7.3)

When variable influent concentrations are considered, Eq. (7.2) can be modified to

$$\frac{S}{S_o} = \frac{1}{1 + kt/S_o}$$ (7.4)

The equation is functionally the same as that employed for aerated lagoons and activated sludge, except that the rate coefficient k includes the effect of biomass concentration. This is because it is generally impractical or impossible to effectively measure the biomass concentration (VSS) in waste stabilization basins.

When multiple ponds are to be considered [Eq. (7.3)], k is assumed to be the same for all ponds. For complex wastewaters, this is probably not true, since the more readily degradable compounds will be removed in the initial ponds. For these cases, an experimental study would need to be conducted for the wastewater in question to define the change in k.

Data from an organic chemicals plant showed k to be 0.05 d^{-1} under anaerobic conditions at 20°C and 0.5 d^{-1} under aerobic conditions.

In a stabilization basin, the average k_m can be computed:

$$k_m = \frac{k_{\text{aerobic}} \times D_{\text{aerobic}} + k_{\text{anaerobic}} \times D_{\text{anaerobic}}}{D_{\text{total}}}$$ (7.5)

As in all biological processes, the biological activity in the pond will be a function of temperature, and the rate coefficient k can be corrected through the application of Eq. (6.37). Evaluation of the 30-d average performance for a kraft pulp and paper mill showed a θ value of 1.053, as shown in Fig. 7.4. During winter operation in colder climates, the pond will ice over, resulting in anaerobic conditions and reduced performance. (Note that an ice cover will act as an insulator, maintaining higher temperature in the liquid.)

Certain design considerations are important to the successful operation of stabilization basins. These have been discussed by Hermann and Gloyna[4] and Marais.[5] Embankments should be constructed of impervious material with maximum slope between 3:1 and 4:1 and minimum slope of 6:1. A minimum freeboard of 3 ft (0.91 m) should be maintained in the basin. Provision should be made for protecting the bank from erosion. Wind action is important for pond mixing and is effective with a fetch of 650 ft (198 m) in a pond with a depth of 3 ft (0.91 m).

Meat waste has been treated in shallow aerobic ponds only 18 in (0.66 m) deep at loadings of 214 lb BOD/(acre · d) [0.024 kg/(m^2 · d)]. The waste was presettled

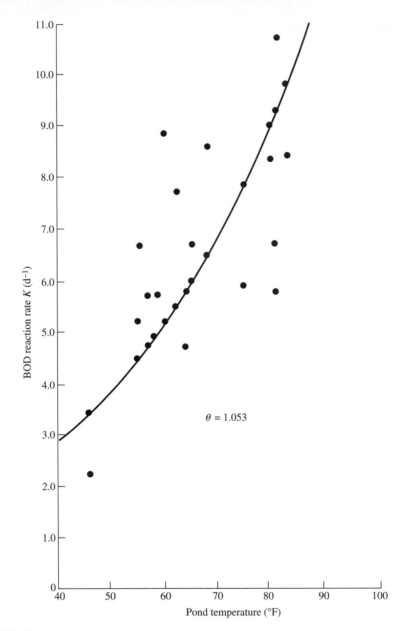

FIGURE 7.4
Temperature effect on 30-d average performance of a pond treating a pulp and paper mill effluent.

and had a concentration of 175 mg/l BOD. BOD reductions of 96 percent in the summer and 70 percent in the winter were obtained.

Use of an aerobic pond treating meat-packing waste after the anaerobic contact process yielded 80 percent BOD removal at a loading of 410 lb BOD/(acre · d)

[0.046 kg/(m^2 · d)]. The concentration of BOD entering the pond was 129 mg/l. Removals at various loadings have been summarized by Steffen.[6]

Packing-house waste loadings to anaerobic ponds 8 to 17 ft (2.4 to 5.2 m) deep have been reported to vary from 0.011 to 0.015 lb BOD/(ft^3 · d) [0.176 to 0.240 kg/(m^3 · d)], with high BOD removals. Treatment of corn waste with an initial BOD of 2936 mg/l in a lagoon with a 9.6-d detention yielded 59 percent BOD reduction. The organic loading was 0.184 lb BOD/(ft^3 · d) [2.95 kg/(m^3 · d)].

In order to improve the efficiency of lagoon operation, nitrogen and phosphorus must be added to nutrient-deficient wastes. The requirements for aerobic and anaerobic processes are discussed on pages 215 and 402, respectively.

Lagoons treating wastes in which aerobic conditions are not maintained frequently emit odors and provide a breeding ground for insects. The odor problems can frequently be eliminated by the addition of sodium nitrate at a dosage equal to 20 percent of the applied oxygen demand, as shown in Fig. 7.5. Surface sprays can be used to reduce the fly and insect nuisance and in some cases the odor.

High efficiencies can frequently be attained by the use of anaerobic ponds followed by aerobic ponds. An anaerobic pond with a 6-d detention and 14-ft (4.3 m) depth loaded at 0.014 lb BOD/(ft^3 · d) [0.224 kg/(m^3 · d)], followed by a 3-ft- (0.9-m-) deep aerobic pond with a 19-d detention loaded at 50 lb BOD/(acre · d) [0.0056 kg/(m^2 · d)], yielded an overall reduction in BOD from 1100 to 67 mg/l.[7] Performance data for aerobic facultative and anaerobic ponds are summarized for various industrial wastewaters in Table 7.1. The design of facultative ponds is illustrated by Example 7.1.

EXAMPLE 7.1. An industrial wastewater with a BOD of 500 mg/l is to be treated in a pond or series of ponds with a total retention time of 50 d with a depth of 6 ft. The anaerobic k at 20°C is 0.05 d^{-1} and the aerobic k is 0.51 d^{-1}. Assume the oxygen relationships as shown in Fig. 7.2 apply and the pond temperature is 20°C.

Solution.

1. For one pond the applied loading is:

$$\text{Loading} = \frac{2.7DS_o}{t}$$

$$= \frac{2.7(6)(500)}{50}$$

$$= 162 \text{ lb BOD/(acre} \cdot \text{d)}$$

Dissolved oxygen will exist to a depth of 0.8 ft. The mean k can be computed:

$$k = \frac{(0.8)(0.5) + 5.2(0.05)}{6} = 0.11 \text{ d}^{-1}$$

The effluent BOD is:

$$S_e = \frac{S_o}{1 + kt}$$

$$= \frac{500}{1 + 0.11(50)}$$

$$= 77 \text{ mg/l}$$

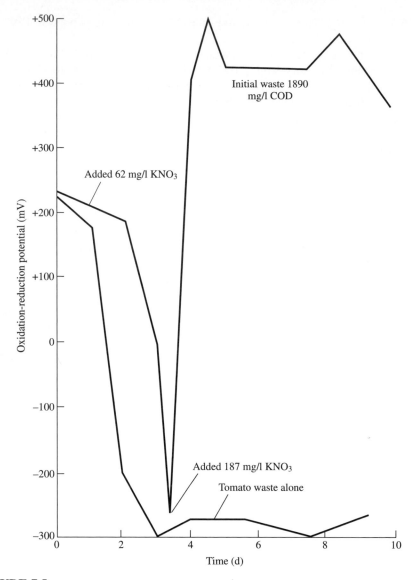

FIGURE 7.5
ORP control of nitrate addition to pond treatment of tomato wastewaters.

2. For four ponds in series, the retention time in each pond will be 12.5 d.

   The loading to the first pond is:

   $$\frac{2.7(6)(500)}{12.5} = 648 \text{ lb BOD/(acre} \cdot \text{d)}$$

   and is anaerobic.

   The effluent from the first pond is:

   $$S_e = \frac{500}{1 + 0.05\,(12.5)} = 308 \text{ mg/l}$$

TABLE 7.1
Performance of lagoon systems

Summary of average data from aerobic and facultative ponds

Industry	Area, acres	Depth, ft	Detention, d	Loading, lb/(acre · d)	BOD removal, %
Meat and poultry	1.3	3.0	7.0	72	80
Canning	6.9	5.8	37.5	139	98
Chemical	31	5.0	10	157	87
Paper	84	5.0	30	105	80
Petroleum	15.5	5.0	25	28	76
Wine	7	1.5	24	221	
Dairy	7.5	5.0	98	22	95
Textile	3.1	4.0	14	165	45
Sugar	20	1.5	2	86	67
Rendering	2.2	4.2	4.8	36	76
Hog feeding	0.6	3.0	8	356	
Laundry	0.2	3.0	94	52	
Miscellaneous	15	4.0	88	56	95
Potato	25.3	5.0	105	111	

Summary of average data from anaerobic ponds

Industry	Area, acres	Depth, ft	Detention, d	Loading, lb/(acre · d)	BOD removal, %
Canning	2.5	6.0	15	392	51
Meat and poultry	1.0	7.3	16	1260	80
Chemical	0.14	3.5	65	54	89
Paper	71	6.0	18.4	347	50
Textile	2.2	5.8	3.5	1433	44
Sugar	35	7.0	50	240	61
Wine	3.7	4.0	8.8		
Rendering	1.0	6.0	245	160	37
Leather	2.6	4.2	6.2	3000	68
Potato	10	4.0	3.9		

Summary of average data from combined aerobic-anaerobic ponds

Industry	Area, acres	Depth, ft	Detention, d	Loading, lb/(acre · d)	BOD removal, %
Canning	5.5	5.0	22	617	91
Meat and poultry	0.8	4.0	43	267	94
Paper	2520	5.5	136	28	94
Leather	4.6	4.0	152	50	92
Miscellaneous industrial wastes	140	4.1	66	128	

Note:
ft = 0.3048 m
lb/(acre · d) = 1.121×10^{-4} kg/(m² · d)
acre = 4.0469×10^3 m²

The loading to the second pond is:

$$\frac{2.7(6)(308)}{12.5} = 399 \text{ lb BOD/(acre} \cdot \text{d)}$$

$$S_e = \frac{308}{1 + (0.05)(12.5)} = 190 \text{ mg/l}$$

The loading to the third pond is 246 lb BOD/(acre · d) and the effluent is 117 mg/l. The loading to the fourth pond is 152 lb BOD/(acre · d) and the aerobic depth is 1 ft. The adjusted k is:

$$\frac{0.5 \, (1) + 0.05 \, (5)}{6} = 0.125 \text{ d}^{-1}$$

The effluent is:

$$S_e = \frac{117}{1 + (0.125) \, (12.5)} = 45.7 \text{ mg/l}$$

Four ponds in series with the same total retention time will produce a superior effluent. This assumes the reaction rate k does not change through the series of basins. This is probably not true in many cases and would have to be determined experimentally.

7.2
AERATED LAGOONS

An aerated lagoon is a basin of significant depth, e.g., 8 to 16 ft (2.4 to 4.9 m) deep, in which oxygenation is accomplished by mechanical or diffused aeration units and through induced surface aeration.

There are two types of aerated lagoons:

1. The aerobic lagoon in which dissolved oxygen and suspended solids are maintained uniformly throughout the basin.
2. The aerobic-anaerobic or facultative lagoon, in which oxygen is maintained in the upper liquid layers of the basin, but only a portion of the suspended solids is maintained in suspension. These basin types are shown in Fig. 7.6. A photo of a typical aerated lagoon is shown in Fig. 7.7.

In the aerobic lagoon, all solids are maintained in suspension, and this system may be thought of as a "flow-through" activated sludge system, i.e., without solids recycle. Thus, the effluent suspended solids concentration will be equal to the aeration basin solids concentration.

In the facultative lagoon, a portion of the suspended solids settle to the bottom of the basin, where they undergo anaerobic decomposition. The anaerobic by-products are subsequently oxidized in the upper aerobic layers of the basin. The facultative lagoon can also be modified to yield a more highly clarified effluent by the inclusion of a separate postsettling pond or a baffled settling compartment.

Aerobic and facultative lagoons are primarily differentiated by the power level employed in the basin. In the aerobic lagoon, the power level is sufficiently high to maintain all solids in suspension and may vary from 14 to 20 hp/million gal (2.8

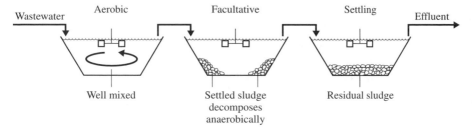

FIGURE 7.6
Aerated lagoon types.

FIGURE 7.7
Aerated lagoon treating pulp and paper mill wastewater.

to 3.9 W/m^3) of basin volume, depending on the nature of the suspended solids in the influent wastewater. Field data demonstrated that 14 hp/million gal (2.8 W/m^3) was generally sufficient to maintain pulp and paper solids in suspension, while 20 hp/million gal (3.9 W/m^3) was required for domestic wastewater treatment.

In the facultative lagoon, the power level employed is sufficient only to maintain dispersion and mixing of the dissolved oxygen. Experience in the pulp and paper industry showed that a minimum power level employing low-speed mechanical surface aerators is 4 hp/million gal (0.79 W/m^3). The use of other kinds of aeration equipment might require different power levels to maintain uniform dissolved oxygen in the basin.

Aerobic Lagoons

At a constant basin detention time, the equilibrium biological solids concentration and the overall rate of organic removal can be expected to increase as the influent organic concentration increases. For a soluble industrial wastewater, the equilibrium biological solids concentration X_v can be predicted from the relationship:

$$X_v = \frac{aS_r}{1 + bt} \tag{7.6}$$

When nondegradable volatile suspended solids are present in the wastewater, Eq. (7.6) becomes

$$X_v = \frac{aS_r}{1 + bt} + X_i \tag{7.7}$$

in which X_i = influent volatile suspended solids not degraded in the lagoon. Combining Eq. (7.6) with the kinetic relationship [Eq. (6.24)], the effluent soluble organic concentration can be computed:

$$\frac{S}{S_o} = \frac{1 + bt}{aKt} \tag{7.8}$$

From Eq. (7.8), it can be concluded that the fraction of effluent soluble organic concentration remaining is independent of the influent organic concentration. For lagoons with fixed detention times, this conclusion is justified because higher influent organic concentrations result in higher equilibrium biological solids levels and, therefore, higher overall BOD removal rates. This relationship is shown in Fig. 7.8 for several wastewaters.

The specific organic reaction rate coefficient K is temperature-dependent, and can be corrected for temperature by Eq. (6.35). The oxygen requirements for an aerobic lagoon are computed using the same relationship employed for activated sludge [Eq. (6.18)]. Aerobic lagoons are employed for pretreatment of high-strength industrial wastewaters prior to discharge to a joint or municipal treatment system, or as the first basin in a two-basin aerated lagoon series, followed by a facultative basin. It should be noted that, while in an aerobic lagoon the soluble organic content is reduced, there is an increase in the effluent suspended solids through synthesis. The relationship obtained for an aerobic lagoon pretreating a brewery wastewater is shown in Fig. 7.9.

Power requirements should generally be designed for summer operation, since the rates of organic removal will be the greatest during this period of operation.

Facultative Lagoons

In a facultative lagoon, the biological solids level maintained in suspension is a function of the power level employed in the basin. Results from the pulp and paper industry are shown in Fig. 7.10. Solids deposited in the bottom of the facultative lagoon will undergo anaerobic degradation, which results in a feedback of soluble

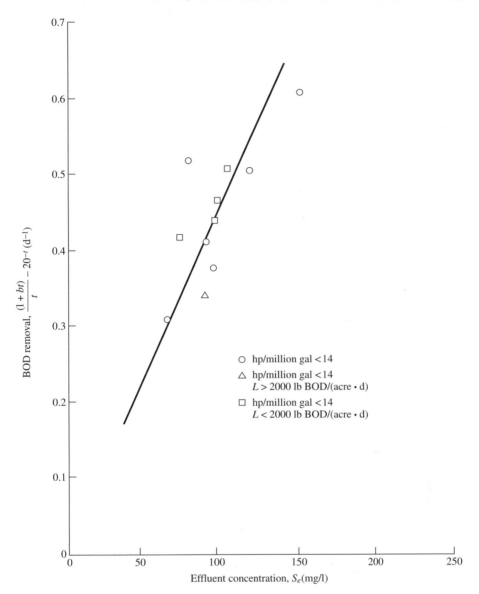

FIGURE 7.8
Kinetic relationships in aerobic lagoons.

organics to the upper aerobic layers. Under these conditions Eq. (6.5) should be modified to

$$\frac{S}{FS_o} = \frac{FS_o}{FS_o + KX_vt} \tag{7.9}$$

in which F is a coefficient that accounts for organic feedback due to anaerobic activity in the deposited sludge layers. The degree of anaerobic activity is highly temperature-dependent and the coefficient F may be expected to vary from 1.0

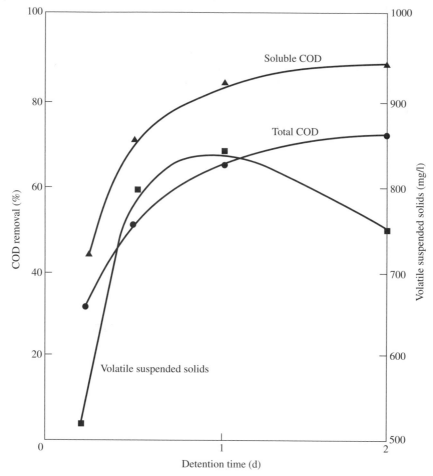

FIGURE 7.9
COD removal from brewery wastewater through an aerobic lagoon.

to 1.4 under winter and summer conditions, respectively, depending on the geographical location of the plant.

In a facultative lagoon, biological solids are maintained at a lower level than in an aerobic lagoon, and soluble organics are fed back to the liquid as anaerobic degradation products. It is therefore not possible to directly compute the oxygen requirements by employing Eq. (6.18). In this case, the oxygen requirements can be related empirically to organic removal and estimated from

$$R_r = F'S_r \tag{7.10}$$

in which F' is an overall oxygen-utilization coefficient for facultative lagoons. Results obtained for various industrial wastewaters would indicate that the coefficient F' is a function of the degree of organic feedback, which in turn is a function of influent settleable solids and temperature. In general, depending on the geographic location of the plant, F' can be estimated to vary from 0. 8 to 1.1 during

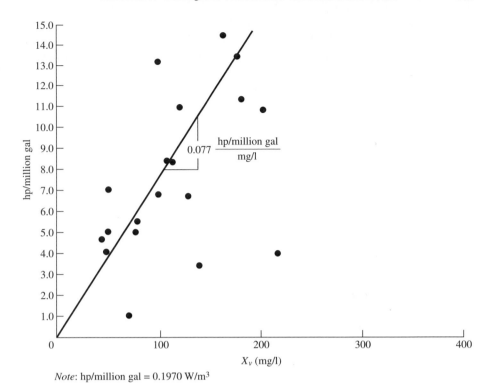

Note: hp/million gal = 0.1970 W/m³

FIGURE 7.10
Correlation for mixing power input versus MLSS concentrations.

winter operation when anaerobic activity in the basin is low, and from 1.1 to 1.5 during summer operation when anaerobic activity in the bottom of the basin is at a maximum. The value selected will depend on the geographical location of the plant.

Nutrient requirements in aerated lagoons are computed in a similar way to the activated sludge process. In the case of a facultative lagoon, however, anaerobic decomposition of sludge deposited in the bottom of the basin will feed back nitrogen and phosphorus. This is usually sufficient for organic removal occurring in this type of basin, and no additional nitrogen and phosphorus needs to be added. The relationships for the design of aerated lagoon systems are summarized in Table 7.2.[9]

Temperature Effects in Aerated Lagoons

The performance of aerated lagoons is significantly influenced by changes in basin temperature. In turn, the basin temperatures are influenced by the temperature of the influent wastewater and the ambient air temperature. Although heat is lost through evaporation, convection, and radiation, it is gained by solar radiation. While several formulas have been developed to estimate the temperature in aerated

TABLE 7.2
Aerated lagoon design relationships[9]

	Aerobic	Facultative	Settling
Relationship	hp/million gal 14 to 20	hp/million gal $> 4 < 10$	
Kinetics	$\dfrac{S_1}{S_o} = \dfrac{1 + bt}{aKt}$	$S_2 = \dfrac{F^2 S_1^{2\dagger}}{KX_v t^\S + FS_1}$	$t < 2\text{d}^\ddagger$
Oxygen requirements	$O_2/d = a'S_r + b'X_v t$	$O_2/d = \begin{cases} 0.8 - 1.1\, S_r\,(\text{winter})^\P \\ 1.1 - 1.4\, S_r\,(\text{summer}) \end{cases}$	
Nutrient requirements	$N = 0.11\,\Delta X_v$ $P = 0.02\,\Delta X_v$	None required††	
Sludge yield	$X_v = \dfrac{aS_r}{1 + bt}$	$0.1\,aS_r + \text{inerts}^{\ddagger\ddagger}$ 3 to 7 percent weight concentration	

† BOD feedback from benthal decomposition varies from 1.0 in winter to 1.4 in summer.
‡ To minimize algae growth.
§ Depends on power level in basin (see Fig. 7.10).
¶ Factors employed depend on geographical location.
†† Benthal decomposition feeds back N and P.
‡‡ Most or biovolatile solids will break down through benthal decomposition.
Note: hp/million gal = 1.98×10^{-2} kW/m^3

lagoons, the following equation will usually give a reasonable estimate for engineering design purposes:

$$\frac{t}{D} = \frac{T_i - T_w}{f(T_w - T_a)} \qquad (7.11)$$

where t = basin detention time, d
       D = basin depth, ft
       T_i = influent wastewater temperature, °F
       T_a = mean air temperature, °F (usually taken as the mean weekly temperature)
       T_w = basin temperature, °F

Equation (7.11) was developed for surface aeration and hence cannot be used for subsurface aeration.

The coefficient f is a proportionality factor containing the heat-transfer coefficients, the surface area increase from aeration equipment, and wind and humidity effects. f has an approximate value of 90 for most aerated lagoons employing surface aeration equipment.

A general temperature model has been developed by Argaman and Adams[8] that considers an overall heat balance in the basin, including heat gained from solar radiation, mechanical energy input, biochemical reaction, and heat lost by long-wave radiation, evaporation from the basin surface, conduction from the basin surface, evaporation and conduction from the aerator spray, and conduction through the basin walls. Their final equation is

$$T_w = T_a + \left[\frac{Q}{A}(T_i - T_a) + 10^{-6}(1 - 0.0071C_C^2)H_{s,o} + 6.95(\beta - 1) \right.$$

$$+ 0.102(\beta - 1)T_a - e^{0.0604T_a}\left(1 - \frac{f_a}{100}\right)1.145A^{-0.05}V_w$$

$$\left. + \frac{126NFV_w}{A} + \frac{10^{-6}H_m}{A} + \frac{1.8S_r}{A} \right]$$

$$\bigg/ \left[\frac{Q}{A} + 0.102 + (0.068e^{0.0604T_a} + 0.118)A^{-0.05}V_w \right.$$

$$\left. + \frac{4.32NFV_w}{A}(3.0 + 1.75e^{0.0604T_a}) + \frac{10^{-6}UA_w}{A} \right] \qquad (7.12)$$

where
T_w = basin water temperature, °C
T_a = air temperature, °C
T_i = influent waste temperature, °C
Q = flow rate, m³/d
A = surface area, m²
C_c = average cloud cover, tenths
$H_{s,o}$ = average daily absorbed solar radiation under clear sky conditions, cal/(m² · d)
U = heat-transfer coefficient, cal/(m² · d · °C)
β = atmospheric radiation factor
f_a = relative humidity, percent
N = number of aerators
F = aerator spray vertical cross-sectional area, m²
V_w = wind speed at tree top, m/s
H_m = $15.2 \times 10^6 p$, where p = aeration power, hp
S_r = organic removal rate, kg COD removed/d
A_w = effective wall area, m²

Equation (7.12) can be used to predict the temperature of diffused air systems by substituting

$$NFV_w = 2Q_A$$

where Q_A = air flow, m³/s.

Aerated Lagoon Systems

Multiple basins may be most effectively employed in aerated lagoon systems under proper conditions. A comparison of single-stage and multistage operation for a pulp and paper mill is shown in Fig. 7.11. As can be seen, multistage operation is more efficient in terms of organic removal. In addition, series operation may be

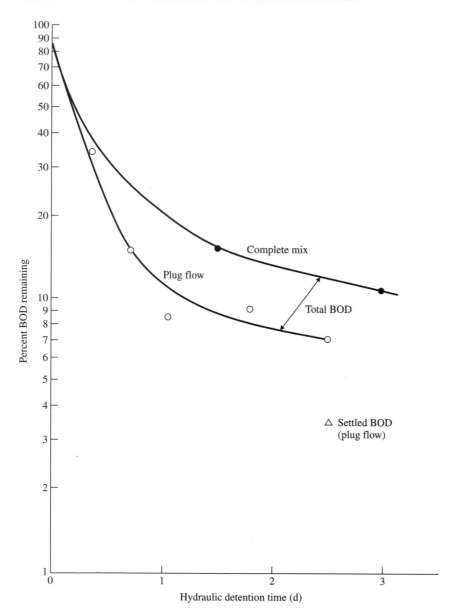

FIGURE 7.11
Pilot plant results for single- and multistage operation.

desired where land availability is a concern. In considering a thermal balance, a minimum total basin volume can be obtained by employing two basins in series. The first basin volume is minimized to maintain a high temperature, a high biological solids level, and a resulting high BOD reaction rate in an aerobic lagoon. The second basin is a facultative basin at lower power (mixing) levels that permit solids settling and decomposition in the bottom of the basin. An optimization procedure can be employed to determine the smallest total basin volume and the lowest aeration horsepower for a specified effluent quality. Where low effluent suspended solids are desired, a final settling basin can be employed. The settling basin should have:

1. A sufficiently long detention period to effect the desired suspended solids removal
2. Adequate volume for sludge storage
3. Minimal algal growth
4. Minimal odors from anaerobic activity

Unfortunately, these design objectives are not always compatible. Frequently short retention times are required to inhibit algal growth, which are too short for proper settling. Also, adequate volume must remain above the sludge deposits at all times to prevent the escape of the odorous gases of decomposition.

In order to achieve these objectives, a minimum detention time of 1 d is usually required to settle the majority of the settleable suspended solids. Where algal growths pose potential problems, a maximum detention time of 3 to 4 d is recommended. For odor control, a minimum water level of 3 ft (0.9 m) should be maintained above the sludge deposits at all times. Parker[10] has shown that a scale-up factor needs to be applied in aerated lagoon process design. The effect of scale on aerated lagoon treatment of a petrochemical wastewater is shown below:

Scale	t, d	K_{20}
Bench	0.83	10.0
Pilot	0.83	5.3
Full	1.3	2.8

Parker[10] attributes this to wall effects in the bench scale and to differences in fluid shearing intensity. In any event, caution should be exercised in scaling up a design from laboratory data.

One problem with aerated lagoons, as well as other technologies, is the effect of shock loads on effluent quality. In one pulp and paper mill effluent, a dissolved oxygen probe at the head end of the basin was used to trigger a diversion to a spill pond as shown in Fig. 7.12. In this case when the dissolved oxygen dropped below 2 mg/l, indicating a shock load, a portion of the wastewater was diverted. The spill basin contents were then pumped at a controlled rate to the aerobic lagoon under dissolved oxygen control. The variability of one aerated lagoon performance is shown in Fig. 7.13.[11]

The design of an aerated lagoon system is illustrated by Example 7.2. Where temperature is not a consideration, a simplified procedure can be employed as shown in Example 7.3.

EXAMPLE 7.2. Design a two-stage aerated lagoon system of 12 ft (3.66 m) depth to treat an 8.5 million gal/d (32,170 m³/d) industrial wastewater with the following characteristics:

Influent: BOD_5 = 425 mg/l
Temperature = 85°F (29°C)
SS = 0 mg/l

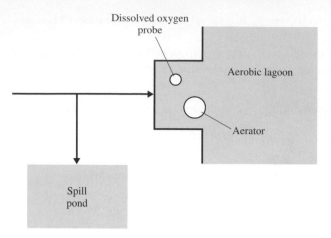

FIGURE 7.12
Spill diversion control.

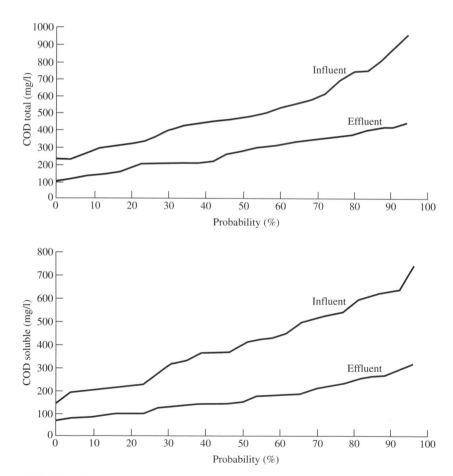

FIGURE 7.13
Performance of aerated ponds—COD removal.[11]

Ambient temperature: summer $=$ 70°F (21°C)
winter $=$ 34°F (1°C)

Kinetic variables:
$K = 63/d$ at 20°C
$a = 0.5$
$b = 0.2/d$ at 20°C
$a' = 0.52$
$b' = 0.28/d$ at 20°C
$\theta = 1.035$ for BOD reaction rate
$\theta = 1.024$ for oxygen-transfer efficiency
$F = 1.0$ (winter)
$= 1.4$ (summer)
$F' = 1.5$ (summer)
$N_o = 3.2$ lb O_2/(hp · h) [1.95 kg/(kW · h)]
$\alpha = 0.85$
$\beta = 0.90$
$C_L = 1.0$ mg/l

The final effluent should have a maximum soluble BOD_5 of 20 mg/l for the summer and 30 mg/1 for the winter.

Solution.

In general, the required detention time of the lagoon system for reaching a terminal BOD_5 is controlled by winter temperatures while the oxygen requirements and power level will usually be controlled by summer conditions.

(a) Design of basin based on minimum detention time

1. Compute basin volume by assuming a detention time.
   For $t = 2$ d:

   $$V = Qt$$
   $$= 8.5 \times 2$$
   $$= 17 \text{ million gal} \qquad (64,350 \text{ m}^3)$$

2. Calculate winter water temperature in basin:

   $$\frac{t}{D} = \frac{T_i - T_w}{f(T_w - T_a)}$$

   or

   $$T_w = \frac{DT_i + ftT_a}{D + ft}$$

   $$= \frac{12 \times 85 + 1.6 \times 2 \times 34}{12 + 1.6 \times 2}$$

   $$= 74.3°F \quad \text{or} \quad 23.5°C$$

3. Correct the BOD reaction rate for winter conditions:

   $$K_{(T_2)} = K_{(T_1)} \theta^{T_2 - T_1}$$

   $$K_{23.5} = 6.3 \times 1.035^{23.5 - 20}$$

   $$= 7.11/d$$

4. Calculate the winter effluent soluble BOD_5:

$$\frac{S_e}{S_o} = \frac{1 + bt}{aKt}$$

$$S_e = \frac{1 + 0.2 \times 2}{0.5 \times 7.11 \times 2} 425$$

$$= 83.7 \text{ mg/l}$$

5. Repeat steps 2 through 4 for summer conditions:

$$T_w = \frac{12 \times 85 + 1.6 \times 2 \times 70}{12 + 1.6 \times 2}$$

$$= 81.8°F \text{ or } 27.7°C$$

$$K_{27.7} = 6.3 \times 1.035^{27.7-20}$$

$$= 8.21/d$$

$$S_e = \frac{1 + 0.2 \times 2}{0.5 \times 8.21 \times 2} 425$$

$$= 72.5 \text{ mg/l}$$

6. Calculate average volatile suspended solids concentration under summer conditions:

$$X_v = \frac{aS_r}{1 + bt}$$

$$= \frac{0.5(425 - 72.5)}{1 + 0.2 \times 2}$$

$$= 126 \text{ mg/l}$$

7. Calculate oxygen requirement:

$$R_r = a'S_r + b'X_v$$

$$= 0.52 \times (425 - 72.5) \times 8.5 \times 8.34 + 0.28 \times 126 \times 8.5 \times 2 \times 8.34$$

$$= 17,996 \text{ lb/d} \quad (8170 \text{ kg/d})$$

8. Compute horsepower requirement:

$$N = N_o \frac{\beta C_s - C_L}{C_{s(20)}} a\theta^{T-20}$$

$$= 3.2 \frac{0.90 \times 7.96 - 1.0}{9.2} 0.85 \times 1.024^{27.7-20}$$

$$= 2.19 \text{ lb/(hp} \cdot \text{h)} \quad (1.33 \text{ kg/kW} \cdot \text{h})$$

$$\text{hp} = \frac{R_r}{N}$$

$$= \frac{17,996}{2.19 \times 24} = 342 \text{ hp} \quad (257 \text{ kW})$$

9. Check power level:

$$PL = \frac{hp}{V}$$

$$= \frac{342}{17} = 20.1 \text{ hp/million gal} \qquad (0.40 \text{ kW/m}^3)$$

For a conservative design, the minimum power level should be 14 hp/million gal (0.28 kW/m³). If the power level is significantly less than 14 hp/million gal, then 14 hp/million gal should be used.

10. Repeat steps 1 through 9 using different detention times. The results can be tabulated as follows:

t, d	Winter or summer	T_w °F	°C	K, d⁻¹	S_e, mg/l	X_v, mg/l	R_r, lb/d	hp required	PL, hp/million gal
1.0	W	79.0	26.1	7.77	131				
	S	83.2	28.4	8.41	121	127	13,727	261	30.7
2.0	W	74.3	23.5	7.11	83.7				
	S	81.8	27.7	8.21	72.5	126	17,996	342	20.1
3.0	W	70.4	21.3	6.59	68.8				
	S	80.7	27.1	8.04	56.4	115	20,436	389	153
4.0	W	67.3	19.6	6.21	61.6				
	S	79.8	26.6	7.91	48.4	105	22,219	476 (423)[†]	14 (12.4)[†]
5.0	W	64.6	18.1	5.90	57.6				
	S	79.0	26.1	7.77	43.8	95	23,480	595 (449)	14 (10.6)
6.0	W	62.3	16.8	5.64	55.3				
	S	78.3	25.7	7.66	40.7	87	24,528	714 (469)	14 (9.2)

† Minimum power level applied.

(b) Design of basin based on minimum detention time
It is assumed that the minimum power level in a facultative lagoon should be 4 hp/million gal (0.079 kW/m³) to maintain 50 mg/l of volatile suspended solids in suspension. The design of this basin is based on selecting an effluent BOD concentration from the first basin at given detention times to be the influent to the second basin.

1. Calculate the temperature in this basin and adjust the BOD reaction rate by assuming a detention time. For example, the effluent BOD concentrations at a 2-d detention time for the first basin are 83.7 and 72.5 mg/l for winter and summer, respectively. Basin temperature:

$$T_w = \frac{DT_i + ftT_a}{D + ft}$$

Corrected influent concentration:

$$S_o' = FS_o$$

Corrected BOD reaction rate:

$$K_{20} = \frac{6.3}{425} S_o'$$

Assume $t = 5$ d.

Winter:
$$T_w = \frac{12 \times 74.3 + 1.6 \times 5 \times 34}{12 + 1.6 \times 5}$$

$$= 58.2°F \quad \text{or} \quad 14.6°C$$

$$K_{20} = \frac{6.3}{425} 1.0 \times 83.7$$

$$= 1.24/\text{d}$$

$$K_{14.6} = 1.24 \times 1.035^{14.6-20}$$

$$= 1.03/\text{d}$$

Summer:
$$T_w = \frac{12 \times 81.8 + 1.6 \times 5 \times 70}{12 + 1.6 \times 5}$$

$$= 77.1°F \quad \text{or} \quad 25.1°C$$

$$K_{20} = \frac{6.3}{425} 1.4 \times 72.5$$

$$= 1.50/\text{d}$$

$$K_{25.1} = 1.50 \times 1.035^{25.1-20}$$

$$= 1.79/\text{d}$$

2. Compute the detention time required to reduce the soluble BOD to the prescribed level:

$$S_o' = FS_o$$

$$t = \frac{S_o'(S_o' - S_e)}{KX_v S_e}$$

Winter:
$$S_o' = 83.7 \text{ mg/l}$$

$$t = \frac{83.7(83.7 - 30)}{1.03 \times 50 \times 30}$$

$$= 2.91 \text{ d}$$

Summer:
$$S_o' = 1.4 \times 72.5$$

$$= 102 \text{ mg/l}$$

$$t = \frac{102(102 - 20)}{1.79 \times 50 \times 20}$$

$$= 4.67 \text{ d}$$

Now summer conditions are controlling.

3. Repeat steps 1 and 2 until the computed detention time in step 2 is close enough to the assumed value in step 1. Final results are as follows:

$$t = 4.67 \text{ d}, \quad V = 39.7 \text{ million gal } (150{,}300 \text{ m}^3)$$

Winter:
$$T_w = 58.8°F \quad \text{or} \quad 14.9°C$$
$$K_{14.9} = 1.04/\text{d}$$
$$S_e = 21.5 \text{ mg/l}$$

Summer:
$$T_w = 77.3°F \quad \text{or} \quad 25.2°C$$
$$K_{25.2} = 1.79/\text{d}$$
$$S_e = 20.0 \text{ mg/l}$$

4. Calculate oxygen requirement:
$$R_r = F'S_r$$
$$= 1.5 \times (1.4 \times 72.5 - 20.0) \times 8.5 \times 8.34$$
$$= 8666 \text{ lb/d} \quad (3934 \text{ kg/d})$$

5. Calculate horsepower requirement:
$$N = N_o \frac{\beta C_s - C_L}{C_s(20)} a\theta^{T-20}$$
$$= 3.2 \frac{0.90 \times 8.36 - 1.0}{9.2} 0.85 \times 1.024^{25.2-20}$$
$$= 2.18 \text{ lb/(hp} \cdot \text{h)} \quad [1.33 \text{ kg/(kW} \cdot \text{h)}]$$

$$\text{hp} = \frac{R_r}{N}$$
$$= \frac{8666}{2.18 \times 24}$$
$$= 166 \text{ hp} \quad (125 \text{ kW})$$

6. Check power level:
$$\text{PL} = \frac{\text{hp}}{V}$$
$$= \frac{166}{39.7}$$
$$= 4.2 \text{ hp/million gal } (0.083 \text{ kW/m}^3)$$

7. Repeat steps 1 through 6 using different detention times in the first basin with the adequate detention time in the second basin to meet the desired effluent quality. The results are tabulated below:

S_o, mg/l	T_w °F	°C	K, d^{-1}	t, d	S_e, mg/l	X_v, mg/l	R_r, lb/d	hp required	PL, hp/million gal
131	55.0	12.8	1.51	8.57	22.1	50			
121	76.2	24.6	2.94	8.57	20.0	50	10,740	291 (205)[†]	4.0 (2.8)[†]
83.7	58.8	14.9	1.04	4.67	21.5	50			
72.5	77.3	25.2	1.79	4.67	20.0	50	8,666	166	4.2
68.8	59.2	15.1	0.862	3.33	22.3	50			
56.4	77.4	25.2	1.40	3.33	20.0	50	6,274	120	4.2
61.6	58.4	14.7	0.761	2.72	23.0	50			
48.4	77.2	25.1	1.19	2.72	20.0	50	5,083	97	4.2
57.6	57.3	14.1	0.697	2.34	23.8	50			
43.8	76.9	24.9	1.08	2.34	20.0	50	4,392	84	4.2
55.3	56.1	13.4	0.653	2.12	24.6	50			
40.7	76.5	24.7	0.993	2.12	20.0	50	3,934	75	4.2

[†] Minimum power level applied

(c) Optimum lagoon system

This two-stage lagoon system can be optimized to minimize either the total detention time or the total horsepower to be installed.

The following figure shows the summarized design results of the two-stage lagoon system. The minimum total required detention time is 6.33 d, with a corresponding total

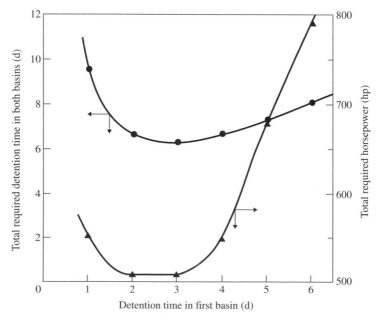

Effect of detention time in first basin on total required time and horsepower.

installed horsepower of 509 hp (382 kW). The minimum total horsepower is 508 hp (381 kW) with a total detention time of 6.67 d. The second alternative may be the optimum system since the increased basin size will be more tolerable to the influent fluctuation in both flow rate and constituent concentration. It will also be justified by an economic gain realized by operating with less power input, although the difference in the horsepower installed is minimal. In this alternative, the detention time in the first basin is 2 d and the detention time in the second basin 4.67 d. The results are tabulated below:

	t, d			hp installed	
First basin	Second basin	Total	First basin	Second basin	Total
1.0	8.57	9.57	261	291	552
2.0	4.67	6.67	342	166	508
3.0	3.33	6.33	389	120	509
4.0	2.72	6.72	476	97	573
5.0	2.34	7.34	595	84	679
6.0	2.12	8.12	714	75	789

EXAMPLE 7.3. Design an aerated lagoon system consisting of an aerobic and a facultative lagoon for the following conditions:

Flow = 4 million gal/d
S_o = 450 mg/l
S_e = 20 mg/l (soluble)
$K = 6 \text{ d}^{-1}$
$b = 0.1 \text{ d}^{-1}$
$a = 0.6$
$a' = 0.5$

Solution.

Design the aerobic lagoon:

$$\frac{S_e}{S_o} = \frac{1 + bt}{aKt}$$

t, d	S_e/S_o	S_e
1	0.3	135
2	0.17	77
3	0.12	54
4	0.097	44

Use a 2-d detention time in the aerobic lagoon.

$$X_v = \frac{aS_r}{1 + bt}$$

$$= \frac{0.6 \cdot 373}{1.2} = 187 \text{ mg/l}$$

Oxygen requirement:

$$O_2 = a'S_r + 0.14\,X_v t$$

$$= 0.5 \cdot 373 + 0.14 \cdot 187 \cdot 2$$

$$= 239 \text{ mg/l}$$

$$\text{lb O}_2/\text{d} = 239 \cdot 4 \cdot 8.34 = 7968$$

$$\text{hp} = \frac{7968}{24 \cdot 1.5 \text{ lb O}_2/(\text{hp} \cdot \text{h})} = 22$$

$$\text{hp/million gal} = \frac{221}{8} = 27.7$$

Nutrient requirements:

$$N = 0.11\,\Delta X_v = 0.11 \cdot 187 = 20.6 \text{ mg/l}$$

$$= 20.6 \cdot 4 \cdot 8.34 = 687 \text{ lb/d}$$

$$P = 0.02\,\Delta X_v = 0.02 \cdot 187 = 3.74 \text{ mg/l}$$

$$= 3.74 \cdot 4 \cdot 8.34 = 125 \text{ lb/d}$$

Facultative lagoon: Reduce K to consider the lower reaction rate in the facultative lagoon [Eq. 6.28].

$$K_1 = K\frac{S_1}{S_o} = 6 \cdot \frac{77}{450}$$

$$= 1.02 \text{ d}^{-1}$$

Use a feedback factor of 1.2. Assume $X_v = 100$ mg/l.

$$\frac{(FS_1)^2 - S_e(FS_1)}{S_e K X_v} = t$$

$$t = 3.27 \text{ d}$$

Oxygen requirement:

$$(77 \cdot 1.2 - 20) \cdot 4 \cdot 8.34 = 2400 \text{ lb/d}$$

$$\text{hp} = \frac{2400}{24 \cdot 1 \cdot 5 \text{ lb O}_2/(\text{hp} \cdot \text{h})} = 67$$

$$\text{hp/million gal} = \frac{67}{3.27 \cdot 4} = 5$$

7.3
ACTIVATED SLUDGE PROCESSES

The objective of the activated sludge process is to remove soluble and insoluble organics from a wastewater stream and to convert this material into a flocculent microbial suspension that is readily settleable and will permit the use of gravitational solid-liquid separation techniques. A number of different modifications or variants of the activated sludge process have been developed since the original experiments of Arden and Lockett in 1914. These variants, to a large extent, have been developed out of necessity or to suit particular circumstances that have arisen. For the treatment of industrial wastewater, the common generic flowsheets are shown in Fig. 7.14. As discussed in Chap. 6, the nature of the wastewater will dictate the process type. Figure 7.15 illustrates the types of reaction that will occur and hence dictate the candidate process.

Plug Flow Activated Sludge

The plug flow activated sludge process uses long, narrow aeration basins to provide a mixing regime that approaches plug flow. Wastewater is mixed with a biological culture under aerobic conditions. The biomass is then separated from the liquid stream in a secondary clarifier. A portion of the biological sludge is wasted and the remainder returned to the head of the aeration tank with additional incoming waste. The rate and concentration of activated sludge returned to the basin determine the mixed liquor suspended solids concentration. As discussed in Chap. 6, a plug flow regime promotes the growth of a well-flocculated, good settling sludge. If the wastewater contains toxic or inhibiting organics, they must be

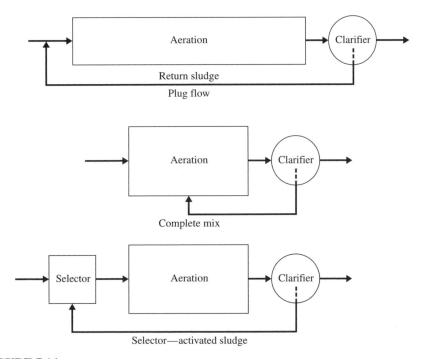

FIGURE 7.14
Types of activated sludge processes.

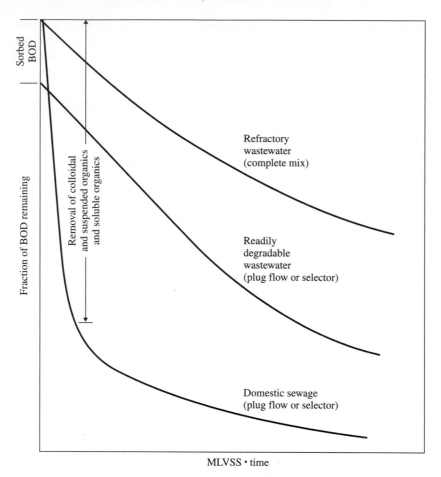

FIGURE 7.15
Removal of BOD in the activated sludge process.

removed or equalized prior to entering the head end of the aeration basin. The oxygen utilization rate is high at the beginning of the aeration basin and decreases with aeration time. Where complete treatment is achieved, the oxygen utilization rate approaches the endogenous level toward the end of the aeration basin.

The dispersion number N_D can be used to express the degree of longitudinal mixing in an activated sludge plant. The dispersion number is a dimensionless number

$$N_D = \frac{D}{UL} \tag{7.13}$$

where N_D = dispersion number
        U = mean velocity of flow, m/s
        L = total length of aeration tank, m
        D = coefficient of axial dispersion, m²/s

A value of 0.068 m^2/s for D was determined by Boon et al.[12] in 24 activated sludge plants. Values of N_D less than about 0.1 should ensure good plug flow hydraulics in practice.

Equation (7.13) can be reexpressed:

$$N_D = \frac{Dt}{L^2} \tag{7.13a}$$

in which t = hydraulic detention time based on $Q + R$. The axial dispersion coefficient N_D is related to the airflow, increasing by a factor of 2 with an airflow increase from 20 to 100 std ft^3/min per 1000 ft^3 of tank volume.[13]

Modification of the way in which wastewater and recycle sludge are brought into contact in a plug flow system can have a number of benefits. For readily degradable wastewaters, a baffled inlet section will ensure sorption. The provisions of a separate zone at the inlet, with a volume of about 15 percent of the total aeration volume, together with a low-energy subsurface mechanical stirrer, can achieve controlled anoxic conditions such that nitrate associated with the recycled sludge fed to the zone can partially satisfy the BOD fed to the zone. In cases where nitrification occurs, recycle of nitrified mixed liquor from the end of the aeration basin to the anoxic zone at the head end can achieve significant denitrification. A typical plug flow process is shown in Fig. 7.16.

FIGURE 7.16
Plug flow activated sludge process.

Complete Mix Activated Sludge

To obtain complete mixing in the aeration tank, proper choices of tank geometry, feeding arrangement, and aeration equipment are required. Through the use of complete mixing, with either diffused or mechanical aeration, it is possible to establish a constant oxygen demand as well as a uniform mixed liquor suspended solids (MLSS) concentration throughout the basin volume. Hydraulic and organic load transients are dampened in these systems, giving a process that is very resistant to upset from shock loadings. Influent wastewater and recycled sludge are introduced to the aeration basin at different points. An activated sludge plant is shown in Fig. 7.17. Readily degradable wastewaters such as food-processing wastes will tend toward filamentous bulking in a complete mix system, as has previously been discussed.

Such conditions may be minimized by the inclusion of a precontacting zone to effect a high level of substrate availability to the recycled mixed liquor. The precontacting zone should have a retention time in the order of 15 minutes to maximize biosorption. Design parameters for this contacting zone appear to be waste-specific and require bench-scale trials for their assessment. By contrast, complex chemical wastewaters do not support filamentous growth and complete mix processes work very effectively in them. Performance data from the treatment of several industrial wastewaters are shown in Table 7.3.

FIGURE 7.17
Complete mix activated sludge plant.

TABLE 7.3
CMAS treatment performance for selected industrial wastewaters

Wastewater	Influent		Effluent		T, °C	F/M		SRT, d	MVLSS, mg/l	HRT, d	SVI, ml/g	ZSV, ft/h
	BOD, mg/l	COD, mg/l	BOD, mg/l	COD, mg/l		BOD, d^{-1}	COD, d^{-1}					
Pharmaceutical	2950	5840	65	712	10.4	0.11	0.19		4970	5.4		
	3290	5780	23	561	20.8	0.11	0.18		5540	5.4		
Coke and by-products chemical plant	1880	1950	65	263		0.18	0.21		2430	4.1	42.4	26
Diversified chemical industry	725	1487	6	257	21	0.41	0.71		2874	0.61	119	4.54
Tannery	1020	2720	31	213	21	0.18	0.45	16	1900	3		
	1160	4360	54	561	21	0.15	0.49	20	2650	3		
Alkylamine manufacturing	893	1289	12	47		0.146	0.21		1977	3.1	133	4.2
ABS	1070	4560	68	510	33.5	0.24	0.94	6	2930	1.5	23	28.7
Viscose rayon	478	904	36	215		0.30	0.47		2759	0.57	117	4.7
Polyester and nylon fibers	207	543	10	107	13.1	0.18	0.40		1689	0.664	116	7.9
	208	559	4	71	22.4	0.20	0.48		1433	0.712	144	8.6
Protein processing	3178	5355	10	362	10	0.054	0.08		2818	21	180	2.9
	3178	5355	5.3	245	26.2	0.100	0.16		2451	12.7	215	2.7
Propylene oxide	532	1124	49	289	20	0.20	0.31		2969	1	51	12.5
	645	1085	99	346	37	0.19	0.25		2491	1.4	32	3.7
Paper mill	375	692	8	79	9.3	0.111	0.19	18.9	1414	2.38	63	22
	380	686	7	75	23.3	0.277	0.45	5.2	748	1.83	504	10
Vegetable oil	3474	6302	76	332		0.57	1.00		1740	3.5	49.2	30
Organic chemicals	453	1097	3	178	20.3	0.10	0.21		2160	2.02	111	6.9

Extended Aeration

In this process, sludge wasting is minimized. This results in low growth rates, low sludge yields, and relatively high oxygen requirements by comparison with the conventional activated sludge processes. The trade-off is between a high-quality effluent and less sludge production. Extended aeration is a reaction-defined mode rather than a hydraulically defined mode, and can be nominally plug flow or complete mix. Design parameters typically include a food/microorganisms (*F/M*) ratio of 0.05 to 0.15, a sludge age of 15 to 35 days, and mixed liquor suspended solids concentrations of 3000 to 5000 mg/l. The extended aeration process can be sensitive to sudden increases in flow due to resultant high MLSS loadings to the final clarifier, but is relatively insensitive to shock loads in concentration due to the buffering effect of the large biomass volume. While the extended aeration process can be used in a number of configurations, a significant number are installed as loop-reactor systems where aerators of a specific type provide oxygen and establish a unidirectional mixing to the basin contents. The use of loop-reactor systems and modifications thereof in wastewater treatment have been significant over recent years.

Oxidation Ditch Systems

A number of loop-reactor or ditch system variants are now available. In any ditch system it is necessary to adequately match basin geometry and aerator performance in order to yield an adequate channel velocity for mixed liquor solids transport. The key design factors in these systems relate to the type of aeration that is to be provided. It is normal to design for a 1 ft/s (0.3 m/s) midchannel velocity in order to prevent solids deposition. The ditch system is particularly amenable to those cases where both BOD and nitrogen removal are desired. Both reactions can be achieved in the same basin by alternating aerobic and anoxic zones, as shown in Fig. 7.18.

A typical oxidation ditch aeration basin consists of a single channel or multiple interconnected channels, as shown in Fig. 7.19.

Intermittently Aerated and Decanted Systems

In the intermittent treatment approach, a single vessel is used to accommodate all of the unit processes and operations normally associated with conventional activated sludge treatment, i.e., primary settlement, biological oxidation, secondary settlement, and sludge digestion, as well as nitrification and the ability for substantial denitrification. In using a single vessel, these processes and operations are simply timed sequences. The sequencing batch reactor (SBR) operates on timed cycles—fill, react, settle, decant—as shown in Fig. 7.20. The feed rate can be adjusted to a batch mode in the case of readily degradable wastewaters to avoid filamentous bulking. The treatment cycle for a pulp and paper mill wastewater is shown in Fig. 7.21. Denitrification is achieved through anoxic cycles. Performance data are shown in Tables 7.4 and 7.5. An SBR plant is shown in Fig. 7.22. A design example is given in Example 7.4.

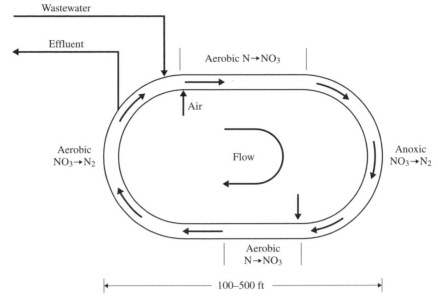

Note: ft = 0.3048 m

FIGURE 7.18
Oxidation ditch with nitrification and denitrification.

FIGURE 7.19
Oxidation ditch.

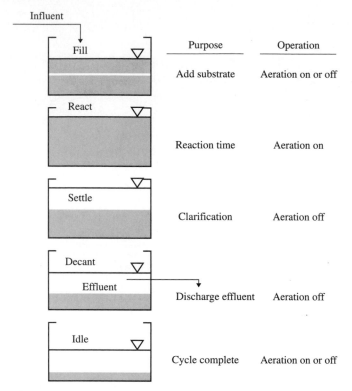

FIGURE 7.20
SBR operation sequence.

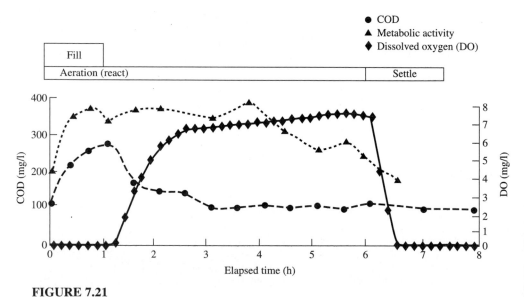

FIGURE 7.21
COD and metabolic activity of the activated sludge during an SBR cycle of reactor for pulp and paper mill wastewater (20-d sludge age, 8-h cycle time, 6-h aerated react time).[14]

TABLE 7.4
Treatment of dairy wastewater by an intermittent activated sludge system

	Influent			Effluent		
	Flow, million gal/d	COD, mg/l	TSS, mg/l	COD, mg/l	BOD, mg/l	TSS, mg/l
Mean	0.094	2400	315	106	15	21
Standard deviation	0.023	903	164	26	104	17
Minimum	0.033	870	73	68	5	3
Maximum	0.133	5636	1030	172	70	80

TABLE 7.5
SBR treatment of a chemical wastewater[†]

Sample point	TOC	TOX	Phenol	Benzoic acid	o-CBA[‡]	m-CBA	p-CBA
Feed, mg/l	8135	780	1650	2475	840	240	285
Effluent, mg/l	409	240	<1	7	3	<2	6

[†] 24-h cycle; MLSS = 10,000 mg/l with HRT = 10 d (10 % feeding over 4-h fill period).
[‡] Chlorobenzoic acid.

FIGURE 7.22
Sequencing batch reactor (SBR) plant.

EXAMPLE 7.4. Design an SBR to treat a wastewater having a flow rate of 0.50 million gal/d and a BOD of 500 mg/l. Assume an $X_v t$ of 1250 (mg · d)/l and a feed plus aeration period of 10 h, with 1 h settling and 1 h effluent decant periods (12 h total cycle time).

Solution.

$$X_v = \frac{1250}{10/24} = 3000 \text{ mg/l}$$

The volume treated during one cycle is 0.25 million gal, and the MLVSS is $(0.25)(3000)(8.34) = 6255$ lb.

At an SVI of 150 ml/g, the volume required for storage of the settled sludge is

$$\left(150 \frac{\text{ml}}{\text{g}}\right)\left(454 \frac{\text{g}}{\text{lb}}\right)\left(3.53 \times 10^{-5} \frac{\text{ft}^3}{\text{ml}}\right) = 2.4 \text{ ft}^3/\text{lb MLSS}$$

Correcting for MLVSS/MLSS = 0.8, the volume required is

$$V = (6255)(1/0.8)(2.4)(7.48)/10^6 = 0.141 \text{ mg}$$

Provide 3 ft of freeboard between the settled sludge blanket and the supernatant water level at the end of the decant cycle.

The volume for aeration and settled sludge is

$$0.25 \text{ mg} + 0.141 \text{ mg} = 0.391 \text{ mg}$$

Select a side water depth (SWD) of 16 ft. The area is

$$\frac{391,000}{(7.48)(16)} = 3267 \text{ ft}^2$$

and the tank diameter is 65 ft. Use a total tank depth of 19 ft to include freeboard.

The MLVSS under aeration will be

$$\frac{6255}{(3267 \cdot 16 \cdot 7.48/106)(8.34)} = 1920 \text{ mg/l}$$

$$\text{MLSS} = \frac{1}{0.8} \times 1920 = 2400 \text{ mg/l}$$

Figure 7.23 is a diagrammatic representation of the operating sequences for a continuous-flow intermittently aerated and decanted activated sludge system.[15] Each sequence (t_0-t_1, t_1-t_2, t_2-t_3) of the cycle t_0-t_3 is initiated by a time-base controller. The treatment cycle begins after the end of decantation from the previous cycle. Aeration begins at time t_0 and continues until time t_1, during which time influent wastewater increases the volume of mixed liquor for aeration. At time t_1, aeration stops and is followed by a nonaeration sequence in which the mixed liquor undergoes settlement and anoxic processes can occur. Following the settlement/anoxic sequence t_1-t_2, treated effluent is discharged during the period t_2-t_3; at the completion of t_3, the same sequence of events is repeated. Operational sequences are developed to optimize specific functions within the main process. For example,

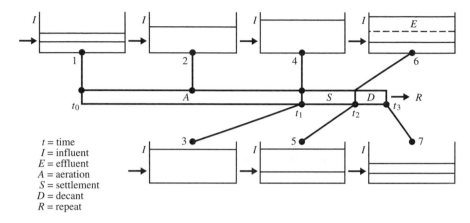

t = time
I = influent
E = effluent
A = aeration
S = settlement
D = decant
R = repeat

FIGURE 7.23
Diagrammatic representation of continuous-flow sequentially aerated activated sludge.

a denitrification cycle will require sufficient aeration to provide for total carbonaceous and nitrogenous oxidation within the time period t_0-t_1, and the period t_1-t_3 must be adequate to effect the reduction of nitrate.

An important feature of these plants is their ability to accept prolonged high-flow conditions without loss of mixed liquor solids. The hydraulic capacity of conventional continuous systems is limited by the operational capacity of the secondary settlement unit.

The decanting device in these systems is located at the vessel end opposite the inlet. The movable weir of the unit is positioned out of the mixed liquor during aeration and settlement. During the decant sequence, a hydraulic ram is activated, which drives the weir through the surface layer of the vessel to the design bottom water level. In this way, a surface layer of treated effluent is skimmed continuously during the decant sequence from the vessel and discharged out of the vessel by gravity via the carrier system of the decanter.

Plants can be designed for an average *F/M* ratio of 0.05 to 0.20 lb BOD/(lb MLSS · d) [0.05 to 0.20 kg BOD/(kg MLSS · d)], depending on the quality of effluent that is specified, at a bottom water level suspended solids concentration of up to 5000 mg/l. In calculating the volume occupied by the sludge mass, an upper sludge volume index of 150 ml/g is used. To ensure solids are not withdrawn during decant, a buffer volume is provided, the depth of which is generally in excess of 1.5 ft (0.5 m) between bottom water level and top sludge level after settlement. Batch activated sludge is similar to the intermittent system except that it is usually employed for high-strength, low-volume industrial wastewaters. Wastewater is added over a short time period to maximize biosorption and flocculent sludge growth. Aeration is then continued for up to 20 h. The mixed liquor is then settled and the treated effluent decanted. A typical batch activated sludge system is shown in Fig. 7.24. Performance data for a high-strength chemical wastewater is shown in Table 7.6. A design example is shown by Example 7.5.

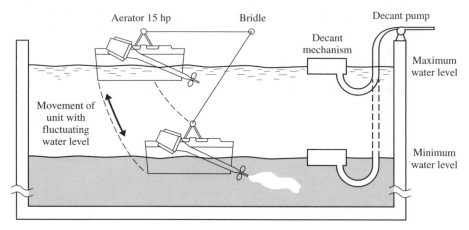

Note: hp = 0.7456 kW

FIGURE 7.24
Batch activated sludge system.

TABLE 7.6
Batch activated sludge treatment performance for a specialty chemicals wastewater

Parameter	Monthly average value							
	March	April	May	June	July	August	September	October
Influent TBOD, mg/l	5,734	5,734	5,734	5,734	7,317	7,317	7,317	7,317
Effluent SBOD, mg/l	43	57	49	119	39	156	91	391
Influent COD, mg/l	10,207	10,207	10,207	10,207	15,242	15,242	15,242	15,242
Effluent COD, mg/l	920	1,992	1,456	2,067	705	2,023	1,682	2,735
Effluent TSS, mg/l	386	828	640	940	250	657	640	700
MLSS, mg/l	9,246	2,430	5,520	3,108	10,300	2,025	9,761	4,572
HRT, d	16.8	16.7	15.6	7.7	14.9	14.6	6.7	6.5
SRT, d	50	50	50	50	30	30	30	30
PAC, mg/l	1,500	—	500	—	2,000	—	2,000	—
Feed time, h	4	4	4	4	4	4	4	4
Aeration time, h	23	23	23	23	23	23	23	23
Settling time, h	1	1	1	1	1	1	1	1
SVI (ml/g)	19	157	32	74	65	75	19	74
F/M (COD basis)	0.11	0.3	0.19	0.19	0.21	0.60	0.51	0.68

EXAMPLE 7.5. Design a batch activated sludge plant for the following wastewater:

$Q = 50,000$ gal/d (190 m³/d)
$BOD(S_o) = 500$ mg/l
$TKN = 2$ mg/l
$a = 0.6$
$a' = 0.55$
$b = 0.1$ d^{-1}
$F/M = 0.1$ d^{-1}
$S = 10$ mg/l (soluble)

The plant will be operated with 20-h aeration, 2-h sedimentation, and 2-h decant.

Solution.

The BOD removal will be

$$S_rQ = (500 - 10) \text{ mg/l} (8.34) \frac{\text{lb/million gal}}{\text{mg/l}} (0.05) \frac{\text{million gal}}{d} = 204 \text{ lb/d} \quad (93 \text{ kg/d})$$

At an F/M of 0.1 d^{-1}, the required MLVSS is

$$X_vV = \frac{QS_o}{F/M} = \frac{0.05 \times 500 \times 8.34}{0.1}$$

$$= 2085 \text{ lb VSS} \quad (946 \text{ kg})$$

and at 85 percent volatile solids the MLSS is

$$\frac{2085}{0.85} = 2453 \text{ lb SS} \quad (1113 \text{ kg})$$

Assuming the sludge has an SVI of 100 ml/g, the sludge volume will be

$$\frac{100 \text{ ml}}{\text{g SS}} \times \frac{454 \text{ g}}{\text{lb}} \times 3.53 \times 10^{-5} \frac{\text{ft}^3}{\text{ml}} = 1.6 \text{ ft}^3/\text{lb}$$

and the volume required for the settled sludge is

$$2453 \text{ lb} \times 1.6 \text{ ft}^3/\text{lb} = 3925 \text{ ft}^3 \text{ or } 29,360 \text{ gal} \quad (111 \text{ m}^3)$$

If sludge is to be wasted twice per month, storage must be provided for accumulated sludge.

At an estimated degradable fraction of the VSS of 0.4, the daily accumulation of VSS is

$$\Delta X_v = aS_rQ - bX_dX_vV$$

$$= 0.6(204) - 0.1 \times 0.4 \times 2085$$

$$= 39 \text{ lb VSS/day} \quad (18 \text{ kg/d})$$

Storage for 15 d will be

$$\frac{39 \text{ lb VSS/d}}{0.85 \text{ VSS/SS}} \times 15 \text{ d} \times 1.6 \text{ ft}^3/\text{lb SS} = 1100 \text{ ft}^3 \text{ or } 8240 \text{ gal} \quad (31 \text{ m}^3)$$

The total volume of basin (excluding freeboard) will be 50,000 + 29,360 + 8240 = 87,600 gal (332 m³). This will be a basin 35.25 ft (10.7 m) in diameter and 12 ft (3.7 m) deep. If 3 ft (0.9 m) is provided for freeboard, the operational basin dimensions will be 35.25 ft (10.7 m) in diameter by 15 ft (4.6 m) depth.

The oxygen requirements can be calculated:

$$O_2/d = a'S_rQ + 1.4b \times X_dX_vV$$

$$= 0.55(204) + 1.4 \times 0.1 \times 0.4 \times 2085$$

$$= 229 \text{ lb/d or } 9.54 \text{ lb/h} \quad (4.3 \text{ kg/h})$$

The required power at 1.5 lb $O_2/(hp \cdot h)$ is

$$\frac{9.54}{1.6} = 6.4 \text{ hp (use 7.5 hp)} \qquad [0.91 \text{ kg}/(kW \cdot h)]$$

This is equivalent to 86 hp/million gal (17 W/m^3) of basin volume, which should provide adequate mixing. The nutrient requirement will be

$$N = 0.123 \frac{X_d}{0.8} \Delta X_v + 0.07 \left(\frac{0.8 - X_d}{0.8} \right) \Delta X_v$$

$$= 0.123 \times \frac{0.4 \times 39}{0.8} + 0.07 \times \frac{0.8 - 0.4}{0.8} \times 39$$

$$= 3.8 \text{ lb/d as N} \qquad (1.7 \text{ kg/d})$$

$$P = 0.026 \times \frac{0.4 \times 39}{0.8} + 0.01 \times \frac{0.8 - 0.4}{0.8} \times 39$$

$$= 0.7 \text{ lb/d as P} \qquad (0.32 \text{ kg/d})$$

Oxygen Activated Sludge

The high-purity oxygen system is a series of well-mixed reactors employing concurrent gas-liquid contact in a covered aeration tank, as shown in Fig. 7.25. The process has been used for the treatment of municipal, pulp and paper mill, and organic chemical wastewaters. Feed wastewater, recycle sludge, and oxygen gas are introduced into the first stage. Gas-liquid contacting can be done by submerged turbines or surface aeration.

Oxygen gas is automatically fed to either system on a pressure demand basis with the entire unit operating, in effect, as a respirator; a restricted exhaust line from the final stage vents the essentially odorless gas to the atmosphere. Normally the system will operate most economically with a vent gas composition of about 50 percent oxygen. For economic considerations, about 90 percent of oxygen utilization with on-site oxygen generation is desired. Oxygen may be generated by a traditional cryogenic air-separation process for large installations (75 million gal/d) ($2.8 \times 10^5 \text{ m}^3/\text{d}$) or a pressure-swing adsorption (PSA) process for smaller installations. The power requirements for surface or turbine aeration equipment vary from 0.08 to 0.14 hp/thousand gal (0.028 kW/m^3). At peak load conditions, the oxygen system is usually designed to maintain 6.0 mg/l dissolved oxygen in the mixed liquor.

Since high dissolved oxygen concentrations are maintained in the mixed liquor, the system can usually operate at high F/M levels (0.6 to 1.0) without filamentous bulking problems. The maintenance of an aerobic floc with high zone settling velocities also permits high MLSS concentrations in the aeration tank. Solids levels will usually range from 4000 to 9000 mg/l, depending on the BOD of the wastewater and design volume of the system.

Pure oxygen also is employed in open aeration basins in which oxygen under high pressure is mixed with the influent wastewater. When introduced into the aer-

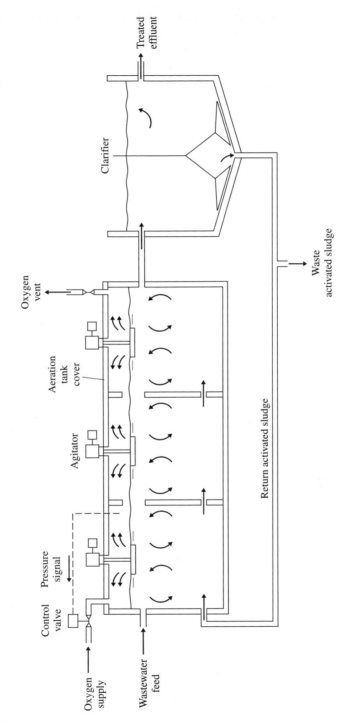

FIGURE 7.25
Schematic diagram of three-stage oxygen system.

355

ation basin, the supersaturated gas comes out of solution in the form of microscopic bubbles. This process is shown in Fig. 7.26.

A membrane filtration unit, shown in Fig. 7.27, permits operation at high MLSS (10,000 to 40,000 mg/l) and is relatively independent of sludge quality. It also produces a high-quality effluent that can be readily disinfected and is frequently suitable for process recycle. This type of system with pure oxygen can have an exceptionally small footprint.

Yamamoto,[16] using hollow fiber membranes, used the following specifications: pore size of 0.1 μ, suction of 13k Pa (1.3 m of water head), and an organic loading

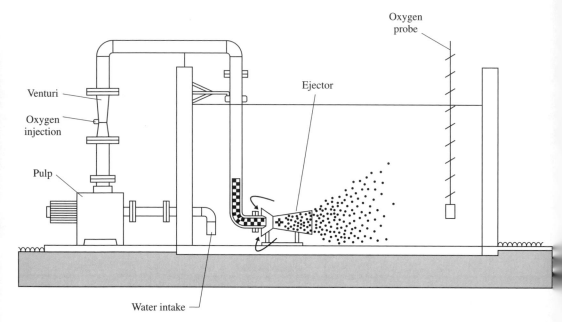

FIGURE 7.26
OXY-DEP[TM] process for biological wastewater treatment using pure oxygen.

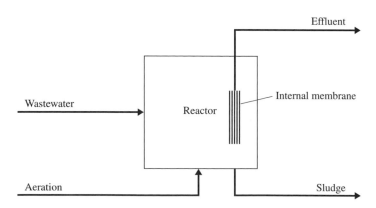

FIGURE 7.27
Diagram of membrane biological reactor with ZeeWeed membranes.

of 3 to 4 kg COD/(m³ · d) for nitrogen removal. Membrane clogging was inevitable to some extent but intermittent suction and low-pressure operation could prevent unrecoverable clogging without cleaning. Design data for an MBR process is shown in Table 7.7.

Deep-Shaft Activated Sludge

The deep-shaft activated sludge process operates at an F/M of 1 to 2 d^{-1} (BOD basis) using a mixing energy level of 800 to 1500 hp/million gal of aeration basin volume. The shaft depth varies from 150 to 400 ft. The operating mixed liquor dissolved oxygen concentration varies from 10 mg/l to 20 mg/l since the increasing shaft depth increases the saturation concentration. The MLSS levels vary from 8000 to 12,000 mg/l. Solids-liquid separation is provided by dissolved air flotation at high MLSS concentrations (greater than 10,000 mg/l) and by vacuum degasification and conventional gravity clarification at lower MLSS levels. A block flow diagram for the deep-shaft process is shown in Fig. 7.28. Performance data for a brewery wastewater is shown in Table 7.8.

Biohoch Process

The Biohoch® reactor consists of an aeration section divided by a perforated plate into lower and upper zones and a cone-shaped final clarifier surrounding the aera-

TABLE 7.7
GM Windsor full-scale MBR system design criteria[17]

Operating and Performance Criteria	Design Value
Feed and Reactor Operating Conditions	
Mean wastewater flow, m³/d	864
Mean feed COD, mg/l	6000
Reactor HRT, d	1.47
Reactor SRT, d	63
Reactor VSS, g/l	23.5
COD volumetric load, kg/m³ d	4.1
TOG volumetric load, kg/m³ d	0.80
Net solids yield coefficient, g VSS/g COD	0.11
Performance Expectations	
Effluent values, mg/l	
COD	≤700
TSS	≤10
TOG	≤45
THOG†	≤5
Membrane flux, l/m² h	51

†THOG represents total hydrocarbon oil and grease.

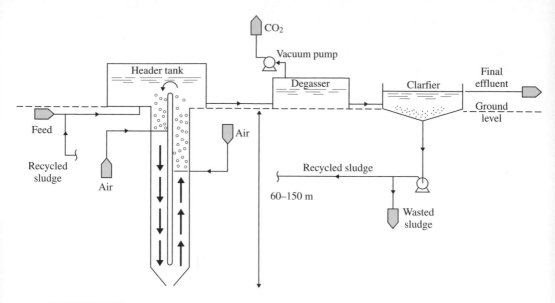

FIGURE 7.28
Flowsheet for deep-shaft activated sludge process.

TABLE 7.8
Brewery wastewater treatment in the deep-shaft process[18]

Parameter	Performance
Average flow	0.65 million gal/d
Average BOD_5	2,400 mg/l
MLSS	12,000 mg/l
MLVSS	7,920 mg/l
F/M	1.51
Hydraulic detention time	0.2 d
Clarifier loading	618 gal/(d · ft^2)
Recycle solids	4 percent
Effluent BOD_5	78 mg/l
TSS	91 mg/l

tion section. The air is supplied to the reactor by means of radial flow jets installed at the reactor bottom.

The untreated wastewater is pumped into the reactor through the radial flow jets or through separate pipes. Turbulence in the lower zone is sufficient to provide completely mixed conditions. In the upper chamber, the stabilizing and degassing zone, bubbles adhering to the activated sludge are removed, since they impede sedimentation in the final clarifier. The depth of the aeration zone is approximately 65 ft.

The Biohoch reactor is illustrated in Fig. 7.29 and performance data for treatment of an organic chemicals wastewater are shown in Table 7.9.

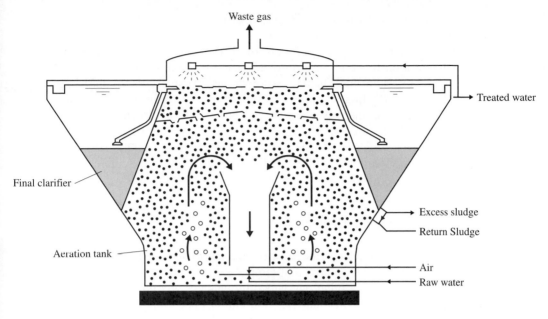

FIGURE 7.29
Flow scheme of a Biohoch reactor.

TABLE 7.9
Treatment of an organic chemicals
wastewater in the Biohoch reactor

Parameter	Value
Flow rate	0.63 million gal/d
Influent BOD	5,000 mg/l
Effluent BOD	40 mg/l
Influent COD	6,000 mg/l
Effluent COD	750 mg/l
F/M (BOD basis)	0.43 lb/(lb MLSS · d)
MLSS	3,500 mg/l
SRT	5.4 d
HRT	80 h
Temperature	95°F

Integrated Fixed Film Activated Sludge (IFAS)

An innovation that has been implemented in recent years is the integration of fixed film media into activated sludge reactors to improve performance and in some cases to minimize expansion of existing facilities. In plants where nitrification and denitrification is practiced, nitrification is usually the rate-limiting step and the media is placed in the aerobic zone to enhance nitrification at low temperatures.

The most common media is Ringlace® ropelike media and small floating plastic sponges such as Captor® and Linpur®.

Randall[19] reviewed performance data of IFAS systems. In one case, using Ringlace nitrification was three times greater than the control reactor and denitrification was 2.5 times greater. The integration of media also decreased the SVI and the solids loading to the final clarifier. Floating sponges have been used in several ways. In one case the sponges were placed before the activated sludge zone. Sponges have also been used after the activated sludge aeration basin for enhanced nitrification. Randall added sponges to the activated sludge aeration basin and reduced the volume requirements for nitrification by 20 percent. Sponges may be used for refractory organics requiring long sludge ages as a second stage treatment.

Thermophilic Aerobic Activated Sludge

Thermophilic aerobic activated sludge offers the advantages of rapid degradation rates and low sludge yield. The optimal temperature for thermophilic oxidation is 55 to 60°C, but common terminology generally includes any process operating at temperatures of 45°C or higher. The reaction rate has been reported to be 3 to 10 times greater than mesophilic operation and the endogenous rate 10 times greater, thereby greatly reducing the net sludge yield. A COD removal of 20,000 to 40,000 mg/l coupled with an oxygen transfer efficiency of 10 to 20 percent are necessary for autoheating to thermophilic temperatures. One disadvantage is that thermophilic bacteria fail to flocculate, making biomass separation in the effluent a problem.

Final Clarification

The final clarifier is a key element in the activated sludge process, and failure to consider major design concepts will result in process upset. Activated sludge exhibits zone settling, which is discussed on p. 494. Batch settling of activated sludge can be characterized by the relationship:

$$V_Z = V_o e^{-KX} \tag{7.14}$$

where
$$
\begin{aligned}
V_Z &= \text{zone settling velocity} \\
X &= \text{sludge concentration at } V_Z \\
V_o \text{ and } K &= \text{empirical constants}
\end{aligned}
$$

The settling properties of activated sludge are commonly related to the sludge volume index (SVI). Daigger and Roper,[20] among others, showed that the zone settling velocity (Eq. 7.14) of municipal activated sludges could be related to SVI:

$$V_Z = 7.80 e^{-(0.148 + 0.0021\text{SVI})X} \tag{7.14a}$$

In Eq. (7.14a), V_z, X, and SVI are in meters per hour, grams per liter, and milliliters per gram, respectively. The clarifier functions primarily as a thickener in which the flux to the clarifier is related to the MLSS concentration, the influent flow rate, the underflow rate, the available clarifier surface area, and the sludge settling characteristics (see p. 494). Figure 7.30 illustrates these relationships.

In Fig. 7.30 the overflow line has a slope equal to the overflow rate, Q/A. The underflow rate operating line has a slope equal to $-R/A$. The intersection of these two lines occurs at the mixed liquor suspended solids concentration X_a. The applied flux G_A (equal to the solids loading rate) is indicated where the underflow rate operating line intersects the solids flux axis. The return sludge concentration X_r is indicated where this line intersects the solids concentration axis.

The maximum flux G_{max} that can be successfully transmitted to the bottom of the clarifier and out in the return sludge stream is determined by drawing the underflow rate operating line tangent to the settling flux curve. The maximum return sludge solids concentration that can be achieved, $X_{r(max)}$, is indicated where this line intersects the solids concentration axis. When the underflow rate operating line is tangent to the settling flux curve, the system is said to be underloaded with respect to thickening. Operation below this limit is desired; prolonged operation above this limit will fill the clarifier with solids, which may result in a gross loss of solids in the effluent.

The amount of solids that can be processed through a clarifier is dictated by the sludge settling characteristics. This relationship is shown in Fig. 7.30: as the sludge settleability decreases (SVI rises from 100 to 150 ml/g), the maximum flux decreases ($G_{max1} < G_{max2}$); the maximum return sludge solids concentration also decreases ($X_{r(max)1} < X_{r(max)2}$). Repeating the calculation procedure for various underflow concentrations results in a series of clarifier underflow, i.e., recycle, concentrations with their corresponding maximum solids flux rates in which each represents a limiting design or operating condition. Figure 7.31 represents these calculations for a range of SVI sludges for municipal wastewater. Acceptable operating

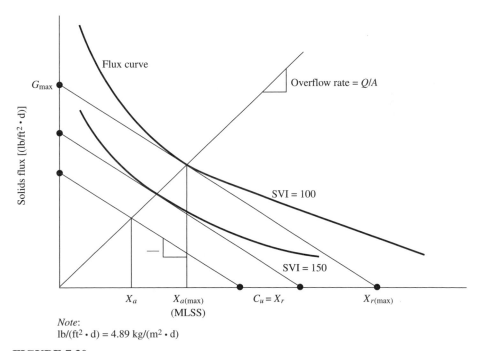

Note:
lb/(ft² · d) = 4.89 kg/(m² · d)

FIGURE 7.30
Final clarifier relationships.

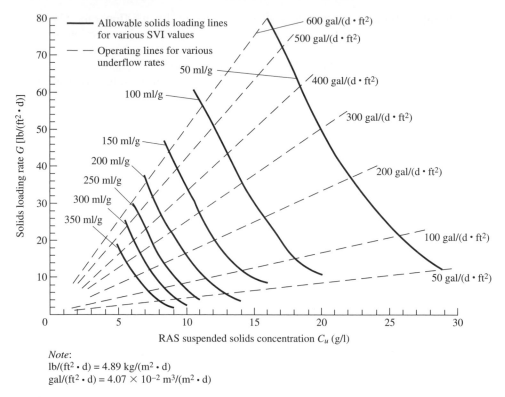

Note:
lb/(ft^2 · d) = 4.89 kg/(m^2 · d)
gal/(ft^2 · d) = 4.07 × 10^{-2} m^3/(m^2 · d)

FIGURE 7.31
Clarifier design and operation diagram (*After Daigger and Roper*[17]).

points on Fig. 7.31 are those which lie below the line for the particular SVI. The dashed lines or operating lines are derived from clarifier mass balances assuming effluent suspended solids and sludge wastage are small relative to the clarifier solids feed. The state point on a given operating line is defined by the solids flux G and the corresponding clarifier underflow or recycle solids concentration X_u. The clarifier will not be solids-limiting, provided the state point lies below the limiting solids flux line for the particular SVI. This is denoted as $X_{a(\text{max})}$ in Fig. 7.30. This analysis is illustrated by Example 7.6. A clarifier design procedure is shown in Example 7.7.

> **EXAMPLE 7.6.** A final clarifier is operating under the following conditions:
>
> Surface area A = 2000 ft^2 (186 m^2)
> Influent flow Q = 1.2 million gal/d (4540 m^3/d)
> Sludge recycle R = 0.6 million gal/d (2270 m^3/d)
> Mixed liquor suspended solids X_a = 3000 mg/l
>
> Compute:
>
> The overflow rate Q/A
> The underflow rate R/A
> The solids flux G
> The recycle suspended solids X_r
> The maximum SVI for clarifier operation

Solution.

The overflow rate is

$$\frac{1.2 \times 10^6}{2000} = 600 \text{ gal/(d} \cdot \text{ft}^2) \qquad [24.4 \text{ m}^3/(\text{m}^2 \cdot \text{d})]$$

The underflow rate is

$$\frac{0.6 \times 10^6}{2000} = 300 \text{ gal/(d} \cdot \text{ft}^2) \qquad [12.2 \text{ m}^3/(\text{m}^2 \cdot \text{d})]$$

The solids flux is

$$G = \frac{X_a(Q + R)}{A} = \frac{3000(1.2 + 0.6)8.34}{2000} = 22.5 \text{ lb/(ft}^2 \cdot \text{d}) \qquad [110 \text{ kg(m}^2 \cdot \text{d})]$$

The recycle suspended solids is

$$\frac{R}{Q} = \frac{X_a}{X_r - X_a}$$

or

$$X_r = \frac{X_a(1 + R/Q)}{R/Q} = \frac{3000(1 + 0.6/1.2)}{0.6/1.2} = 22.5 \text{ lb/(ft}^2 \cdot \text{d}) \qquad [110 \text{ kg/(m}^2 \cdot \text{d})]$$

From Fig. 7.31, with a recycle suspended solids of 9000 mg/l and a solids flux of 22.5 lb/(ft$^2 \cdot$ d) [110 kg/(m$^2 \cdot$ d)], the maximum permissible SVI is 200 ml/g.

EXAMPLE 7.7. State point analysis. Given the following data from an activated sludge plant:

Influent flow Q = 12 million gal/d
Number of clarifiers = 2
Surface area of each, A = 9000 ft^2
Raw activated sludge (RAS) flow from each, R = 3.5 million gal/d
MLSS concentration = 3 gal/l
V_o settling parameter = 564 ft/d
K settling parameter = 0.41/g

Compute (a) the overflow rate operating line and (b) the underflow operating line. (c) Construct the solids flux curve. (d) Find the maximum loading to the clarifier.

Solution.

(a) The slope of the overflow rate operating line equals the surface overflow rate Q/A. The overflow rate is

$$\frac{Q}{A} = \frac{12 \text{ million gal/d}}{2 \times 9000} = 666 \text{ gal/(d} \cdot \text{ft}^2)$$

Converting Q/A to ft/d gives

$$\frac{666}{7.48} = 89 \text{ ft/d}$$

But the overflow rate in ft/d is equal to

$$\frac{16 \cdot G \; \text{lb}/(\text{ft}^2 \cdot \text{d})}{\text{MLSS} \; (\text{g}/\text{l})}$$

Therefore, for an MLSS of 3 g/l, G can be computed:

$$\frac{16 \cdot G}{3} = 89$$

$$G = 16.7 \; \text{lb}/(\text{ft}^2 \cdot \text{d})$$

In Fig. 7.32, the overflow rate operating line is the line passing through the origin and the point defined by $G = 16.7$ lb/d, MLSS $= 3$ g/l.

(*b*) The slope of the underflow rate operating line equals the underflow rate. The underflow rate is

$$\frac{2R}{2A} = \frac{2 \times 3.5 \; \text{million gal}/\text{d}}{2 \times 9000 \; \text{ft}^2} = 389 \; \text{gal}/(\text{d} \cdot \text{ft}^2)$$

Converting the underflow rate to ft/d gives

$$\frac{389}{7.48} = 52 \; \text{ft}/\text{d}$$

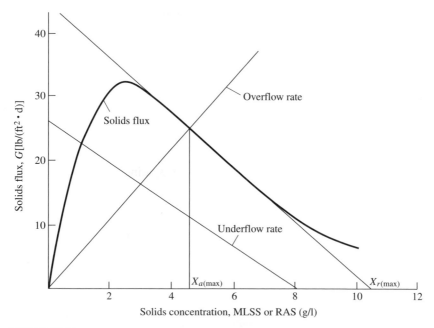

FIGURE 7.32
Plot of overflow and underflow operating lines and solids flux curve.

The underflow rate in ft/d is equal to

$$\frac{16G \ (\text{lb/d} \cdot \text{ft}^2)}{\text{RAS} \ (\text{g/l})}$$

The flux G to the clarifier is

$$G = \frac{(Q + R)X_a}{2A}$$

$$= \frac{(12 + 7)3 \cdot 8.34 \cdot 10^3}{18,000} = 26.4 \ \text{lb/(ft}^2 \cdot \text{d)}$$

The RAS concentration is computed

$$\frac{16 \cdot 26.4 \ \text{lb/(ft}^2 \cdot \text{d)}}{\text{RAS(g/l)}} = 52 \ \text{ft/d}$$

$$\text{RAS} = 8.1 \ \text{g/l}$$

The underflow rate operating line is now plotted on Fig. 7.32 as the line extending from 8.1 g/l on the RAS axis to 26.4 lb/(ft^2 · d) on the G axis.

(c) The solids flux curve can be constructed by modifying Eq. (7.14)

$$V_z X = G = V_o X e^{-KX}$$

For $V_o = 564$ ft/d and $K = 0.4$ l/g, the solids flux curve is plotted on Fig. 7.32.

(d) The maximum loading to the clarifier is obtained by drawing a tangent to the inflection point of the solids flux curve as shown in Fig. 7.32. This yields a flux of 44 lb/(ft^2 · d) with an underflow RAS of 10.4 g/l. The MLSS as shown in Fig. 7.32 is 4500 mg/l. The underflow rate is increased from 389 gal/(d · ft^2) to 563 gal/(d · ft^2).

Flocculation and Hydraulic Problems

Wahlberg et al.[21] devised a series of tests, called the dispersed suspended solids (DSS) and flocculated suspended solids (FSS) tests, to differentiate between flocculation and hydraulic problems in secondary clarifiers. DSS is operationally defined as the supernatant suspended solids concentration after 30 minutes of settling in a Kemmerer sampler. The use of a Kemmerer sampler allows the sample to be collected and settled in the same container, thereby sparing the biological flocs any unquantifiable aggregation or breakup effects during an intermediate transfer step.

FSS is operationally defined as the supernatant suspended solids concentration after 30 minutes of flocculation and 30 minutes of settling. The FSS concentration is particularly useful because it is a measure of the effluent TSS concentration that would be possible if the mixed liquor entering a secondary clarifier is optimally flocculated and the flow characteristics in the clarifier are ideal. Differences between the FSS concentration and the effluent TSS concentration reflect any inefficiencies caused by poor flocculation or poor hydraulics. Most activated sludges will produce an effluent less than 10 mg/l in TSS if properly flocculated and ideally settled.

The existence of density currents in secondary clarifiers is well known. These currents result from the underwater waterfall that results when the solids-laden mixed liquor enters the relatively solids free secondary clarifier. High velocity jets also can occur as a result of improperly designed secondary clarifier inlets that fail to dissipate the kinetic energy of the influent flow. Poor secondary clarifier designs resulting in chronic hydraulic problems have plagued the wastewater treatment industry. Hydrodynamic models are being used successfully to identify hydraulic problems and to design modifications to correct them. Increasingly, these models are being used to design new secondary clarifiers as well.

Treatment of Industrial Wastewaters in Municipal Activated Sludge Plants

Municipal wastewater is unique in that a major portion of the organics are present in suspended or colloidal form. Typically, the BOD in municipal sewage will be 50 percent suspended, 10 percent colloidal, and 40 percent soluble. By contrast, most industrial wastewaters are almost 100 percent soluble. In an activated sludge plant treating municipal wastewater, the suspended organics are rapidly enmeshed in the flocs, the colloids are adsorbed on the flocs, and a portion of the soluble organics are adsorbed. These reactions occur in the first few minutes of aeration contact. By contrast, for readily degradable wastewaters, i.e., food processing, a portion of the BOD is rapidly sorbed and the remainder removed as a function of time and biological solids concentration. Very little sorption occurs in refractory wastewaters. These phenomena are shown in Fig. 7.15. The kinetics of the activated sludge process will therefore vary, depending on the percentage and type of industrial wastewater discharged to the municipal plant, and must be considered in the design calculations.

The percentage of biological solids in the aeration basin will also vary with the amount and nature of the industrial wastewater. For example, municipal wastewater without primary clarification will yield a sludge that is 47 percent biomass at a 3-d sludge age. Primary clarification will increase the biomass percentage to 53 percent. Increasing the sludge age will also increase the biomass percentage as volatile suspended solids undergo degradation and synthesis. Soluble industrial wastewater will increase the biomass percentage in the activated sludge.

As a result of these considerations, there are a number of phenomena that must be considered when industrial wastewaters are discharged to municipal plants:

1. *Effect on effluent quality.* Soluble industrial wastewaters will affect the reaction rate K, as shown in Table 6.5. Refractory wastewaters such as tannery and chemical will reduce K, while readily degradable wastewaters such as food processing and brewery will increase K.
2. *Effect on sludge quality.* Readily degradable wastewaters will stimulate filamentous bulking, depending on basin configuration, while refractory wastewaters will suppress filamentous bulking.
3. *Effect of temperature.* An increased industrial wastewater input, i.e., soluble organics, will increase the temperature coefficient θ, thereby decreasing efficiency at reduced operating temperatures.
4. *Sludge handling.* An increase in soluble organics will increase the percentage of biological sludge in the waste sludge mixture. This generally will decrease

dewaterability, decrease cake solids, and increase conditioning chemical requirements. An exception is pulp and paper mill wastewaters, in which pulp and fiber serves as a sludge conditioner and will enhance dewatering rates.

It should be noted that most industrial wastewaters are nutrient deficient; i.e., they lack nitrogen and phosphorus. Municipal wastewater, with a surplus of these nutrients, will provide the required nutrient balance. However, wastewaters from industries such as poultry processing can be very high in both nitrogen and phosphorus relative to organic matter, and substantially increase the costs of nutrient removal at municipal plants.

Pretreatment of an industrial wastewater in a municipal plant is shown in Example 7.8.

EXAMPLE 7.8. A wastewater is to be pretreated in an activated sludge process and then blended with a domestic wastewater for final treatment in a second activated sludge basin as shown below:

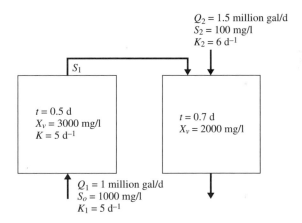

Solution.

The effluent SBOD from the first aeration basin is, from Eq. (6.24),

$$(S_e)_1 = \frac{1000^2}{1000 + (5)(3000)(0.5)}$$

$$= 118 \text{ mg/l}$$

The reaction rate K_2 of the pretreated industrial wastewater in the second basin is, from Eq. (6.29),

$$K_2 = 5\left(\frac{118}{1000}\right) = 0.59 \text{ d}^{-1}$$

The influent concentration to the second basin is

$$(S_o)_2 = \frac{(118)(1.0) + (100)(1.5)}{(1.0 + 1.5)}$$

$$= 107 \text{ mg/l}$$

The average rate coefficient, $\overline{K}$, after blending the two wastewaters is, from Eq. (6.28),

$$\frac{1}{\overline{K}} = \frac{\dfrac{1}{0.59}(118)(1.0) + \dfrac{1}{6}(100)(1.5)}{(118)(1.0) + (100)(1.5)}$$

$$\overline{K} = 1.2 \ d^{-1}$$

The SBOD from the second basin will be

$$(S_e)_2 = \frac{107^2}{107 + (1.2)(2000)(0.7)}$$

$$= 6.5 \ mg/l$$

Effluent Suspended Solids Control

Carryover of suspended solids in the secondary clarifier effluent can have several causes:

- Floc shear due to high aeration basin power levels
- Poor clarifier hydraulics
- High wastewater TDS concentration
- Low or high mixed liquor temperature
- Rapid change in mixed liquor temperature
- Low mixed liquor surface tension

High mixed liquor turbulence levels created by turbine-type or mechanical surface aerators can cause floc breakup that results in high-effluent suspended solids. This problem can frequently be solved by reducing the aeration basin power level and/or by installing a flocculation zone between the aeration basin and the final clarifier. Results of flocculation of mixed liquor from an activated sludge plant treating a pulp and paper mill wastewater at two aeration basin power levels are shown in Table 7.10. These data indicate that flocculation times of 1 to 3 min were

TABLE 7.10
Floc shear test results for pulp and paper mill wastewaters

Flocculation time, min	Settled TSS,[†] mg/l at	
	690 hp/million gal	360 hp/million gal
0	81	64
1	30	28
3	28	22
5	26	19
7	27	21

[†] Supernatant TSS following flocculation and 15-min settling period.

effective at reducing the settled effluent TSS concentration. The effect of turbulence on effluent suspended solids when fine bubble diffusers are used is shown in Fig. 7.33.

High-effluent suspended solids levels will frequently result from a poor clarifier hydraulic design that causes density currents and/or short-circuiting. These conditions result in an upwelling of floc solids at the clarifier peripheral weir. This problem can be minimized by installing a Stamford baffle, which redirects the upflow of solids away from the effluent weir. A Stamford baffle and its performance are illustrated by Fig. 7.34.

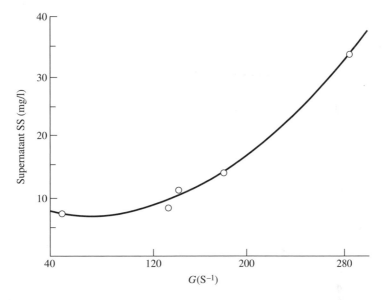

G is defined as the mean velocity gradient, which is a measure of the mixing intensity. For example what is G for an air flow of 20 scfm/1,000 ft^3 in an aeration basin of 0.5 mg, a 26 ft SWD at 20°C

$$G = \sqrt{\frac{Qa\gamma h}{V\mu}}$$

in which Qa = air flow, m^3/sec
         γ = liquid specific weight, N/m^3
         h = liquid depth, m
         V = aeration basin volume, m^3
         μ = absolute viscosity, N sec/m^2

$$G = \sqrt{\frac{0.63 \text{ m}^3/\text{sec} \cdot 9{,}790 \text{ N/m}^3 \cdot 7.47 \text{ m}}{1{,}894 \text{ m}^3 \cdot 1.003 \times 10^{-3} \dfrac{\text{N} \cdot \text{sec}}{\text{m}^2}}}$$

$$= 155 \text{ sec}^{-1}$$

FIGURE 7.33
The effect of G on effluent SS concentration for fine-bubble aerated plants.[22]

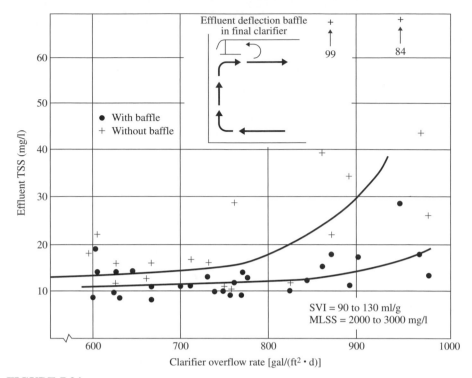

FIGURE 7.34
Effect of Stamford baffle on clarifier effluent TSS concentration.

In the case of industrial wastewaters, however, high effluent suspended solids of a dispersed character can result from one of several causes:

1. High total dissolved salt (TDS) may cause an increase in nonsettleable suspended solids. While the specific cause of floc dispersion has not been defined, increasing the TDS has generally resulted in an increase in nonsettleable, dispersed solids. High TDS will also increase the specific gravity of the liquid, thereby reducing the settling rate of the biological sludge. The salt content appears to have little effect on the kinetics of the process under acclimated conditions.

2. Dispersed suspended solids increase as the aeration basin temperature decreases. For example, an aerated sludge plant in West Virginia treating an organic chemicals wastewater had an effluent suspended solids concentration of 42 mg/1 during summer operation and 104 mg/1 during winter operation. The change in coagulant dosage with respect to temperature at a Tennessee organic chemicals plant is shown in Fig. 7.35.

3. Dispersed suspended solids increase with a decrease in surface tension. At one deinking mill the effluent suspended solids was directly related to the surfactant usage in the mill.

4. The nature of the organic content may increase the effluent suspended solids. While this effect is ill-defined, some plants will consistently generate high effluent suspended solids.

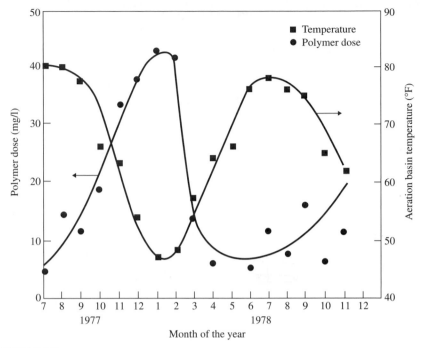

FIGURE 7.35

Polymer addition for effluent suspended solids control.[23]

The effluent suspended solids can be reduced by the addition of a coagulant prior to the final clarifier. It is important that there be sufficient time for flocculation to occur. This can be achieved by a flocculation chamber between the aeration basin and the clarifier or by a flocculation well within the clarifier, as shown in Fig. 7.36.

Cationic polyelectrolytes, or alum or iron salts, may be used as a coagulant. Choice of coagulant depends on a test program to select the most economical solution. Data reported by Paduska[23] for a cationic polyelectrolyte are shown in Fig. 7.35. It is important when using a cationic polymer to avoid overdosing, which will cause a charge reversal and a redispersion of the solids. The temperature effect as previously noted is also shown in Fig. 7.35. An activated sludge design is illustrated by Example 7.9.

EXAMPLE 7.9. Given the following data:

$Q = 4$ million gal/d (15,140 m^3/d)
$S_o = 610$ mg/l
$S_e = 40$ mg/l
$SS_{effl} = 40$ mg/l
mg BOD/mg SS $= 0.3$
$K = 3.0$ d^{-1} at 20°C
$b = 0.1$ d^{-1} at 20°C
$X_v = 3000$ mg/l
$a = 0.55$
$a' = 0.50$
$\theta_b = 1.04$
$\theta_K = 1.065$

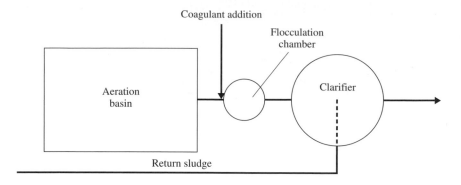

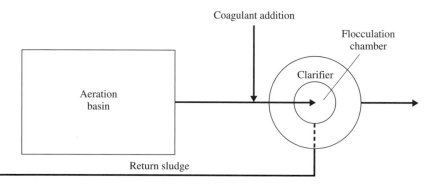

FIGURE 7.36
Coagulant addition for suspended solids control.

Compute:
(a) The sludge age and F/M to meet this effluent quality
(b) The N and P requirements at 20°C
(c) The MLVSS at 10°C to meet the same effluent quality
(d) The excess sludge at 10°C
(e) The oxygen requirements at 30°C

Solution.

(a) Food to microorganisms ratio F/M and sludge age θ_c:

F/M: The effluent soluble BOD is

$$S = S_e - 0.3SS_{\text{effl}} = 40 - 0.3 \times 40 = 28 \text{ mg/l}$$

The detention time is obtained by rearranging Eq. (6.24):

$$t = \frac{(S_o - S)S_o}{KX_vS} = \frac{(610 - 28) \text{ mg/l} \times 610 \text{ mg/l}}{3.0/\text{d} \times 3000 \text{ mg/l} \times 28 \text{ mg/l}}$$

$$= 1.41 \text{ d}$$

$$F/M = \frac{S_o}{(X_v t)} = \frac{610 \text{ mg/l}}{3000 \text{ mg/l} \times 1.41 \text{ d}} = 0.144/\text{d}$$

Sludge age: The BOD removed is

$$S_r = S_o - S = 610 - 28 = 582 \text{ mg/l}$$

The degradable fraction is given by

$$(*)X_d = \frac{aS_r + bX_vt - [(aS_r + bX_vt)^2 - (4bX_vt)(0.8aS_r)]^{\frac{1}{2}}}{2bX_vt}$$

$$= 0.47$$

The sludge age can be calculated from Eq. (6.11).

$$\theta = \frac{X_vt}{aS_r - bX_dX_vt}$$

$$= \frac{3000 \text{ mg/l} \times 1.41 \text{ d}}{0.55 \times 582 \text{ mg/l} - 0.1/\text{d} \times 0.47 \times 3000 \text{ mg/l} \times 1.41 \text{ d}}$$

$$= 34.9 \text{ d} \approx 35 \text{ d}$$

Note that θ_c could have been calculated first using the following equation [obtained by combining Eqs. (6.5) and (6.11)]:

$$\theta_c = \frac{-(aS_r - bX_vt) + [(aS_r - bX_vt)^2 + 4(abX_n'S_r)(X_vt)]^{\frac{1}{2}}}{2abX_n'S_r}$$

with $X_n' = 0.2$, this equation gives $\theta_c = 35$ d; then X_d can be computed from Eq. (6.5):

$$X_d = \frac{X_d'}{(1 + bX_n'\theta_c)}$$

$$= \frac{0.8}{1 + 0.1 \times 0.2 \times 35} = 0.47$$

(b) Nitrogen N and phosphorus P requirements:

The excess volatile sludge produced at 20°C is computed from Eq. (6.10):

$$\Delta X_{v20} = (aS_r - bX_dX_vt)Q$$

$$= (0.55 \times 582 \text{ mg/l} - 0.1/\text{d} \times 0.47 \times 3000 \text{ mg/l}$$

$$\times 1.41 \text{ d}) (4 \text{ million gal/d}) \times [8.34 \text{ (lb/million gal)}/(\text{mg/l})]$$

$$= 4046 \text{ lb/d} \quad (1837 \text{ kg/d})$$

The nitrogen and phosphorus requirements are given by Eqs. (6.19) and (6.20), therefore

$$N = 0.123\left(\frac{X_d}{0.8}\right)\Delta X_v + 0.07\left(\frac{0.8 - X_d}{0.8}\right)\Delta X_v$$

$$= \left[0.123\left(\frac{0.47}{0.8}\right) + 0.07\left(\frac{0.8 - 0.47}{0.8}\right)\right](4046) \text{ lb/d}$$

$$= 409 \text{ lb/d} \quad (186 \text{ kg/d})$$

$$P = 0.026\left(\frac{X_d}{0.8}\right)\Delta X_v + 0.01\left(\frac{0.8 - X_d}{0.8}\right)\Delta X_v$$

$$= \left[0.026\left(\frac{0.47}{0.8}\right) + 0.01\left(\frac{0.8 - 0.47}{0.8}\right)\right](4046) \text{ lb/d}$$

$$= 79 \text{ lb/d} \qquad (36 \text{ kg/d})$$

(c) Mixed liquor volatile suspended solids at 10°C, X_{v10}:
The kinetic coefficients at 10°C are computed:

$$K_{10°C} = K_{20°C}\theta_K{}^{10-20} = 3.0/\text{d} \times 1.065^{-10}$$

$$= 1.6/\text{d}$$

$$b_{10°C} = b_{20°C}\theta_b{}^{10-20} = 0.1/\text{d} \times 1.04^{-10}$$

$$= 0.068/\text{d}$$

From Eq. (6.26), rearranged,

$$X_{v10} = \frac{S_r S_o}{K_{10°C}St} = \frac{582 \text{ mg/l} \times 610 \text{ mg/l}}{1.6/\text{d} \times 28 \text{ mg/l} \times 1.41 \text{ d}}$$

$$= 5624 \text{ mg/l}$$

(d) Excess sludge at 10°C, ΔX_{v10}:
With the new values of K, b, and X_v in Eq. (6.12), the degradable fraction at 10°C is

$$X_{d10} = 0.40$$

The excess sludge is computed from Eq. (6.10):

$$\Delta X_{v10} = (0.55 \times 582 - 0.068 \times 0.40 \times 5624 \times 1.41) \times (4)(8.34) \text{ lb/d}$$

$$= 3483 \text{ lb/d} \qquad (1581 \text{ kg/d})$$

(e) Oxygen requirements at 30°C, R_{30}

The kinetic coefficients at 30°C are

$$K_{30°C} = 3.0/\text{d} \times 1.065^{30-20}$$

$$= 5.6/\text{d}$$

$$b_{30°C} = 0.1/\text{d} \times 1.04^{10}$$

$$= 0.15/\text{d}$$

Then

$$X_{v30} = \frac{582 \times 610}{5.63 \times 28 \times 1.41}$$

$$= 1597 \text{ mg/l}$$

and

$$X_{d30} = 0.54$$

$$R_{30} = (a'S_r + 1.4b_{30°C} X_{d30}X_{v30}t)Q$$

$$= (0.50 \times 582 \text{ mg/l} + 1.4 \times 0.15/\text{d} \times 0.54 \times 1597 \text{ mg/l}$$

$$\times 1.41 \text{ d})(4 \text{ million gal/d})[8.34(\text{lb/million gal})/(\text{mg/l})]$$

$$= 18,226 \text{ lb/d} \qquad (8275 \text{ kg/d})$$

7.4
TRICKLING FILTRATION

A trickling filter is a packed bed of media covered with slime growth over which wastewater is passed. As the waste passes through the filter, organic matter present in the waste is removed by the biological film.

Plastic packings are employed in depths up to 40 ft (12.2 m), with hydraulic loadings as high as 4.0 gal/(min · ft²) [0.16 m³/(min · m²)]. Depending on the hydraulic loading and depth of the filter, BOD removal efficiencies as high as 90 percent have been attained on some wastewaters. In one industrial plant, a minimum hydraulic loading of 0.5 gal/(min · ft²) [0.02 m³/(min · m²)] was required in order to avoid the generation of filter flies (psychoda). Figure 7.37 shows an installation of a plastic-packed filter.

FIGURE 7.37
Plastic-packed trickling filter.

Theory

As wastewater passes through the filter, nutrients and oxygen diffuse into the slimes, where assimilation occurs, and by-products and CO_2 diffuse out of the slime into the flowing liquid. As oxygen diffuses into the biological film, it is consumed by microbial respiration, so that a defined depth of aerobic activity is developed. Slime below this depth is anaerobic, as shown in Fig. 7.38.

As in BOD removal by activated sludge, BOD removal through a trickling filter is related to the available biological slime surface and to the time of contact of wastewater with that surface.

The mean time of contact of liquid with the filter surface is related to the filter depth, the hydraulic loading, and the nature of the filter packing:

$$t = \frac{CD}{Q^n} \tag{7.15}$$

where
$$\begin{aligned}
t &= \text{mean detention time} \\
D &= \text{filter depth, ft} \\
Q &= \text{hydraulic loading, gpm/ft}^2 \\
C \text{ and } n &= \text{constants related to the specific surface and the} \\
&\quad \text{configuration of the packing}
\end{aligned}$$

There are a number of commercial packings available. These include vertical-flow, random-packed, and cross-flow media. The properties of these media are summarized in Table 7.11. Investigations of several types of packings showed that

$$C = C'A_v^m \tag{7.16}$$

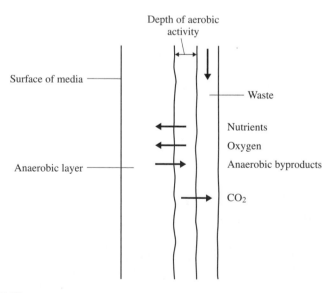

FIGURE 7.38
Mode of operation of a trickling filter.

TABLE 7.11
Comparative physical properties of common trickling filter media

Medium type	Nominal size, in	Unit weight, lb/ft^3	Specific surface area, ft^2/ft^3	Void space, %	Application
Bundle	24 × 24 × 48	2–5	27–32	>95	C, CN, N
(sheet)	24 × 24 × 48	4–6	42–45	>94	N
Rock	1–3	90	19	50	N
Rock	2–4	100	14	60	C, CN, N
Random	Varies	2–4	25–35	>95	C, CN, N
(Dump)	Varies	3–5	42–50	>94	N
Wood	48 × 48 × 1$\frac{7}{8}$	10.3	14		C, CN

C = CBOD$_{5R}$
CN = CBOD$_{5R}$ and NOD$_R$
N = tertiary NOD$_R$

1 in = 25.4 min
1 lb/ft^3 = 16.05 kg/m^3
1 ft^2/ft^3 = 0.305 m^2/m^3

in which A_v is the specific surface expressed in square feet per cubic foot (*note:* ft^2/ft^3 = 3.28 m^2/m^3). For spheres, rock, and Polygrid plastic media without slime, C' has a value of 0.7 and m a value of 0.75. This relationship will vary for others that have a different general configuration.

The exponent n in Eq. (7.15) has been observed to decrease with decreasing specific surface A_v. Equations (7.15) and (7.16) can be combined to yield a general expression for the mean retention time through any type of filter packing:

$$\frac{t}{D} = \frac{C'A_v^m}{Q^n} \tag{7.17}$$

The mean retention time increases considerably in the presence of filter slime, in some cases being as much as 4 times that prevailing with the nonslimed surface.

In order to avoid filter plugging, a maximum specific surface of 30 ft^2/ft^3 (98 m^2/m^3) is recommended for the treatment of carbonaceous wastewaters, Specific surfaces in excess of 100 ft^2/ft^3 (328 m^2/m^3) can be used for nitrification because of the low yield of biological cellular material.

Recent data have shown improved performance with cross-flow media with the same specific surface as other media because of the increased time of contact of the wastewater passing through the filter.

In Chap. 6 it is shown that soluble BOD removal in the plug flow activated sludge process could be expected to follow the relationship

$$\frac{S}{S_o} = e^{-K_b X_v t / S_o}$$

These equations may be analogously applied to trickling filters under the following conditions:

1. The specific surface must remain constant. This is true for any specific filter medium but varies from medium to medium.

2. The medium must have a uniform, thin slime cover. This does not always occur, particularly in the case of rock filters. It is essential therefore, that the entire surface be wetted. A minimum hydraulic loading of 0.50 gal/(min · ft^2) is recommended.

The second condition also requires uniform distribution of the hydraulic loading to the filter. Albertson[24] has shown the dosing cycle should be optimized to ensure media wetting and flushing of excessive slime growth. Albertson and Eckenfelder[25] have shown that, with adequate hydraulic loading for media wetting, the BOD removal efficiency is independent of depth and that the value of K is depth-dependent as follows:

$$K_2 = K_1 \left(\frac{D_1}{D_2} \right)^{0.5} \tag{7.18}$$

This condition is violated when heavy slime buildup short-circuits the filter.

For the retention time as defined by Eq. (7.15) and assuming that the available bacterial surface to be proportional to the specific surface A_v,

$$\frac{S}{S_o} = e^{-KA_vD/Q^nS_o} \tag{7.19}$$

In Eq. (7.18) the hydraulic loading of q includes both the forward flow and the recirculated flow. It has been shown that in many cases BOD removal can be increased by recirculation of filter effluent around the filter. The recirculated flow serves as a diluent to the influent waste. When recirculation is used, the applied BOD to the filter, S_o, becomes

$$S_o = \frac{S_a + NS}{1 + N} \tag{7.20}$$

where S_o = BOD of waste applied to the filter after mixing with the recirculated flow
S_a = BOD of the influent
S = BOD of filter effluent
N = recirculation ratio, R/Q.

Filter performance in accordance with Eq. (7.19) is shown in Fig. 7.39. Combining Eqs. (7.19) and (7.20) yields the BOD removal relationship:

$$\frac{S}{S_a} = \frac{e^{-KA_vD/Q^nS_o}}{(1 + N) - Ne^{-KA_vD/Q^nS_o}} \tag{7.21}$$

When the BOD of the waste exhibits a decreasing removal rate with decreasing concentration, the BOD in the recirculated flow will be removed at a lower rate than that prevailing in the influent. In this case, one must apply a coefficient of retardation to the recirculated flow.

It has been shown that BOD removal through the filter can be related to the applied organic loading expressed as lb BOD/(1000 ft^3 · d):

$$\frac{S}{S_o} = e^{-kA_v/L} \tag{7.22}$$

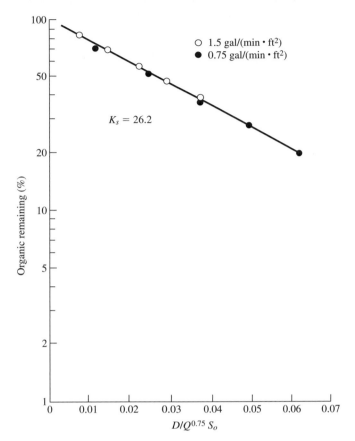

FIGURE 7.39
Treatment of dilute black liquor (S_o = 400 mg/l) on plastic packing.

in which L is expressed as lb BOD/(1000 ft^3 · d) (kg of BOD/m^3 · d). Data correlated in accordance with Eq. 7.22 are shown in Fig. 7.40. It should be noted that when the exponent n in Eq. (7.22) is unity, Eq. (7.22) and Eq. (7.19) become mathematically the same. Filter performance characteristics for various wastewaters, in accordance with Eq. (7.22), are shown in Table 7.12.

Oxygen Transfer and Utilization

Oxygen is transferred from air passing through the filter to the films of liquid passing over the slimes. Since the rate of oxygen transfer is related to fluid mixing and turbulence, it can be expected that hydraulic loading and media configuration will influence the transfer rate. Experimentation on various types of filter media has shown that the following relationship applies:[28]

$$\frac{-dC}{dD} = K_o(C_s - C_L) \tag{7.23}$$

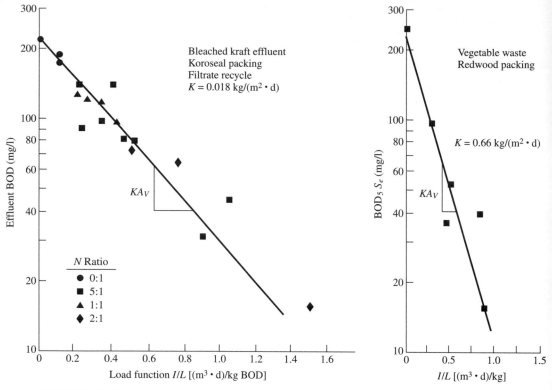

FIGURE 7.40
Data correlation for two wastewaters in accordance with Eq. (7.22).

where D = filter depth
   C_L = concentration of dissolved oxygen in the liquid passing over the filter
   K_o = transfer rate coefficient

This integrates to

$$\frac{(C_s - C_L)_1}{(C_s - C_L)_2} = e^{-K_o(D_2 - D_1)} \qquad (7.24)$$

The transfer rate coefficient K_o is related to the hydraulic loading to the filter. The oxygen transfer relationship is shown in Fig. 7.41.

The oxygen transfer can be expressed in mass of oxygen per hour per unit volume of filter medium by the following relationship

$$N = 5.0 \times 10^{-4} K_o (C_s - C_L) Q \qquad (7.25)$$

where N = lb $O_2/(ft^3 \cdot h)$
   C_s = oxygen saturation, mg/l
   C_L = dissolved oxygen concentration, mg/l
   Q = hydraulic loading, gal/(min $\cdot$ ft^2)
   K_o = transfer rate coefficient, ft^{-1}

TABLE 7.12
Trickling filter performance

Type of waste	Type of medium	Mean value, S_o or S_a, mg/l	Rate coefficient, kg/m²
Pharmaceutical	Vinyl core	5248	0.2160
Phenolic	Vinyl core	340	0.0210
Sewage	6 parallel: basalt, slag, Surfpac I and II, Flocor, Cloisonyle	280	0.0480
Sewage	4 parallel: slag 6 in and 4 in, Flocor, Surfpac	332	0.0480
Sewage	4 parallel: slag 6 in and 4 in, Flocor, Surfpac	215	0.0500
Kraft mill sludge recycle	Vinyl core	210	0.0160
Kraft mill filtrate recycle	Vinvl core	220	0.0180
Vegetable	Del Pak	235	0.0660
Fruit canning	Surfpac	2200	0.0930
Kraft mill	Surfpac	130	0.005
Fruit processing	Surfpac	3200	0.001
Pulp and paper	Vinyl core	280	0.016

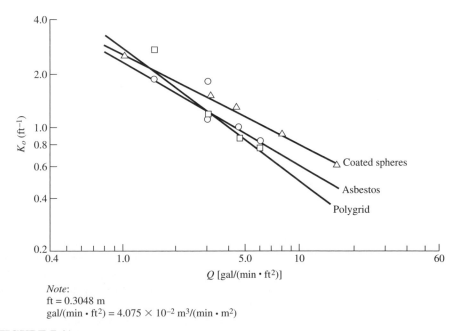

Note:
ft = 0.3048 m
gal/(min · ft²) = 4.075 × 10⁻² m³/(min · m²)

FIGURE 7.41
Relationship between the oxygen transfer rate coefficient and hydraulic loading.

or
$$N = 0.06K_o(C_s - C_L)Q$$

where N = kg $O_2/(m^3 \cdot h)$
Q = $m^3/(m^2 \cdot min)$
K_o = m^{-1}

The ability of a filter to remove BOD is limited by the aerobic activity of the filter film. This in turn is limited by the amount of oxygen transferred to the slime from the flowing liquid.

The activity of the filter films is measured by the surface oxygen-utilization rate. British studies[29] of the treatment of domestic sewage showed an oxygen-utilization rate of 0.028 mg $O_2/(cm^2 \cdot h)$. Studies of the treatment of dilute black liquor at a filter loading of 400 lb BOD/(1000 ft$^3 \cdot$ d) [6.41 kg/(m$^3 \cdot$ d)] yielded a utilization rate of 0.0434 mg $O_2/(cm^2 \cdot h)$.[27]

The total quantity of aerobic film can be estimated from the relationship:

$$h = \sqrt{\frac{2D_L C_L}{K_r \rho}} \tag{7.26}$$

where h = depth of film to which oxygen penetrates
D_L = diffusivity of oxygen through the filter film
C_L = concentration of dissolved oxygen in the liquid passing the film surface
ρ = density of the filter film
K_r = unit oxygen-utilization rate of the filter film

It is possible to estimate the maximum BOD which can be assimilated by the filter film from Eqs. (7.25) and (7.26). Albertson[24] has estimated oxygen is not limiting up to organic loadings of 160 lb $O_2/(1000$ ft$^3 \cdot$ d) for media containing 30 ft^3/ft^3. In order to ensure adequate oxygen ventilation, fans should be installed.

In treating high-concentration industrial wastewaters, a maximum influent concentration must be maintained in order to avoid anaerobic conditions prevailing in the filter slimes and resultant escape of anaerobic products and odors to the atmosphere. Depending on the degradability of the wastewater this may vary from 600 to 1200 mg/l. Higher influent BOD concentrations require recirculation for dilution of the influent strength.

Effect of Temperature

The performance of trickling filters will be affected by changes in the temperature of the filter films and the liquid passing over the films. It is usually assumed that these two temperatures will be essentially the same when only the aerobic portion of the film is considered. A decrease in temperature results in a decrease in respiration rate, a decrease in oxygen-transfer rate, and an increase in oxygen saturation. The combined effect of these factors results in an increase of aerobic film at a lower activity level, yielding a somewhat reduced efficiency at lower temperatures. The relationship of efficiency and temperature can be expressed as[26]

$$E_T = E_{20} \times 1.035^{T-2} \tag{7.27}$$

where E = filter efficiency and T = temperature, °C

Trickling Filter Applications

In most cases, the reaction rate K for soluble industrial wastewaters is relatively low, and hence filters are not economically attractive for high treatment efficiency (85 percent BOD reduction) of such wastewaters. Plastic-packed filters, however, have been employed as a pretreatment for high-strength wastewaters in which BOD removals in the order of 50 percent have been achieved at hydraulic and organic loadings of greater than 4 gal/(min · ft²) [0.16 m³/(min · m²)] and 500 lb BOD/(thousand ft³ · d) [8.0 kg/(m³ · d)]. Performance characteristics for several industrial wastewaters are shown in Table 7.13 and Figure 7.42. Example 7.10 presents a trickling filter design.

> **EXAMPLE 7.10.** One million gallons per day (3785 m³/d) of an industrial waste-water with a BOD of 900 mg/l is to be pretreated to an effluent BOD of 300 mg/l. A plastic tower with a specific surface of 30 ft²/ft³ (98 m²/m³) and a depth of 20 ft (6.1 m) is to be used. Determine the hydraulic loading, the recirculation ratio, and the filter area required. The maximum BOD applied to the filter is 600 mg/l. The reaction rate coefficient for this wastewater is 1.0. The media has a coefficient n of 0.5.

TABLE 7.13
High-rate trickling filtration performance

Waste	Hydraulic loading, MGAD	Depth, ft	Raw BOD	Recycle ratio	BOD removal (clarified), %	Temperature	BOD loading, lb/1000 ft³
Sewage	126	21.6	145	3	88		54
	252		131	3	82		110
	252		175	1	70		250
	63		173	0	78		95
	126		152	0	76		67
	189	10.8	166	0	45		549
	135		165	0	57		390
	95		185	0	51		304
Citrus	72	21.6	542	3	69		199
	189		464	2	42		612
Citrus and sewage	189	21.6	328	2	53		384
Kraft mill	365	18.0	250	0	10	34	
	185	18.0	250	0	24	36	
	200	21.6	250	0	23	40	
	90	21.6	250	0	31	33	
Black	47	18	400	0	73	24	200
Liquor	95	18	400	0	58	29	380
	189	18	400	0	58	35	780

Note:
MGAD = 0.9354 m³/(m² · d)
ft = 0.3048 m
lb/1000 ft³ = 16 × 10⁻³ kg/m³

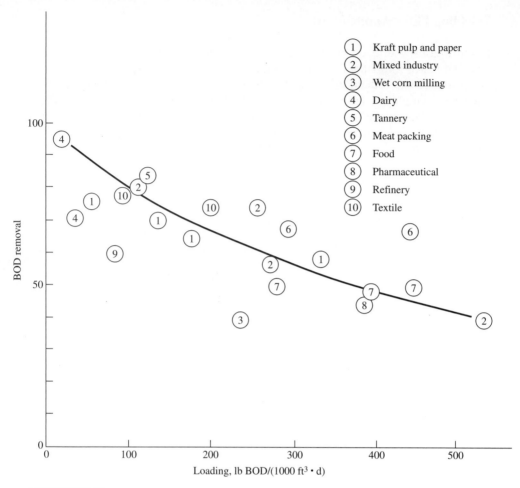

FIGURE 7.42

Pretreatment of organic wastewater by high-rate trickling filters using plastic media
($Av = 30$ ft²/ft³).

Solution.

Calculate the required recycle ratio:

$$S_o = \frac{S_a + NS_e}{1 + N}$$

$$600 = \frac{900 + 300N}{1 + N}$$

$$N = 1.0$$

Calculate the hydraulic loading:

$$\frac{S}{S_o} = e^{-KA_v D/Q^n S_o}$$

$$\frac{300}{600} = e^{-(1 \times 30 \times 20)/(Q_o^{0.5} \times 600)} = e^{-x}$$

$$x = 0.69 = \frac{1}{Q_o^{0.5}}$$

$$Q_o = 2 \text{ gal}/(\text{min} \cdot \text{ft}^2) \qquad [0.082 \text{ m}^3/(\text{min} \cdot \text{m}^2)]$$

Since $N = 1.0$, $R = Q$, and the wastewater flow is 1.0 gal/(min · ft²) [0.041 m³/(min · m²)],

$$A = \frac{10^6 \text{ gal/d} \times 6.94 \times 10^{-4} \text{ d/min}}{1.0 \text{ gal}/(\text{min} \cdot \text{ft}^2)} = 694 \text{ ft}^2 \qquad (64.5 \text{ m}^2)$$

The filter volume $V = AD = 694 \times 20 = 13,880$ ft³ (393 m³).

Check for oxygen requirements:
The BOD removed through the filter is

$$2 \text{ million gal/d} \times (600 - 300)(\text{mg/l}) \times 8.34 \frac{\text{lb/million gal}}{\text{mg/l}} = 5000 \text{ lb/d } (2270 \text{ kg/d})$$

Assuming 0.8 lb O_2/lb BOD removed, the oxygen requirement is 4000 lb/d (1816 kg/d).
From Fig 7.41 at a hydraulic loading of 2 gal/(min · ft²), the oxygen-transfer rate coefficient K_o is 2.0/ft.
The oxygen transfer can be calculated from Eq. 7.25:

$$N = 5.0 \times 10^{-4} \times 2.0/\text{ft} \times (9 - 1) \text{ mg/l} \times 2 \text{ gal}/(\text{min} \cdot \text{ft}^2)$$

$$= 160 \times 10^{-4} \text{ lb } O_2/(\text{ft}^3 \cdot \text{h}) \qquad [0.26 \text{ kg}/(\text{m}^3 \cdot \text{h})]$$

Assuming an α value of 0.85,

$$N = 160 \times 10^{-4} \times 0.85 = 136 \times 10^{-4} \text{ lb } O_2/(\text{ft}^3 \cdot \text{h}) \qquad [0.22 \text{ kg}/(\text{m}^3 \cdot \text{h}]$$

$$\text{lb } O_2/\text{d} = 136 \times 10^{-4} \text{ lb}/(\text{ft}^3 \cdot \text{h}) \times 24 \text{ h/d} \times 13,880 \text{ ft}^3 = 4341 \text{ lb/d} \qquad (2055 \text{ kg/d})$$

Nitrification in a trickling filter is related to the organic loading. Since heterotropic growth is considerably greater than that of nitrifiers, the nitrifiers compete very poorly for film growth. Therefore, only under low organic loadings will significant nitrification occur. This is shown in Fig. 7.43.

Tertiary Nitrification

Nitrification in a trickling filter in the absence of carbonaceous organics follows a zero-order reaction for effluent NH_3^-N concentrations of 3.5 to 10 mg/l. Results, reported by Jiumm et al[30] are shown in Fig. 7.44 for a synthetic wastewater activated sludge effluent. Maximum ammonia removal rates for municipal wastewater have been reported to vary from 1.2 to 1.8 g $NH_3^-N/(\text{m}^2 \cdot \text{d})$. This rate must be corrected for industrial effluents. Reported rates are based on a soluble BOD applied to the filter of less than 20 mg/l. In order to ensure wetting of the filter surface, hydraulic loadings should exceed 0.8 gal/(min · ft²). Jiumm et al.[30] have

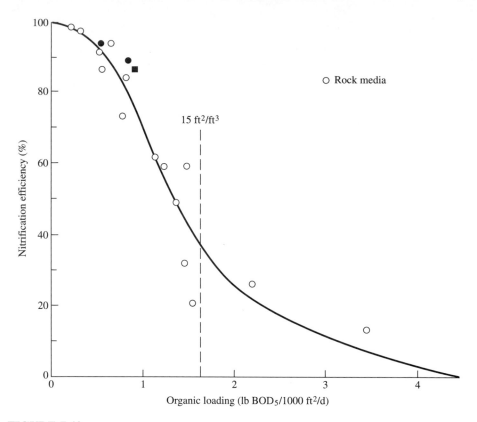

FIGURE 7.43
Combined carbon oxidation–nitrification performance.

shown that alkalinity requirements are less than the theoretical value of 7.2 mg alkalinity per mg of N oxidized. They showed a variation of 6.7 mg alkalinity/mg N oxidized at low nitrogen loadings to 4.6 at mg/1 at high loadings. This decrease in alkalinity requirement may be attributed to denitrification in the biofilm, which generates some alkalinity. Tertiary nitrification in a trickling filter is illustrated by Example 7.11.

EXAMPLE 7.11. It is desired to reduce NH_3^-N from 100 mg/l to 20 mg/l through a 20-ft-deep trickling filter using media with a specific surface of 44 ft^2/ft^3. No inhibition is expected. The wastewater flow is 1 million gal/d.

Compute:

    The filter volume required
    The hydraulic loading on the filter
    The alkaline loading on the filter

Use a NH_3^-N removal rate of 0.28 lb $NH_3^-N/(1000\ ft^2 \cdot d)$

Solution.

The NH_3^-N removed is

$$(100 - 20)(1)(8.34) = 667\ \text{lb/d}$$

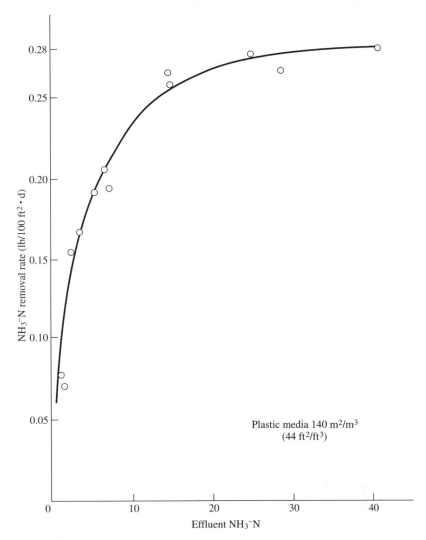

FIGURE 7.44
Tertiary nitrification through a trickling filter.

The packing area required is

$$\frac{667 \text{ lb NH}_3^- \text{N/d}}{0.28 \text{ lb NH}_3^- \text{N}/(1000 \text{ ft}^2 \cdot \text{d})} = 2383 \times 10^3 \text{ ft}^2$$

The filter volume, for a specific surface area of 44 ft²/ft³, is

$$\frac{2383 \times 10^3 \text{ ft}^2}{44 \text{ ft}^2/\text{ft}^3} = 54 \times 10^3 \text{ ft}^3$$

The surface area of the filter is therefore

$$\frac{54 \times 10^3}{20 \text{ ft}} = 2700 \text{ ft}^2$$

and the filter diameter is thus 52 ft.

The hydraulic loading is based on the need to maintain 0.8 gal/(min · ft²) across the filter surface. This will require 2160 gal/min, or a recycle flow of 2160 − 690 = 1470 gal/min.

For the alkalinity requirement, assume 7.14 lb alkalinity/lb NH_3^-N removed. Then

$$(667)(7.14) = 4762 \text{ lb/d}$$

7.5
ROTATING BIOLOGICAL CONTACTORS

The rotating biological contactor consists of large-diameter plastic media mounted on a horizontal shaft in a tank, as shown in Fig. 7.45. The contactor is slowly rotated with approximately 40 percent of the surface area submerged. A 1- to 4-mm layer of slime biomass is developed on the medium. (This would be equivalent to 2500 to 10,000 mg/l in a mixed system.) As the contactor rotates, it carries a film of wastewater through the air, resulting in oxygen and nutrient transfer. Additional removal occurs as the contactor rotates through the liquid in the tank. Shearing forces cause excess biomass to be stripped from the media, as in a trickling filter. This biomass is removed in a clarifier. The attached biomass is shaggy with small filaments, providing a high surface area for organic removal to occur. The present

FIGURE 7.45
Rotating biological contactor (*Courtesy of Envirex Inc.*).

medium consists of high-density polyethylene with a specific surface of 37 ft²/ft³ (121 m²/m³). Single units are up to 12 ft (3.7 in) in diameter and 25 ft (7.6 m) long, containing up to 100,000 ft² (9290 m²) of surface in one section.

The primary variables affecting treatment performance are:

1. Rotational speed
2. Wastewater retention time
3. Staging
4. Temperature
5. Disk submergence

In the treatment of low-strength wastewaters (BOD to 300 mg/l), performance increased with rotational speed up to 60 ft/min (18 m/min), with no improvement noted at higher speeds. Increasing rotational speed increases contact, aeration, and mixing, and would therefore improve efficiency for high-BOD wastewaters. However, increasing rotational speed rapidly increases power consumption, so that an economic evaluation should be made of the trade-off between increased power and increased area.

In the treatment of domestic wastewater, performance increases with liquid volume to surface area ratios up to 0.12 gal/ft² (0.0049 m³/m²). No improvement was noted above this value.

In many cases, significant improvement was observed by increasing from two to four stages with no significant improvement with greater than four stages. Several factors could account for these phenomena. The reaction kinetics would favor plug flow or multistage operation. With a variety of wastewater constituents, acclimated biomass for specific constituents may develop in different stages. Nitrification will be favored in the later stages, where low BOD levels permit a higher growth of nitrifying organisms on the media. In the treatment of industrial wastewaters with high BOD levels or low reactivity, more than four stages may be desirable. For high-strength wastewaters, an enlarged first stage may be employed to maintain aerobic conditions. An intermediate clarifier may be employed where high solids are generated to avoid anaerobic conditions in the contactor basins. RBC performance treating minewater tailings is shown in Table 7.14.

Design for industrial wastewaters will usually require a pilot plant study. A kinetic model similar to that developed for activated sludge may be employed:

$$\frac{Q}{A}(S_o - S) = kS \tag{7.28}$$

where Q = flow rate
  A = surface area
  S_o = influent substrate concentration
  S = effluent substrate concentration
  k = reaction rate

Or, for wastewaters with a highly variable influent strength,

$$\frac{Q}{A}(S_o - S) = K\frac{S}{S_o} \tag{7.29}$$

TABLE 7.14
RBC performance on goldmine tailings wastewater

	SCN⁻, mg/l	CN$_t^-$, mg/l	CN$_c^-$, mg/l	Cu, mg/l	NH$_3^-$N, mg/l
Influent	51.9	6.94	4.60	0.99	5.5
Effluent	< 1.0	0.35	0.05	0.03	0.5

CN$_c^-$ = cyanide amenable to chlorination
CN$_t^-$ = total cyanide

Results across four stages:

	Influent	Stage 1	Stage 2	Stage 3	Stage 4
SCN⁻, mg/l	51.9	20.3	3.0	< 1.0	< 1.0
NH$_3^-$N, mg/l	2.2	8.4	3.4	0.9	0.3

Equation (7.26) for several industrial wastewaters is shown in Fig. 7.46. For high-BOD wastewaters, performance can be improved by surrounding the media with enriched oxygen to enhance oxygen transfer and BOD removal.

Several factors become apparent from Fig. 7.46. The maximum BOD removal rate $Q/A(S_o - S)$ for a given operating condition (rotational speed, gas oxygen content, etc.) will relate to both the concentration of influent BOD and the biodegradability of the wastewater. The performance of multiple contactors in series can be defined by the relationship:

$$\frac{S}{S_o} = \left(\frac{1}{1 + kA/Q}\right)^n \tag{7.30}$$

in which n is the number of stages. Since at some loading oxygen will be limiting, it is important that the loading for each stage be checked for a multistage system.

As previously noted, the economics of the alternatives should be evaluated for each application. A process design is shown in Example 7.12.

EXAMPLE 7.12. An RBC is to be designed for an effluent soluble BOD of 20 mg/l for a wastewater with an initial soluble BOD of 300 mg/l. The flow is 0.5 million gal/d (1900 m³/d). Compute the total number of stages and the required area. The performance relationship is shown in Fig. 7.47.

Solution.

The maximum loading to an RBC occurs at that point where oxygen becomes limiting, as shown in Fig. 7.47. The maximum hydraulic loading can be computed:

$$\frac{Q}{A} = \frac{[(Q/A)S_r]_{\max}}{S_o - [(Q/A)S_r]_{\max}/k}$$

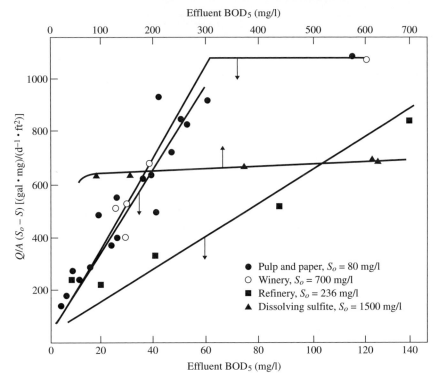

Note: gal/(ft² · d) = 4.08 × 10⁻² m³/(m² · d)

FIGURE 7.46
BOD removal characteristics of an RBC treating industrial wastewaters.

The performance of multiple contactors in series can be defined by the relationship:

$$\frac{S_e}{S_o} = \left(\frac{1}{1 + kA/Q}\right)^n$$

$$\frac{Q}{A} = \frac{k}{(S_o/S_e)^{1/n} - 1}$$

The maximum hydraulic loading can be computed:

$$\left(\frac{Q}{A} S_r\right)_{max} = 1000 \text{ gal}/(d \cdot ft^2) \cdot mg/l$$

$$\frac{Q}{A} = \frac{1000}{300 - 1000/7} = 6.37 \text{ gal}/(d \cdot ft^2) \qquad [0.26 \text{ m}^3/(d \cdot m^2)]$$

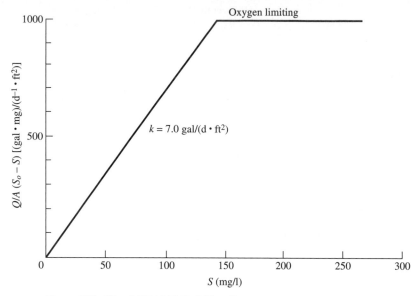

FIGURE 7.47
BOD removal relationship.

The required area for various numbers of stages can be computed as tabulated below.

Number of stages, n	Q/A, gal/(d · ft^2)	$(Q/A)/n$, gal/(d · ft^2 · stage)	$A \times 10^{-3}$, ft^2/stage
1	0.5	0.5	1000
2	2.44	1.22	410
3	4.83	1.61	310
4	7.29	1.82	275
5	10.0	2.0	250

$$\frac{Q}{A} = \frac{7.0}{(S_o/S_e)^{1/n} - 1}$$

The hydraulic loading (Q/A) can be plotted versus the number of stages, as shown in Fig. 7.48. In order not to exceed the maximum loading to the first stage, the maximum number of stages would be 3.5. In this case, the plant would be designed for three stages.

The calculations can also be made graphically, as shown in Fig. 7.49.

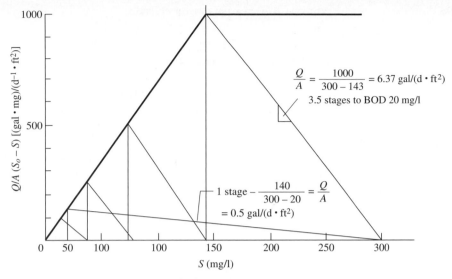

$$\frac{Q}{A} = \frac{1000}{300 - 143} = 6.37 \text{ gal/(d} \cdot \text{ft}^2)$$

3.5 stages to BOD 20 mg/l

$$1 \text{ stage} - \frac{140}{300 - 20} = \frac{Q}{A}$$
$$= 0.5 \text{ gal/(d} \cdot \text{ft}^2)$$

Note: gal/(d $\cdot$ ft^2) = 4.075 $\times$ 10^{-2} m^3/(d $\cdot$ m^2)

FIGURE 7.48
Graphical design solution.

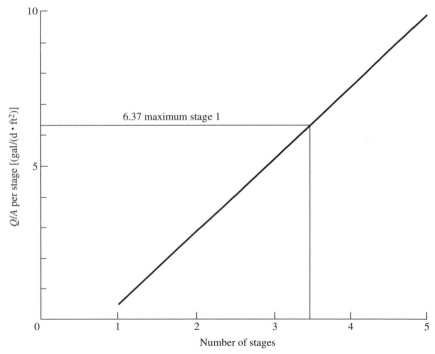

6.37 maximum stage 1

Note: gal/(d $\cdot$ ft^2) = 4.075 $\times$ 10^{-2} m^3/(d $\cdot$ m^2)

FIGURE 7.49
Calculation of number of stages.

7.6
ANAEROBIC DECOMPOSITION

Anaerobic decomposition involves the breakdown of organic wastes to gas (methane and carbon dioxide) in the absence of oxygen. Although the process kinetics and material balances are similar to those of aerobic systems, certain basic differences require special consideration.

The conversion of organic acids to methane gas yields little energy; hence the rate of growth is slow and the yield of organisms by synthesis is low. The kinetic rate of removal and the sludge yield are both considerably less than in the activated sludge process. The quantity of organic matter converted to gas will vary from 80 to 90 percent. Since there is less cell synthesis in the anaerobic process, the nutrient requirements are correspondingly less than in the aerobic system. High process efficiency requires elevated temperatures and the use of heated reaction tanks. The methane gas produced by the reaction can be used to provide this heat. Wastes of low COD or BOD concentration will not provide sufficient methane for heating, and a supplementary source of heat is necessary.

The anaerobic process is operated in one of several ways (shown in Fig. 7.50):

1. The *anaerobic filter reactor* establishes growth of the anaerobic organisms on a packing medium. The filter may be operated upflow, as shown in Fig. 7.50, or downflow. The packed filter medium, while retaining biological solids, also provides a mechanism for separating the solids and the gas produced in the digestion process. Jennett and Dennis,[34] treating a pharmaceutical wastewater, were able to achieve a 97 percent removal of COD at a loading of 3.5 kg COD/(m^3 · d) using plastic media (37°C). Sachs et al.[35] achieved an 80 percent COD reduction at a loading of 0.56 kg COD/(m^3 · d) (35°C) and a 36-h HRT for a synthetic organic chemicals wastewater. Obayashi and Roshanravan[36] obtained a 70 percent COD reduction at a loading of 2 kg COD/(m^3 · d) on a rendering wastewater. Start-up periods may vary from 3 to 9 months, depending on substrate and OLR differences (Colleran et al.).[37]

2. The *anaerobic contact process*[31,32] provides for separation and recirculation of seed organisms, thereby allowing process operation at retention periods of 6 to 12 h. A degasifier is usually needed to minimize floating solids in the separation step. For high-degree treatment, the solids retention time has been estimated at 10 d at 90°F (32°C); the estimate doubles for each 20°F (11°C) reduction in operating temperature. Steffen and Bedker reported a full-scale anaerobic contact process (30 to 35°C) that achieved 90 percent COD removed while treating a meat-packing wastewater at a loading of 2.5 kg COD/(m^3 · d) and an HRT of 13.3 h. The SRT was about 13.3 d. Speece[33] has shown that with a six percent rate of synthesis the days required for a twofold and a tenfold increase in biomass is 12 and 40 days, respectively.

3. In the fluidized-bed reactor (FBR) wastewater is pumped upward through a sand bed on which microbial growth has been developed. Biomass concentrations exceeding 30,000 mg/l have been reported. Effluent is recycled to mix with the feed in quantities dictated by the strength of the wastewater and the fluidization

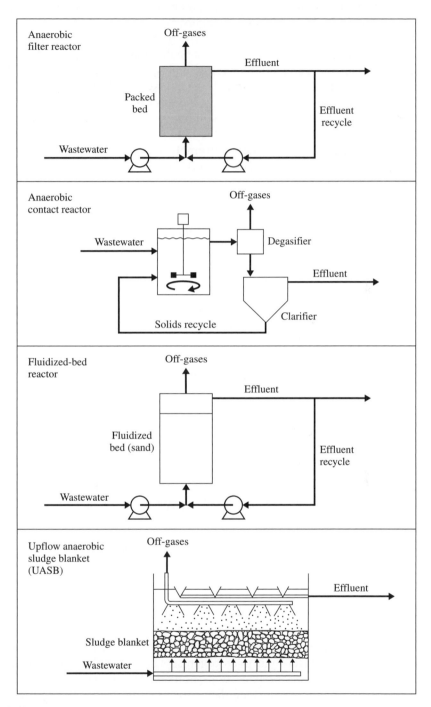

FIGURE 7.50
Anaerobic wastewater treatment processes.

velocity. Organic removal efficiencies of 80 percent were achieved at loadings of 4 kg COD/(m^3 · d) on dilute wastewaters.[40]

4. In the *upflow anaerobic sludge blanket* (UASB) process, wastewater is directed to the bottom of the reactor, where it must be distributed uniformly. The wastewater flows upward through a blanket of biologically formed granules which consume the waste as it passes through the blanket. Methane and carbon dioxide gas bubbles rise and are captured in the gas dome. Liquid passes into the settling portion of the reactor, where solids-liquid separation takes place. The solids return to the blanket area while the liquid exits over the weirs. Formation of granules and their maintenance is extremely important in the operation of the process. Palns et al.[38] hypothesized that favorable granule formation requires a plug flow reactor configuration with a neutral pH, a zone of high H$_2$ partial pressure, a nonlimiting NH$_3$$^-$N source, and a limited source of cysteine, and a substrate that produces H$_2$ as an intermediate. With carbohydrate wastewaters the alkalinity requirement is 1.2 to 1.6 g alkalinity as CaCO$_3$/g influent COD, which is sufficient to maintain the pH above 6.6. Guiot et al.[39] found the addition of trace metals enhanced biomass activity. To keep the blanket in suspension, an upflow velocity of 2 to 3 ft/h (0.6 to 0.9 m/h) has been used. Waste stabilization occurs as the waste passes through the sludge bed, with solids concentrations in the sludge bed having been reported to reach as high as 100 to 150 g/l. Loadings up to 600 lb COD/(100 ft^3 · d) [96 kg/(m^3 · d)] have been used successfully for certain wastewaters. In pilot studies, organic loadings of 15 to 40 kg COD/(m^3 · d) at liquid residence times of 3 to 8 h successfully treated high-strength wastewaters. A full-scale plant treating beet sugar wastewater achieved 80 percent removal with loadings of 10 kg COD/(m^3 · d) and an HRT of 4 h.

5. The ADI-BVF process is a low-rate anaerobic reactor with intermittent mixing and sludge recycle. The reactor has two zones: a reaction zone at the inlet end and a clarification zone at the outlet end. The reactor can be an aboveground tank or a lined earthen basin with a floating insulated membrane cover for gas recovery and temperature and odor control. Because of the large reactor volume, equalization requirements are minimized. A typical unit is shown in Fig. 7.51.

Performance data for the various anaerobic processes is shown in Table 7.15. Performance of the ADI-BVF process treating a variety of industrial wastewaters is shown in Table 7.16.

Mechanism of Anaerobic Fermentation

In anaerobic fermentation, roughly four groups of microorganisms sequentially degrade organic matter. Hydrolytic microorganisms degrade polymer-type material such as polysaccharides and proteins to monomers. This reduction results in no reduction of COD. The monomers are then converted into fatty acids (VFA) with a small amount of H$_2$. The principal acids are acetic, propionic, and butyric with small quantities of valeric. In the acidification stage there is minimal reduction of COD. Should a large amount of H$_2$ occur, some COD reduction will result. This

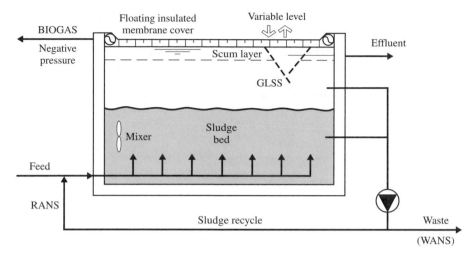

(a)

(b)

FIGURE 7.51
ADI-BVF® reactor (*Courtesy of ADI Systems Inc.*).

seldom exceeds 10 percent. All acids higher than acetic acid are converted into acetate and H_2 by acetogenic microorganisms. The conversion of propionic acid is

$$C_3H_6O_2 + 2H_2O \rightarrow C_2H_4O_2 + CO_2 + 3H_2$$

In this reaction, COD reduction does occur in the form of H_2. This reaction will occur only if the concentration of H_2 is very low. The breakdown of organic acids

TABLE 7.15
Performance of anaerobic processes

Wastewater	Loading, kg/(m³ · d)	HRT, d	Temperature, °C	Removal, %	Reference
Anaerobic contact process:					
Meat packing	3.2 (BOD)	12	30	95	32
Meat packing	2.5 (BOD)	13.3	35	95	31
Keiring	0.085 (BOD)	62.4	30	59	39
Slaughterhouse	3.5 (BOD)	12.7	35	95.7	39
Citrus	3.4 (BOD)	32	34	87	41
Upflow filter process:					
Synthetic	1.0 (COD)	—	25	90	42
Pharmaceutical	3.5 (COD)	48	35	98	35
Pharmaceutical	0.56 (COD)	36	35	80	34
Guar gum	7.4 (COD)	24	37	60	36
Rendering	2.0 (COD)	36	35	70	43
Landfill leachate	7.0 (COD)	—	25	89	44
Paper mill foul condensate	10–15 (COD)	24	35	77	45
Fluidized bed reactor process:					
Synthetic	0.8–4.0 (COD)	0.33–6	10–3	80	46
Paper mill foul condensate	35–48 (COD)	8.4	35	88	45
USAB process:					
Skimmed milk	71 (COD)	5.3	30	90	46
Sauerkraut	8–9 (COD)	—	—	90	46
Potato	25–45 (COD)	4	35	93	41
Sugar	22.5 (COD)	6	30	94	41
Champagne	15 (COD)	6.8	30	91	34
Sugar beet	10 (COD)	4	35	80	46
Brewery	95 (COD)	—	—	83	47
Potato	10 (COD)	—	—	90	47
Paper mill foul condensate	4–5 (COD)	70	35	87	45
ADI-BFV process:					
Potato	0.2 (COD)	360	25	90	—
Cornstarch	0.45 (COD)	168	35	85	—
Dairy	0.32 (COD)	240	30	85	—
Confectionary	0.51 (COD)	336	37	85	—

TABLE 7.16
Anaerobic treatment of industrial wastewaters
COD, BOD, and SS values for different wastewaters

Wastewater	Raw wastewater				Anaerobic effluent			
	COD, mg/l	BOD, mg/l	BOD/ COD	SS, mg/l	COD, mg/l	BOD, mg/l	BOD/ COD	SS, mg/l
Potato processing	4,263	2,664	0.62	1,888	144	32	0.22	70
Yeast, cane molasses	13,260	6,630	0.50	1,086	4,420	600	0.14	883
Brewery and municipal	9,750	2,790	0.29	4,146	332	179	0.54	168
Clam processing	3,813	1,895	0.50	856	594	337	0.57	130
Corn processing and municipal	5,780				1,210			136
Hardboard mill effluent	12,930	5,990	0.46	486	2,590	740	0.29	507
Dairy wastewater	13,076	7,204	0.55	1,919	596	173	0.29	260
Semichemical pulp mill	6,826	2,221	0.32	851	3,822	524	0.14	881
Brewery	2,692	1,407	0.52	778	295	122	04.1	201
Alcohol stillage—1	90,000	23,000	0.26					
Alcohol stillage—2	120,000	40,000	0.33		57,000	4,700	0.08	
Alcohol stillage—3	98,000	31,000	0.32		54,000	6,000	0.11	
Alcohol stillage—4	80,000	24,000	0.30		36,000	4,100	0.11	
Dairy	3,250	1,970	0.61	252	372	111	0.30	55
Potato processing	1,890	1,090	0.58	341	165	98	0.59	50
Kraft foul condensates	13,960	6,710	0.48	10	1,076	660	0.61	190
Molasses stillage	65,000	25,000	0.38	5,000	15,000	1250	0.08	500
Corn wet milling	3,510	1,700	0.48	1,080	410	133	0.32	64
Pulp and paper	5,349	2,287	0.43	3,792	965	308	0.32	199
Dairy	25,541	20,575	0.81	974	737	190	0.26	337
Dairy	19,200	10,400	0.54	3,400	770	130	0.17	500
Brewery	4,011	2,786	0.69	139	510	306	0.60	105
Industrial and domestic	3,000	1,620	0.54	550	300	105	0.35	120
Dairy	8,830	7,890	0.89	1,670	150	86	0.57	53
Potato processing	8,356	5,300	0.63	5,250	1,113	486	0.44	708
Apple processing	3,994	2,441	0.61	2,573	174	87	0.50	54
Olive processing	13,395	5,550	0.41	289	2,332	786	0.34	212
Beans and pasta processing	2,604	1,200	0.46		1,285	528	0.41	
Pharmaceutical	9,200	4,000	0.43	2,400	3,300	850	0.26	350
Pharmaceutical	7,100	3,300	0.46	1,000	1,490	460	0.31	170
Confectionary	10,560	6,550	0.62	1,050	320	70	0.22	180
Potato processing	12,489	5,978	0.48	9,993	4,692	1573	0.34	2200
Ethanol corn processing	1,155	743	0.64	20	397	204	0.51	162

to CH_4 and CO_2 is shown in Fig. 7.52. The acetic acid and H_2 are converted to CH_4 by methanogenic organisms.

Acetic acid: $$C_2H_4O_2 \rightarrow CO_2 + CH_4$$

$$CH_3COO^- + H_2O \rightarrow CH_4 + HCO_3^-$$

Hydrogen: $$HCO_3^- + 4H_2 \rightarrow CH_4 + OH^- + 2H_2O$$

The specific activity of typical anaerobic processes treating soluble industrial wastewaters is approximately 1 kg COD utilized/(kg biomass · d). There are two classes of methanogens that convert acetate to methane, namely *Methanothrix* and *Methanosarcina*. *Methanothrix* has a low specific activity, so it will predominate in systems with a low steady-state acetate concentration. In highly loaded systems, *Methanosarcina* will predominate with a higher specific activity (3 to 5 times as high as *Methanothrix*) if trace nutrients are available. These trace nutrients are iron, cobalt, nickel, molybdenum, selenium, calcium, magnesium, and microgram per liter levels of vitamin B.[33]

Speece[33] has reported trace mineral requirements as Ca 0.018 mg/g Ac, Fe 0.023 mg/g Ac, Ni 0.004 mg/g Ac, Co 0.003 mg/g Ac, and Zn 0.02 mg/g Ac.

Zehnder et al.[49] found that optimal methanogen growth and the specific rate of methane production required between 0.001 and 1.0 mg/l sulfur as S.

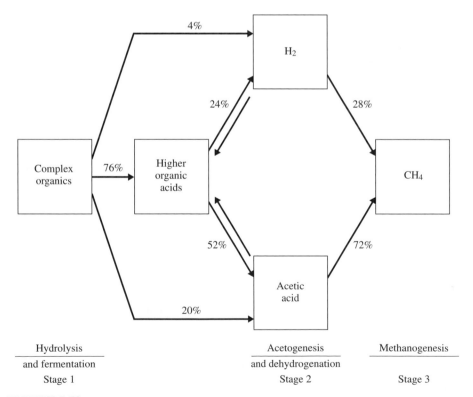

FIGURE 7.52
Three stages of methane fermentation.

The kinetic relationship commonly employed for anaerobic degradation is the Monod relationship:

$$\frac{ds}{dt} = \frac{k_{max}SX}{K_s + S} \tag{7.31}$$

where ds/dt = substrate utilization rate, mg/(l · d)
 k_{max} = maximum specific substrate utilization rate, g COD/(g VSS · d)
 S = effluent concentration, mg/l
 X = biomass concentration, mg/l
 K_S = half saturation concentration, mg/l

Typical values for the coefficients are (Lawrence and McCarty)[50]:

Temperature, °C	K_{max}, d^{-1}	K_S, mg/l
35	6.67	164
25	4.65	930
20	3.85	2130

Reactor staging will frequently result in a reduced volume: 1 lb (0.454 kg) of COD or ultimate BOD removed in the process will yield 5.62 ft^3 (0.16 m^3) of methane at 0°C and 6.3 ft^3 CH$_4$ at 35°C.

The quantity of cells produced during methane fermentation will depend on the strength of the waste, the character of the waste, and the retention of the cells in the system. As in an aerobic system, a portion of the cells produced will be destroyed by endogenous metabolism. The data of McCarty and Vath[51] are shown in Fig. 7.53.

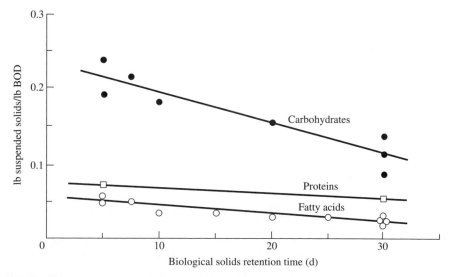

FIGURE 7.53
Biological solids production resulting from methane fermentation (*After McCarty and Vath.*[51]).

A relationship similar to Eq. (6.10) can be used to estimate cell yield. The relationships obtained by McCarty and Vath are:

Amino and fatty acids: $A = 0.054F - 0.038M$

Carbohydrates: $A = 0.46F - 0.088M$

Nutrient broth: $A = 0.076F - 0.014M$

where A = biological solids accumulated, mg/l
       M = mixed liquor volatile suspended solids, mg/l
       F = COD utilized, mg/l

McCarty[48] has estimated the cell composition as $C_5H_9NO_3$, in which the nitrogen requirement is 11 percent of the net cell weight. The phosphorus requirement has been estimated as 2 percent of the biological cell weight. The COD of the cells is 1.21 mg/mg volatile suspended solids.

Biodegradation of Organic Compounds under Anaerobic Conditions

Anaerobic processes can break down a variety of aromatic compounds. It is known that anaerobic breakdown of the benzene nucleus can occur by two different pathways: photometabolism and methanogenic fermentation. It has been shown that benzoate, phenylacetate, phenylpropionate, and annamate are completely degraded to CO_2 and CH_4. Lower fatty acids were detected as reaction intermediates. While long acclimation periods were required to initiate gas production, the time required could be reduced by adapting the bacteria to an acetic acid substrate before adapting them to the aromatic.

Chmielowski et al.[52] showed that phenol, p-cresol, and resorcinol were completely converted to CH_4 and CO_2. A tabulation of organic compounds mineralized under anaerobic conditions is shown in Table 7.17.[49] As described by Speece,[33] fortuitous metabolism is defined as biodegradation of refractory organics by microbes from which they derive no energy. If a cosubstrate is added along with a refractory organic such as chloroform, the biomass is able to metabolize it fortuitously. Cosubstrates include sugar, methanol, and lactate.

As in aerobic processes, soluble microbial products (SMP) are also generated under anaerobic conditions. Kuo et al.[54] found SMP production ranged from 0.2 to 1.0 percent for acetate and 0.6 to 2.5 percent for glucose.

Food-processing and brewery wastes are readily broken down anaerobically with BOD removals in the range of 85 to 95 percent. In the anaerobic treatment of distillery waste, special precautions must be taken to reduce the inhibitory tendency of sulfates contained in the waste by either dilution or adjustment of loading rates. While animal wastes are readily treated anaerobically, ammonia toxicity may be a problem in dealing with fresh waste that contains a large amount of urine. Vath[55] showed that linear anionic and nonionic ethoxylated surfactants underwent degradation as observed by a loss of surfactant properties. Numerous investigators have shown that a variety of pesticides including lindane and isomers of benzene hexachloride will degrade under anaerobic conditions.

TABLE 7.17
Organics mineralized under anaerobic conditions[53]

Acetylsalicylic acid	Phloroglucinol
Acrylic acid	Phthalic acid
p-Anisic acid	Polyethelene glycol
Benzoic acid	Pyrogallol
Benzyl alcohol	p-Aminobenzoic acid
2,3-Butanediol	Butylbenzylphthalate
Catechol	4-Chloroacetanilide
m-Cresol	m-Chlorobenzoic acid
p-Cresol	Diethylphthalate
Di-n-butylphthalate	Geraniol
Dimethylphthalate	4-Hydroxyacetinilide
Ethyl acetate	p-Hydroxybenzyl alcohol
2-Hexanone	2-Octanol
o-Hydroxybenzoic acid	Propionanilide
p-Hydroxybenzoic acid	Butylbenzylphthalate
3-Hydroxybutanone	m-Chlorobenzoic acid
1-Octanol	m-Methoxyphenol
Phenol	o-Nitrophenol
	p-Nitrophenol

Many of the higher-molecular-weight hydrocarbons that make up oil are decomposable by anaerobic bacteria. Shelton and Hunter[56] demonstrated that anaerobic decomposition of oil occurs naturally in bottom deposits.

An excellent review of the anaerobic treatment of industrial wastewaters by anaerobic processes has been presented by Obayashi and Gorgan.[57] The BOD/COD characteristics of effluents from anaerobic treatment of various wastewaters are shown in Table 7.18.

Factors Affecting Process Operation

The anaerobic process will effectively function over two temperature ranges: the mesophilic range of 85 to 100°F (29 to 38°C) and the thermophilic range of 120 to 135°F (49 to 57°C). Although the rates of reaction are much greater in the thermophilic range, the maintenance of higher temperatures is usually not economically justifiable.

The methane organisms function over a pH range of 6.6 to 7.6 with an optimum near pH 7.0. When the rate of acid formation exceeds the rate of breakdown to methane, a process unbalance results in which the pH decreases, gas production falls off, and the CO_2 content of the gas increases.

pH control is therefore essential to ensure a high rate of methane production. Lime is commonly used to raise the pH of an anaerobic system when there is a

TABLE 7.18
**BOD/COD characteristics of effluents from
anaerobic treatment of various wastewaters**

Wastewater	BOD, mg/l	COD, mg/l
Sugar	50–500	250–1500
Dairy	150–500	250–1200
Maize starch	—	500–1500
Potato	200–300	250–1500
Vegetable	100	700
Wine	3,500	—
Pulp	350–900	1,400–8000
Fibre board	2500–5500	8800–14900
Paper mill	100–200	280–300
Landfill leachate	—	500–4000
Digester supernatant	400	800–1400
Brewery	—	200–350
Distillery	—	320–400

process imbalance. Caution must be taken, since excess application of lime will result in precipitation of calcium carbonate. Sodium bicarbonate can alternatively be used for pH adjustment. It is desirable to have a bicarbonate alkalinity in the range of 2500 to 5000 mg/l in order to provide a buffer capacity to handle the increase in volatile acids with a minimal decrease in pH. If sufficient amounts of alkalinity are not present in the feed, then the alkalinity can be controlled by reducing the feed rate or adding alkalinity to the wastewater.

Alkalinity is required to neutralize the volatile fatty acids (VFA) generated and the H_2CO_3 formed, which results from the high partial pressure of CO_2 in the reactor. These alkalinity requirements are shown in Fig. 7.54.

In a well-operated complete mix reactor, the VFA will be in the range of 20 to 200 mg/l. In plug flow systems, however, elevated concentrations of VFA in the inlet region will reduce additional alkalinity.

Inorganic salts in low concentration may provide stimulation; in high concentration, they may be toxic.[58] In some cases, adaptation will increase the tolerance level of the organisms. The stimulatory and inhibitory concentrations of some common cations are summarized in Table 7.19.

The presence of antagonistic ions may sharply reduce the inhibitory effect of specific cations. Kugelman and McCarty[59] have shown that 300 mg/l potassium will reduce by 80 percent the inhibitory effect of 7000 mg/l sodium. The inhibitory effect will be completely eliminated by the addition of 150 mg/l calcium, but the calcium without the potassium will exert no beneficial effect. Ammonia is toxic at concentrations in excess of 3000 mg/l and inhibitory at levels greater than 1500 mg/l. The toxicity is related to the pH, which indicates how much ammonia is present in gaseous form. Gaseous ammonia is more inhibitory than the ammonium ion.

The maximum nontoxic concentration of soluble sulfide in an anaerobic system is 200 mg/l. Since sulfide is present as H_2S (gaseous), HS^-, and precipitated

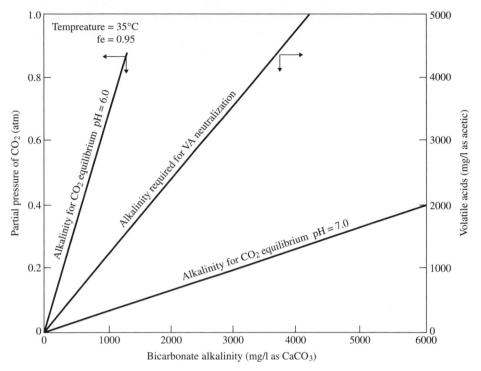

FIGURE 7.54
Design diagram for alkalinity requirements.

TABLE 7.19
Cation toxicity to anaerobic decomposition
Maximum ion concentration, mol/l

Ion	Other ions < 10 mg/l		Antagonist present	
	Slug	**Acclimated**	**Slug**	**Acclimated**
Na^+	0.2	0.3	0.25–0.3	>0.35
NH^4	0.1	0.15–0.18		
K^+	0.09	0.15	0.15–0.2	>0.35
Ca^{2+}	0.07	>0.2	0.13	>0.2
Mg^{2+}	0.05	0.075	0.1	>0.14

sulfide, the total sulfide concentration may be considerably higher. The loss of sulfide in the gas would also permit higher concentrations of either sulfate or sulfide in the waste feed to the system.

The heavy metals, such as copper, zinc, and nickel, are toxic at low concentrations. In the presence of sulfides, however, some of these metals are precipitated. The toxicity to organisms is related to their concentration in the digester itself.

When toxic materials are present, they must be removed before treatment or diluted below the toxic levels. The proper approach will depend principally on the overall waste composition.

In order to avoid a decrease in pH, sufficient alkalinity must be present to compensate for the high CO_2 content. At a 30 percent CO_2 content in the digester gas, 1500 mg/l of alkalinity is necessary.

7.7
LABORATORY EVALUATION OF ANAEROBIC TREATMENT

Owen et al.[60] developed an anaerobic biotransformation assay termed the biochemical methane potential (BMP). This assays the concentration of organic pollutants in a wastewater which can be anaerobically converted to CH_4 and evaluates the potential anaerobic process efficiency. The test procedure has been described by Speece[33] as follows.

The procedure for implementing the BMP assay involves placing an aliquot of the effluent sample, normally 50 ml, in a 125-ml serum bottle with an anaerobic inoculum. In many cases the reactor effluent already contains an adequate inoculum. In other cases an acclimated inoculum can be taken directly from an anaerobic reactor.

The headspace in the serum bottle should be purged with CO_2 at 30 to 50 percent CO_2 composition for pH control and N_2 or CH_4. The serum bottle is then incubated at 35°C and CH_4 production recorded after a prescribed number of days (usually 5 days). The gas production is measured by inserting a hypodermic needle, connected to a calibrated fluid reservoir, through the serum cap. At this temperature, 395 ml of CH_4 production is equivalent to 1 g COD reduction, a stoichiometric relationship which allows calculation of the COD reduction in the liquid phase.

It is important that CO_2 production be excluded because CO_2 does not represent COD reduction under anaerobic conditions. For example, if 2000 mg/l (COD equivalent) of biodegradable organic pollutant remains in the effluent, a BMP assay would indicate that after a period of time 39.5 ml of CH_4 net gas production would result from a 50 ml sample of effluent.

A cardinal rule of the BMP assay, as with the BOD, is that the biomass must be acclimated to the pollutants. Care must be exercised to run a control with only the anaerobic inoculum to ensure adequate time and acclimation for the biomass to metabolize the pollutant. Whereas a 20-day BOD is considered to represent the ultimate demand aerobically, the BMP may be extended to 30 or 60 days to accommodate acclimation of the biomass to toxic and/or unusual pollutants occurring in some industrial wastewaters.

Since COD conversion is normally proportional to the product of biomass and time, the relative amount of biomass inoculum will affect the rate of conversion, but not the net ultimate value. A BMP analysis of a degradable wastewater is shown in Fig. 7.55. The corresponding COD reduction is shown in Fig. 7.56.

Owen et al.,[60] as described by Speece,[33] also developed a very useful and simple assay procedure to evaluate the potential toxicity of a wastewater sample to the

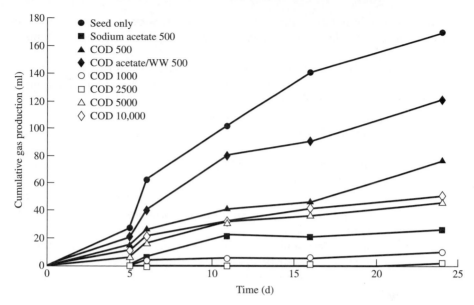

FIGURE 7.55
Cumulative gas production for nontoxic wastewaters.

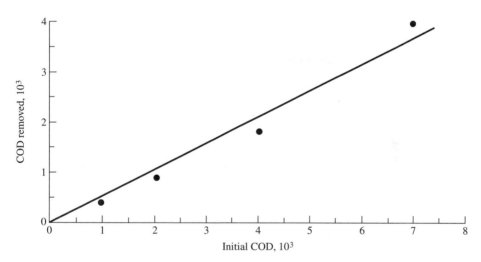

FIGURE 7.56
COD removal relative to initial COD.

anaerobic biomass, the anaerobic toxicity assay (ATA). The biomass to be evaluated is placed in a serum bottle and gassed with 50 percent CO_2 and 50 percent CH_4; then the wastewater sample is injected in increasing volumes into successive bottles. This procedure results in a range of dilution of the wastewater with the initial inocula of biomass. Excess substrate is also added initially to the serum bottles to avoid substrate limitation. If there is toxicity in the wastewater sample, it will be reflected in a reduced initial rate of gas production in proportion to the volume of wastewater added.

Because the aceticlastic methanogens are commonly the most sensitive to toxicity in the consortium, this characteristic can be assayed by adding a surplus of acetate (10,000 mg/l of calcium acetate salt is recommended). More complex substrates such as glucose, ethanol, propionate, or other complex substrates can be added in excess to assay toxicity to members in the consortium other than methanogens.

The significant difference between the BMP and the ATA assays is that the ATA is flooded with acetate (or other simple substrate noted above) as well as the wastewater sample, whereas the BMP is not. Also it must be borne in mind that in the ATA assay, the initial rate of gas production is of primary interest, while in the BMP it is the total amount of gas production which is important. Acclimation phenomena can be observed in both assays as the biomass demonstrates the ability to acclimate to the toxicity. In the BMP, if gas production rate (corrected for the control) per unit volume of wastewater decreases as the amount injected into the bottles increases, this change is also an indication of inherent toxicity in the wastewater, as shown in Fig. 7.57.

Methane production can be estimated by the relationship

$$G = 5.62(S_r - 1.42\,\Delta X_v) \qquad (7.32)$$

where G = CH$_4$ produced per day, ft^3/d
         S_r = BOD removed, lb/d
         ΔX_v = VSS produced, lb/d

or $$G = 0.351(S_r - 1.42\,\Delta X_v) \qquad (7.33)$$

where G is in m^3/d and S_r and ΔX_v are in kg/d. One cubic foot (0.0283 m^3) of methane has a net heating value of 960 Btu (1.01 × 10^6 J).

A design example is shown in Example 7.13.

> **EXAMPLE 7.13.** Design an anaerobic contact process to achieve 90 percent removal of COD from a wastewater flow of 100,000 gal/d (379 m^3/d).
>
> Total influent COD = 10,300 mg/l
> Nonremovable COD = 2200 mg/l
> Removable COD (COD$_R$) = 8100 mg/l
> COD to be removed = 90 percent
>
> Process parameters:
>
> SRT = 20 days minimum
> Temperature = 35°C
> a = 0.136 mg VSS/mg COD$_R$
> b = 0.021 mg VSS/(mg VSS · d)
> k = 0.0004 l/(mg · d)
> X_v = 5000 mg/l

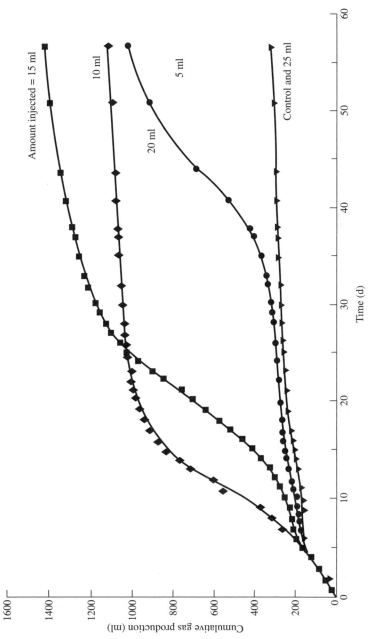

FIGURE 7.57
Cumulative gas production for inhibitory wastewater.

Solution.

Digester volume

From the kinetic relationship:

$$t = \frac{S_r}{X_v k S}$$

$$= \frac{(8100)(0.9)}{(5000)(0.0004)(810)} = 4.5 \text{ d}$$

The digester volume is therefore

$$(4.5 \text{ d})(0.1 \text{ million gal/d}) = 0.45 \text{ million gal} \qquad (1700 \text{ m}^3)$$

Check SRT (sludge age):

$$\text{SRT} = \frac{X_v t}{\Delta X_v} = \frac{X_v t}{a S_r - b X_v t}$$

$$= \frac{(5000)(4.5)}{(0.136)(7290) - (0.021)(5000)(4.5)} = 43.4 \text{ d}$$

This is in excess of the recommended SRT of 20 d to ensure the growth of methane formers.

A vacuum degasifier or flash aerator should be provided between the digester and the clarifier to purge gas from the sludge. A flash aerator inhibits further methane production in the clarifier with resulting floating sludge.

Sludge yield

The sludge yield from the process is

$$\Delta X_v = a S_r - b X_v t$$

$$= (0.136)(7290) - (0.021)(5000)(4.5) = 519 \text{ mg/l}$$

$$\Delta X_v = 519 \text{ mg/l} \times 0.1 \text{ million gal/d} \times 8.34 \frac{\text{lb/million gal}}{\text{mg/l}}$$

$$= 433 \text{ lb/d} \qquad (196 \text{ kg/d})$$

Gas production

$$G = 5.62(S_r - 1.42 \Delta X_v)$$

$$= 5.62[(7290)(0.1)(8.34) - (1.42)(433)]$$

$$= 30,700 \text{ ft}^3 \text{ CH}_4/\text{d} \ (869 \text{ m}^3/\text{d})$$

Heat requirements

These can be estimated by calculating the energy required to raise the influent wastewater temperature to 95°F (35°C) and allowing 1°F (0.56°C) heat loss per day of detention time.

Average wastewater temperature = 75°F (23.9°C)
Heat-transfer efficiency = 50 percent

$$\text{Btu} = \frac{W(T_i - T_e)}{E} \times (\text{specific heat})$$

$$\text{Btu}_{\text{required}} = \frac{(0.1 \text{ million gal/d})(8.34 \text{ lb/gal})(95° + 4.5° - 75°)}{0.5} \times \left(\frac{1 \text{ Btu}}{\text{lb} \cdot °F}\right)$$

$$= 40,900,000 \text{ Btu/d } (4.3 \times 10^{10} \text{ J/d})$$

The heat available from gas production is

$$\text{Btu}_{\text{available}} = (30,700 \text{ ft}^3 \text{ CH}_4/\text{d})(960 \text{ Btu/ft}^3 \text{ CH}_4)$$

$$= 29,500,000 \text{ Btu/d} \qquad (3.1 \times 10^{10} \text{ J/d})$$

External heat of $40,900,000 - 29,500,000 = 11,400,000$ Btu/d $(1.2 \times 10^{10} \text{ J/d})$ should be supplied to maintain the reactor at 95°F (35°C).

Nutrient requirements
The nitrogen requirement is

$$N = 0.12 \,\Delta X_v = 0.12 \times 433 \text{ lb/d}$$

$$= 52 \text{ lb/d } (23.6 \text{ kg/d})$$

The phosphorus requirement is

$$P = 0.025 \,\Delta X_v = 0.025 \times 433 \text{ lb/d}$$

$$= 11 \text{ lb/d} \qquad (5 \text{ kg/d})$$

PROBLEMS

7.1. A wastewater of 3 million gal/d (11,350 m³) has a BOD of 2100 mg/l and a VSS content of 400 mg/l. It is to be pretreated in an aerobic lagoon with a retention period of 1.2 d. The K is 8.0/d. Compute:
(*a*) The effluent soluble BOD in mg/l
(*b*) The effluent VSS in mg/l
(*c*) The oxygen required in mass/day if
$\quad a = 0.5$
$\quad a' = 0.55$
$\quad b = 0.15/\text{d}$

7.2. An industrial facility is to treat its waste in a series of aerated lagoons. The final effluent should have a maximum soluble BOD₅ of 20 mg/l for the summer months and 30 mg/l for the winter months.

Flow = 8.5 million gal/d (32,170 m³/d)
BOD₅ inf = 425 mg/l
Temperature of waste = 95°F (35°C)
$K = 6.3\text{d}^{-1}$ at 20°C
$\theta = 1.065$
$a = 0.5$

$a' = 0.52$
$b = 0.20 \text{ d}^{-1}$
$\alpha = 0.88$
$C_1 = 1.0 \text{ mg/l}$
$T_a = 30°\text{F (winter)}$
$\quad = 65°\text{F (summer)}$

Design the lagoons and optimize the system with respect to area. Design and specify the aeration system. Present a diagram showing basin geometries and aerator locations.

7.3. Design a two-stage aerated lagoon to operate at 20°C for the following conditions:

Flow $= 2$ mgd
BOD, $S_o = 500$ mg/l
$S_{e \text{ soluble}} = 20$ mg/l
$K = 10 \text{ day}^{-1}$
$a = 0.6$
$b = 0.1 \text{ d}^{-1}$
$a' = 0.55$

Assume the detention period in the aerobic lagoon is 1 d.

7.4. The parameters for an activated sludge process are

Wastewater

Flow $= 150 \text{ m}^3/\text{h}$ (0.95 million gal/d)
$COD_T = 450$ mg/l
$COD_D = 250$ mg/l
$BOD_5 = 125$ mg/l
$TSS = 42$ mg/l
Eff. $BOD_T = 20$ mg/l
$BOD_S = 10$ mg/l
$COD_D = 30$ mg/l

Process

$a'COD = 0.55$
$a\text{VSS/COD} = 0.32$
$a\text{VSS/BOD} = 0.55$
$\theta_c = 8 \text{ d}$
$F/M_{BOD} = 0.35$
$K_{BOD} = 5 \text{ d} - 1$

Assume the 42 mg/l influent TSS are 80 percent volatile and are nondegradable. Assume X_v to be 3000 mg/l and the sludge age to be 8 d.

Compute the following:

(*a*) Waste activated sludge
(*b*) Oxygen requirements
(*c*) Nutrient requirements
(*e*) Sludge recycle
(*f*) Biological selector
(*g*) Final clarifier area for a flux of 20 lb/(ft² · d)

7.5. A batch activated sludge plant is to be designed for the following conditions:

Q = 40,500 gal/d (153 m³/d)
S_o = 975 mg/l
S_e = 10 mg/l
Period of wastewater discharge = 11.0 h

A laboratory bench-scale study yielded the following operational parameters:

$(F/M)_{av}$ = 0.2/d
a = 0.53
b = 0.16/d
a' = 0.43
Eff. SS = 50 mg/l (85% volatile)
Equilibrium sludge density = 0.8 lb sludge/ft³ (12.8 kg/m³)

The sludge is to be stored for a period of 30 d when the reactor will be emptied and cleaned. A value of 45 lb BOD removed/(hp · d) [27.4 kg/(kW · d)] was also found to be in effect.

(a) Calculate the following:
(i) Volume of active sludge required to oxidize each day's accumulation of degradable solids
(ii) The 30 days sludge storage volume
(iii) The daily volume for the waste
(iv) The freeboard at 5 percent of the waste volume
(v) The total volume and basin dimensions
(b) Calculate the amount of oxygen per day required.
(c) Calculate the effluent BOD total.
(d) Estimate the aeration power requirements.

7.6. A 5 million gal/d (18,930 m³/d) municipal activated sludge plant is presently operating at an F/M of 0.3 d^{-1}. The plant plans to accept 0.5 million gal/d (1890 m³/d) of a brewery wastewater with a BOD of 1200 mg/l. Determine what changes need to be made to avoid filamentous bulking. Use the F/M dissolved oxygen relationship shown in Fig. 6.32. The municipal sewage has a BOD of 200 mg/l. The effluent soluble BOD is 10 mg/l.

7.7. A municipal plant receives fruit-processing wastewaters and has a combined BOD of 600 mg/l. The return sludge has a concentration of 8000 mg/l SS. What recycle is required for a contactor with a floc load of 150 mg BOD/g VSS?

7.8. Compute the required clarifier area for a wastewater flow of 1.0 million gal/d (3785 m³/d), an MLSS of 3000 mg/l, an SVI of 150 ml/g, and an R/Q of 0.5. What is the maximum MLSS if the SVI increases to 250 ml/g and the recycle rate stays the same?

7.9. A trickling filter presently achieves 52 percent BOD reduction from a brewery wastewater under the following conditions:

N = 3:1
Q = 3 gal/(min · ft²) [0.122 m³/(m² · min)]
d = 20 ft (6.1 m)
S_o = 850 mg/l
Flow = 0.8 million gal/d (3030 m³/d)
n = 0.5

In-plant changes will reduce the flow to 0.5 million gal/d and the BOD to 700 mg/l. For the same recycle, what will the new BOD removal efficiency be?

7.10. The following data were obtained on a rotating biological contactor (RBC) for a bleached pulp-mill effluent. The test unit had four stages.

Average flow, gal/(d · ft²)	Influent, mg/l	Stages, mg/l			
		1	2	3	4
2.0	76	26	10	7	6
4.0	72	41	24	16	12
6.0	81	42	25	15	10
8.0	78	52	38	25	18
10.0	83	60	47	36	30

Correlate the data and design a system for 80 percent BOD removal, assuming an influent BOD of 85 mg/l.

7.11. Consider Example 7.9. At present, an air-activated sludge plant follows the anaerobic process. The activated sludge plant operates under the following conditions to produce an effluent with a COD of 30 mg/l (degradable):

$T = 20°C$
$F/M = 0.3 \text{ d}^{-1}$
$X_v = 2500 \text{ mg/l}$

The load to the anaerobic plant is to be increased to 120,000 gal/d (454 m³/d) with the same removable COD.

(a) Compute the new effluent from the anaerobic plant assuming X_v remains the same. What will the new gas production be?

(b) What modifications to the aerobic plant must be made to maintain the same effluent quality? Assume the sludge settling characteristics are the same as originally. Compute the following:

(i) The new sludge recycle and the new MLSS. The volatile content of the sludge is 74 percent.

(ii) The new F/M.

(iii) Calculate the increase in power required assuming 1.5 lb O_2/(hp · h) [0.91 kg/kW · h]

The process parameters are:

$a = 0.5$
$a' = 0.6$
$b = 0.15 \text{ at } 20°C$

REFERENCES

1. Gloyna, E. F.: *Waste Stabilization Pond Concepts and Experiences,* Division of Environmental Health, World Health Organization, Geneva, 1965.
2. Oswald, W. J.: *Advances in Water Quality Improvement* (E. F. Gloyna and W. W. Eckenfelder, eds.), vol. 1, University of Texas Press, Austin, Texas, 1968.

3. Gellman, I., and H. F. Berger: *Advances in Water Quality Improvement* (E. F. Gloyna and W. W. Eckenfelder, eds.), vol. 1, University of Texas Press, Austin, Texas, 1968.

4. Hermann, E. R., and E. F. Gloyna: *Sewage Ind. Wastes,* vol. 30, p. 963, 1958.

5. Marais, G. V., and V. A. Shaw: *Trans. S. Afr. Inst. Civ. Engrs.,* vol. 3, p. 205, 1961.

6. Steffen, A. J.: *J. WPCF,* vol. 35, pt. 4, p. 440, 1963.

7. Cooper, R. C.: *Dev. Appl. Microbiol.,* vol. 4, pp. 95–103, 1963.

8. Argamon, Y., and C. Adams: *Proc. 8th International Conf. on Water Pollution Research,* Pergamon Press, Oxford, England, 1976.

9. Eckenfelder, W. W., C. Adams, and S. McGee: *Advances in Water Pollution Research,* vol. 2, Pergamon Press, Oxford, 1972.

10. Parker, D.: Personal communication.

11. Galil, N. et al.: *Proc. 2nd Specialized Conf. on Pretreatment of Industrial Wastewaters,* Athens, Greece 1996.

12. Boon, A. G. et al.: Report from Water Resources Research Center, Stevenage, U.K., 1983.

13. Murphy, K. L., and P. L. Timpany: *J. San. Eng. Div.,* ASCE, SA 5, October, 1967.

14. Franta, J. R., and P. A. Wilderer: *Water Sci. Tech.,* vol. 35, no. 1, p. 67, 1997.

15. Goronszy, M.: *J. WPCF,* vol. 41, pt. 2, p. 274, 1979.

16. Yamamoto, K. et al: *Wat. Sci. Technol.,* vol. 21, p. 43, 1989.

17. Mishra, P. N. et al: *89th Meeting Air and Waste Management Assoc.,* Nashville, TN 1996.

18. Cuthbert, D. C., and D. C. Pollock: *Wat. Env. Assoc. of Texas,* May 1995.

19. Randall, C. W.: *Water and Env. Management,* vol. 12, no. 5, p. 375, 1998.

20. Daigger, G. T., and R. E. Roper: *J. WPCF,* vol. 57, p. 859, 1985.

21. Wahlberg, E. J. et al: *Proc. Wat. Env. Fed. 68th Ann. Conf.,* vol. 1, p. 435, 1995.

22. Parker, D. S. et al.: *Wat. Sci. Tech.* vol. 25, no. 6, p. 301, 1992.

23. Paduska, R. A.: *Proc 24th Industrial Waste Conf.,* Purdue University, Lafayette, Ind., 1979.

24. Albertson, O. E.: *WPCF Operations Forum,* p. 15, 1989.

25. Albertson, O. E., and Eckenfelder, W. W.: *Proc. 2nd International Conf. on Fixed Film Biological Processes,* Washington, D.C., 1984.

26. Howland, W. E.: *Proc. 12th Ind. Waste Conf.,* Purdue University, vol. 94, p. 435, 1958.

27. Eckenfelder, W. W., and E. L. Barnhart: *J. WPCF,* vol. 35, p. 535, 1963.

28. Eckenfelder, W. W.: *Proc. ASCE SA,* vol. 4, pt. 2, pp. 33, 860, July 1961.

29. Department of Scientific and Industrial Research: "Water Pollution Research, 1956" (British), Her Majesty's Stationery Office, London, 1957.

30. Jiumm, M. H. et al.: *Proc. 1st International Conf. on Fixed Film Biological Processes,* Kings Island, Ohio, 1982.

31. Steffen, A. J., and M. Bedker: *Proc. 16th Ind. Waste Conf.,* Purdue University, 1961.

32. Schropfer, G. J. et al.: *Sewage Ind. Wastes,* vol. 27, p. 460, 1955.

33. Speece, R. F.: *Anaerobic Biotechnology for Industrial Wastewaters,* Archae Press, Nashville, Tenn., 1996.

34. Jennett, J. C., and N. D. Dennis: *J. WPCF,* vol. 47, p. 104, 1975.

35. Sachs, E. F. et al.: *Proc. 33rd Ind. Waste Conf.,* Purdue University, 1978.

36. Obayashi, A. W., and M. Roshanravan: Unpublished report, Illinois Institute of Technology, Chicago, 1980.

37. Colleran, E. S. et al.: *Proc. 7th International Symposium on Anaerobic Digestion,* South Africa, p. 160, 1994.

38. Palns, S. S. et al.: *Water SA IT,* pp. 47–56, pp. 1991.

39. Guiot, S. et al: *Wat. Sci. Tech.,* vol. 25, p. 1, 1992.

40. Jewell, W. J. et al.: *J. WPCF,* vol. 53, p. 482, 1981.

41. McNary, R. R. et al.: *Proc. 17th Ind. Waste Conf.,* Purdue University, 1962.

42. Young, J. C., and P. L. McCarty: *Technical Report 87*, Department of Civil Engineering, Stanford University, 1968.
43. Witt, E. R. et al.: *Proc. 34th Ind. Waste Conf.,* Purdue University, 1979.
44. Dewalle, F. B., and E. S. K. Chian: *Biotech. Bioengineering,* vol. 18, p. 1275, 1976.
45. Donovan, G.: Personal communication.
46. Lettinga, G., and W. de Zeeuw: *Proc. 35th Ind. Waste Conf.,* Purdue University, 1980.
47. Pette, K. C. et al.: *CSM Suiker,* Amsterdam, Netherlands, 1986.
48. McCarty, P. L.: *Progress in Wat. Tech.,* vol. 7, p. 157, 1975.
49. Zehnder, A. J. et al.: *Anaerobic Digestion,* Elsevier, Amsterdam, 1982.
50. Lawrence, A. W., and McCarty, P. L.: *J. WPCF,* vol. 41, pp. R1–R17, 1969.
51. McCarty, P. L., and C. A. Vath: *Int. J. Air Water Pollution,* vol. 6, p. 65, 1962,
52. Chmielowski, J. et al.: *Zesz. Nauk,* Politech Slaska Inz. (Polish), vol. 8, p. 97, 1965.
53. Shelton, D. R., and J. M. Tiedjc: *Applied and Env. Microbiol.,* vol. 47, pp. 850–857, 1984.
54. Kuo, W. C. et al.: *Water Env. Research,* 1995.
55. Vath, C. A.: *Soap and Chem. Specif.,* March 1964.
56. Shelton, T. B., and J. B. Hunter: *J. WPCF,* vol. 47, pt. 9, p. 2257, 1975.
57. Obayashi, A. W., and J. M. Gorgan: *Management of Industrial Pollutants by Anaerobic Processes,* Lewis Publishers, Chelsea, Mich., 1985.
58. McCarty, P. L. et al.: *J. WPCF,* vol. 35, pt. 1, p. 501, 1963.
59. Kugelman, I. J., and P. L. McCarty: *Proc. 19 Ind. Waste Conf.,* Purdue University, 1964.
60. Owen, W. R. et al.: *Water Res.,* vol. 13, p. 485, 1979.

8

ADSORPTION

Many industrial wastes contain organics which are refractory and which are difficult or impossible to remove by conventional biological treatment processes.
Examples are ABS and some of the heterocyclic organics. These materials can frequently be removed by adsorption on an active-solid surface. The most commonly
used adsorbent is activated carbon.

8.1
THEORY OF ADSORPTION

A solid surface in contact with a solution tends to accumulate a surface layer of
solute molecules because of the unbalance of surface forces. Chemical adsorption
results in the formation of a monomolecular layer of the adsorbate on the surface
through forces of residual valence of the surface molecules. Physical adsorption
results from molecular condensation in the capillaries of the solid. In general, substances of the highest molecular weight are most easily adsorbed. There is a rapid
formation of an equilibrium interfacial concentration, followed by slow diffusion
into the carbon particles. The overall rate of adsorption is controlled by the rate of
diffusion of the solute molecules within the capillary pores of the carbon particles.
The rate varies reciprocally with the square of the particle diameter, increases with
increasing concentration of solute, increases with increasing temperature, and
decreases with increasing molecular weight of the solute. Morris and Weber[1] found
the rate of adsorption to vary as the square root of the time of contact, as shown in
Fig. 8.1.

The adsorptive capacity of a carbon for a solute will likewise be dependent on
both the carbon and the solute.

Most wastewaters are highly complex and vary widely in the adsorbability of
the compounds present. Molecular structure, solubility, etc., all affect the adsorbability. These effects are shown in Table 8.1. The relative adsorbability of organics
on carbon is shown in Table 8.2.

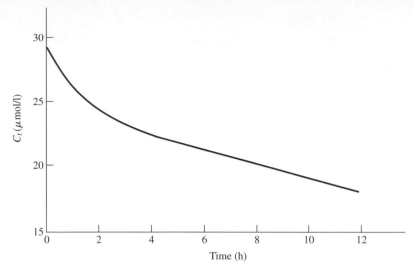

FIGURE 8.1
Rate of adsorption of 2-dodecyl benzene sulfonate by Columbia carbon (30°C, 75 mg/l, 0.273 mm diameter). (*After Morris and Weber.*[1])

TABLE 8.1
Influence of molecular structure and other factors on adsorbability

1. An increasing solubility of the solute in the liquid carrier decreases its adsorbability.

2. Branched chains are usually more adsorbable than straight chains. An increasing length of the chain decreases solubility.

3. Substituent groups affect adsorbability:

Hydroxyl	Generally reduces adsorbability. Extent of decrease depends on structure of host molecule.
Amino	Effect similar to that of hydroxyl but somewhat greater. Many amino acids are not adsorbed to any appreciable extent.
Carbonyl	Effect varies according to host molecule. Glyoxylic acid more adsorbable than acetic but similar increase does not occur when introduced into higher fatty acids.
Double bonds	Variable effects as with carbonyl.
Halogens	Variable effects.
Sulfonic	Usually decreases adsorbability.
Nitro	Often increases adsorbability.

4. Generally, strong ionized solutions are not as adsorbable as weakly ionized ones; i.e., undissociated molecules are in general preferentially adsorbed.

5. The amount of hydrolytic adsorption depends on the ability of the hydrolysis to form an adsorbable acid or base.

6. Unless the screening action of the carbon pores intervenes, large molecules are more sorbable than small molecules of similar chemical nature. This is attributed to more solute carbon chemical bonds being formed, making desorption more difficult.

7. Molecules with low polarity are more sorbable than highly polar ones.

TABLE 8.2
Amenability of selected organic compounds to activated carbon adsorption

Compound	Molecular weight	Aqueous solubility, %	Concentration, mg/l Initial C_o	Final C_f	Adsorbability, g compound/ g carbon	Percent reduction
Alcohols						
Methanol	32.0	∞	1000	964	0.007	3.6
Ethanol	46.1	∞	1000	901	0.020	10.0
Propanol	60.1	∞	1000	811	0.038	18.9
Butanol	74.1	7.1	1000	466	0.107	53.4
Aldehydes						
Formaldehyde	30.0	∞	1000	908	0.018	9.2
Acetaldehyde	44.1	∞	1000	881	0.022	11.9
Propionaldehyde	58.1	22	1000	723	0.057	27.7
Butyraldehyde	72.1	7.1	1000	472	0.106	52.8
Aromatics						
Benzene	78.1	0.07	416	21	0.080	95.0
Toluene	92.1	0.047	317	66	0.050	79.2
Ethyl benzene	106.2	0.02	115	18	0.019	84.3
Phenol	94	6.7	1000	194	0.161	80.6

Formulation of Adsorption

The degree to which adsorption will occur and the resulting equilibrium relationships have been correlated according to the empirical relationship of Freundlich and the theoretically derived Langmuir relationship. For practical application, the Freundlich isotherm usually provides a satisfactory correlation. The Freundlich isotherm is expressed as

$$\frac{X}{M} = kC^{1/n} \tag{8.1}$$

where X = weight of substance adsorbed
M = weight of adsorbent
C = concentration remaining in solution

and k and n are constants depending on temperature, the adsorbent, and the substance to be adsorbed. The Freundlich constants for several priority pollutants are shown in Table 8.3.[2]

The Langmuir equation is based on an equilibrium between condensation and evaporation of adsorbed molecules, considering a monomolecular adsorption layer:

$$\frac{X}{M} = \frac{abC}{1 + aC} \tag{8.2}$$

TABLE 8.3
Summary of Freundlich parameters at neutral pH

Compound	K, mg/g	$1/n$
Hexachlorobutadiene	360	0.63
Anethole	300	0.42
Phenyl mercuric acetate	270	0.44
p-Nonylphenol	250	0.37
Acridine yellow	230	0.12
Benzidine dihydrochloride	220	0.37
n-Butylphthalate	220	0.45
N-Nitrosodiphenylamine	220	0.37
Dimethylphenylcarbinol	210	0.33
Bromoform	200	0.83
β-Naphthol	100	0.26
Acridine orange	180	0.29
α-Naphthol	180	0.31
α-Naphthylamine	160	0.34
Pentachlorophenol	150	0.42
p-Nitroaniline	140	0.27
1-Chloro-2-nitrobenzene	130	0.46
Benzothiazole	120	0.27
Diphenylamine	120	0.31
Guanine	120	0.40
Styrene	120	0.56
Dimethyl phthalate	97	0.41
Chlorobenzene	93	0.98
Hydroquinone	90	0.25
p-Xylene	85	0.16
Acetophenone	74	0.44
1,2,3,4-Tetrahydronaphthalene	74	0.81
Adenine	71	0.38
Nitrobenzene	68	0.43
Dibromochloromethane	63	0.93

This can be reexpressed in linear form as

$$\frac{1}{X/M} = \frac{1}{b} + \frac{1}{ab}\frac{1}{C} \tag{8.2a}$$

where b = amount adsorbed to form a complete monolayer on the surface and a = constant which increases with increasing molecular size.

Since most wastewaters contain more than one substance which will be adsorbed, direct application of the Langmuir equation is not possible. Morris and

Weber[1] have developed relationships from the Langmuir equation for competitive adsorption of two substances:

$$\frac{X_A}{M} = \frac{a_A b_A C_A}{1 + a_A C_A + a_B C_B} \qquad (8.3a)$$

$$\frac{X_B}{M} = \frac{a_B b_B C_B}{1 + a_A C_A + a_B C_B} \qquad (8.3b)$$

More complex relationships could similarly be developed for multicomponent mixtures. It should be noted that although the equilibrium capacity for each individual substance adsorbed in a mixture is less than that of the substance alone, the combined adsorption is greater than that of the individuals alone. In industrial application, contact times of less than 1 h are usually used. Equilibrium is probably closely realized when high carbon dosages are employed, since the rate of adsorption increases with carbon dosage.

8.2
PROPERTIES OF ACTIVATED CARBON

Activated carbons are made from a variety of materials including wood, lignin, bituminous coal, lignite, and petroleum residues. Granular carbons produced from medium volatile bituminous coal or lignite have been most widely applied to the treatment of wastewater. Activated carbons have specific properties depending on the material source and the mode of activation. Property standards are helpful in specifying carbons for a specific application. In general, granular carbons from bituminous coal have a small pore size, a large surface area, and the highest bulk density. Lignite carbon has the largest pore size, least surface area, and the lowest bulk density. Adsorptive capacity is the effectiveness of the carbon in removing desired constituents such as COD, color, phenol, etc., from the wastewater. Several tests have been employed to characterize adsorptive capacity. The phenol number is used as an index of a carbon's ability to remove taste and odor compounds. The iodine number relates to the ability of activated carbon to adsorb low-molecular-weight substances (micropores having an effective radius of less than 2 μm), while the molasses number relates to the carbon's ability to adsorb high-molecular-weight substances (pores ranging from 1 to 50 μm). In general, high iodine numbers will be most effective on wastewaters with predominantly low-molecular-weight organics, while high molasses numbers will be most effective for wastewaters with a dominance of high-molecular-weight organics. Properties of commercial carbons are shown in Table 8.4.

Laboratory Evaluation of Adsorption

In order to evaluate the feasibility and economics of adsorption, a laboratory adsorption study should be conducted. If granular carbon is to be evaluated, the

TABLE 8.4
Properties of commercially available carbons[3]

	NORIT (lignite)	Calgon Filtrasorb 300 (8 × 30) (bituminous)	Westvaco Nuchar WV-L (8 × 30) (bituminous)	Witco 517 (12 × 30) (bituminous)
Physical properties				
Surface area, m²/g (BET)	600–650	950–1050	1000	1050
Apparent density, g/cm³	0.43	0.48	0.48	0.48
Density, backwashed and drained, lb/ft³	22	26	26	30
Real density, g/cm³	2.0	2.1	2.1	2.1
Particle density, g/cm³	1.4–1.5	1.3–1.4	1.4	0.92
Effective size, mm	0.8–0.9	0.8–0.9	0.85–1.05	0.89
Uniformity coefficient	1.7	1.9 or less	1.8 or less	1.44
Pore volume, cm³/g	0.95	0.85	0.85	0.60
Mean particle diameter, mm	1.6	1.5–1.7	1.5–1.7	1.2
Specifications				
Sieve size (U.S. standard series)				
Larger than No. 8 (max. %)	8	8	8	†
Larger than No. 12 (max. %)	†	†	†	5
Smaller than No. 30 (max. %)	5	5	5	5
Iodine No.	650	900	950	1000
Abrasion No., minimum	‡	70	70	85
Ash, %	‡	8	7.5	0.5
Moisture as packed (max. %)	‡	2	2	1

† Not applicable to this size carbon.
‡ No available data from the manufacturer.
Note: lb/ft³ = 16 kg/m³.

carbon must first be ground to pass a 325-mesh screen. Grinding the carbon will not significantly affect its adsorptive capacity, but will increase the rate of adsorption. The time of contact required to approach equilibrium should first be evaluated. A carbon dosage of 500 mg/l is mixed with waste for various periods of time, and the degree of adsorption is determined at selected time intervals. A mixing time sufficient to achieve 90 percent or more of equilibrium should be used for subsequent studies. Usually, a 2-h contact is sufficient to attain greater than 90 percent of equilibrium, though in some cases a longer contact time is required. The initial testing should include a 24-h contact time. If the equilibrium value after 2 h is greater than 90 percent of the 24-h value, the 2-h test can be used.

Various dosages of carbon are then mixed with waste for the time interval selected. The carbon is then filtered off and the concentration remaining in solution is measured. These data are then plotted in accordance with the Freundlich isotherm to determine the adsorption characteristics. This is shown in Fig. 8.2. Note that in most wastewaters there will be a nonsorbable residual. A crude estimate of the carbon required can be made from this plot for any required removal.

Depending on the characteristics of the wastewater, one type of carbon may be superior to another (Fig. 8.3) since the capacity is greater at equilibrium effluent concentrations. For TOC, carbon-2 is better for carbon column operation, since the capacity is greater in equilibrium with the influent (C_o) at column exhaustion, while carbon-1 would be better for batch treatment.

Continuous Carbon Filters

Although batch laboratory adsorption studies provide useful information on the application of adsorption to the removal of specific waste constituents, continuous

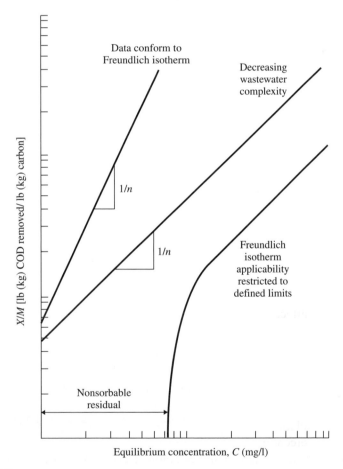

FIGURE 8.2
Freundlich adsorption isotherm.

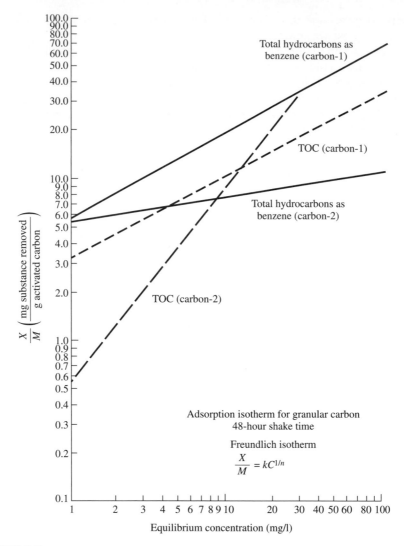

FIGURE 8.3
Freundlich isotherms for total hydrocarbons and total organic carbon.

carbon filters provide the most practical application of this process in waste treatment. The reasons for this are:

1. High capacities in equilibrium with the influent concentration rather than the effluent concentration can be approached, as shown in Fig. 8.4.
2. Biological activity in the presence of degradable organics will affect the apparent carbon capacity.

The carbon filter can be considered as a non-steady-state process in which, as an increasing quantity of water is passed through the bed, the adsorbent is removed in an increasing amount. Consider the system shown in Fig. 8.5. As water initially

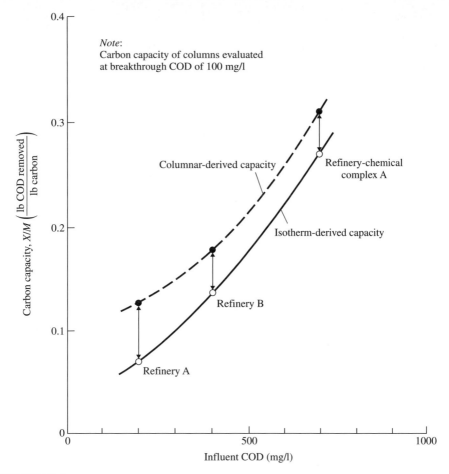

FIGURE 8.4
Carbon capacity from batch and column systems.

passes through the uppermost layers, rapid adsorption occurs in equilibrium with the effluent concentration. As this water passes through the bed, the equilibrium shifts with the decreasing concentration of remaining solute; this results in a substantially solute-free effluent. With continuing flow of water, the adsorption zone, in equilibrium with the influent concentration, moves downward in the bed. As this zone approaches the bottom of the bed, the concentration of solute in the effluent increases. The breakpoint is defined as the volume of water passed through the bed before a maximum effluent concentration is reached. As the adsorption zone falls to the bottom of the bed, the effluent concentration increases until it equals the influent concentration.

The breakthrough point:

1. Decreases with decreased empty bed contact time (EBCT)
2. Decreases with increasing particle size of adsorbent
3. Decreases with increased initial solute concentration

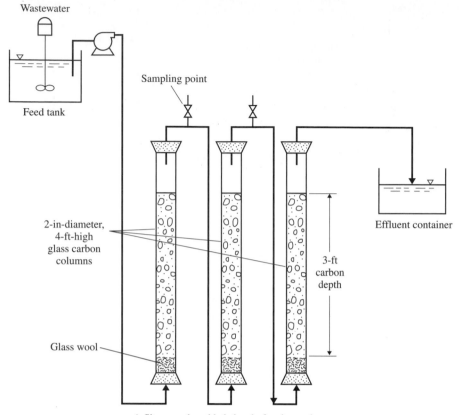

A filter may be added ahead of carbon columns
if needed for suspended solids control.

FIGURE 8.5
Granular activated carbon columns schematic (laboratory scale).

A granular carbon pilot plant for the development of breakthrough curves is shown in Fig. 8.5.

Carbon Regeneration

It is generally feasible to regenerate spent carbon for economic reasons. In the regeneration process, the object is to remove from the carbon pore structure the previously adsorbed materials. The modes of regeneration are thermal, steam, or solvent extraction; acid or base treatment; and chemical oxidation. The methods other than thermal are usually to be preferred when applicable, since they can be accomplished in situ. The difficulty arises that adsorption from multicomponent wastewaters usually does not lend itself to high-efficiency regeneration by these methods. An exception is phenol, which can be treated with caustic to convert it to the more soluble phenate and a single chlorinated hydrocarbon, which can be removed with steam. In most wastewater cases, however, thermal regeneration is required. Thermal regeneration is the process of drying, desorption, and high-temperature heat treatment (1200 to 1800°F; 650 to 980°C) in the presence of limited quantities of water vapor, flue gas, and oxygen. Multiple-hearth furnaces or fluidized-bed furnaces can be used.

Weight losses of carbon result from attrition and carbon oxidation. Depending on the type of carbon and furnace operation, this usually amounts to 5 to 10 percent by weight of the carbon regenerated. There is also a change in carbon capacity through regeneration that may be caused by a change in pore size (usually an increase resulting in a decrease in iodine number) and a loss of pores by deposition of residual materials. In the evaluation of carbons for a wastewater-treatment application, the change in capacity through successive regeneration cycles should be evaluated. In most cases, three to six regeneration cycles will define the maximum capacity loss. This is shown in Fig. 8.6.

Adsorption System Design

In granular carbon columns, the carbon capacity at breakthrough as related to exhaustion is a function of the waste complexity, as shown in Fig. 8.7. A single organic such as dichloroethane will yield a sharp breakthrough curve such that the

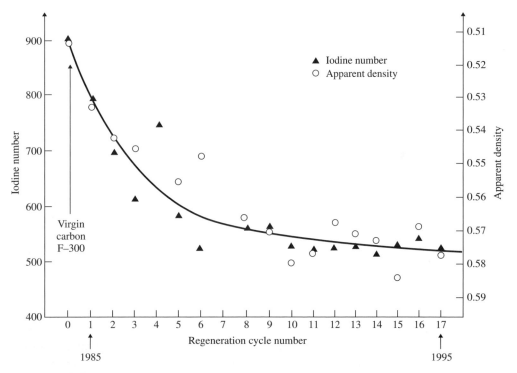

FIGURE 8.6
Regenerated carbon quality trends.[4]

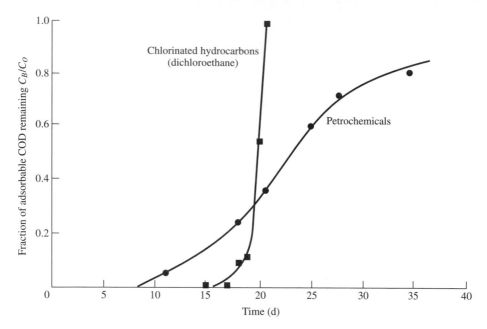

FIGURE 8.7
Continuous carbon column breakthrough curves.[5]

column is greater than 90 percent exhausted when breakthrough occurs. By contrast, a multicomponent petrochemical wastewater shows a drawn-out breakthrough because of varying rates of sorption and desorption, as shown in Fig. 8.8. Depending on the nature of the wastewater one of several modes of carbon column design may be employed:

1. *Downflow.* These are fixed beds in series. When breakthrough occurs in the last column, the first column is in equilibrium with the influent concentration (C_o) in order to achieve a maximum carbon capacity. After carbon replacement in the first column, it becomes the last column in a series, etc. (Fig. 8.9).
2. *Multiple units.* These are operated in parallel with the effluent blended to achieve the final desired quality. The effluent from a column ready for regeneration or replacement, which is high in COD, is blended with the other effluents from fresh carbon columns to achieve the desired quality (Fig. 8.9). This mode of operation is most adaptable to waters in which the capacity at breakthrough/capacity at exhaustion ratio is near 1.0, as in the dichloroethane case previously mentioned.
3. *Upflow.* Expanded beds are used when suspended solids are present in the influent or when biological action occurs in the bed (Fig. 8.9).
4. *Continuous counterflow.* These are column or pulsed beds with the spent carbon from the bottom (in equilibrium with influent solute concentration) sent to regeneration. Since this design cannot be backwashed, residual biodegradable organic content in the influent should be very low to avoid plugging. Regenerated and makeup carbon is fed to the top of the reactor (Fig. 8.9). A granular carbon system is shown in Fig. 8.10.

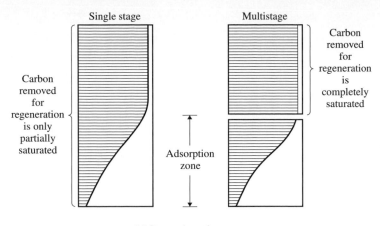

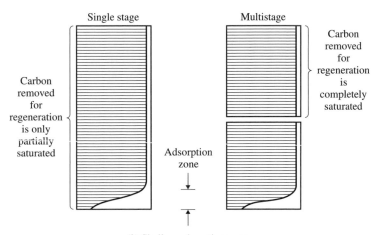

(a) Deep adsorption zone

(b) Shallow adsorption zone

FIGURE 8.8
Adsorption zones for single- and multicomponent wastewaters.

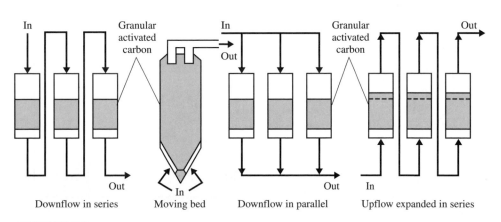

| Downflow in series | Moving bed | Downflow in parallel | Upflow expanded in series |

FIGURE 8.9
Types of GAC column design. *(Calgon Carbon Corporation.)*

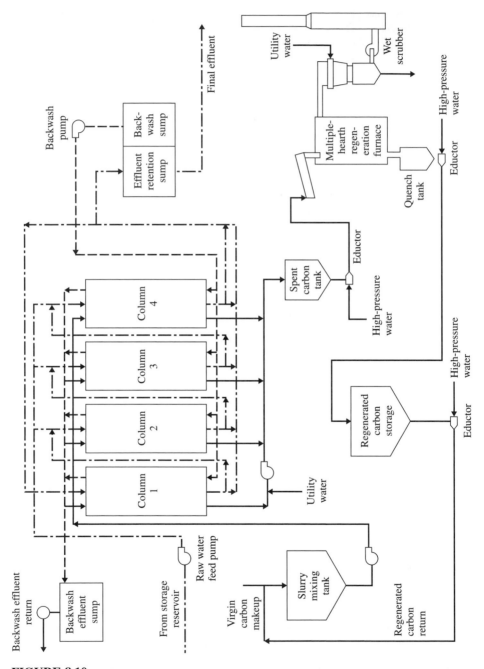

FIGURE 8.10
GAC process flowsheet.

5. *Upflow-downflow.* This concept provides a countercurrent two-bed series system. The two beds are arranged so that the gravity, open-top structures are operated in a series upflow "roughing" contactor and a downflow "polishing" contactor. Once breakthrough occurs, the pair of columns are taken off line, the spent upflow column is regenerated, and the unused capacity of the downflow column is used by reversing the flow and employing it as the upflow reactor, using the former upflow column containing regenerated carbon as the downflow polishing unit.

Bohart and Adams[6] developed a relationship, based on a surface-reaction-rate theory, which can be used to predict the performance of continuous carbon columns:

$$\ln\left(\frac{C_o}{C_B} - 1\right) = \ln(e^{KN_oX/v} - 1) - KC_o t \tag{8.4}$$

Since $e^{KN_oX/v}$ is much greater than unity, Eq. (8.4) can be simplified to

$$t = \frac{N_o}{C_o v}\left[X - \frac{v}{KN_o}\ln\left(\frac{C_o}{C_B} - 1\right)\right] = \frac{N_o}{C_o}\left[\text{EBCT} - \frac{1}{KN_o}\ln\left(\frac{C_o}{C_B} - 1\right)\right] \tag{8.5}$$

where
$$
\begin{aligned}
t &= \text{service time} \\
v &= \text{linear flow rate} \\
X &= \text{depth of bed} \\
K &= \text{rate constant} \\
N_o &= \text{adsorptive capacity} \\
C_o &= \text{influent concentration} \\
C_B &= \text{allowable effluent concentration} \\
\text{EBCT} &= \text{empty bed contact time, } X/v
\end{aligned}
$$

The bed depth that is theoretically just sufficient to prevent penetration of concentration in excess of C_b at zero time is defined as the critical depth and determined from Eq. (8.5) when $t = 0$:

$$X_o = \frac{v}{KN_o}\ln\left(\frac{C_o}{C_B} - 1\right) \tag{8.6}$$

The critical EBCT (EBCT$_o$) is:

$$EBCT_o = \frac{1}{KN_o}\ln\left(\frac{C_o}{C_B} - 1\right) \tag{8.6a}$$

From Eq. (8.5) it can be shown that the adsorptive capacity N_o can be determined from the slope of a linear plot of t versus X or EBCT. The rate constant K is then calculated from the intercept of this plot:

$$b = -\frac{1}{C_o K}\ln\left(\frac{C_o}{C_B} - 1\right) \tag{8.7}$$

The application of the Bohart-Adams equation to the removal of ABS[4] in continuous carbon columns is shown in Example 8.1.

EXAMPLE 8.1. Experiments in a carbon column 1 in (2.54 cm) in diameter were conducted on a 10-ppm ABS solution.[5] The data obtained are given in Table 8.5.

Compute: (a) the coefficients in the Bohart-Adams equation for this carbon and (b) the carbon required per year to treat 100,000 gal (380 m³) of water per week with an ABS content of 10 mg/l to a residual of 0.5 ppm.

The tower will be 5 ft (1.5 m) in depth and 2 ft (0.6 m) in diameter; compute the adsorption efficiency of the carbon.

Solution.

(a)
$$t = \frac{N_o}{C_o v}\left[X - \frac{v}{KN_o}\ln\left(\frac{C_o}{C_B} - 1\right)\right]$$

where
$N_o = M\ ABS/L^3\ C$
$X = L$
$C_o = M/L^3$
$v = L/T$
$K = L^3/MT$
$t = T$

Area of column A_c, (1-in diameter) = 0.00545 ft² (5.06 cm²)
Calculation of N_o and K

$$C_o = 10\text{ mg/l} \times 62.4 \times 10^{-6}\text{ lb/ft}^3 = 0.000624\text{ lb/ft}^3$$

From the equation for t above, N_o is computed from the slope of the t versus X plot (see Fig. 8.11 and Table 8.6). To prepare Table 8.6 the following considerations apply:

$$v(\text{ft/h}) = q[\text{gal}/(\text{min}\cdot\text{ft}^2)] \times 8.02(\text{ft/h})/[\text{gal}/(\text{min}\cdot\text{ft}^2)]$$

$$V(\text{ft}^3) = \text{throughput (gal)} \times 0.134\ (\text{ft}^3/\text{gal})$$

$$t = \frac{V}{vA_c}$$

TABLE 8.5
Column capacity data for the adsorption of ABS from a 10-ppm aqueous solution (capacity to 0.5 ppm breakpoint)

Flow rate, gal/(min · ft²)	Bed depth, ft	Throughput volume, gal
2.5	2.5	363
	5.0	1216
	7.5	2148
5.0	2.5	141
	5.0	730
	10.0	2190
10.0	5.0	332
	10.0	1380
	15.0	2760

Note:
gal/(min · ft²) = 4.07 × 10⁻² m³ /(min · m²)
ft = 0.3048 m
gal = 3.785 × 10⁻³ m³

At 2.5 gal/(min · ft²):

$$N_o = C_o v_a$$

$$= 0.000624 \times 20 \times 445 = 5.55 \text{ lb/ft}^3 \ (89 \text{ kg/m}^3)$$

At 5.0 gal/(min · ft²):

$$N_o = 0.000624 \times 40 \times 170 = 4.24 \text{ lb/ft}^3 \ (67.9 \text{ kg/m}^3)$$

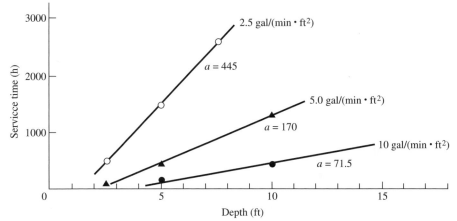

Note:
gal/(min · ft²) = 4.07 × 10⁻² m³/(min · m²)
ft = 0.3048 m

FIGURE 8.11

TABLE 8.6

	Flow		Depth, ft	Volume treated V, ft³	t, h
q, gal/(min · ft²)		v, ft/h			
2.5		20	2.5	49	440
			5.0	162	1480
			7.5	268	2620
5.0		40	2.5	19	87
			5.0	98	440
			10.0	290	1340
10		80	5.0	44	102
			10.0	184	420
			15.0	370	835

Note: gal/(min · ft²) = 8.02 ft/h = 4.07 × 10⁻2 m³/(min · m²)

At 10 gal/(min · ft^2):

$$N_o = 0.000624 \times 80 \times 71.5 = 3.57 \text{ lb/ft}^3 \ (57.2 \text{ kg/m}^3)$$

$$b = -\frac{1}{C_oK} \ln\left(\frac{C_o}{C_B} - 1\right) \quad \text{and} \quad K = -\frac{1}{C_ob} \ln\left(\frac{C_o}{C_B} - 1\right)$$

From Fig. 8.12 the values of b are taken and K can be calculated, giving the values presented in Table 8.7.

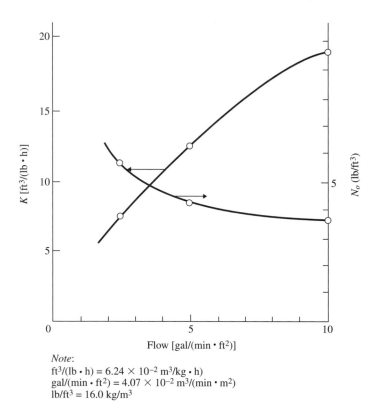

Note:
ft^3/(lb · h) = 6.24 × 10^{-2} m^3/kg · h)
gal/(min · ft^2) = 4.07 × 10^{-2} m^3/(min · m^2)
lb/ft^3 = 16.0 kg/m^3

FIGURE 8.12

TABLE 8.7

Flow, gal/(min · ft^2)	b, h	K, ft^3/(lb · h)
2.5	−630	7.5
5.0	−370	12.7
10.0	−250	18.8

Note:
ft^3/(lb · h) = 6.24 × 10^{-2} m^3/(kg · h)
gal/(min · ft^2) = 4.07 × 10^{-2} m^3/(min · m^3)
lb/ft^3 = 16.0 kg/m^3

Calculation of critical layer [Eq. (8.6)]

$$X_o = \frac{v}{KN_o} \ln\left(\frac{C_o}{C_a} - 1\right)$$

$$X_o = \frac{20}{7.5 \times 5.55} \cdot 2.95 = 1.41 \text{ ft } (0.43 \text{ m}) \quad \text{for } 2.5 \text{ gal}/(\text{min} \cdot \text{ft}^2)$$

In this manner

$$X_o = 2.19 \text{ ft } (0.67 \text{ m}) \quad \text{for } 5.0 \text{ gal}/(\text{min} \cdot \text{ft}^2)$$

$$X_o = 3.51 \text{ ft } (1.07 \text{ m}) \quad \text{for } 10.0 \text{ gal}/(\text{min} \cdot \text{ft}^2)$$

(b) Using a tower of 24 in diameter (0.61 m), 5 ft deep (1.53 m), 3.15 ft² (0.29 m²), 15.75 ft³ (0.45 m³),

$$\frac{100,000 \text{ gal/week}}{5 \text{ d/week} \times 1440 \text{ min/d}} = 13.9 \text{ gal/min } (0.05 \text{ m}^3/\text{min})$$

$$\text{or } 4.4 \text{ gal}/(\text{min} \cdot \text{ft}^2)[0.18 \text{ m}^3/(\text{min} \cdot \text{m}^2)]$$

At 4.4 gal/(min ·ft²), $v = 35$ ft/h, $K = 11.5$ ft³/(lb · h), and $N_o = 4.2$ lb/ft³,

$$t = \frac{N_o}{C_o v}\left[X - \frac{v}{KN_o} \ln\left(\frac{C_o}{C_B} - 1\right)\right]$$

$$= \frac{4.2}{0.000624 \times 35}\left(5 - \frac{35}{11.5 \times 4.2} 2.95\right)$$

$$= 551 \text{ h}$$

At a flow rate of 834 gal/h the total volume treated before breakthrough is (551)(834) = 460,000 gal. For a yearly volume of 5.2 million gallons, 11 carbon charges are required.

Bed efficiency

$$\text{Total ABS adsorbed} = 460,000 \times 8.34 \times 9.5 = 36.4 \text{ lb}$$

$$\text{Total capacity} = 4.2 \text{ lb ABS/ft}^3 \text{ C} \times 15.7 \text{ ft}^3 \text{ C} = 66 \text{ lb}$$

$$\% \text{ efficiency} = \frac{36.4}{66} \times 100 = 55\%$$

Efficiency computed from X_o:
Using Eq. (8.6),

$$X_o = \frac{35}{11.5 \times 4.2} \ln\left(\frac{10}{0.5} - 1\right) = 2.13 \text{ ft}$$

$$\% \text{ efficiency} = \frac{X - X_o}{X} \times 100 = \frac{5 - 2.13}{5} \times 100 = 57\%$$

Hutchins[7] presented a modification of the Bohart-Adams equation which required only three column tests to collect the necessary data. This is called the bed depth service time (BDST) approach. The Bohart-Adams equation can be expressed as

$$t = aX + b$$
$$= a'\text{EBCT} + b \tag{8.8}$$

where $a = \text{slope} = N_o/C_o v$
           $b = \text{intercept} = -1/KC_o[\ln(C_o/C_B - 1)]$
           $a' = va$

If a value of a is determined for one flow rate, values for other flow rates can be computed by multiplying the original slope by the ratio of the original and new flow rates. The b value change is insignificant with respect to changing flow rates. No adjustment for flow rate is needed if the EBCT versus t plot is used. Adjustment for changing initial concentration can be made as follows:

$$a_2 = a_1 \frac{Q_1}{Q_2}$$

$$b_2 = b_1 \frac{C_1}{C_2} \frac{\ln(C_2/C_F - 1)}{\ln(C_1/C_B - 1)}$$

in which C_F and C_B are the effluent concentrations at C_2 and C_1, respectively. In order to develop a BDST correlation, a number of pilot columns of equal depth are operated in series and breakthrough curves plotted for each, as shown in Fig. 8.13.

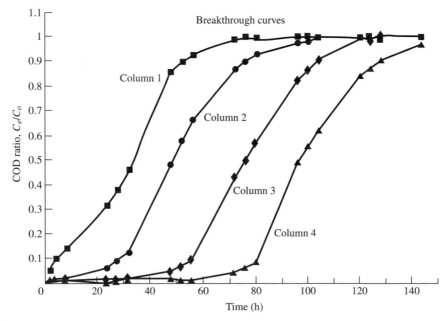

FIGURE 8.13
Column breakthrough curves.

These data are then used to plot a BDST correlation by recording the operating time required to reach a certain removal at each EBCT. A BDST plot of the data from Fig. 8.13 is shown in Fig. 8.14.

The slope of the BDST line is equal to the reciprocal velocity of the adsorption zone and the X intercept is the critical depth defined as the minimum bed depth required to obtain the desired effluent quality at time zero.

If the adsorption zone is arbitrarily defined as the carbon layer through which the liquid concentration varies from 90 to 10 percent of the feed concentration, then this zone is defined by the horizontal distance between these two lines in the BDST plot.

In order to design an adsorption system with maximum carbon utilization, the carbon removed should be near saturation in equilibrium with the influent concentration. In multistage columns, the first run of three columns will be made with all fresh carbon. When the third column breaks through, the first column should be exhausted and a fourth column with fresh carbon is placed at the end of the train. This process is repeated each time the last column breaks through.

The BDST curve should be developed from the breakthrough curves after the third column breaks through. Assuming that 90 percent removal represents exhaustion, the horizontal distance between 90 and the desired breakthrough concentration is taken as the depth of the adsorption zone. This is also the minimum in a pulsed or moving-bed system. For a multistage system, the number of stages and the bed depth in each stage are related to the depth of the adsorption zone:

$$n = \frac{D}{d} + 1 \qquad (8.9)$$

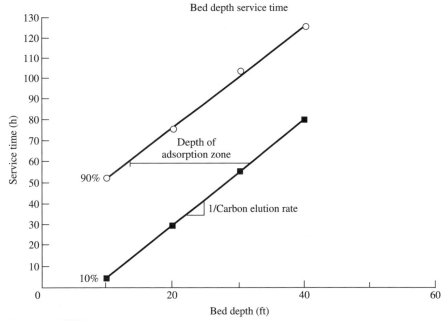

Note: ft = 0.3048 m

FIGURE 8.14
BDST design curves.

where n = number of stages in series
        D = depth of adsorption zone
        d = depth of single stage

Selection of d should be based on practical considerations and should be an integer fraction of D. Selecting a small d will result in small-size equipment with lower carbon inventory, but a high number of stages and consequently more costly equipment. Example 8.2 illustrates a BDST design.

EXAMPLE 8.2. A petrochemical washwater with a flow of 85,000 gal/d (322 m³/d) has to be treated to an effluent standard of 50 mg/l. A four-column pilot plant was operated with a carbon that had a density of 30 lb/ft³ (481 kg/m³). The columns were 10 ft (3 m) long and loaded at a hydraulic rate of 5 gal/(min · ft²) [0.20 m³/(min · m²)]. The pilot plant is operated in series: the effluent from column 1 is passed to the top of column 2 and sequentially to columns 3 and 4.

Calculate the depth of the adsorption zone, the required number of columns, the time required to exhaust a column, the column diameter, the daily carbon use, and the solution carbon adsorption loading.

Solution.

The data should be plotted as sequential breakthrough curves (Fig. 8.13) of the bed effluent concentration as a ratio of the influent concentration of the system (630 mg/l).

The breakthrough curves for the four columns are symmetrical. The adsorption zone is defined as the carbon layer through which the concentration varies from 10 percent to 90 percent of the feed concentration. The breakthrough times are plotted as the service times and are a function of total carbon bed depth (Fig. 8.11).

In this case, the lines are almost parallel and the depth of the adsorption zone is between 18 and 19 ft. The total bed depth will be the adsorption zone plus an additional column:

$$\text{Number of columns} = \frac{19}{10} + 1 = 2.9, \text{round up to 3}$$

The bed depth service time (Fig. 8.14) data fits well in the Bohart-Adams equation:

$$t = \frac{N_o}{C_o v}(X) - \frac{1}{C_o K}\ln\left(\frac{C_o}{C_e} - 1\right)$$

At 10 percent,

$$t\,(\text{h}) = 2.57(X) - 21.5$$

At 90 percent,

$$t\,(\text{h}) = 2.50(X) + 27.0$$

$$\text{Adsorption velocity} = \frac{1}{\text{slope}} = \frac{1}{2.57\ \text{h/ft}}$$

$$= 0.39\ \text{ft/h or } 9.34\ \text{ft/d} \qquad (2.85\ \text{m/d})$$

The rest of the design and operation can be calculated from the above determinations:

$$\text{Time to exhaust a column} = \frac{10\ \text{ft}}{9.34\ \text{ft/d}} \times 24\ \text{h/d} = 26\ \text{h}$$

$$\text{Area required} = \frac{Q}{A} = \frac{85{,}000 \text{ gal/d}}{5 \text{ gal/(min} \cdot \text{ft}^2)} \times \frac{d}{1440 \text{ min}} = 11.8 \text{ ft}^2 \ (1.1 \text{ m}^2)$$

$$\text{Diameter} = 3.88 \text{ ft} \ (1.2 \text{ m})$$

$$\text{Carbon use} = 11.8 \text{ ft}^2 \times 9.34 \text{ ft/d} \times 39 \text{ lb/ft}^3 = 3300 \text{ lb/d} \ (1500 \text{ kg/d})$$

$$\text{Carbon adsorption loading} = \frac{(630 - 50) \text{ mg/L} \times Q}{\text{carbon use}}$$

$$= \frac{580 \text{ mg/L} \times 85{,}000 \text{ gal/d}}{3300 \text{ lb/d}} \times \frac{8.34 \text{ lb/gal}}{10^6 \text{ mg/l}}$$

$$= 0.125 \text{ lb COD/lb carbon}$$

GAC Small Column Tests

The two dominant factors that control the breakthrough in GAC columns are the adsorption capacity and adsorption kinetics. As previously discussed, pilot columns utilize the same reliable predictors of breakthrough behavior in full-scale columns in terms of both capacity and rate of adsorption. However, this approach may require time-consuming and expensive studies. Rapid methods to design GAC columns from small columns have been developed to reduce the study time and cost. Examples of methods using small columns are the short fixed bed, the minicolumns, the high-pressure minicolumns, the dynamic minicolumn adsorption technique, the accelerated column tests, the small-scale columns, and the rapid small-scale column test (RSSCT). The use of the RSSCT, which does not require the use of complicated models, will be illustrated.

The RSSCT method, which was developed by Frick[8] and improved and applied by Crittenden and coworkers,[9,10] is a scaled-down version of a pilot- or full-scale GAC column. If the RSSCT and the GAC column use carbons with the same bulk density and capacity, similitude is maintained through the use of dimensional analysis for the relationship between particle size, column length or empty bed contact time (EBCT), and operation time. These relationships may be summarized by Eqs. (8.10) and (8.11).

$$\frac{\text{EBCT}_S}{\text{EBCT}_L} = \frac{t_S}{t_L} \tag{8.10}$$

where EBCT_S = EBCT of small-particle column
EBCT_L = EBCT of large-particle column
t_S = operation time of small-particle column
t_L = operation time of large-particle column

Equal Reynolds numbers (Re) for the RSSCT and the full-scale GAC column assure hydraulic similarity with equal length of the mass transfer zone (MTZ) to length of the column ratio. This equality in the Re numbers is expressed by the following equation:

$$\mathrm{Re}_S = \frac{d_S\, v_S}{\nu\, \varepsilon_S} = \frac{d_L\, v_L}{\nu\, \varepsilon_L} \qquad (8.11)$$

This is illustrated in Example 8.3.

EXAMPLE 8.3. Design a 100 gal/min (379 l/min) full-scale GAC column using the following information from a small-column study:

$d_S = 0.06$ cm (>0.0077 cm)
$\varepsilon_S = 0.40$ (0.36 to 0.48)
$\rho_b = 0.42$ g/ml (0.32 to 0.42 g/ml)
$\mathrm{ID}_S = 1.5$ cm ($\mathrm{ID}_S/d_S > 25$–50)
$L_S = 5$ cm
$\mathrm{Re}_S = 1$
$t_{bS} = 2$ d (time to breakthrough or $C = C_{to}$)
$t_{eS} = 3.5$ d (time to exhaustion or $C = 0.95\, C_o$)
C_o = influent generation
C_{to} = treatment objective
C = effluent concentration
$T_w = 20°C$

Solution.

1. Small column
   a. Dimensions:
      Surface Area

      $$A_S = \frac{\pi \mathrm{ID}^2}{4} = \frac{\pi(1.5)^2}{4} = 1.77 \text{ cm}^2$$

      Volume of carbon bed

      $$V_{bS} = A_S \cdot L_S = 1.77 \times 5 = 8.84 \text{ cm}^3$$

   b. Mass of carbon in the column

      $$M_{cS} = V_{bS} \cdot \rho b = 8.84 \times 0.42 = 3.71 \text{ g C}$$

   c. Hydraulic loading rate (approaching velocity) using $\mathrm{Re}_S = 1$

      $$V_S = \frac{\mathrm{Re}_S \varepsilon}{d_S} = \frac{1 \times 0.602 \text{ cm}^2/\text{min} \times 0.4}{0.06 \text{ cm}} = 4.01\,\frac{\text{cm}}{\text{min}} \quad \left(\frac{0.98 \text{ gal/min}}{\text{ft}^2}\right)$$

   d. Flow rate

      $$Q_S = V_S A_S = 4.01 \times 1.77 = 7.10 \text{ cm}^3/\text{min}$$

   e. $$\mathrm{EBCT}_S = \frac{V_{bS}}{Q_S} = \frac{8.84}{7.10} = 1.25 \text{ min}$$

   f. Length of mass transfer zone

      $$\mathrm{MTZ}_S = L_S \left[\frac{t_{eS} - t_{bS}}{t_{eS}}\right] = 5 \left[\frac{3.5 - 2}{3.5}\right] = 2.1 \text{ cm } (42\% \text{ of } L_S)$$

2. GAC Adsorber
   a. Select GAC

   $d_L = 0.1$ cm
   $\varepsilon_L = 0.4$
   $\rho_b = 0.42$ g/cm^3 = 420 kg/m^3

   b. Dimensions for an $EBCT_L$ of 15 min and an HLR_L = 4 gal/(min · ft^2) = 16.3 cm/min = V_L

$$V_{bL} = Q_L \times EBCT_L = 100 \times 15 = 1500 \text{ gal} = 5.68 \text{ m}^3$$

$$A_L = \frac{Q_L}{HLR} = \frac{100}{4} = 25 \text{ ft}^2 = 2.33 \text{ m}^2$$

$$ID_L = \left(\frac{4A_L}{\pi}\right)^{0.5} = \left(\frac{4 \times 2.33 \text{ m}^2}{\pi}\right)^{0.5} = 1.72 \text{ m} \qquad (5.65 \text{ ft})$$

$$L_L = \frac{V_{bL}}{A_L} = \frac{5.68}{2.33} = 2.44 \text{ m} \qquad (8.0 \text{ ft})$$

   c. Time to breakthrough

$$t_{bL} = t_{bS} \frac{EBCT_L}{EBCT_S} = 2 \times \frac{15}{1.25} = 24 \text{ d}$$

   d. Mass of carbon in the column

$$M_{cL} = V_{bL} \times \rho_b = 5.68 \text{ m}^3 \times \frac{420 \text{ kg}}{\text{m}^3} = 2390 \text{ kg} \qquad (5260 \text{ lb})$$

   e. Estimate bed volumes of wastewater treated to breakthrough:

$$BV_b = \frac{t_{bL}}{EBCT_L} = \frac{24}{15} \times 1440 = 2300$$

3. Estimate the volume of wastewater treated per unit mass of carbon (specific volume):

$$V_{sp} = \frac{Q_L t_{bL}}{M_{cL}} = \frac{0.379 \times 24}{2390} \times 1440 = \frac{5.5 \text{ m}^3}{\text{kg}} \qquad \left(\frac{660 \text{ gal}}{\text{lb}}\right)$$

4. Estimate the carbon utilization rate:

$$CU_r = \frac{1}{V_{sp}} = 0.000182 \frac{\text{kg}}{1} = 182 \frac{\text{mg}}{1} \qquad \left(\frac{0.00152 \text{ lb}}{\text{gal}} = \frac{1.52 \text{ lb}}{1000 \text{ gal}}\right)$$

5. Estimate the length of the MTZ:

$$MTZ_L = L_L \times 0.42 = 2.44 \times 0.42 = 1.02 \text{ m} \qquad (3.4 \text{ ft})$$

6. Determine configuration of GAC system:

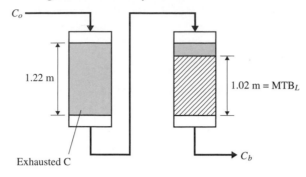

Using a safety factor of 1.5, use 2 columns in series with:

$$ID_L = 1.72 \text{ m}$$

$$L_L = 1.83 \text{ m}$$

$$V_{bL} = \frac{\pi \, 1.72^2}{4} \times 1.83 = 4.25 \text{ m}^3/\text{column}$$

$$EBCT_L = 2 \times \frac{4.25}{0.379} = 22 \text{ min} \quad (11 \text{ min per column})$$

$$t_b = 2 \times \frac{22}{1.25} = 35 \text{ d}$$

Performance of Activated Carbon Systems

Activated carbon columns are employed for the treatment of toxic or nonbiodegradable wastewaters and for tertiary treatment following biological oxidation.

When degradable organics (BOD) are present in the wastewater, biological action provides biological regeneration of the carbon, thus increasing the apparent capacity of the carbon. Biological activity may be an asset or a liability. When the applied BOD is in excess of 50 mg/l, anaerobic activity in the columns may cause serious odor problems while aerobic activity may cause plugging due to the biomass generation by aerobic activity.

Most heavy metals are removed through carbon columns as shown in Table 8.8 for a petroleum refinery wastewater. In order to avoid reduced capacity after regen-

TABLE 8.8
Heavy metals removal on activated carbon from a petroleum refinery wastewater

Parameter	API separator, mg/l	Carbon-treated, mg/l
Chromium	2.2	0.2
Copper	0.5	0.03
Iron	2.2	0.3
Lead	0.2	0.2
Zinc	0.7	0.08

eration, the carbon should be acid-washed prior to reuse. The effectiveness of activated carbon in the treatment of various industrial wastewaters is shown in Table 8.9.

High-molecular-weight compounds such as SMP are strongly adsorbed on carbon and will replace weakly adsorbed compounds, as shown in Fig. 8.15.

8.3
THE PACT® PROCESS

Powdered activated carbon (PAC) can be added to the activated sludge process for enhanced performance (the PACT® process). The flowsheet for this process is shown in Fig. 8.16. The addition of PAC has several process advantages, namely, decreased variability in effluent quality and removal by adsorption of nondegradable organics (principally color), reduction of inhibition in industrial wastewater treatment, and removal of refractory priority pollutants. PAC can be integrated into

TABLE 8.9
Results from adsorption isotherms on various industrial wastewaters

Type of industry	Initial TOC (or phenol), mg/l	Initial color OD	Average reduction, %	Carbon exhaustion rate, lb/1000 gal
Food and kindred products	25–5300	—	90	0.8–345
Tobacco manufacturers	1030	—	97	58
Textile mill products	9–4670	—	93	1–246
	—	0.1–5.4	97	0.1–83
Apparel and allied products	390–875	—	75	12–43
Paper and allied products	100–3500	—	90	3.2–156
	—	1.4	94	3.7
Printing, publishing, and allied industries	34–170	—	98	4.3–4.6
Chemicals and allied products	19–75,500	—	85	0.7–2905
	(0.1–5325)	—	99	1.7–185
	—	0.7–275	98	1.2–1328
Petroleum refining and related industries	36–4400	—	92	1.1–141
	(7–270)	—	99	6–24
Rubber and miscellaneous plastic products	120–8375	—	95	5.2–164
Leather and leather products	115–9000	—	95	3–315
Stone, clay, and glass products	12–8300	—	87	2.8–300
Primary metal industries	11–23,000	—	90	0.5–1857
Fabricated metal products	73,000	—	25	606

Note: lb/1000 gal = 0.120 kg/m^3

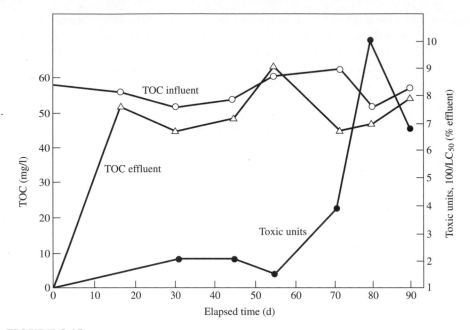

FIGURE 8.15
TOC and toxicity reduction by granular carbon columns.

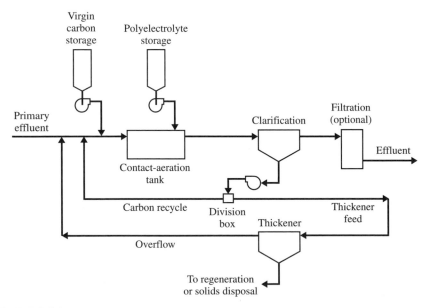

FIGURE 8.16
PACT® wastewater-treatment system general process diagram.

existing biological treatment facilities at minimum capital cost. Since the addition of PAC enhances sludge settleability, conventional secondary clarifiers will usually be adequate, even with high carbon dosages. In some industrial waste applications, nitrification is inhibited by the presence of toxic organics. The application of PAC has been shown to reduce or eliminate this inhibition. Batch isotherm screening tests are used on the biological effluent in order to select the optimal carbon. Bench-scale continuous reactors as shown in Fig. 6.70 can be used to develop process design criteria; several reactors are run in parallel: a control with no PAC and several with varying dosages of PAC.

The PAC dosage and the PAC mixed liquor solids concentration are related to the sludge age:

$$X_p = \frac{X_i \theta_c}{t}$$
(8.12)

where X_p = equilibrium PAC MLSS content
X_i = PAC dosage
t = hydraulic retention time

The sludge age affects the PAC efficiency with higher sludge ages enhancing the organic removal per unit of carbon; affects the molecular configuration of the adsorbate depending on varying biological uptake patterns and end products; and establishes the equilibrium biological solids level in the aeration basin. There is some evidence that the attached biomass degrades some of the low-molecular-weight compounds that are adsorbed, as demonstrated by superior TOC removal rates for PAC when added to an aeration basin as opposed to isotherm predictions of adsorption capacity. Figure 8.17 was developed by evaluating the difference in TOC removal for biological units operated in parallel with and without PAC. As can be seen, the performance of the carbon in the bioreactor was significantly greater than predicted by the isotherm only. The mechanisms felt to be responsible for this phenomena include:

1. Additional biodegradation of organics due to decreased biological toxicity or inhibition via activated carbon.
2. Degradation of normally nondegradable substances due to increased exposure time to the biomass through adsorption on the carbon. The carbon with adsorbed material remains in the system for one sludge age, typically 10 to 30 d, while without carbon the substances would remain in the system for only one hydraulic retention time, typically 6 to 36 h.
3. Substitution/adsorption phenomena, replacement of low-molecular-weight compounds with high-molecular-weight compounds, resulting in improved adsorption efficiency and lower toxicity.

Elimination of nitrification inhibition by the addition of PAC for a coke plant wastewater is shown in Table 8.10.

When there is a small or intermittent application of PAC, the carbon is disposed of with the excess sludge. Continuous application at larger plants, however, requires regeneration of the carbon. This can be accomplished by the use of wet air oxidation (WAO).

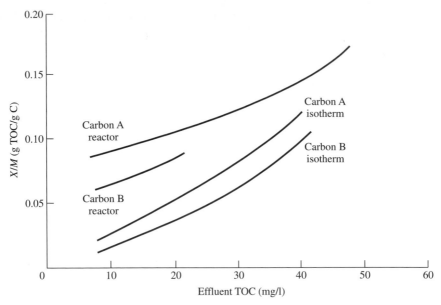

FIGURE 8.17
Performance relationship of PACT® reactors with isotherm data.

TABLE 8.10
Effect of PAC on nitrification of coke plant wastewaters

PAC feed, mg/l	SRT, d	TOC, mg/l	TKN, mg/l	NH_3^-N, mg/l	NO_2^-N, mg/l	NO_3^-N, mg/l
0	40	31	72	68	4.0	0
33	30	20	6.3	1	4.0	9.0
50	40	26	6.4	1	1.0	13.0

Influent conditions: TOC = 535 mg/l, TKN = 155 mg/l, NH_3^-N = 80 mg/l; pH = 7.5.

In the WAO process, the biological carbon-sludge mixture is treated in a reactor at 450°C and 750 lb/in² (51 atm) for 1 h in the presence of oxygen. The biological sludge is oxidized and solubilized under these conditions and the carbon regenerated. The exothermic reaction will provide energy for the reaction provided that the influent solids content exceeds 10 percent. The decant liquor from the reactor will contain 5000 mg/l BOD, which is recycled back to the aeration basins. In some cases, there is an ash buildup which must be removed from the system. Depending on the characteristics of the wastewater and the type of carbon used, there may be significant losses in carbon capacity through regeneration. This phenomenon should be evaluated by a pilot plant study for any specific application.

Carbon dosages may vary from 20 to 200 mg/l, depending on the results desired. Since the carbon is abrasive, equipment selection should consider this fact. The effect of PAC dosage on the removal of chlorinated benzenes is shown in Fig. 8.18. Performance of the PACT® process for an organic chemicals wastewater is shown in Table 8.11. Activated sludge process performance with and without PAC is summarized in Table 8.12.

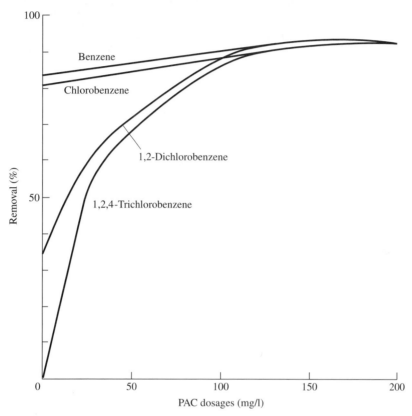

FIGURE 8.18
Effect of PAC dosage on chlorinated benzenes.[11]

TABLE 8.11
**Larger doses of powdered activated carbon result in greater removal
of organic carbon, color, and heavy metals**

| | **Wastewater composition, mg/l** | | | | | | | |
	BOD	**TOC**	**TSS**	**Color**	**Cu**	**Cr**	**Ni**	**Bioassay[†] LC$_{50}$**
Influent	320	245	70	5365	0.41	0.09	0.52	
Biotreatment	3	81	50	3830	0.36	0.06	0.35	11
+ 50 mg/l PAC	4	68	41	2900	0.30	0.05	0.31	25
+ 100 mg/l PAC	3	53	36	1650	0.18	0.04	0.27	33
+ 250 mg/l PAC	2	29	34	323	0.07	0.02	0.24	>75
+ 500 mg/l PAC	2	17	40	125	0.04	<0.02	0.23	>87

[†] Percentage of wastewater in which 50 percent of aquatic organisms survive for 48 h.

TABLE 8.12
Suspended growth systems comparison
(with and without PAC)

	Industrial					
	Organic chemicals[13]		Textile finishing[13]		Berndt-Polkowski[12]	
Operating conditions	PACT®[†]	Activated sludge	PACT®	Activated sludge	PACT®[†]	Air activated sludge
Aeration d, d	6	6	2.4	2.4	3.2–4.5	[‡]
SRT, d	25	45[†]				
Temperature, °C	25	25			13	16
Performance results					20	20
Influent charac- teristics, mg/l						
BOD$_5$	4,035	4,035	660	Unknown		
TOC	2,965	2,965			134	128
COD	10,230	10,230	1362	1590	364	320
TKN	120	120	Unknown	106	39.4	32.0
NH$_3^-$N	76	76	74	31	19.5	19.8
Chlorinated hydrocarbons	5–67	5				
Phenol	8.1	8.1				
Effluent charac- teristics, mg/l						
BOD$_5$	11	17	5	26	1.2	24
TOC	25	65			49.8	63
COD	102	296	116	270	3.7	6.4
TKN	4	—	6	29	0.2	3.6
NH$_3^-$N	0.8	—	3.6	15		
Chlorinated hydrocarbons	0.1	0.9				
Phenol	0.01	0.22	0.44	1.6		
Color, APHA	94	820	240	600		
Detergents			0.8	11.4		

[†] Includes wet oxidation carbon regeneration.
[‡] Contact stabilization facility; aeration d_t = 4.2 h, stabilization d = 6.2 h.

PROBLEMS

8.1. Given the following data for the adsorption of napthlene on Filtersorb 400 carbon:

Carbon dose, mg/l	Effluent C_f, mg/l
0	9.94
11.2	5.3
22.3	3.0
56.1	0.71
168.3	0.17
224.4	0.06

(a) Develop the adsorption isotherm and the Freundlich parameters.
(b) What is the adsorption capacity at an initial concentration of 10 mg/l?
(c) What is the carbon dose required to reduce the concentration from 1 to 0.01 mg/l?

8.2. The breakthrough curves for a pilot plant operation are shown in Fig. P8.2. The pilot columns were 3.5 ft (1.07 m) long. Design a carbon column system to a breakthrough TOC concentration of 20 mg/l. The influent wastewater has a flow of 3100 gal/min (11.7 m³/min) and an influent TOC of 60 mg/l. Use a hydraulic loading of 4 gal/ (min · ft²) [0.163 m³/(min · m²)] and a carbon density of 28 lb/ft³ (449 kg/m³).

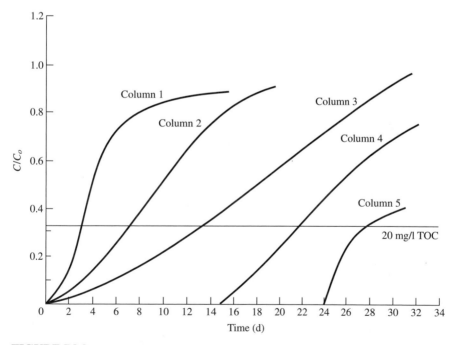

FIGURE P8.2
Breakthrough curves for GAC columns.

REFERENCES

1. Morris, J. C., and W. J. Weber: "Adsorption of Biochemically Resistant Materials from Solution," Environmental Health Series AWTR-9, May 1964.
2. EPA: "Carbon Adsorption Isotherms for Toxic Organics," EPA-600/8-80-023, April 1980.
3. U.S. EPA: *Process Design Manual for Carbon Adsorption,* Technology Transfer, 1973.
4. Roll, R. R., and Crocker, D. N.: *Proc. WEF,* vol. 1, Dallas, 1996.
5. Argamon, Y., and W. W. Eckenfelder: *Water 1975* Symposium, Series 72, p. 151, Association of Industrial Chemical Engineers, New York, 1976.
6. Dale, J., J. Malcolm, and I. M. Klotz: *Ind. Eng. Chem.,* vol. 38, pt. 1, p. 289, 1946.
7. Hutchins, R. A.: *Chem. Engineering,* vol. 80, pt. 19, p. 133, 1973.
8. Calgon Carbon Corp., Pittsburgh.
9. Crittenden, J. et al.: *J. Env. Eng. ASCE,* vol. 113, no. 2, p. 243, 1987.
10. Crittenden, J. et al.: *J. AWWA,* vol. 77, p. 87, 1991.
11. Weber, W. J., and B. E. Jones: EPA NTIS PB86-182425/AS, 1983.
12. Berndt, C., and L. Polkowski: "A Pilot Test of Nitrification with PAC," 50th Ann. Control States WPCA, 1977.
13. US Filter/Zimpro, Inc. Rothchild, Wisc.

9

ION EXCHANGE

9.1
THEORY OF ION EXCHANGE

Ion exchange can be used for the removal of undesirable anions and cations from a wastewater. Cations are exchanged for hydrogen or sodium and anions for hydroxyl ions.

Ion exchange resins consist of an organic or inorganic network structure with attached functional groups. Most ion exchange resins used in wastewater treatment are synthetic resins made by the polymerization of organic compounds into a porous three-dimensional structure. The degree of crosslinking between organic chains determines the internal pore structure, with higher crosslink density giving smaller pore sizes. From a kinetic viewpoint, a low degree of crosslinking would enhance diffusion of ions through larger pores. However, physical strength decreases and swelling in water increases as crosslink density is lowered. The functional ionic groups are usually introduced by reacting the polymeric matrix with a chemical compound containing the desired group. Exchange capacity is determined by the number of functional groups per unit mass of resin.

Ion exchange resins are called *cationic* if they exchange positive ions and *anionic* if they exchange negative ions. Cation exchange resins have acidic functional groups, such as sulfonic, whereas anion exchange resins contain basic functional groups, such as amine. Ion exchange resins are often classified by the nature of the functional group as strong acid, weak acid, strong base, and weak base. The strength of the acidic or basic character depends upon the degree of ionization of the functional groups, as with soluble acids or bases. Thus, a resin with sulfonic acid groups would act as a strong-acid cation exchange resin. The most common strong-acid ion exchange resin is prepared by copolymerizing styrene and divinylbenzene followed by sulfonation of the copolymer. The degree of crosslinking is controlled by the fraction of divinylbenzene in the initial mixture of monomers.

The types of ion exchange resins are

1. *Strong-acid cation resins.* Strong-acid resins are so named because their chemical behavior is similar to that of a strong acid. The resins are highly ionized in both the acid ($R\text{-}SO_3H$) and salt ($R\text{-}SO_3Na$) form, over the entire pH range.
2. *Weak-acid cation resins.* In a weak-acid resin, the ionizable group is a carboxylic acid (—COOH) as opposed to the sulfonic acid group (SO_3H^-) used in strong-acid resins. These resins behave like weak organic acids that are weakly dissociated.
3. *Strong-base anion resins.* Like strong-acid resins, strong-base resins are highly ionized and can be used over the entire pH range. These resins are used in the hydroxide (OH) form for water deionization.
4. *Weak-base anion resins.* Weak-base resins are like weak-acid resins, in that the degree of ionization is strongly influenced by pH.
5. *Heavy-metal selective chelating resins.* Chelating resins behave like weak-acid cation resins but exhibit a high degree of selectivity for heavy-metal cations. Chelating resins tend to form stable complexes with the heavy metals. In fact, the functional group used in these resins is an EDTA compound. The resin structure in the sodium form is expressed as R-EDTA-Na.

The reactions that occur depend upon chemical equilibria situations in which one ion will selectively replace another on the ionized exchange site. Cation exchange on the sodium cycle can be illustrated by the following reaction:

$$Na_2 \cdot R + Ca^{2+} \rightleftharpoons Ca \cdot R + 2Na^+ \tag{9.1}$$

where R represents the exchange resin. When all the exchange sites have been substantially replaced with calcium, the resin can be regenerated by passing a concentrated solution of sodium ions through the bed. This reverses the equilibrium and replaces the calcium with solium.

A 5 to 10 percent brine solution is usually used for regeneration:

$$2Na^+ + Ca \cdot R \rightleftharpoons Na_2 \cdot R + Ca^{2+} \tag{9.2}$$

Similar reactions occur for cation exchange on the hydrogen cycle:

$$Ca^{2+} + H_2 \cdot R \rightleftharpoons CaR + 2H^+ \tag{9.3}$$

Regeneration with 2 to 10 percent H_2SO_4 yields

$$Ca \cdot R + 2H^+ \rightleftharpoons H_2 \cdot R + Ca^{2+} \tag{9.4}$$

Anion exchange similarly replaces anions with hydroxyl ions:

$$SO_4^{2-} + R \cdot (OH)_2 \rightleftharpoons R \cdot SO_4 + 2OH^- \tag{9.5}$$

Regeneration with 5 to 10 percent sodium hydroxide will renew the exchange sites:

$$R \cdot SO_4 + 2OH^- \rightleftharpoons R \cdot (OH)_2 + SO_4^{2-} \tag{9.6}$$

In addition to the factors of concentration, the nature of the exchanger, and the exchanging ions, such factors as temperature and the particle size of the exchanger are also of considerable importance to the kinetics of ion exchange. The degree of exchange depends on several factors:

1. The size and valence (charge) of the ions entering into the exchange
2. The concentration of ions in the water or solution
3. The nature (both physical and chemical) of the ion exchange substance
4. The temperature

The following sequence shows the selectivity and ease of exchange of cations (Clifford et al.[1]): $Ra^{2+} > Ba^{2+} > Sr^{2+} > Ca^{2+} > Ni^{2+} > Cu^{2+} > Co^{2+} > Zn^{2+} > Mn^{2+} > UO_2^{2+} > Ag^+ > Cs^+ > K^+ > NH_4^+ > Na^+ > Li^+$. Thus, radium is the most preferred and lithium is the least preferred cation. The following sequence shows the selectivity of exchange on anions: $HCRO_4^- > CrO_4^{2-} > ClO_4^- > SeO_4^{2-} > SO_4^{2-} > NO_3^- > Br^- > HPO_4^-, HAsO_4^-, SeO_3^{2-} > CO_3^{2-} > CN^- > NO_2^- > Cl^- > H_2PO_4^-, H_2AsO_4^-, HCO_3^- > OH^- > CH_3COO^- > F^-$. The least preferred anion has the shortest retention time and appears first in the effluent, whereas the most preferred anion has the longest retention time and is eluted last.

In certain cases a nonpreferred ion can be converted into a polyvalent complex that is highly preferred by the resin, thereby making ion exchange a feasible decontamination process. For example, the removal of UO_2^{2+} from acid solutions is generally not practical because of competition from the highly preferred polyvalent cations Fe^{3+} and Al^{3+}. However, UO_2^{2+} ion undergoes a stepwise formation of anion complexes with sulfate as follows (Dorfiner[2]):

$$UO_2^{2+} + nSO_4^2 \rightleftharpoons UO_2(SO_4)n^{2-2n} \; (n = 1, 2, 3)$$

In this case, the resulting two-valent ($n = 2$) and four-valent ($n = 3$) complex anions are highly preferred by strong-base anion exchangers.

The performance and economics of ion exchange are related to the capacity of the resin to exchange ions and to the quantity of regenerant required. Since exchange occurs on an equivalent basis, the capacity of the bed is usually expressed as equivalents per liter of bed volume. In some cases capacity has been expressed as kilograms of $CaCO_3$ per unit of bed volume or as mass of ions per unit volume of bed. In like manner, the quantity of ions to be removed in the wastewater is expressed as equivalents per liter of wastewater to be treated.

In a fixed-bed exchanger, there is a relationship between the operating capacity of the bed and the quantity of regenerant employed. Resin utilization is defined as the ratio of the quantity of ions removed during treatment to the total quantity of ions that could be removed at 100 percent efficiency. The regenerant efficiency is the quantity of ions removed from the resin compared to the quantity of ions present in the volume of regenerant used. The resin utilization will increase as the regenerant efficiency decreases. Typical performance curves for a cation exchange resin are shown in Fig. 9.1. The shape of these curves will vary, depending upon the characteristics of the resin and the concentration of the regenerant used.

Treatment of a wastewater by ion exchange involves a sequence of operating steps. The wastewater is passed through the resin until the available exchange sites are filled and the contaminant appears in the effluent. This process is defined as the *breakthrough*. At this point treatment is stopped and the bed is backwashed to

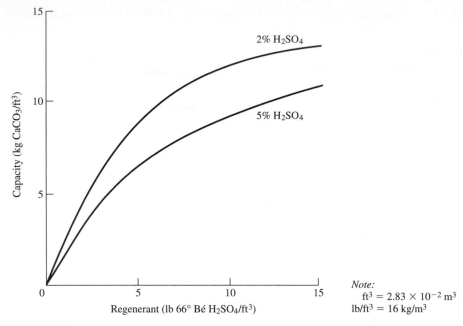

FIGURE 9.1
Performance of a cation exchange resin.

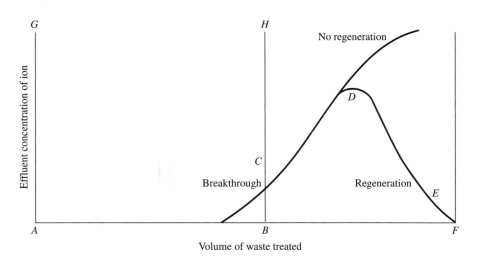

FIGURE 9.2
Treatment and regeneration cycle of an ion exchange resin.

remove dirt and to regrade the resin. The bed is then regenerated. After regeneration, the bed is rinsed with water to wash out residual regenerant. The bed is then ready for another treatment cycle.

The treatment and regeneration cycle is shown in Fig. 9.2. In this figure, area *ABHG* is the quantity of ions in the volume of solution treated before break-

through. Area *ABC* is the quantity of ions leaking through the column and area
ACHG is the quantity of ions removed by the exchange resin. The resin utilization
is therefore area *ACHG/K*, where *K* is the ultimate capacity of the resin. Area
BCDEF is the quantity of ions removed from the bed during regeneration. The
regeneration efficiency is therefore area *BCDEF/R,* where *R* is equal to the concentration of the regenerant times its volume.

To ensure contact of liquid with the resin and to minimize leakage, the minimum bed depth is 24 to 30 in (61 to 76 cm). The treatment flow rate can vary
between 2 and 5 gal/(min · ft^3) [0.27 to 0.67 m^3/(min · m^3)], although breakthrough
will occur more quickly at the higher flow rates. The regenerant flow rate is 1 to 2
gal/(min · ft^3) [0.13 to 0.27 m^3/(min · m^3)]. A rinsewater volume of 30 to 100
gal/ft^3 (4.0 to 13.4 m^3/m^3), applied at a flow rate of 1 to 1.5 gal/(min · ft^3) [0.13 to
0.20 m^3/(min · m^3)], will usually be sufficient to flush a bed of residual regenerant.

Experimental Procedure

It is frequently necessary to operate a laboratory ion exchange column to develop
the necessary design criteria for the removal of ions from complex industrial
wastes. A typical laboratory assembly is shown in Fig. 9.3. A suggested procedure
based on the apparatus shown in Fig. 9.3 is detailed as follows:

1. Rinse the column for 10 min with deionized water at a flow rate of 50 ml/min.
2. Adjust the flow rate to the column to 50 ml/min of solution containing the waste
   to be treated.

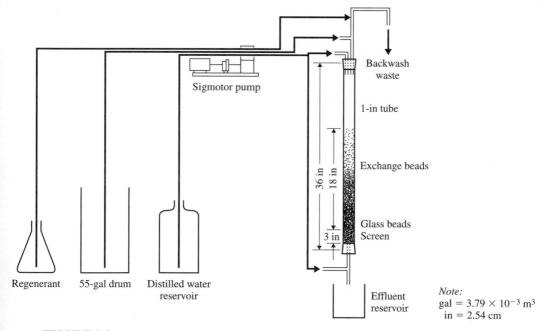

FIGURE 9.3
Laboratory ion exchange column.

3. Measure the initial volume of solution to be treated.
4. Start the treatment cycle. Develop the breakthrough curve until the ion concentration reaches the maximum effluent limit.
5. Backwash to 25 percent bed expansion for 5 to 10 min. (Use distilled water for the backwash operation.)
6. Regenerate at a flow rate of 6 ml/min, using the concentration and volume recommended for the resin. Collect the spent regenerant and measure the recovered ions.
7. Rinse the column with distilled water.

When several runs are made, it is possible to develop a relationship between resin utilization and regenerant efficiency and to select the optimum operating level for the system.

Macroreticular resins are employed for the removal of specific nonpolar organic compounds. These resins are highly specific and can be formulated to remove one compound or a class of compounds. The resins are solvent-regenerated. Treatment results of selected compounds using macrorecticular resins are shown in Table 9.1.

Arsenic (v) has been removed with a strong base anion exchange resin.[3] As (v) is present as the divalent anion $HAsO_4 2^-$ appears to be preferred on strong-base resins over monovalent anions in wastewaters.

Selenium can be removed by ion exchange under the following conditions:

Oxidation of all aqueous to selenate $SeO_4 2^-$ anion.
Strong-base anion exchange removal of selenate anion. The amount of selenate that can be removed is dependent on sulfate and nitrate concentrations in the water.[3]

TABLE 9.1
Macroreticular resin treatment of selected compounds

Compound	Influent, $\mu g/l$	Effluent, $\mu g/l$
Carbon tetrachloride	20,450	490
Hexachloroethane	104	0.1
2-Chloronaphthalene	18	3
Chloroform	1,430	35
Hexachlorobutadiene	266	<0.1
Hexachlorocyclopentadiene	1,127	1.5
Napthalene	529	<3
Tetrachloroethylene	34	0.3
Toluene	2,360	10
Aldrin	84	0.3
Dieldrin	28	0.2
Chlordane	217	<0.1
Endrin	123	1.2
Heptachlor	40	0.8
Heptachlor epoxide	11	<0.1

Ammonia can be removed by ion exchange using a natural inorganic zeolite clinoptilolite, which has an unusual selectivity for ammonium ions.[3] This unusual selectivity, which makes it attractive for ammonium ions, is caused by structurally related ion sieve properties. While the total exchange capacity of clinoptilolite is somewhat less than that of synthetic organic resins, its selectivity for the ammonium ion compensates. Regeneration is accomplished using a 3 to 6 percent NaCl, which is reused following NH_3 removal from the spent regenerant by air stripping or break-point chlorination.

9.2
PLATING WASTE TREATMENT

One of the major applications of ion exchange in industrial waste treatment has been in the plating industry, where chrome recovery and water reuse have often resulted in considerable savings.[4-7]

For recovery of the spent chromic acid in plating baths, the chromic acid is passed through a cation exchange resin to remove other ions (Fe, Cr^{3+}, Al, etc.). The effluent can be returned to the plating bath or to storage. Since the maximum concentration of CrO_3 that can be passed through some resins to avoid deterioration is 14 to 16 oz/gal (105 to 120 kg/m^3) as CrO_3, the bath may require dilution and the recovered solution may require makeup to strength.

The rinsewaters are first passed through a cation exchanger to remove metal ions. The effluent from this unit is passed through an anion exchanger to remove chromate and to obtain demineralized makeup water. It is desirable to pass the rinsewater through the cation unit first to avoid precipitation of metal hydroxides on the exchange resin. The anion exchanger is regenerated with sodium hydroxide; this results in a mixture of Na_2CrO_4 and NaOH in the spent regenerant. This mixture is passed through a cation exchanger to recover H_2CrO_4, which is returned to the plating bath. The recovered chromic acid from the spent regenerant will average 4 to 6 percent concentration. The spent regenerant from the cation exchanger will require neutralization and possibly precipitation of metallic ions before it is discharged to the sewer.

Since most of the metal ions are eluted in the first 70 percent of the regenerant volume, neutralization requirements can be reduced by reuse of the last portion of the acid regenerant for the subsequent regeneration.[4] In like manner, the last portion of the caustic regeneration can be used for neutralization of the spent cation regenerant.

In cation units, the regeneration requirements are higher than in water purification because of the competition with the H^+ present in the waste solution. For water reuse, 4 to 5 lb/ft^3 (64 to 80 kg/m^3) H_2SO_4 may be required for regeneration, but for recovery of H_2CrO_4 as much as 25 lb/ft^3 (400 kg/m^3) H_2SO_4 may be required to reduce leakage of sodium ions. A flow diagram of an ion exchange process for a plating plant is shown in Fig. 9.4. Ion exchange performance is shown in Table 9.2.

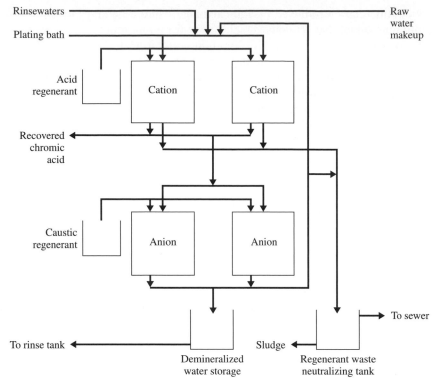

FIGURE 9.4
Ion exchange system for chromate removal and water reuse.

TABLE 9.2
Ion exchange performance in hexavalent chromium removal[8]

Wastewater source	Chromium, mg/l		Resin capacity[†]
	Influent	**Effluent**	
Cooling tower blowdown	17.9	1.8	5–6
	10.0	1.0	2.5–4.5
	7.4–10.3	1.0	—
	9.0	0.2	2.5
Plating rinsewater	44.8	0.025	1.7–2.0
	41.6	0.01	5.2–6.3
Pigment manufacture	1210	<0.5	—

[†] lb. chromate/ft^3 resin.

EXAMPLE 9.1. A general plating plant operates 16 h/d, 5 d/week. The total discharge of rinsewaters has the following characteristics:

Copper	22mg/l as Cu
Zinc	10 mg/l as Zn
Nickel	15 mg/l as Ni
Chromium	130 mg/l as CrO_3

The rate of flow is 50 gal/min (0.19 m³/min), and in-plant separation is not feasible. Design an exchanger system to include water and chromium recovery. The operating characteristics of the cation exchanger are given in Table 9.3.

Solution.

Anion exchanger
In the anion exchanger, CrO_3 is exchanged for OH.

$$130 \text{ mg/l} \times 50 \text{ gal/min} \times 60 \text{ min/h} \times 16 \text{ h/d} \times 8.34$$

$$\times 10^{-6} \frac{\text{lb/gal}}{\text{mg/l}} = 52 \text{ lb/d}$$

For a resin capacity of 3.8 lb CrO_3/ft_3 at a regeneration level of 4.8 lb NaOH/ft³ and a daily regeneration,

$$\text{Volume of resin} = \frac{52}{3.8} = 13.7 \text{ ft}^3 \quad (0.39 \text{ m}^3)$$

Treatment flow rate is 3.6 gal(min · ft³) [0.48 m³/(min · m³)], for a resin depth of 30 in (0.76 m), 2 units, 2 ft (0.61 in) diameter by 30 in (0.76 in) deep, plus 50 percent for bed expansion.

Regeneration

$$\text{NaOH required} = 4.8 \times 13.7 = 66 \text{ lb/reg} \quad (30 \text{ kg/reg})$$

TABLE 9.3

	Exchanger	
	Cation	**Anion**
Regenerant	H_2SO_4	NaOH
Dosage, lb/ft³	12	4.8
Concentration, %	5	10
Flow rate	0.5 gal/(min · ft³)	
Operating capacity	1.5 equiv wt/l	3.8 lb CrO_3/ft^3

Note:
lb/ft³ = 16.0 kg/m³
gal/(min · ft³) = 0.134 m ³/(min · m³)

$$\text{Regenerant tank volume} = 66 \text{ lb NaOH} \times \frac{1}{0.10} \times \frac{1}{9.6} \quad \text{lb reg/gal}$$

$$= 68 \text{ gal} \quad (0.26 \text{ m}^3)$$

$$\text{Rinse requirement at } 100 \text{ gal/ft}^3 = 1370 \text{ gal} \quad (5.2 \text{ m}^3)$$

Cation exchanger
The cations to be removed are:

$$\text{Zn} \quad \frac{10 \text{ mg/l}}{32.7 \text{ mg/meq}} = 0.306 \text{ meq/l}$$

$$\text{Cu} \quad \frac{22 \text{ mg/l}}{31.8 \text{ mg/meq}} = 0.693 \text{ meq/l}$$

$$\text{Ni} \quad \frac{15 \text{ mg/l}}{29.4 \text{ mg/meq}} = 0.511 \text{ meq/l}$$

In the cation unit, Cu, Zn, and Ni are exchanged for H^+.
    The total daily equivalents are

$$(0.306 + 0.693 + 0.511) \times 10^{-3} \times 50 \times 60 \times 16 \times 3.78 \text{ l/gal} = 273 \text{ equiv wt/d}$$

For an operating capacity of 1.5 equiv wt/l at a regeneration level of 12 lb H_2SO_4/ft^3 (5 percent), the resin required for a 2-d regeneration is

$$\frac{273 \times 2}{1.5 \times 28.3 \text{ l/ft}^3} = 13.0 \text{ ft}^3 \text{ resin} \quad (0.36 \text{ m}^3)$$

The treatment flow rate is 3.8 gal/(min · ft³) [0.51 m³/(min · m³)]. Use 2 units, 2 ft (0.61 m) diameter by 30 in (0.76 m) deep plus 50 percent for bed expansion.

Regeneration
Using 5 percent H_2SO_4 at 12 lb/ft³, H_2SO_4 required is

$$12 \times 13 = 156 \text{ lb} \quad (71 \text{ kg})$$

$$\text{Regenerant tank} = 156 \times \frac{1}{0.05} \times \frac{1}{1.0383 \times 8.34 \text{ lb/gal}}$$

$$= 360 \text{ gal} \quad (1.36 \text{ m}^3)$$

where 1.0383 is the specific gravity of 5% H_2SO_4.

$$\text{Rinse requirement} = 120 \text{ gal/ft}^3 \times 13 \text{ ft}^3 = 1560 \text{ gal} \quad (5.9 \text{ m}^3)$$

Anion regenerant capacity for chromium recovery:

$$\text{Sodium} = \frac{66 \text{ lb NaOH} \times 453 \text{ g/lb}}{40 \text{ g/equiv wt}} = 750 \text{ equiv wt}$$

If it is assumed that 70 percent of anion exchanger regenerant will pass through the cation unit, 525 equiv wt must be exchanged, which is compatible with the capacity of the cation units.

PROBLEM

9.1. A metal finishing plant operates 8 h/d, 6 d/week. The total discharge of reuse waters has the following characteristics:

    Zn   15 mg/l
    Ni   12 mg/l
    Cr   190 mg/1 as CrO_3

The flow rate is 100 gal/min (0.38 m³/min). Design an ion exchange system to include water and chromium recovery. The operating characteristics of the resins are given in Example 9.1.

REFERENCES

1. Clifford, D. et al.: *J. ES&T*, vol. 20, no. 11, p. 1072, 1986.
2. Dorfner, K.: *Ion Exchange Properties and Applications,* Ann Arbor Science Pub., 1973.
3. Bin, Luo et al.: *Toxicity Reduction: Evaluation and Control,* Technomic Pub. Co., Lancaster, Pa., 1998.
4. Fadgen, T. J.: *Proc. 7th Ind. Waste Conf.,* 1952, Purdue University.
5. Paulson, C. F.: *Proc. 7th Ind. Waste Conf.,* 1952, Purdue University.
6. Rich, L. G.: *Unit Processes of Sanitary Engineering,* John Wiley, New York, 1963.
7. Keating, R. J., R. Dvorin, and V. J. Calise: *Proc. 9th Ind. Waste Conf.,* 1954, Purdue University.
8. Patterson, J.: *Industrial Wastewater Treatment Technology,* 2nd ed., Butterworth Pub., Stoneham, Mass., 1985.

10

CHEMICAL OXIDATION

10.1
INTRODUCTION

Chemical oxidation generally refers to the use of oxidizing agents such as ozone, O_3, hydrogen peroxide, H_2O_2, permanganate, MnO_4^-, chloride dioxide, ClO_2, chlorine, Cl_2 or $HOCl$, or even oxygen, O_2, without the need for microorganisms for the reactions to proceed. These reactions frequently require one or more catalysts in order to increase the rate of reaction to acceptable levels. Catalysts include simple pH adjustment, transition metal cations, enzymes, and a variety of proprietary catalysts of unreported composition.

Chemical oxidation is typically applied to situations where organic compounds are nonbiodegradable (refractory), toxic, or inhibitory to microbial growth. However, chemical oxidation is also effective for the destruction of many inorganic compounds and the elimination of odorous compounds, e.g., oxidation of sulfides ($H_2S \rightarrow SO_4^{2-}$).

While oxygen serves as a readily available and extremely economical oxidant for biological treatment processes, other chemical oxidants are relatively expensive and cannot compete economically with aerobic biological treatment. However, it is not necessary to carry chemical oxidation to the fullest extent of reaction (conversion of organic carbon to CO_2). Partial oxidation of compounds may be sufficient to render specific compounds, such as priority pollutants, more amenable to subsequent biological treatment. On a general basis, the oxidation of specific compounds may be characterized by the extent of degradation of the final oxidation products:[1]

1. *Primary degradation.* A structural change in the parent compound.
2. *Acceptable degradation (defusing).* A structural change in the parent compound to the extent that toxicity is reduced.
3. *Ultimate degradation (mineralization).* Conversion of organic carbon to inorganic CO_2.
4. *Unacceptable degradation (fusing).* A structural change in the parent compound resulting in an increase in toxicity.

With the exception of unacceptable by-products, chemical oxidation can result in reduced toxicity or inhibitory behavior, and greatly increased biodegradability of the parent compounds at dosages far less than required for ultimate degradation, i.e., the "stoichiometric dosage." Therefore, coupled chemical/biological oxidation processes, with chemical oxidation used for pretreatment of difficult wastes is frequently considered as a treatment alternative.[2]

10.2
STOICHIOMETRY

It is important to define the stoichiometric relationship between an oxidant and the compounds to be treated so that required oxidant dosages may be estimated, and experiments may be designed within reasonable limits. A general approach may be taken so that one can easily convert the stoichiometry for one particular compound from one oxidant to another. For convenience, the half-reactions for each oxidant can be expressed in terms of "free reactive oxygen," $O^{\cdot}$, derived for each oxidant, or, using simple oxygen as an example:

$$O_2 \rightarrow 2O^{\cdot} \tag{10.1}$$

This may be developed from the electrochemical half-reaction for any oxidant by balancing the electrons with the equivalent free reactive oxygens based on water, or:

$$H_2O \rightarrow O^{\cdot} + 2e^- + 2H^+ \tag{10.2}$$

where the free reactive oxygen is half a diatomic oxygen.

Then, as an example we may consider one of the possible half-reactions for permanganate:

$$MnO_4^- + 2H_2O + 3e^- \rightarrow MnO_2 + 4OH^- \tag{10.3a}$$

Balancing electrons with Eq. (10.2) gives

$$\begin{array}{r} 2MnO_4^- + 4H_2O + 6e^- \rightarrow 2MnO_2 + 8OH^- \\ + \qquad\qquad 3H_2O \rightarrow 3O^{\cdot} + 6e^- + 6H^+ \\ \hline MnO_4^- + H_2O \rightarrow 2MnO_2 + 3O^{\cdot} + 2OH^- \end{array} \tag{10.3b}$$

Equation (10.3b) is the equivalent half-reaction to Eq. (10.3a) except the electrons are replaced by the free reactive oxygens. The same procedure can be completed for any oxidant using Eq. (10.2) and the appropriate half-reaction. Then, any oxidation reaction may be expressed in terms of the equivalent free oxygen and related back to a specific oxidant by the stoichiometry of the half-reaction. The half-reactions for a variety of oxidants are presented in Table 10.1.

For the ultimate conversion of an organic compound to CO_2 and H_2O, a general stoichiometric equation may be derived for reactive oxygens:

$$C_aH_bO_c + dO^{\cdot} \rightarrow aCO_2 + (b/2)H_2O \tag{10.4}$$

TABLE 10.1
Oxidant half-reactions

Half-reaction	Equivalent reactive oxygens	
	Moles [O˙] per mole oxidant (n)	Moles [O˙] per kg oxidant
Chlorine:		
$Cl_2 + H_2O \rightarrow O˙ + 2Cl^- + 2H^+$	0.5	14.1
$HOCl \rightarrow O˙ + Cl^- + H^+$	1.0	19.0
Chlorine dioxide:		
$2ClO_2 + H_2O \rightarrow 5O˙ + 2Cl^- + 2H^+$	2.5	37.0
Hydrogen peroxide:		
$H_2O_2 \rightarrow O˙ + H_2O$	1.0	29.4
Permanganate: pH < 3.5		
$2MnO_4^- + 6H^+ \rightarrow 2Mn^{2+} + 5O˙ + 3H_2O$	2.5	15.8
3.5 < pH < 7.0		
$2MnO_4^- + 2H^+ \rightarrow 2MnO_2 + O˙ + 3H_2O$	1.5	9.5
7.0 < pH < 12.0		
$2MnO_4^- + H_2O \rightarrow 3O˙ + 2MnO_2 + 2OH^-$	1.5	9.5
12.0 < pH < 13.0		
$2MnO_4^- + H_2O \rightarrow 2MnO_4^{-2} + O˙ + 2H^+$	0.5	3.2
Ozone: high pH		
$O_3 \rightarrow O˙ + O_2$	1.0	20.8
low pH		
$O_3 \rightarrow 3O˙$	3.0	61.4

where $d = 2a + b/2 - c$. Then, the equation may be balanced for any oxidant by adding the half-reaction for the oxidant times the number of free reactive oxygens required (d) divided by the stoichiometric number of free reactive oxygens produced (see half-reactions in Table 10.1).

EXAMPLE 10.1. For phenol (C_6H_5OH), balance the half-reactions with peroxide (H_2O_2) and permanganate (MnO_4^-) as oxidants.

Solution.

From Eq. (10.4),

$$C_6H_5OH + 14O˙ \rightarrow 6CO_2 + 3H_2O$$

H_2O_2:

$$C_6H_5OH + 14O^{\cdot} \rightarrow 6CO_2 + 3H_2O$$
$$+ \quad (14/1)\,(H_2O_2 \rightarrow O^{\cdot} + H_2O)$$
$$\overline{= C_6H_5OH + 14H_2O_2\ 6CO_2 + 17H_2O}$$

MnO_4^-:

$$C_6H_5OH + 14O^{\cdot} \rightarrow 6CO_2 + 3H_2O$$
$$+ \quad (14/3)\,(2MnO_4^- + H_2O \rightarrow 3O^{\cdot} + 2MnO_2 + 2OH^-)$$
$$\overline{= C_6H_5OH + (28/3)\ MnO_4^- + (5/3)\ H_2O\ 6CO_2 + (28/3)\ MnO_2 + (28/3)\ OH^-}$$

In many cases, a wastewater will consist of a wide variety of compounds rather than one or a few specific known compounds. Therefore, the use of the theoretical stoichiometry cannot be applied to specific compounds without some error introduced by the background oxidizable compounds. However, the general approach developed using the free reactive oxygens can be adapted to the surrogate chemical oxygen demand, COD. This readily measurable parameter can be converted to the total stoichiometric requirement for an arbitrary wastewater as follows:

$$\text{Oxidant demand (mg oxidant/l)} = (2/n)\,(MW/32)\,\text{COD} \qquad (10.5)$$

where
n = moles O per mole oxidant (see Table 10.1)
MW = molecular weight of oxidant (g/mole)
COD = chemical oxygen demand (mg O_2/l)

For H_2O_2 and $KMnO_4$ as examples:

H_2O_2 stoichiometric demand $= (2/1)\,(34/32)\,\text{COD} = 2.13\,\text{COD}$

$KMnO_4$ stoichiometric demand $= (2/1.5)\,(158/32)\,\text{COD} = 6.58\,\text{COD}$

For the stoichiometric requirements and the yield of free reactive oxygens, i.e., moles O per kg (from Table 10.1), the cost for the stoichiometric dosage of each oxidant may be determined (based on $/kg oxidant). Then, a ranking of candidate oxidants based on the potential oxidant cost can be established, and their actual effectiveness and cost determined from laboratory data.

10.3
APPLICABILITY

Apart from oxygen, most oxidants are expensive and not competitive with biological wastewater treatment for high-strength, large-volume wastewaters. However, chemical oxidation processes are generally designed for wastewaters that are not amenable to biological treatment, i.e., toxic, inhibitory, and/or refractory compounds. In addition, significant savings can be obtained through the coupling of chemical oxidation with biological treatment by pretreating toxic/refractory wastes to improve biological treatment performance. The partial oxidation of many wastes can be economically achieved, yielding a variety of readily biodegradable organic acids.

The average oxidation state (OX) of organic carbon in the waste mixture may be expressed as:[3]

$$OX = \frac{4\,(TOC - COD)}{TOC} \tag{10.6}$$

The COD/TOC ratio can be used as a primary parameter to determine the extent of the overall reaction. Figure 10.1 shows a typical case. The initial TOC before reaction is point A. Point B identifies the initial reaction products, and point C identifies the products after the reaction is substantially complete. An evaluation for toxicity and for biodegradability is made at points B and C to determine both the dosage and the time of contact.

The efficiency of treatment is based on the ability of the oxidant to produce an acceptable organic by-product without significant ultimate conversion of organic carbon to carbon dioxide. This can be expressed as:

$$f = \frac{OD_T}{OX_T} \tag{10.7}$$

where f = fraction of ultimate oxidant demand
         OD_T = oxidant consumed, mg/l
         OX_T = total stoichiometric oxidant demand (can be based on COD), mg/l

Results for the oxidation of 2,4-dichlorophenol (DCP) and 2,4-dinitro-*ortho*-cresol (DNOC) are shown in Fig. 10.2. The results show that H_2O_2 was not used for ultimate conversion of the organic carbon, but rather the original compounds were drastically altered, leaving by-products in which the mean oxidation state of the

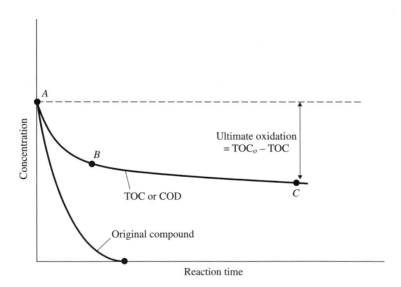

FIGURE 10.1
Change in original compound and TOC or COD over the course of reaction with an oxidant.

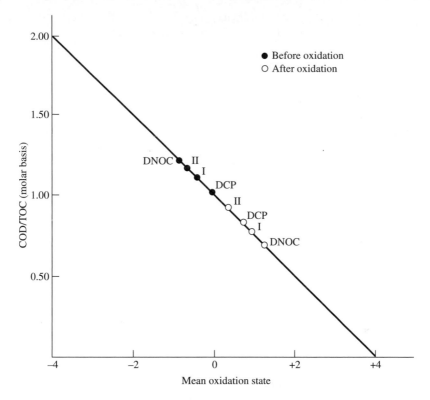

FIGURE 10.2
The COD/TOC ratio versus the mean oxidation state of carbon in the system. Note that a +4 oxidation state corresponds to inorganic carbon (CO_2) and a −4 oxidation state corresponds to methane. I and II refer to different times of oxidation.

carbon was more highly oxidized (positive instead of negative). DNOC exhibited the greatest change (OX = +2.22) while DCP showed the least (OX = +0.867).

10.4
OZONE

Ozone, O_3, is a powerful oxidant ($E_H > 2.0$ volts) that is commonly used for disinfection and wastewater treatment. Ozone is a metastable gas at normal temperatures and pressures and must be generated on site. At high pressures decomposition is rapid. Therefore, generation and mass transfer operations are carried out at low pressure, typically less than 20 lb/in^2 gauge.[4]

Ozone generation is the key to economical operation. Generators operate on electrical current and produce ozone from either air or pure oxygen gas streams. Moisture retards the process and air must be dehumidified prior to ozone generation (dew point < −60°C). Recent advances in pure oxygen generation have greatly reduced the cost of ozone production. A comparison of air and pure O_2 for ozone production is presented in Table 10.2.

TABLE 10.2
Typical systems for generating ozone[4]

Feature	Typical operation
Principle of generation	Corona discharge
Electrical frequency	Low: 50 or 60 Hz
	Median: 60 to 1000 Hz
	High: > 1000 Hz
Operating pressure	7 to 14 lb/in^2 gauge (low frequency)
	20 lb/in^2 gauge (high frequency)
Cooling	Water or oil
Ozone output	Air: 1 to 6%
	Pure O$_2$: 8 to 14%
Energy consumption, kWh/lb	Air: 6.4 to 9.1
	Pure O$_2$: 3.2 to 6.8

Ozone decomposes in water, especially at high pH values, to produce free radicals: $O_3 \rightarrow O^{\cdot} + O_2$ ($n = 1$, see Table 10.1). The rate of decomposition can be written as a function of pH (represented by OH$^-$) in terms of a first-order decay:[5]

$$\frac{dO_3}{dt} = r_d = -9.811 \times 10^7 \, [OH^-]^{0.123} [O_3] \exp(-5606/T) \qquad (10.8)$$

where
$$r_d = \text{rate of O}_3 \text{ self-decomposition, M/min}$$
$$[OH^-], [O_3] = \text{concentrations of OH}^- \text{ and O}_3 \text{ in aqueous solution, M}$$
$$T = \text{temperature, K}$$

The ozone reaction mechanism is then uniquely dependent on the rate of mass transfer into an aqueous system, and three operating regions of reactivity exist:

Region 1: High pH and/or low ozone dose rate, implying that the major reaction mechanism with solutes entails $^{\cdot}$OH. Little selectivity will be exhibited.

Region 2: Approximately equal rates of decomposition and mass transfer. Both direct O_3 and $^{\cdot}$OH reactions are important.

Region 3: Low pH and/or extremely high ozone dose rate, implying that direct oxidation of solutes by O_3 is the controlling reaction. Ozone may exhibit a high degree of selectivity.

The mechanisms of ozone oxidation of organics are:

1. Oxidation of alcohols to aldehydes and then to organic acids:

$$RCH_2OH \xrightarrow{O_3} RCOOH$$

2. Substitution of an oxygen atom onto an aromatic ring
3. Cleavage of carbon double bonds

Ozonation can be employed for the removal of color and residual refractory organics in effluents. In one case, while there was a decrease in TOC in the final filtered effluent, the soluble BOD increased from 10 to 40 mg/l because of the conversion from long-chain biologically refractory organics to biodegradable compounds. Similar results were obtained from the ozonation of a secondary effluent from low- and high-rate activated sludge units treating a tobacco-processing wastewater, as shown in Table 10.3. TOC will not be reduced until the organic carbon has been oxidized to CO_2, while COD will generally be reduced with any oxidation.

Oxidation of unsaturated aliphatic or aromatic compounds causes a reaction with water and oxygen to form acids, ketones, and alcohols. At a pH greater than 9.0 in the presence of redox salts such Fe, Mn, and Cu, aromatics may form some hydroxyaromatic structures (phenolic) which may be toxic. Many of the by-products of ozonation are readily biodegradable.

For many of the priority pollutants there is a rapid, first-order reduction in COD to some level followed by a slow or even zero removal, indicating the non-reactive nature of the by-products.

COD reduction by ozonation of an aerated lagoon effluent treating an organic chemicals wastewater is shown in Fig. 10.3.

Phenol can be oxidized with ozone, producing as many as 22 intermediate products between phenol and CO_2 and H_2O. The reaction is first order with respect to phenol and proceeds optimally over a pH of 8 to 11. The ozone consumption is 4 to 6 mole O_3 consumed/mole phenol oxidized. This requires in the order of 25 mole O_3/mole phenol to be generated in the gas phase.

Ozone may be used in combination with UV radiation to catalyze the oxidation of nonreactive organics, such as saturated hydrocarbons and highly chlorinated organics. Ozone reacts with UV light (253.7 nm) in aqueous systems to produce hydrogen peroxide:[6]

$$O_3 + H_2O \xrightarrow{\;uv\;} O_2 + H_2O_2 \qquad (10.9)$$

These two oxidants, O_3 and H_2O_2, react with one another, providing O_3 and hydroxyl radical, $\cdot OH$, at low pH values. In some cases, H_2O_2 may be added directly to complement the coupled oxidation process, in advanced oxidation processes (AOPs). In addition, UV light may react directly with some of these organics, further promoting reaction of the first by-products with O_3.

TABLE 10.3
Ozonation results on secondary clarifier effluent

Parameter	Low F/M (0.15) time, min[†]		High F/M (0.60) time, min[†]	
	0	60	0	60
BOD, mg/l	27	22	97	212
COD, mg/l	600	154	1100	802
pH	7.1	8.3	7.1	7.6
Org-N, mg/l	25.2	18.9	40	33
NH_3^-N, mg/l	3.0	5.8	23	25
Color (Pt-Co)	3790	30.0	5000	330

[†] Loading of 155 mg O_3/min.

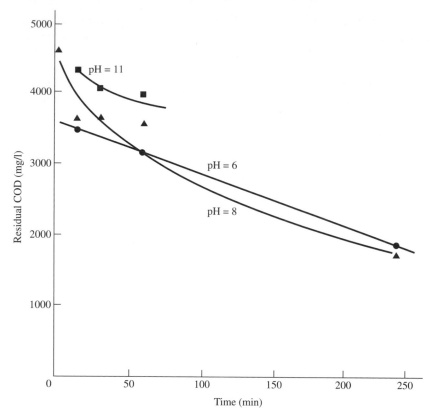

FIGURE 10.3
Effect of pH on COD removal by ozonation-aerated lagoon effluent.

10.5
HYDROGEN PEROXIDE

Hydrogen peroxide, H_2O_2, is commercially available in a variety of grades, with 30 or 50 percent (by weight) solutions being most common for wastewater applications. Inhibitors, typically phosphates, are added to prolong storage times. Hydrogen peroxide has a long history of use for oxidation of sulfides in sewer lines and wastewater treatment plants, and more recently has been widely applied to toxic and refractory organics. Sulfide oxidation with H_2O_2 is summarized in Table 10.4.

Alkaline peroxidation (pH 9.5) will oxidize formaldehyde: $2CH_2O + H_2O_2 + 2OH^- \rightarrow 2HCOO^- + H_2 + 2H_2O$.

Alkaline peroxidation (pH 10 to pH 12) is an effective means of providing total cyanide destruction. The reactions are $CN^- + H_2O_2 \rightarrow OCN^- + H_2O$ $OCN^- + 2H_2O \rightarrow NH_4^+ + CO_3^{2-}$ This process was evaluated for treatment of a wastestream discharged from specialty polymers manufacturing. The wastewater was discharged at 40°C and pH 7.5 and contained 4700 mg/l COD and 96 mg/l CN. Hydrogen peroxide was added to the wastestream and allowed to react at pH 10.7 (sodium hydroxide addition was required) for 6 h. Following 1 h of reaction, 5600

TABLE 10.4
Sulfide oxidation with hydrogen peroxide

Acidic or neutral pH

$$H_2O_2 + H_2S \rightarrow 2H_2O + S$$

Reaction time: 15–45 min

Catalyst: Fe^{2+}

pH: 6.0–7.5

Reaction time: seconds

Basic pH

$$4H_2O_2 + S^{2-} \rightarrow SO_4^{2-} + 4H_2O$$

Reaction time: 15 min

---

mg/l H_2O_2 had been consumed and the residual CN concentration was 8.3 mg/l. Following 6 h of reaction time, 6000 mg/l H_2O_2 had been consumed and the residual CN concentration was 2.5 mg/l.

Reactions with H_2O_2 alone are slow, and a catalyst is generally required. A wide variety of reaction schemes are possible, using high pH (alkaline catalysis); metals such as ferrous sulfate (Fenton's reagent), complexed Fe (Fe-EDTA or Heme), Cu, or Mn; or natural enzymes such as horseradish peroxidase. By far, the most common catalyst is ferrous iron ($FeSO_4$ or Fenton's reagent), at a pH of about 3.5. A series of reactions with Fe are hypothesized leading to the generation of free radicals, $\cdot OH$ and $HO_2 \cdot$ and regeneration of Fe (II):

$$Fe^{2+} + H_2O_2 \rightarrow Fe^{3+} + OH^- + \cdot OH \tag{10.10a}$$

and

$$Fe^{3+} + H_2O_2 \rightarrow Fe^{2+} + HO_2 \cdot + H^+ \tag{10.10b}$$

A chain reaction then occurs between the hydroxyl radical and an organic compound R:

$$RH + \cdot OH \rightarrow R \cdot + H_2O \tag{10.11a}$$

$$R \cdot + O_2 \rightarrow ROO \cdot \tag{10.11b}$$

$$ROO \cdot + RH \rightarrow ROOH + R \cdot \tag{10.11c}$$

The reaction proceeds optimally at pH between 2.0 to 4.0, primarily so that the Fe (III) produced stays in solution as Fe^{3+} rather than precipitating as the ferrihydroxides, $Fe(OH)_3$ or FeOOH. Some processes make use of this, precipitating the Fe (III) after the reaction is complete and recycling the ferrihydroxides back to the process for reuse. A typical catalyzed H_2O_2 system is shown in Fig. 10.4.

Other catalysts include UV radiation (UV-H_2O_2). It is hypothesized that the H_2O_2 molecule can be split directly into hydroxyl radicals by UV light:

$$H_2O_2 \xrightarrow{UV} 2 \cdot OH \tag{10.12}$$

Ultraviolet light is derived from lamps that have been developed for a high quantum yield in the appropriate wavelengths for hydroxyl radical generation, based on

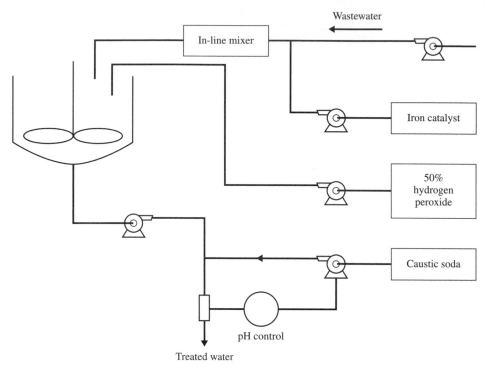

Note: A pH of ≤ 5 is required for this reaction.

FIGURE 10.4
Wastewater treatment system with catalyzed hydrogen peroxide.

decomposition of ferrioxalate as an indicator. Additionally, heat generated by the lamps can increase the rates of reaction significantly. The UV-H_2O_2 process is generally applied to aqueous wastes of lower color, turbidity, and concentration, such as contaminated groundwaters. However, proprietary additives are available for higher-strength wastes. Many compounds are effectively treated by UV-H_2O_2 including benzene, toluene, xylene, trichloroethylene, and perchloroethylene, but some are refractory to treatment, including chloroform, acetone, trinitrobenzene, and *n*-octane. Performance is shown in Fig. 10.5.

Several catalysts are compared in Table 10.5. It should be noted that catalysts must be evaluated experimentally for each wastewater and/or specific pollutant.

Hydrogen peroxide has been shown to be useful in many cases for removal of specific undesirable pollutants such as priority pollutants, reduction of toxicity, and/or improvements in biodegradability (both rate and extent of degradation). Table 10.6 summarizes the results of tests on a variety of aromatic compounds oxidized which indicate reactivity with these compounds, COD and TOC reduction, and a significant improvement in toxicity (except for 2,6-dichlorophenol). Oxidation with H_2O_2 can then be coupled with biological treatment to further treat the COD to an acceptable level. In many cases oxidation of refractory COD will result in an increase of BOD as shown in Fig. 10.6.

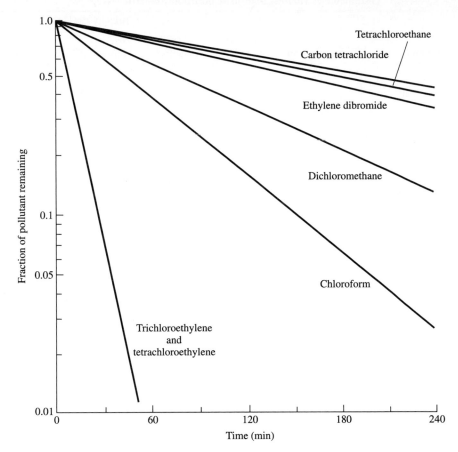

FIGURE 10.5
Comparison of rates of reaction of halogenated aliphatics at 20°C with UV and hydrogen peroxide.[6]

TABLE 10.5
Catalysts for hydrogen peroxide treatment of wastes

Catalyst	Optimum pH	Conditions required	Comments
Fenton's reagent: FeSO$_4$	2–4	H$_2$O$_2$:Fe ≈ 10:1	Generation of Fe(OH)$_3$ sludge
Alkaline pH	> 9	_____	_____
Complexed iron: Fe-EDTA,[†] Heme	5–10	H$_2$O$_2$:Fe ≈ 5:1	Complexed iron remains in solution
UV, radiation	2–10[‡]	Low turbidity Low UV absorbance Low organic contaminant	Electrical costs dominate

[†] As ferrous or ferric ethylenediaminetetraacetate.
[‡] pH plays little role.

TABLE 10.6
Oxidation of various aromatic compounds with H_2O_2[a, b]

Compound	% Destruction	% COD reduction	% TOC reduction[c]	Toxicity (EC$_{50}$)[d]	
				Original compound	After oxidants
Nitrobenzene	>99	72.4	37.3	62	217
Benzoic acid	>99	75.8	48.8	289	>292
Phenol	>99	76.1	44.1	69	NT
o-Cresol	>99	75.0	55.6	29	NT
m-Cresol	>99	73.3	38.2	16	NT
p-Cresol	>99	71.8	40.0	5	NT
o-Chlorophenol	>99	75.1	47.9	52	—
m-Chlorophenol	>99	75.0	41.3	18	NT
p-Chlorophenol	>99	75.7	21.7	3	NT
2,3-Dichlorophenol	>99	70.2	52.6	10	NT
2,4-Dichlorophenol	>99	68.9	50.3	6	—
2,5-Dichlorophenol	>99	74.2	41.8	18	NT
2,6-Dichlorophenol	>99	61.1	32.5	54	26[e]
3,5-Dichlorophenol	>99	69.1	48.9	5	NT
2,3-Dinitrophenol	>99	80.1	50.7	86	NT
2,4-Dinitrophenol	>99	72.5	51.0	22	NT
Aniline	>99	76.5	43.4	394	NT

[a] Stoichiometric dose of H_2O_2, 5×10^{-3} M pollutant, 50 mg/l FeSO$_4$ (as Fe) for catalyst; samples reacted in batch for 24 h at 20°C ($\pm$2°).
[b] Based on original compound only.
[c] Corresponds to ultimate conversion.
[d] EC$_{50}$ = concentration which causes a 50 percent reduction in activity. Based on Microtox® toxicity; values are as mg DOC/l; NT = nontoxic.
[e] Shows increased toxicity.

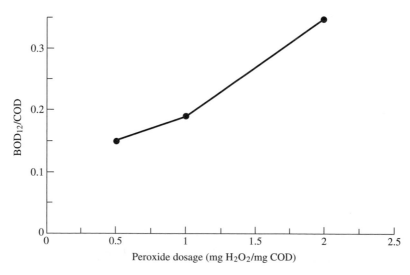

FIGURE 10.6
BOD$_{12}$/COD ratio at different peroxide dosages.[7]

10.6
CHLORINE

Chlorine has a long history of use as an oxidant in water and wastewater treatment, and has been especially successful for color removal where organic dyes are present. However, recent concerns regarding the formation of chlorinated by-products, such as chloroform, have greatly reduced the applications of chlorine in wastewater. Chlorination of an activated sludge effluent for color removal is shown in Fig. 10.7. While chlorine is highly successful for color removal, chlorinated organics must be accounted for. Similar effluents have had difficulty meeting discharge permit requirements where zero- and low-flow receiving streams were involved.

Chlorine is still frequently applied for the oxidation of cyanides in metal finishing operations. The oxidation of cyanide by chlorine proceeds through

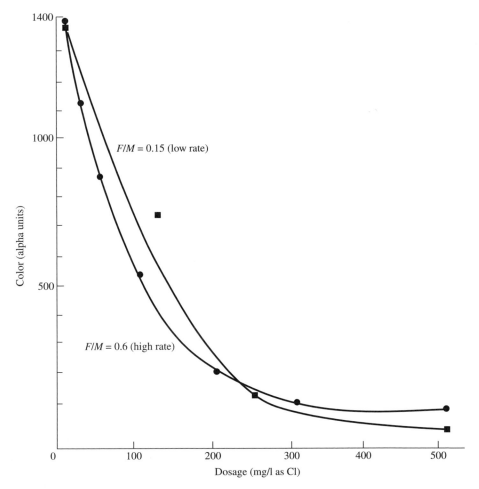

FIGURE 10.7
Color removal by chlorination.

several reactions that are highly pH-dependent, and a two-step process is typically practiced:

Step 1
Reaction 1: Reaction of cyanide with hypochlorite to form cyanogen chloride (all pH values):

$$CN^- + OCl^- + H_2O \xrightarrow{slow} CNCl + 2OH^- \tag{10.13a}$$

Reaction 2: Hydrolysis of cyanogen chloride to form cyanate, CNO^- (minimum pH 9 to 10; pH 11.5 recommended):

$$CNCl + 2OH^- \xrightarrow{fast} CNO^- + Cl^- + H_2O \tag{10.13b}$$

Step 2
Reaction 3: Oxidation of cyanate to form bicarbonates and nitrogen gas (pH 8.0 to 8.5):

$$2CNO^- + 3HOCl \rightarrow 2HCO_3^- + N_2 + 3Cl^- + H^+ \tag{10.13c}$$

Cyanogen chloride is a toxic gas and must be eliminated immediately. It is not stable and hydrolyzes rapidly to CNO^- at high pH. Therefore, Reactions 1 and 2 are carried out simultaneously in step 1 at the recommended pH of 11.5.

At this point, CNO^- is about 1000 times less toxic than CN^-, and frequently the reaction is stopped at this point. However, complete treatment requires step 2, where CNO^- is oxidized to HCO_3^- and N_2 (gas) at pH 8 to 8.5. While a lower pH (< 7.6) would be preferred, the potential for HCN evolution in the event of a process malfunction places a lower limit on pH in this process.

In practice, the dosages of chlorine depend heavily on the identity of the metals complexed with CN^- and the presence of other background constituents in the wastewater. Since these concentrations change continually, the system is usually operated on the basis of oxidation-reduction potential (ORP). A summary of typical dosages and operating parameters is presented in Table 10.7. Chlorine is typically applied as sodium hypochlorate (15 percent NaOCl) for operation at under 20 gal/min and chlorine gas (Cl_2) for larger operations. As an example problem consider the example on the next page.

TABLE 10.7
Summary of cyanide oxidation practice with chlorine

Reaction step	Chlorine dose[†]	Operating ORP, mV
Step 1: $CN^- \rightarrow CNO^-$	Stoichiometric (Cu, Zn),	~350[‡]
	2 × Stoichiometric (Fe, Co, Ni)	
Step 2: $CNO^- \rightarrow HCO_3^- + N_2$	Stoichiometric	500–800

[†] This must be tuned experimentally for each wastewater.
[‡] Stoichiometric dosages based on Eqs. (10.10a) and (10.10c).

EXAMPLE 10.2. How much Cl_2 must be supplied to oxidize 130 mg/l cyanide (as CN) given a flow of 10,000 l/d?

(*a*) Consider oxidation to CNO^-.
(*b*) Consider complete oxidation to HCO_3^- and N_2.

First, consider that Cl_2 reacts with water to produce HOCl and HCl:

$$Cl_2 + H_2O \longleftrightarrow HOCl + HCl$$

Therefore, only half of our chlorine is effective. Next, all calculations must be done on a molar basis, or molecular weight of CN = 12 + 14 = 26 g/mole. Then

$$\frac{130 \text{ mg/l cyanide}}{(26 \text{ g/mole})(1000 \text{ mg/g})} = 5.0 \times 10^{-3} \text{ M}$$

Solution.

(*a*) Combine Reactions 1 and 2 [Eqs. (10.10*a*) and (10.10*b*)] to give an overall reaction of CN^- to CNO^-:

$$\text{Reaction 1} + \text{Reaction 2} = CN^- + OCL^- \rightarrow CNO^- + Cl^-$$

Therefore, 1 mole of OCL^- is required for each CN^-:

$$5 \times 10^{-3} \text{ MCN}^- \left(\frac{1 \text{ mole OCl}^-}{1 \text{ mole CN}^-}\right) = 5 \times 10^{-3} \text{ M OCl}^- \text{ required}$$

$$= \frac{\text{stoichiometric dosage}}{5 \times 10^{-3} \text{ M Cl}_2 \text{ required}}$$

Finally, for a 10,000 l/d flow:

$$(5 \times 10^{-3} \text{ mole Cl}_2/\text{l})(10,000 \text{ l/d}) = 50 \text{ moles Cl}_2/\text{d}$$

then Cl_2 = 71 g/mole, or

$$\frac{(50 \text{ moles Cl}_2/\text{d})(71 \text{ g/mole})}{1000 \text{ g/kg}} = 3.55 \text{ kg Cl}_2/\text{d}$$

and, since dosage is 1 to 2 times stoichiometric, depending on the metal complexed with CN^- (see Table 10.5), the actual required dose = 3.55 to 7.10 kg Cl_2/d.

(*b*) Consider Reaction 3 [Eq. (10.10*c*)]:

$$ZCNO^- + 3HOCl \rightarrow 2HCO_3^- + N_2 + 3Cl^- + H^+$$

Note that 3 moles HOCl produced per 2 moles CNO^- and 1 mole CNO^- produced per mole CN^- are initially present (100 percent conversion). Therefore,

$$5 \times 10^{-3} \text{ MCNO}^- = \left(\frac{3 \text{ moles HOCl}}{2 \text{ moles CNO}^-}\right)(10,000 \text{ l/d})\left(\frac{71 \text{ g/mole}}{1000 \text{ g/kg}}\right) = 5.33 \text{ kg Cl}_2/\text{d}$$

and the total requirement is step 1 + step 2, or

$$Cl_2 = (3.55 \text{ to } 7.10 \text{ kg/d}) + 5.33 \text{ kg/d} = 8.88 \text{ to } 12.43 \text{ kg Cl}_2/\text{d}$$

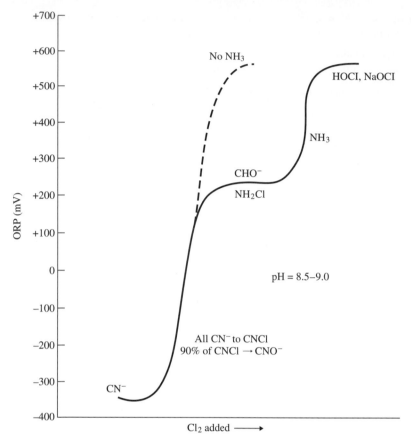

FIGURE 10.8
Electrode potential relationships in the alkaline chlorination of cyanide waste.

Electrode potential relationships in the alkaline chlorination of cyanide wastewater are shown in Fig. 10.8.

10.7
POTASSIUM PERMANGANATE

Permanganate, MnO_4^-, is a powerful oxidizing agent ($E_h = 1.68$ volts) that is reactive over a wide pH range (although rates are reportedly faster at high pH) for a
variety of organic and inorganic compounds. This is supplied in stable form as a
solid (96.5 to > 99 percent purity) or in concentrated aqueous form. Applications
of $KMnO_4$ traditionally include odor control (oxidation of inorganic and organic
sulfides) and textile, tannery, steel processing, metal finishing, pulp and paper, and
oil refinery wastes. Oxidation of sulfides proceeds differently under acidic or alkaline conditions:

Acidic

$$3H_2S + 2KMnO_4 \rightarrow 3S° + 2H_2O + 2KOH + 2MnO_2(s)$$

Alkaline

$$3H_2S + 8KMnO_4 \rightarrow 3K_2SO_4 + 2H_2O + 2KOH + 8MnO_2(s)$$

Unlike other oxidants, a solid reaction by-product, $MnO_2(s)$, is formed and must be disposed of as a waste sludge, along with other precipitates/solids already understood as part of the waste itself. This sludge can be considerable for concentrated wastewaters. Data shows that $KMnO_4$ can be effective in the destruction of specific compounds and in toxicity reduction for phenolics and other aromatic compounds, and even certain chlorinated aliphatics, such as trichloroethylene, perchloroethylene, and trichloroethane.

10.8
OXIDATION OVERVIEW

The chemical oxidation of various aromatic compounds by H_2O_2, $KMnO_4$, and O_3 is shown in Table 10.8. The toxicity before and after oxidation of these compounds is shown in Table 10.9.

Application of chemical oxidation for the treatment of different types of dyes is shown in Table 10.10.

TABLE 10.8
Chemical oxidation of various aromatic compounds using H_2O_2, $KMnO_4$, and O_3[8]

Compound	Initial oxidation state	TOC removal, %[a,b,c]			COD removal, %[d]		
		H_2O_2	$KMnO_4$	O_3	H_2O_2	$KMnO_4$	O_3
Pyrrolidine	−1.76	34.9	NR[e]	32.1	72.1	NR[e]	58.5
Sulfanilic acid	−0.84	46.3	NR	57.5	74.9	NR	57.4
Naphthalene	−0.80	46.2	NR	0.0	80.4	NR	>99.0
Diphenylamine	−0.66	69.4	NR	30.6	87.7	NR	90.0
Skatole	−0.66	0.0	NR	0.0	39.0	NR	38.1
Benzaldehyde	−0.57	78.6	67.6	74.4	93.5	79.1	74.2
Indole	−0.50	62.3	60.3	60.9	95.5	91.0	77.2
Catechol	−0.33	57.0	52.2	22.0	80.5	66.3	30.7
Hydroquinone	−0.33	30.7	27.3	17.2	78.5	71.2	45.0
Resorcinol	−0.33	56.5	27.8	29.1	79.8	73.1	50.1
Vanillin	−0.25	70.3	53.4	63.6	87.8	55.2	63.6
Pyrogallol	0.00	45.4	22.1	28.5	75.1	78.2	48.5
Salicylic acid	0.00	28.6	31.6	31.2	74.6	49.8	41.6
Coumarin	+0.22	25.9	NR	NR	65.3	NR	NR
Phthalic acid	+0.25	37.0	NR	31.1	71.2	NR	52.0
Average[f]	−0.44	45.9	42.8	36.8	77.1	70.5	55.9

[a] Arranged in order of lowest oxidation state of organic carbon.
[b] Mean oxidation state of the organic carbon calculated from molecular structure.
[c] Corresponds to the percentage converted to CO_2, i.e., ultimate oxidation.
[d] All compounds were initially at a concentration of 5×10^{-3} M (as the original compound) unless limited by solubility (naphthalene and diphenylamine).
[e] NR = nonreactive (no measurable reduction in COD).
[f] Average based on reactive compounds only.

TABLE 10.9
Toxicity of compounds before and after oxidation[8]

Compound	Original EC$_{50}$, %^{a}	EC$_{50}$ after chemical oxidation, %^{a,b}		
		H$_2$O$_2$	KMnO$_4$	O$_3$
Pyrrolidine	NT^c	NT	NR^d	NT
Sulfanilic acid	14.0 (49.6)	62.0 (117.8)	NR	70.0 (107.1)
Naphthalene	41.0 (5.3)	NT (7.0)^e	NR	NT (14.0)^e
Diphenylamine	14.0 (5.0)	83.0 (9.1)	NR	88.0 (22.0)
Skatole	0.9 (2.6)	4.2 (12.6)	NR	8.0 (24.2)
Benzaldehyde	4.8 (20.2)	54.0 (48.6)	11.0 (15.0)	39.0 (42.1)
Indole	1.3 (6.2)	56.0 (100.8)	25.0 (47.5)	45.0 (84.6)
Catechol	8.1 (28.4)	30.0 (45.3)	49.5 (95.5)	39.0 (108.0)
Hydroquinone	0.3 (1.0)	1.6 (3.9)	63.0 (172.0)	48.0 (143.0)
Resorcinol	55.0 (194.7)	64.0 (98.6)	42.0 (107.9)	48.0 (120.5)
Vanillin	22.0 (103.8)	53.0 (74.2)	52.0 (236.4)	65.0 (113.1)
Pyrogallol	2.8 (9.4)	61.0 (111.6)	50.0 (130.5)	42.0 (102.1)
Salicylic acid	27.0 (110.4)	25.0 (73.0)	26.0 (90.2)	28.0 (98.3)
Coumarin	2.0 (12.4)	NT (460.0)^e	NR	NR
Phthalic acid	NT	NT	NR	NT

^{a} Based on Microtox® toxicity; % volume (or concentration) to reduce the light produced from luminescence by 50%.
^{b} Values in parentheses represent the EC$_{50}$ on a TOC basis (as mg/l).
^{c} NT = nontoxic, i.e., no detectable light reduction.
^{d} NR = nonreactive compound for this particular oxidant.
^{e} Value in parentheses for nontoxic compounds represents the highest TOC value at no dilution.

TABLE 10.10
Process recommendations for different types of dyes

Type of dye	Fenton (FSR)	UV/H$_2$O$_2$	O$_3$/H$_2$O$_2$
Reactive	+	+	+
Direct	+	+	+
Metal complex	+	+	+
Pigment	+	−	−
Disperse	−	+	+
Vat	0	−	−
Mixtures	0	+	0

Recommended +
Not recommended −
Applicable 0

10.9
HYDROTHERMAL PROCESSES

Hydrothermal processes refer to aqueous treatment of wastewaters at elevated temperatures and pressures. There are three basic operating regimes that are in use or have been investigated in the laboratory:

1. *Wet air oxidation (WAO).* A commonly used process for more concentrated wastes, especially those that are toxic and/or biologically refractory. These processes use an oxidant, primarily O_2 from air, to partially oxidize organics, yielding a variety of low-molecular-weight organic acids (readily biodegradable). Usual temperatures range from 150 to 320°C and pressures from 150 to 3000 lb/in^2 gauge (1.0 to 20.7 Mpa). This process is shown in Fig. 10.9.
2. *Hydrothermal hydrolysis.* Hydrolysis of organic compounds can occur at elevated temperatures and pressures, for example, $CN^- \rightarrow HCOO^-$ or $CCL_4 \rightarrow$ HCl. To date, all investigations have been on a laboratory scale. Temperatures proposed for hydrothermal hydrolysis range from 200 to 374°C and pressure from 220 to 3200 lb/in^2 gauge.
3. *Supercritical water oxidation (SCWO).* Aqueous oxidation of organics takes place to completion, at even higher temperature and pressures than wet air oxidation. SCWO takes place beyond the critical point of water (about 374°C and 218 atm). Typical operating conditions are 400 to 650°C and 3500 to 5000 lb/in^2 gauge (24.1 to 34.5 Mpa). At these temperatures and pressures, materials of construction become critical and solubility of salts can decrease dramatically, causing fouling.

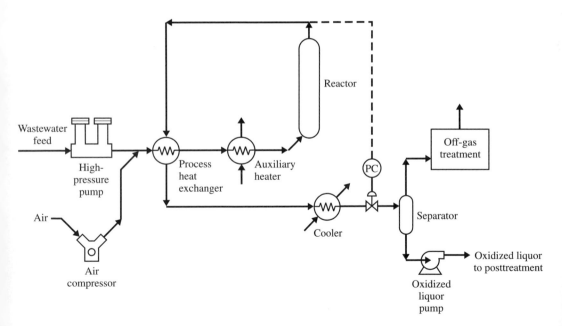

FIGURE 10.9
Wet air oxidation flow scheme.

To date, only wet air oxidation has achieved any degree of commercial success. Hydrothermal hydrolysis and SCWO have been mainly demonstrated in the laboratory, and only a few operations have been run continuously in the field. However, WAO is a commercial process, with at least three commercial suppliers of equipment (US Filter/Zimpro, Kenox Corporation, and Nippon Petrochemical) and over 250 units sold to customers. Field units operate a roughly 2.5 to 300 gal/min capacity and treat a variety of wastes from spent caustic solutions to pharmaceuticals. The fate of wastewater components is shown in Table 10.11. A summary treatment performance for several case studies is presented in Table 10.12. These systems all operate with air as the source of oxygen. The high temperatures, pressures, and corrosivity of these wastes require annual maintenance and inspection as well as frequent cleaning to remove scale deposit from the boilers and heat exchangers. With regular maintenance, a WAO unit is expected to be in service 80 to 85 percent of the time, while having secondary pumps and compressors available can increase in-service time to 90 to 95 percent.

TABLE 10.11
Fate of wastewater components during wet air oxidation

Influent soluble, colloidal, or suspended species	Wet air oxidation products	
	Partially oxidized	**Completely oxidized**
Complex organics	Low-molecular-weight organics (carboxylic acids, aldehydes, ketones, hydrocarbons)	CO_2, H_2O, HX
Inorganics	NH_4, N_2, SO_3	SO_4^{2-} PO_4^{3-}, NO^{3-}

TABLE 10.12
Summary of operating conditions and destruction efficiencies at WAO installations[†]

Waste compound[‡]	Treatment parameters		Destruction, %
	Temperature, °C	**Residence time, min**	
Acenaphthene	275	60	99.99
Acenaphthene	275	60	99.0
Carbon tetrachloride	275	60	99.7
Chloroform	275	60	99.5
Dibutylphthalate	275	60	99.5
Malathion	250	60	99.9
Mercaptons	200	—	>99.99
4-Nitrophenol	275	—	99.6
Phenols	200	—	97.7–98.2

[†] Data from US Filter/Zimpro, Rothchild, Wisconsin.
[‡] Influent concentrations ranged from 287 to 11,800 mg/l.

PROBLEM

10.1. A wastewater has a flow of 40,000 gal/d (151 m^3/d) containing 35 mg/l CN$^-$. Compute the chemical requirements for the alkaline chlorination of this wastewater for CN$^-$ removal.

REFERENCES

1. Lyman, W. J., W. F. Reehl, and D. H. Rosenblatt: *Handbook of Chemical Property Estimation Methods*, American Chemical Society, Washington, D.C., 1990.
2. Lankford, P. W., and W. W. Eckenfelder: *Toxicity Reduction in Industrial Effluents*, Van Nostrand Reinhold, New York, 1990.
3. Stumm, W., and J. J. Morgan: *Aquatic Chemistry*, John Wiley & Sons, New York.
4. Langlais, B., D. A. Reckhow, and D. R. Brink, eds.: *Ozone in Water Treatment: Application and Engineering*, Lewis Publishers, Chelsea, Mich., 1991.
5. Sullivan, D. E., and J. A. Roth: "Kinetics of Ozone Self-Decomposition in Aqueous Solution," *Water 1979*, vol. 76, no. 197; pp. 142–149.
6. Sundstrom, D. W. et al.: presented at the Association of Industrial Chemical Engineers Spring Meeting, April 1994.
7. Albers, H., and Kayser, R.: *42nd Purdue Ind. Waste Conf.*, p. 893, 1988.
8. Bowers, A.: *Chemical Oxidation,* Technomic Pub. Co., Lancaster, Pa., 1997.

11

SLUDGE HANDLING AND DISPOSAL

Most of the treatment processes normally employed in industrial water pollution control yield a sludge from a solids-liquid separation process (sedimentation, flotation, etc.) or produce a sludge as a result of a chemical coagulation or a biological reaction. These solids usually undergo a series of treatment steps involving thickening, dewatering, and final disposal. Organic sludges may also undergo treatment for reduction of the organic or volatile content prior to final disposal. Sludges contain free water, capillary water, and bound water. Free water is removed by thickening of the sludge. Capillary water is removed through dewatering. Bound water can be removed only by chemical or thermal means. These mechanisms are shown in Fig. 11.1.

In general, gelatinous-type sludges such as alum or activated sludge yield lower concentrations, whereas primary and inorganic sludges yield higher concentrations in each process sequence.

Conventional sludge handling alternatives are shown in Fig. 11.2. The processes selected depend primarily on the nature and characteristics of the sludge and on the final disposal method employed. For example, activated sludge is more effectively concentrated by flotation than by gravity thickening. Final disposal by incineration desires a solids content that supports its own combustion. In some cases, the process sequence is apparent from experience with similar sludges or by geographical or economic constraints. In other cases, an experimental program must be developed to determine the most economical solution to a particular problem.

11.1
CHARACTERISTICS OF SLUDGES FOR DISPOSAL

The physical and chemical characteristics of sludges dictate the most technically and economically effective means of disposal. For thickening, the concentration ratio C_u/C_o (the concentration of the underflow divided by the concentration of the influent) is related to the mass loading [lb solids/(ft$^2 \cdot$ d) or kg/(m$^2 \cdot$ d)], which indicates the feasibility of gravity thickening.

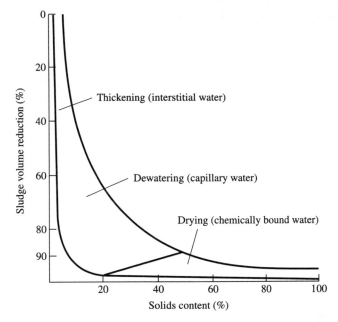

FIGURE 11.1
Mechanisms of sludge dewatering.

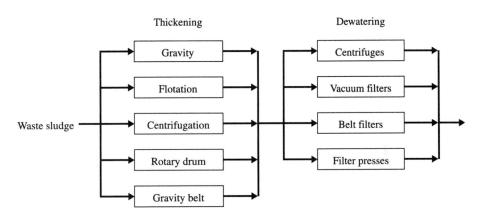

FIGURE 11.2
Sludge handling process alternatives.

The dewaterability of a sludge by filtration is related to the specific resistance. While the specific resistance of a sludge can be reduced by the addition of coagulants, economic considerations may dictate alternative dewatering methods.

Ultimate disposal usually considers land disposal or incineration. In incineration, the heat value of the sludge and the concentration attainable by dewatering dictate the economics of the operation. Land disposal may use the sludge as a fertilizer or soil conditioner, as in the case of waste activated sludges or in a confined landfill for hazardous sludges. It is important, if a sludge is to be used for land disposal, that heavy metals be removed by pretreatment.

The toxicity characteristic leaching procedure (TCLP) is a protocol developed by the U.S. Environmental Protection Agency (USEPA) (Method 1311) to evaluate the propensity of a waste material to release hazardous constituents in a mismanagement scenario, such as placement of a hazardous waste into a municipal solid waste landfill. The procedure involves chemical analyses of an extract of the waste. If the waste contains less than 0.5 percent filterable solids, the filtrate from the Method 1311 protocol is considered to be the extract. Wastes with higher solids contents are leached with an acetic acid solution to produce an extract. If any of the extract concentrations from a waste exceed the regulatory levels established by USEPA, that waste is considered to exhibit the hazardous waste characteristic of toxicity. At present, USEPA has promulgated regulatory levels for 40 elements and compounds, including heavy metals, volatile and semivolatile organics, pesticides, and herbicides (see 40 CFR 261.24), as shown in Table 11.1.

11.2
AEROBIC DIGESTION

Aerobic digestion when applied to excess biological sludges involves the oxidation of cellular organic matter through endogenous metabolism. The oxidation of cellular organics has been found to follow first-order kinetics when applied to the degradable volatile suspended solids.[1] Under batch or plug flow conditions,

$$\frac{(X_d)_e}{(X_d)_o} = e^{-k_d t} \tag{11.1}$$

where $(X_d)_e$ = degradable volatile solids after time t
$(X_d)_o$ = initial degradable volatile solids
k_d = reaction rate coefficient, d^{-1}
t = time of aeration, d

The kinetic parameters can be determined from a batch oxidation in the laboratory as shown in Fig. 11.3.

If the total volatile suspended solids are considered, Eq. (11.1) becomes

$$\frac{X_e - X_n}{X_o - X_n} = e^{-k_d t} \tag{11.1a}$$

where X_o = initial VSS
X_e = effluent VSS
X_n = nondegradable VSS

For a completely mixed reactor, the relationship is modified to

$$\frac{X_e - X_n}{X_o - X_n} = \frac{1}{1 + k_d t} \tag{11.2}$$

TABLE 11.1
Maximum concentration of contaminants for the toxicity characteristic

Contaminant	Regulatory level (mg/l)
Arsenic	5.0
Barium	100.0
Benzene	0.5
Cadmium	1.0
Carbon tetrachloride	0.5
Chlordane	0.03
Chlorobenzene	100.0
Chloroform	6.0
Chromium	5.0
o-Cresol	200.0[†]
m-Cresol	200.0[†]
p-Cresol	200.0[†]
Cresol	200.0[†]
2,4-D	10.0
1,4-Dichlorobenzene	7.5
1,2-Dichloroethane	0.5
1,1-Dichloroethylene	0.7
2,4-Dinitrotoluene	0.13[‡]
Endrin	0.02
Heptachlor (and its epoxide)	0.008
Hexachlorobenzene	0.13[‡]
Hexachlorobutadiene	0.5
Hexachloroethane	3.0
Lead	5.0
Lindane	0.4
Mercury	0.2
Methyoxychlor	10.0
Methyl ethyl ketone	200.0
Nitrobenzene	2.0
Pentrachlorophenol	100.0
Pyridine	5.0
Selenium	1.0
Silver	5.0
Tetrachloroethylene	0.7
Toxaphene	0.5
Trichloroethylene	0.5
2,4,5-Trichlorophenol	400.0
2,4,6-Trichlorophenol	2.0
2,4,5-TP (Silvex)	1.0
Vinyl chloride	0.2

[†] If *o*-, *m*-, and *p*-cresol concentrations cannot be differentiated, the total cresol concentration is used. The regulatory level of total cresol is 200 mg/l.
[‡] Quantification limit is greater than the calculated regulatory level. The quantification limit therefore becomes the regulatory level.

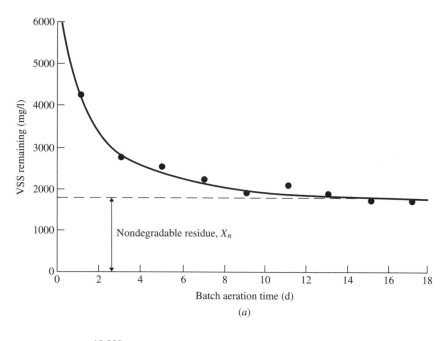

(a)

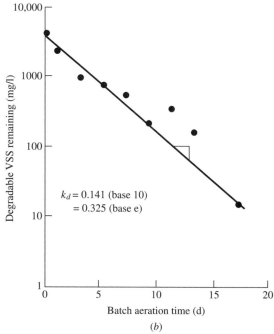

$k_d = 0.141$ (base 10)
$= 0.325$ (base e)

(b)

FIGURE 11.3
Kinetics of aerobic sludge digestion. (a) Chronological destruction of VSS in batch reactor; (b) correlation of degradable VSS with detention time.

and the required retention time is

$$t = \frac{X_o - X_e}{k_d(X_e - X_n)} \qquad (11.3)$$

For n mixed reactors in series,

$$\frac{X_e - X_n}{X_o - X_n} = \frac{1}{(1 + k_d t_n)^n} \qquad (11.4)$$

In accordance with the kinetic relationship, mixed reactors in series are more efficient than one mixed reactor. For example, to achieve 90 percent removal of degradable solids at 20°C would require 9.7 days in a single-stage digester and 7.2 days in a three-stage digester.

The oxygen requirements for aerobic digestion can be estimated as 1.4 lb of oxygen consumed for each pound of VSS destroyed (1.4 kg O_2/kg VSS destroyed). Nitrogen and phosphorus will be released by the oxidation process. Under mesophilic digestion conditions, with long sludge ages and a nitrifier seed, nitrification will usually occur. If nitrification does not occur there will be a buildup of ammonia and nondegradable COD. Oxygen and alkalinity must be available for this oxidation. Under thermophilic operation, nitrification will be inhibited.

Temperature will affect the rate coefficient k_d. The temperature relationship is shown in Fig. 11.4. Conventional aerobic digestion design employs secondary clarifier underflow (0.5 to 1.5 percent solids) in one or more completely mixed aeration basins. Power levels of 15 to 20 standard ft³/(min · thousand ft³) [15 to 20 std m³/(min · thousand m³)] using diffused air or 100 hp/million gal (0.02 kW/m³) using surface mechanical aerators are usually adequate for providing both mixing

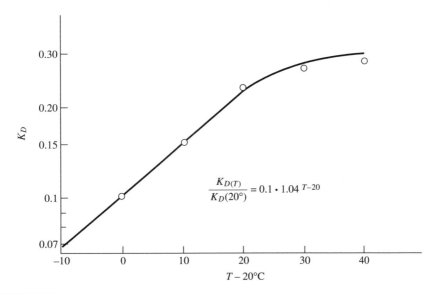

FIGURE 11.4
Temperature effect on aerobic digestion.

and oxygen requirements. Prethickening the sludge offers a number of advantages, in particular, reducing the basin volume requirements and increasing the temperature due to the exothermic heat of reaction. Andrews and Kambhu[2] have estimated the heat of combustion as 9000 Btu/lb (2.1×10^7 J/kg) VSS destroyed. Aerobic digestion designs are illustrated in Example 11.1. Aerobic digestion requirements will depend on the operating sludge age in the aeration process. As the sludge age is increased, more of the degradable biomass is oxidized in the aeration basin and hence less will be oxidized in the aerobic digester. Consider two cases as shown in Example 11.1.

EXAMPLE 11.1. Design an aerobic digester to yield a final degradable fraction of 0.37 for a waste activated sludge from a system with a 10-d SRT and a 30-d SRT. The following data apply for the wastewater:

$S_r = 690$ mg/l
$Q = 5$ million gal/d (18,925 m³/d)
$a = 0.6$
$b = 0.1$ d^{-1}

Solution.

Equation (6.11) can be rearranged to calculate ΔX_v.

$$\Delta X_v = \frac{aS_r}{1 + bX_d\theta_c}$$

For the 10-d SRT, $X_d = 0.67$.

$$\Delta X_v = \frac{0.6 \cdot 690}{(1 + 0.1 \cdot 0.67 \cdot 10)}$$

$$= 248 \text{ mg/l} \qquad (10{,}337 \text{ lb/d; } 4693 \text{ kg/d})$$

For the 30-d SRT, $X_d = 0.5$.

$$\Delta X_v = \frac{0.6 \cdot 690}{(1 + 0.1 \cdot 0.5 \cdot 30)}$$

$$= 166 \text{ mg/l} \qquad (6905 \text{ lb/d; } 3135 \text{ kg/d})$$

A degradable fraction of 0.37 yields a degradable VSS of 2000 lb/d (908 kg/d).

	10-d SRT	30-d SRT
X_o	10,337 lb/d	6905 lb/d
X_D	6926 lb/d (3144 kg/d)	3494 lb/d (1586 kg/d)
X_N	3411 lb/d	3411 lb/d (1549 kg/d)
X_{DN}	2000 lb/d	2000 lb/d
X_e	5411 lb/d	5411 lb/d (2457 kg/d)

Required detention time:

$$t = \frac{X_o - X_e}{k_d(X_e - X_n)}$$

$$k_d = 0.155 \text{ d}^{-1}$$

For a 10-d SRT:

$$t = \frac{10{,}337 - 5411}{0.155(5411 - 3411)}$$

$$= 15.9 \text{ d}$$

For a 30-d SRT:

$$t = \frac{6905 - 5411}{0.155(5411 - 3411)}$$

$$= 4.8 \text{ d}$$

Oxygen requirements:
For VSS destruction,

$$\text{Oxygen} = (X_D - X_{DN}) \cdot 1.4$$

For a 10-d SRT,

$$\text{Oxygen} = (6926 - 2000) \cdot 1.4$$

$$= 6896 \text{ lb/d} \quad (3131 \text{ kg/d})$$

For a 30-d SRT,

$$\text{Oxygen} = (3494 - 2000) \cdot 1.4$$

$$= 2092 \text{ lb/d} \quad (950 \text{ kg/d})$$

Nitrification (see Fig. 6.11):
It is assumed that the $NH_3\text{-}N$ released through endogenous metabolism is nitrified.

$$N_{ox} = N_o X_o - N_e X_e$$

For a 10-d SRT,

$$N_{ox} = 0.091 \cdot 10{,}337 - 0.072 \cdot 5411$$

$$= 551 \text{ lb/d} \quad (250 \text{ kg/d})$$

For a 30-d SRT,

$$N_{ox} = 0.076 \cdot 6905 - 0.072 \cdot 5411$$

$$= 135 \text{ lb/d} \ (61 \text{ kg/d})$$

The oxygen requirements for nitrification are

$$551 \cdot 433 = 2386 \text{ lb/d} \quad (1083 \text{ kg/d}) \quad \text{10-d SRT}$$

$$135 \cdot 4.33 = 585 \text{ lb/d} \quad (266 \text{ kg/d}) \quad \text{30-d SRT}$$

The total oxygen requirements are

$$6896 \text{ lb/d} + 2386 \text{ lb/d} = 9282 \text{ lb/d} \quad (4214 \text{ kg/d}) \quad \text{10-d SRT}$$

$$2092 \text{ lb/d} + 585 \text{ lb/d} = 2677 \text{ lb/d} \quad (1215 \text{ kg/d}) \quad \text{30-d SRT}$$

The alkalinity requirements are

$$551 \cdot 7.14 = 3934 \text{ lb/d} \quad (1786 \text{ kg/d}) \quad \text{10-d SRT}$$

$$135 \cdot 7.14 = 964 \text{ lb/d} \quad (438 \text{ kg/d}) \quad \text{30-d SRT}$$

Denitrification:
Assume sludge at 10,000 mg/l. For the 10-d SRT system at 10,337 lb/d, the sludge flow is 0.124 million gal/d. The volume of the digester is

$$0.124 \text{ million gal/d} \cdot 15.9 \text{ d} = 19.7 \text{ million gal}$$

The oxygen-uptake rate for VSS oxidation is

$$\frac{6896 \text{ lb/d}}{1.97 \cdot 8.34 \cdot 24} = 17.5 \text{ mg/(l·h)}$$

The denitrification rate mg $NO_3^-N/(l \cdot h)$ is estimated as

$$17.5 \cdot 0.25 = 4.4 \text{ mg NO}_3/(l \cdot h)$$

$$NO_3^-N = 551 \text{ lb/d} \quad (250 \text{ kg/d})$$

or 33.5 mg/l in the aeration volume. Then

$$t_{DN} = 33.5/4.4 = 7.6 \text{ h}$$

Recently a focus has been on autothermal aerobic digestion (ATAD). In this process the exothermic heat of combustion of the volatile solids increases the temperature in the reactor to the thermophilic range, i.e., 55°C. Deeny et al.[18] showed that an influent VSS of 3 percent was sufficient to maintain a digester temperature of 55 to 60°C. This is illustrated in Example 11.2.

EXAMPLE 11.2. Determine the temperature achieved in an aerobic digester with feed solids of 3 percent at 22°C. The daily sludge volume is 50,000 gal (189 m³). Volatile suspended solids reduction is 40 percent and the endogenous rate k_d is 0.3 d^{-1}. The degradable fraction of the sludge is 0.67.

Solution.

From Eq. (11.3),

$$t = \frac{X_o - X_e}{k_d (X_e - X_n)}$$

in which

$$X_o = 30{,}000 \text{ mg/l}$$

$$X_e = 30{,}000 \cdot 0.6 = 18{,}000 \text{ mg/l}$$

$$X_n = 30{,}000 \cdot 0.33 = 9900 \text{ mg/l}$$

$$t = \frac{30{,}000 - 18{,}000}{0.3(18{,}000 - 9900)} = 4.9 \text{ d}$$

The volume of the digester is

$$50,000 \cdot 4.9 = 245,000 \text{ gal} \qquad (927 \text{ m}^3)$$

The VSS reduced is

$$30,000 \text{ mg/l} \cdot 0.4 \cdot 0.05 \text{ million gal/d} \cdot 8.34 = 5004 \text{ lb/d} \qquad (2272 \text{ kg/d})$$

The heat generated is

$$9300 \text{ Btu/lb} \cdot 5004 \text{ lb/d} = 46.5 \times 10^6 \text{ Btu/d} \qquad (49.1 \text{ kJ/d})$$

The water per day is

$$50,000 \text{ gal/d} \cdot 8.34 \text{ lb/gal} = 0.417 \times 10^6 \text{ lb/d} \qquad (189,000 \text{ kg/d})$$

The temperature increase will be

$$46.5 \times 10^6 \text{ Btu/d} = \Delta T \cdot 1 \text{ Btu/(lb} \cdot {}^\circ\text{F)} \cdot 0.417 \times 10^6 \text{ lb/d}$$

$$\Delta T = 111.5 {}^\circ\text{F} \qquad (44 {}^\circ\text{C})$$

Considering heat losses, this should be sufficient to maintain a digester temperature in excess of 55°C.

The oxygen requirements are

$$5004 \text{ lb/d} \cdot 1.4 = 7006 \text{ lb/d} \qquad (3181 \text{ kg/d})$$

and the oxygen uptake rate will be

$$\frac{7006 \text{ lb/d}}{0.245 \text{ mg} \cdot 8.34 \cdot 24} = 142 \text{ mg/(l} \cdot \text{h)}$$

11.3
GRAVITY THICKENING

Gravity thickening is accomplished in a tank equipped with a slowly rotating rake mechanism that breaks the bridge between sludge particles, thereby increasing settling and compaction. Gravity thickening is usually applied to primary and chemical sludges which thicken well by gravity settling. A typical gravity thickener is shown in Fig. 11.5.

The primary objective of a thickener is to provide a concentrated sludge underflow. The area of thickener for a specified underflow concentration is related to the mass loading [lb/(ft$^2 \cdot$ d), kg/(m$^2 \cdot$ d)] or to the unit area [ft^2(lb/d), m^2/(kg/d)].

The mass loading can be computed from a stirred laboratory cylinder test. For municipal sewage, the mass loading might be expected to vary from 4 lb/(ft$^2 \cdot$ d) [19.5 kg/(m$^2 \cdot$ d)] for waste activated sludge to 22 lb/(ft$^2 \cdot$ d) [107 kg/(m$^2 \cdot$ d)] for primary sludge.

A procedure for the design of gravity thickeners has been developed by Dick.[3] The most important criterion in thickener design and operation is the mass loading or solids flux expressed as pounds of solids fed per square foot per day (or

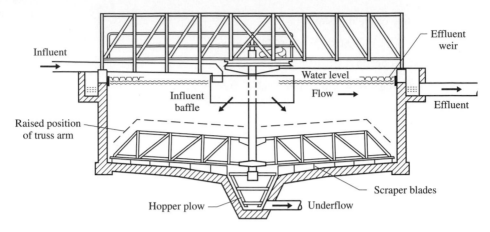

FIGURE 11.5
Gravity thickener (*Courtesy of Link Belt, FMC Company*).

kilograms per square meter per day). The limiting flux that produces the desired underflow for a given area must equal the solids loading rate to the thickener:

$$G_L = \frac{C_o Q_o}{A} = \frac{M}{A} \tag{11.5}$$

where Q_o = influent flow, ft^3/d (m^3/d)
   C_o = influent solids, lb/ft^3 (kg/m^3)
   M = solids loading, lb/d (kg/d)
   G_L = limiting solids flux, lb/(ft^2 · d) [kg/(m^2 · d)]
   A = area, ft^2 (m^2)

The limiting flux can be obtained from the following rationale.
   The capacity of a thickener for removing solids under batch conditions is

$$G_B = C_i V_i \tag{11.6}$$

where G_B = batch flux, lb/(ft^2 · d) [kg/(m^2/d)]
   C_i = solids concentration, lb/ft^3 (kg/m^3)
   V_i = settling velocity at C_i, ft/d (m/d)

A relationship can be developed between C_i and V_i that is usually linear on a log scale over a wide range of concentrations, as shown for an activated sludge in Fig. 11.6.
   In a continuous thickener the solids are removed both by gravity and by the velocity resulting from the removal of sludge from the tank bottom:

$$G = C_i V_i + C_i U \tag{11.7}$$

where G = continuous solids flux, lb/(ft^2 · d) [kg/(m^2/d)], and U = downware sludge velocity due to sludge removal, ft/d (m/d). G can be varied by controlling

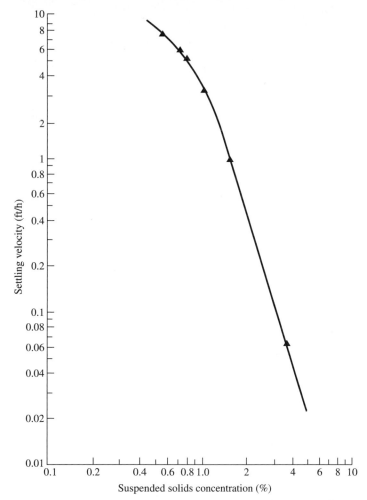

Note: ft/h = 0.305 m/h

FIGURE 11.6
Sludge settling characteristics.

U since this is determined by the underflow pumping rate. Assuming total solids removal from the bottom:

$$U = \frac{Q_u}{A} = \frac{C_u Q_u}{C_u A} = \frac{M}{C_u A} = \frac{G_L}{C_u} \tag{11.8}$$

where Q_u = underflow, ft³/d (m³/d), and C_u = underflow concentration, lb/ft³ (kg/m²). It is important to note from Eq. (11.8) that by increasing U, the withdrawal rate decreases the underflow concentration C_u. A batch flux curve as shown in Fig. 11.7 can be employed to determine the limiting flux G_L for a given underflow concentration C_u. This is because the slope of any line connecting G_L on the y axis with

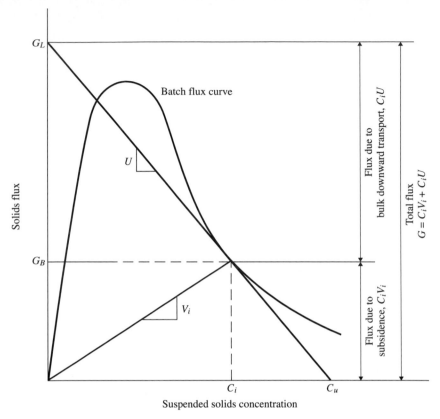

FIGURE 11.7
Batch flux plot illustrating how to determine limiting flux for a continuous thickener.

C_u on the x axis on the batch flux curve, as shown in Fig. 11.7, is obtained by plotting G_B [computed from Eq. (11.6)] against its corresponding concentration C_i.

The required thickener area, A, is then computed from Eq. (11.5). It should be noted that the selected underflow concentration C_u must be less than the ultimate concentration attainable, C_∞. C_∞ is determined from thickening studies as

$$C_\infty = \frac{C_o H_o}{H_\infty} \qquad (11.9)$$

where C_o = initial solids concentration
         H_o = initial height
         C_∞ = final or ultimate concentration
         H_∞ = final height

Example 11.3 illustrates a gravity thickener design.

EXAMPLE 11.3. Waste sludge from a chemical coagulation process is to be gravity-thickened from 0.5 to 4 percent. The average sludge volume is 550,000 gal/d (2082 m³/d) with a variation of 450,000 to 700,000 gal/d (1703 to 2650 m³/d). Determine the thickener area required and the underflow solids concentration at minimum flow.

Solution.

The relationship between zone settling velocity and suspended solids concentration is shown in Table 11.2. The batch flux curve is developed by plotting the flux G versus its corresponding concentration. For example, at 2 percent solids,

$$G = 0.02 \times 62.4 \text{ lb/ft}^3 \times 0.50 \text{ ft/h} \times 24 \text{ h/d}$$

$$= 15.0 \text{ lb/(ft}^2 \cdot \text{d)} \qquad [73.3 \text{ kg/(m}^2 \cdot \text{d)}]$$

The batch flux curve is shown in Fig. 11.8. For the desired underflow concentration of 4 percent, the limiting flux is found from the batch flux curve by constructing a tangent which extends from 4 percent solids to the limiting flux of $G_L = 26 \text{ lb/(ft}^2 \cdot \text{d)}$ [127 kg/(m^2 · d)]. The required thickener area is calculated:

$$A = \frac{C_o Q_o}{G}$$

$$= \frac{(0.7 \text{ million gal/d})(5000 \text{ mg/l})(8.34)(\text{lb/million gal})(\text{mg/l})}{26 \text{ lb/(ft}^2 \cdot \text{d)}}$$

$$= 1123 \text{ ft}^2 \qquad (104 \text{ m}^2)$$

When the sludge flow to the thickener is 0.45 million gal/d, the solids flux will be

$$G = \frac{(0.45 \text{ million gal/d})(5000 \text{ mg/l})(8.34)(\text{lb/million gal})/(\text{mg/l})}{1123 \text{ ft}^2}$$

$$= 16.7 \text{ lb/(ft}^2 \cdot \text{d)} \qquad [81.6 \text{ kg (m}^2 \cdot \text{d)}]$$

From the batch flux curve (Fig. 11.8), the underflow concentration at this loading will be 4.9 percent.

TABLE 11.2
Batch setting data

Solids concentration, %	Settling velocity, ft/h
0.50	7.5
0.75	5.5
1.00	4.2
1.25	3.1
1.50	1.5
2.00	0.50
4.00	0.075
6.00	0.030

Note: ft/h = 0.3048 m/h.

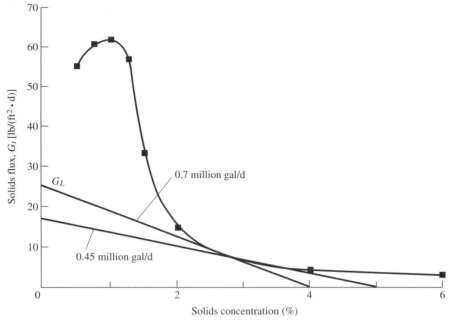

FIGURE 11.8

Note: $lb/(ft^2 \cdot d) = 4.89 \ kg/(m^2 \cdot d)$

11.4
FLOTATION THICKENING

Thickening through dissolved air flotation is particularly applicable to gelatinous sludges such as activated sludge. In flotation thickening, small air bubbles released from solution attach themselves to and become enmeshed in the sludge flocs. The air-solid mixture rises to the surface of the basin, where it concentrates and is removed. The primary variables are recycle ratio, feed solids concentration, air/solids (A/S) ratio, and solids and hydraulic loading rates. Pressures between 50 and 70 lb/in^2 (345 to 483 kPa or 3.4 to 4.8 atm) are commonly employed. Recycle ratio is related to the air/solids ratio and the feed solids concentration. The float solids are related to the A/S ratio, as shown in Fig. 11.9. Typical design criteria are:

Solids loading rate	50 to 100 kg/m^2 d
Pressurized effluent recycle	15 to 120 percent
Pressure requirement	3 to 5 bars
A/S ratio	0.005 to 0.060 kg/kg

Experience has shown that in some cases dilution of the feed sludge to a lower concentration increases the concentration of the floated solids. A flotation thickener is shown in Fig. 11.10. Performance data for the thickening of excess activated sludge is shown in Table 11.3. The use of polyelectrolytes will usually increase the solids capture.

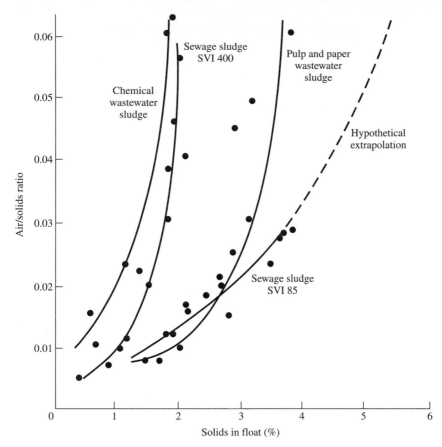

FIGURE 11.9
Influence of air/solids ratio on float solids content.

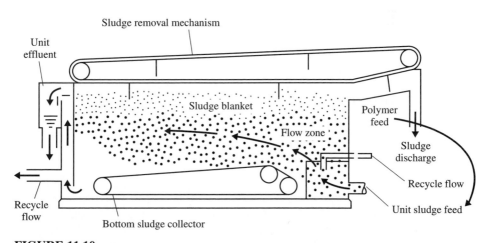

FIGURE 11.10
Flow diagram of a flotation unit (*Courtesy of Komline-Sanderson Engineering Company*).

TABLE 11.3
Thickening and dewatering of wastewater sludges

Equipment	Type of sludge	Loading	Cake solids, %	Chemicals, lb/ton polymer	Reference
Thickening					
Gravity	WAS	5–6 lb/(ft^2 · d)	2.5–3.0	None	9
Gravity	Pulp and paper 53% P, 47% WAS	25	4	None	9
	67% P, 33% WAS	25	6	None	—
	100% P	25	9	None	—
DAF	WAS	2.9–4.5 lb/(ft^2 · h)	4–5.7	Low dosage	9
Solid bowl centrifuge	WAS	75–100 gal/min	5–7	None	9
Basket centrifuge	Citrus, WAS	25–40 gal/min	9–10	10–20	10
Gravity belt	WAS	315 gal/min	5.5	39	—
Solid bowl centrifuge	Paper mill, WAS	100 gal/min	11	10	11
Solid bowl centrifuge	Chemical, WAS	—	7–9	5–10	12
Dewatering					
Basket centrifuge	Citrus, WAS	25–40 gal/min	9–10	10–20	10
Basket centrifuge	Paper mill, WAS	60 gal/min	11	5	11
Belt press	Citrus, WAS	40 gal/(min · m)	18	10–20	10
Belt press	Paper mill, WAS	70 gal/(min · m)	16	6.5	11
Belt press	Chemical, WAS	—	13–15	10–20	12
Belt press	Organic chemical, WAS	190 gal/min	15	25	13
Belt press	Deinking primary	500 l/m	37	4	14
Belt press	Bleached and unbleached kraft; 67 percent P, 33 percent WAS	240 l/m	27	12	14
Belt press	Kraft linerboard	75 l/m	19	25	14

Note:
lb/(ft^2 · d) = 4.88 kg/(m^2 · d)
gal/min = 3.78 × 10^{-3} m^3/min
WAS = waste activated sludge
P = primary sludge

The quality of the sludge has a significant effect on its ability to thicken by flotation. For example, a filamentous, bulking activated sludge may not achieve 2 percent solids, compared to 4 to 5 percent with a well-flocculated sludge. This in turn will affect all other solids handling operations.

11.5
ROTARY DRUM SCREEN

A rotary drum screen consists of a stainless steel or nonferrous wire mesh screen cloth. Screen openings typically vary from 6 to 20 mm. The drum revolves at about 4 rev/min around a horizontal axis and operates slightly less than half submerged. The wastewater flows in one end of the drum and outward through the screen cloth. Solids are raised above the liquid level by rotation of the screen and are backflushed into receiving troughs by high-pressure jets. With the finer mesh, effluent may be used for spray water. A typical rotary drum screen is shown in Fig. 11.11.

Typical design criteria for waste activated sludge are

Loading rate	33 l/min/m^2
Polymer consumption	5 to 9 kg/ton of solids
Capture rate	95 to 99 percent

Low SVI sludge will thicken to 6 to 10 percent with a filtrate clarity of 100 to 500 mg/l.

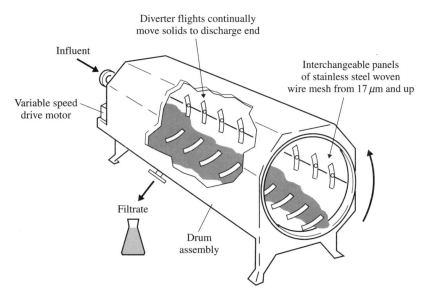

FIGURE 11.11
Rotary drum screen thickener.

11.6
GRAVITY BELT THICKENER

The gravity belt thickener reduces the volume of water in the sludges by using gravity forces to remove water that has been freed by polymer or chemical conditioning. Polymer or chemical is injected into the sludge feed line and mixed by an adjustable in-line mixer. Conditioned sludge enters a stainless steel tank and is then distributed evenly across the width of the belt without shearing the flocculated sludge particles. A series of free-floating plows furrow and roll the sludge to expose ponded water to open areas on the belt while a low-wear grid system supports the belt and shears capillary water from its underside. An adjustable-angle discharge dam rolls the sludge for maximum thickened solids concentration. Thickened sludge is continuously removed by an adjustable spring-tensioned blade. The drainage belt then passes through a high-pressure/low-volume shower assembly to remove any particles trapped in the belt. Typically a hydraulic loading rate up to 35 m³/h per meter of width is employed. Thickened sludge solids concentration of 4 to 7 percent is achievable. A typical gravity belt thickener is shown in Fig. 11.12.

11.7
DISK CENTRIFUGE

In the disk-nozzle separator, the feed enters at the top and is distributed between a multitude of channels, or spaces between the stacked conical disks. Solid particles settle through the layer of liquid flowing in these channels to the underside of the disk, and then slide down to a sludge compaction zone. The thickened sludge is flushed out of the bowl with a portion of the wastewater, thus limiting the solids concentration to 10 to 20 times the feed rate. The disk-nozzle separator finds its major application in the thickening of activated and similar sludges. They are very efficient in thickening waste activated sludge at high feed rates without the addition of polymers. In an industrial waste treatment plant in Germany, excess activated sludge has been thickened from 1 percent to 8 to 10 percent solids.

11.8
BASKET CENTRIFUGE

In the basket-type centrifuge, feed is introduced in the bottom of the basket. At equilibrium, solids settle out of the annular moving liquid layer to the layer that builds up on the bowl wall, while the contrate overflows the lip at the top. When solids have filled the basket, feed is stopped, the basket speed is reduced, and a knife moves into the cake, discharging it from the bottom of the casing. Cycles are automated and cake unloading requires less than 10 percent of the cycle time. Chemical addition is generally not required for high solids recovery. However, the unit operates at low centrifugal forces and has a discontinuous cake discharge and a fairly low solids handling capacity.

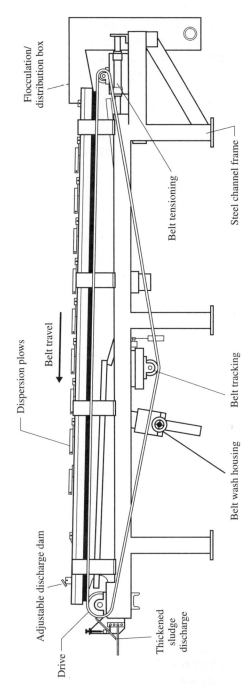

Flocculation/
distribution box

Belt tensioning

Steel channel frame

Dispersion plows

Belt travel

Belt tracking

Belt wash housing

Adjustable discharge dam

Drive

Thickened
sludge
discharge

FIGURE 11.12
Gravity belt thickener.

Activated sludge can be thickened to 4.5 to 8.0 percent total solids by using up to 2.5 g of cationic polymer per kilogram waste. A low-SVI sludge will yield the higher cake density.

A solid bowl centrifuge is also used for sludge thickening. It can usually obtain over 90 percent capture without the use of polymers.

11.9
SPECIFIC RESISTANCE

Sludge dewaterability can be defined in terms of its specific resistance. While this test has usually been applied to drum filters, it does give a measure of dewaterability and coagulant requirements.

The rate of filtration of sludges has been formulated according to Poiseuille's and Darcy's laws by Carman and by Coackley and Jones:[4]

$$\frac{dV}{dt} = \frac{PA^2}{\mu(rcV + R_m A)} \tag{11.10}$$

where V = volume of filtrate, ml
   t = cycle time (approximate form time in continuous drum filters), s
   P = vacuum, in Hg
   A = filtration area, cm^2
   μ = filtrate viscosity, s^2/g
   r = specific resistance, gr · s^2/g^2
   c = weight of solids/unit volume of filtrate, g/ml

$$c = \frac{1}{C_i/(100 - C_i) - C_f/(100 - C_f)} \quad \text{units} \tag{11.11}$$

where C_i = initial moisture content, %, and C_f = final moisture content, %.

R_m, the initial resistance of the filter media, can usually be neglected, as it is small compared to the resistance developed by the filter cake. The specific resistance r is a measure of the filterability of the sludge and is numerically equal to the pressure difference required to produce a unit rate of filtrate flow of unit viscosity through a unit weight of cake.

Integration and rearrangement of Eq. (11.10) yields

$$\frac{t}{V} = \left(\frac{\mu rc}{2PA^2}\right)V + \frac{\mu R_m}{PA} \tag{11.12}$$

From Eq. (11.12), a linear relationship will result from a plot of t/V versus V. The specific resistance can be computed from the slope of this plot:

$$r = \frac{2bPA^2}{\mu c} \tag{11.12a}$$

where b = slope of the t/V versus V plot.

Although specific resistance has limited value for the design of a sludge dewatering device, it provides a valuable tool for the evaluation of the relative filterability of sludges. Typical values are given in Table 11.4.

TABLE 11.4
Specific resistance of sludges [†]

Description		Specific resistance, $(gr \cdot s^2/g^2) \times 10^{-7}$	Coefficient of compressibility
Domestic activated sludge		2800	
Activated (digested)		800	
Primary (raw)		1310–2110	
Primary (digested)		380–2170	
Primary (digested)		1350	25.00
Primary (digested)			
Detention time	Stage		
7.5 days	1	1590	
10.0 days	1	1540	
15.0 days	1	1230	
20.0 days	1	530	
30.0 days	1	760	
15.0 days	2	400	
20.0 days	2	400	
30.0 days	2	480	
Activated sludge + 13.5% $FeCl_3$		45	
Activated sludge + 10.0% $FeCl_3$		75	
Activated sludge + 125% (by weight) newsprint		15	
Activated digested sludge + 6% $FeCl_3$ + 10% CaO		5	
Activated digested sludge + 125% newsprint + 5% CaO		4.5	
Vegetable-processing sludge		46	7.00
Vegetable tanning		15	20.00
Lime neutralization acid mine drainage		30	10.50
Alum sludge (water works)		530	14.50
Neutralization of sulfuric acid with lime slurry		1–2	
Neutralization of sulfuric acid with dolomitic lime slurry		3	0.77
Aluminum processing		3	0.44
Paper industry		6	
Coal (froth flotation)		80	1.60
Distillery		200	1.30
Mixed chrome and vegetable tannery		300	
Chemical wastes (biological treatment)		300	
Petroleum industry (from gravity separators)			
Refinery A		10–100	0.50
Refinery B		100	0.70

[†] All values were recorded at 500 gr/cm² pressure.

Most wastewater sludges form compressible cakes in which the filtration rate (and the specific resistance) is a function of the pressure difference across the cake:

$$r = r_o P^s \tag{11.13}$$

where s = coefficient of compressibility. The greater the value of s, the more compressible is the sludge. When $s = 0$, the specific resistance is independent of pressure and the sludge is incompressible.

Some generalizations on filtration characteristics can be made. Filterability is influenced by particle size, shape, and density, and by the electrical charge on the particle. Smaller particles exert a greater chemical demand than larger particles. The larger the particle size, the higher is the filter rate and the lower the cake moisture. Municipal and industrial sludges filter very poorly, and coagulants must be added; lime and ferric salts have been the most common coagulating agents used in the past. Polyelectrolytes have proved effective coagulants in many applications. Frequently, the dual use of anionic and cationic polymers is the most economic and effective procedure. The cationic polymer affects charge neutralization and the anionic affects polymer particle bridging and agglomeration of the particles. Note that excessive coagulant dosages result in a charge reversal and in an increase in specific resistance.

Laboratory Procedures

Filtration characteristics of sludges can be obtained in the laboratory by the Büchner funnel test. The Büchner funnel test can be used to determine specific resistance. It is usually possible, however, to determine the sludge compressibility s and the optimum coagulant dosage from a series of Büchner funnel tests.

The procedure for the Büchner funnel test is as follows (see Fig. 11.13):

1. Prepare the Büchner funnel by placing a wire mesh or screen under the filter paper to ensure drainage.
2. Moisten the filter paper with water and adjust the vacuum to obtain a seal.
3. Condition the sludge if necessary; mix it and permit it to stand 30 s to 1 min; use 200 ml samples. The relationship between specific resistance and coagulant dosage is shown in Fig. 11.14.
4. Transfer it to the Büchner funnel, allow sufficient time for a cake to form (5 to 10 s), and apply the vacuum.
5. Record the milliliters of filtrate after selected time intervals (usually 5 to 10 s).
6. Continue filtration until the vacuum breaks.
7. Determine the initial and final solids in the feed sludge and cake.
8. Record the data obtained and calculate the specific resistance in accordance with Equation (11.12a).

Calculations for specific resistance are shown in Example 11.4.

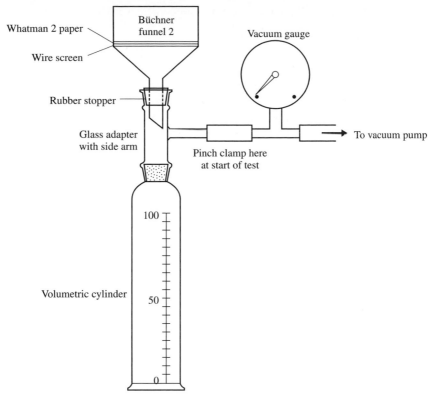

FIGURE 11.13
Büchner funnel assembly for determination of sludge filterability.

EXAMPLE 11.4. Calculate the specific resistance given the values in the table and the data in Fig. 11.15,

where A = 176.5 cm²
T = 84°F (29°C)
P = 20 in Hg = 704 g/cm²
C_i = 97.6%
C_j = 77.4%
μ = 0.01 s²/g (viscosity at 29°C)
b = 0.0007

Time, s	Volume, ml	t/V
5	78	0.064
10	114	0.088
15	142	0.106
25	178	0.140
35	212	0.165
45	224	0.201
50	228	0.220

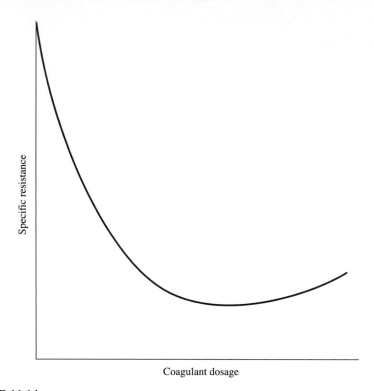

FIGURE 11.14
Relation between specific resistance and coagulant dosage.

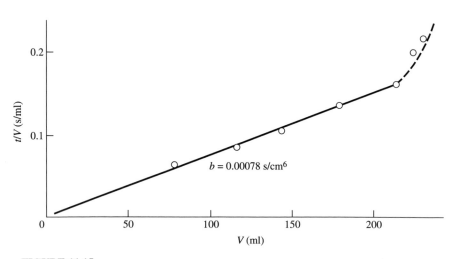

$b = 0.00078$ s/cm^6

FIGURE 11.15

Solution.

$$r = \frac{2PbA^2}{\mu c}$$

$$c = \frac{1}{C_i/(100 - C_i) - C_f/(100 - C_f)}$$

$$= \frac{1}{97.6/2.4 - 77.4/22.6} = 0.0269 \text{ g/ml}$$

$$r = \frac{(2)(0.00078)(704)(176.5)^2}{(0.0269)}$$

$$= 12.7 \times 10^7 \text{ gr} \cdot \text{s/g}^2$$

Capillary Suction Time Test

An evaluation technique based on the capillary suction time (CST) has been found to be a rapid, easy, inexpensive, and reproducible method of characterizing the dewaterability of a given sludge. The assembly developed by the Water Research Center (WRC) at Stevanage Laboratory, England, is shown in Fig. 11.16. Filtrate is withdrawn from the sludge sample by capillary adsorbent filter paper. Filterability is measured by observing the time for an area of paper to become wetted. The CST can be correlated to specific resistance. The optimum coagulant and its respective dosage can be determined from a CST versus coagulant dosage relationship.

11.10
CENTRIFUGATION

Centrifuge performance is affected by both machine and process variables. The significant machine variables for the solid bowl decanter are bowl speed, pool volume, and conveyor speed. Process variables include the feed rate of solids to the machine, solids characteristics, chemical addition, and temperature.

The solid bowl decanter consists of an imperforated cylindrical-conical bowl with an internal helical conveyor as shown in Fig. 11.17. The feed sludge enters the cylindrical bowl through the conveyor discharge nozzles. Centrifugal forces compact the sludge against the bowl wall, and the internal scroll or conveyor, which rotates slightly slower than the bowl, conveys the compacted sludge along the bowl wall toward the conical section (beach area) and out.

When the feed rate to a centrifuge is increased, the retention time in the unit is decreased and the recovery decreases. Flow rates are usually limited to 0.5 to 2.0 gal/(min · hp) [3.65 to 14.6 m³/(d · kW)] to obtain satisfactory solids recovery. Since the lower recovery results in the removal of only larger particles, a drier cake is produced. Increasing the feed solids concentration reduces the liquid overflow from the machine, resulting in an increased recovery of solids.

Chemical flocculants (polyelectrolytes) are used to increase recovery. The flocculants both increase the structure strength of the solids and flocculate fine particles.

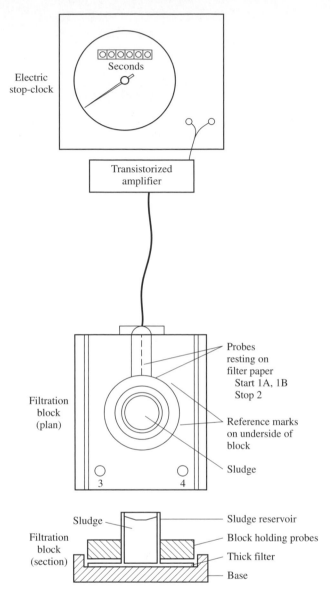

FIGURE 11.16
Capillary suction time apparatus.

Because of the increased removal of the fine particles, chemical addition usually lowers the cake dryness. Centrifuge performance characteristics are shown in Fig. 11.18.

Bernard and Englande[5] correlated centrifuge performance data in accordance with the relationship

$$R = \frac{C_1\,(C_2 + P)^m}{Q^n} \tag{11.14}$$

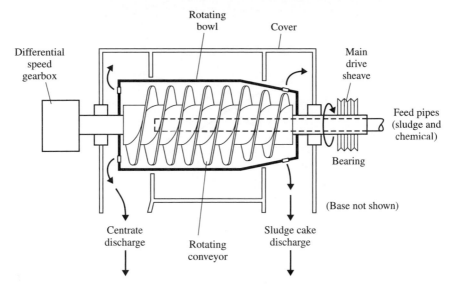

FIGURE 11.17
Continuous countercurrent solid bowl conveyor discharge centrifuge.

where R = recovery, %
              P = polymer dosage, lb/ton dry solids feed (kg/ton)
              Q = feed rate, gal/(min · ft²) [m³/(min · m²)]
      C_1, C_2 = constants
        m, n = exponents

Centrifuge performance in accordance with Eq. (11.14) is shown in Fig. 11.19. Centrifuge design is illustrated in Example 11.5.

> **EXAMPLE 11.5.** Determine the polymer required and size of centrifuge with respect to surface area for dewatering 10,000 lb/d (4536 kg/d) of sludge previously thickened to 4 percent. The centrifuge will operate over 8 h with a 95 percent solids recovery. The pilot plant data yielded the following relationship:
>
> $$R(\%) = \frac{48(0.47 + P)^{0.37}}{Q^{0.52}}$$
>
> where R = solid recovery efficiency, %
> P = polymer dosage, lb/ton of sludge
> Q = hydraulic loading, gal/(min · ft²)
>
> ***Solution.***
>
>   For 4 percent solids concentration, the sludge flow is
>
> $$Q' = 10,000 \times \frac{1}{0.04} \times \frac{1}{8.34}$$
>
> $$= 30,000 \text{ gal/d} \quad (114 \text{ m}^3/\text{d})$$

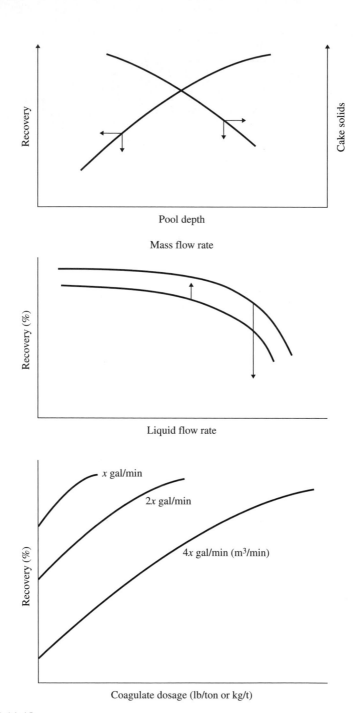

FIGURE 11.18
Centrifuge operating relationships.

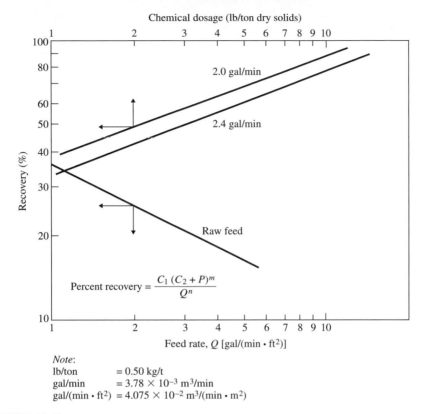

FIGURE 11.19
Solids recovery for digested activated sludge with cationic polymer.

If the centrifuge will operate for 8 h per d, the total sludge flow to the centrifuge is

$$Q' = 30,000 \times \frac{24}{8}$$

$$= 90,000 \text{ gal/d or } 62.5 \text{ gal/min} \quad (341 \text{ m}^3/\text{d or } 0.237 \text{ m}^3/\text{min})$$

At 95 percent solids recovery, the relationship between polymer requirement and hydraulic loading can be calculated by

$$95 = \frac{48(0.47 + P)^{0.37}}{Q^{0.52}} \quad \text{(plot } P \text{ versus } Q \text{ in Fig. 11.20)}$$

The centrifuge size is computed by

$$A = \frac{62.5}{Q} \text{ ft}^2 \quad \text{(plot } A \text{ versus } Q \text{ in Fig. 11.20)}$$

From Fig. 11.20,

$$A = 46 \text{ ft}^2 \quad (4.3 \text{ m}^2)$$

$$P = 46 \text{ lb/d} \quad (21 \text{ kg/d})$$

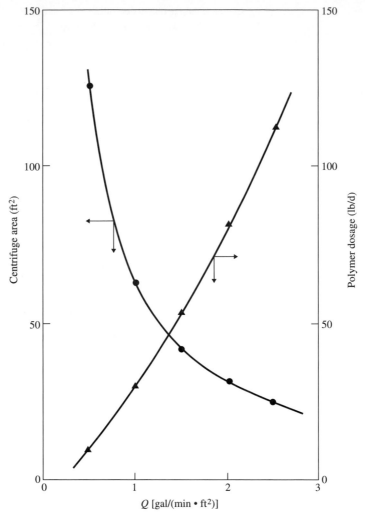

FIGURE 11.20
Required centrifuge area and polymer dosage for various hydraulic loadings.

11.11
VACUUM FILTRATION

Vacuum filtration was one of the most common methods for dewatering waste-water sludges. Vacuum filtration dewaters a slurry under applied vacuum by means of a porous medium, which retains the solids but allows the liquid to pass through. Media used include cloth, steel mesh, or tightly wound coil springs.

In vacuum filter operation, a rotary drum passes through a slurry tank in which solids are retained on the drum surface under applied vacuum. The drum submergence can vary from 12 to 60 percent. As the drum passes through the slurry, a cake is built up and water is removed by filtration through the deposited solids and the filter media. The time the drum remains submerged in the slurry is the form time t_f. As the drum emerges from the slurry tank, the deposited cake is further dried by liquid transfer to air drawn through the cake by the applied vacuum. This period of the drum's cycle is called the dry time t_d. At the end of the cycle, a knife edge scrapes the filter cake from the drum to a conveyor. The filter medium is usually washed with water sprays prior to again being immersed in the slurry tank. A vacuum filter is schematically shown in Fig. 11.21.

The variables that influence the dewatering process are solids concentration, sludge and filtrate viscosity, sludge compressibility, chemical composition, and the nature of the sludge particles (size, shape, water content, etc.).

The filter operating variables are vacuum, drum submergence and speed, sludge conditioning, and the type and porosity of the filter medium.

Equation (11.12) can be modified to express filter loading (neglecting the initial resistance of the filter medium):

$$L_f = 35.7 \left(\frac{P^{1-s}}{\mu R_o} \right)^{1/2} \frac{c^m}{t_f^{\,n}} \tag{11.15}$$

where

$R_o = r_o \times 10^{-7}$, gr · s²/g²
P = vacuum, lb/in²
c = solids deposited per unit volume filtrate, g/ml
μ = filtrate viscosity, centipoise
t_f = form cycle time, min
L_f = filter loading, lb/(ft² · h)
m, n = constants related to the sludge characteristics
s = coefficient of compressibility

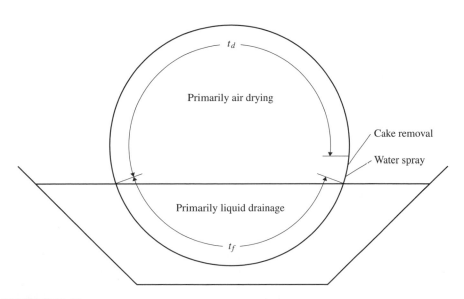

FIGURE 11.21
Mechanism of vacuum filtration.

For routine calculations C_i is used in Eq. (11.15) as c.

Equation (11.15) is in terms of form time and is conventionally converted to cycle time by

$$L_c = L_f \frac{\% \text{ submergence}}{100} \times 0.8$$

The factor 0.8 compensates for the area of the filter drum where the cake is removed and the medium washed. The total cycle time on a filter may vary from 1 to 6 min. Submergence of the drum may vary from 10 to 60 percent, resulting in a maximum spread of form time of 0.1 to 3.5 min. This also yields a maximum spread of dry time of 2.5 to 4.5 min. In general, the filter yield from highly compressible cakes is unaffected by increases in form vacuum varying from 12 to 17 in (30 to 43 cm) of mercury.

Diatomaceous earth is composed of fossil skeletons of microscopic diatoms and is used as a filter aid in rotary vacuum precoat filtration and pressure filtration. Typically 2 to 6 in (5 to 15 cm) of diatomaceous earth or perlite is applied prior to sludge application. To maintain a porous cake, filter aid may also be added continuously to the sludge slurry prior to vacuum filtration. Collected solids plus a few thousandths of an inch of precoat are scraped off by a knife as the drum advances at rates of 0.5 to 5 minutes per revolution. Depending on the waste stream and filter aid, flow rates can vary from 2 to 50 gal/(h · ft²) [81 to 2035 liters/(h · m²)]. Operating costs with diatomaceous earth or perlite precoat are high because of the need to replenish expended filter aid. Precoating is especially required for dewatering of gelatinous-type solids such as alum sludge, clarification of oily wastes, and for sludges characterized by a high percentage of fines. Precoating is also generally employed during pressure filtration. As with the vacuum filter operation, it both protects the filter medium against frequent blinding and provides a thin, nonadherent parting plane between the cake and the medium which minimizes cake discharge difficulties. The optimal type and quantity of precoat must be established by laboratory evaluation. Vacuum filtration design is shown in Example 11.6.

EXAMPLE 11.6. A combined primary and activated sludge from a pulp-and-paper mill is to be dewatered. The sludge flow is 100 gal/min (0.38 m³/min) of 6 percent solids. Design a vacuum filter to operate 16 h/d, 7 d/week, using 15 in Hg (381 mm Hg) vacuum (7.35 lb/in²) and 30 percent submergence. Laboratory and pilot studies have shown that:

1. The coefficient $m = 0.25$.
2. The coefficient $n = 0.65$.
3. An optimum cake solids of 28 percent is obtained at a 3-min dry time.
4. The coefficient of compressibility is 0.85.
5. The specific resistance r_o is 1.3×10^7 gr · s²/g².

Solution.

$$L_f = 35.7 \left(\frac{P^{1-s}}{\mu R_o} \right)^{1/2} \frac{c^m}{t^n}$$

The cycle time is $3/0.7 = 4.3$ min.

$$t_f = 4.3 - 3 = 1.3 \text{ min}$$

$$c = \frac{1}{94/6 - 72/28} = 0.077 \text{ g/cm}^3$$

$$L_f = 35.7 \left[\frac{7.35^{0.15}}{(1)(1.3)} \right]^{1/2} \frac{0.077^{0.25}}{1.3^{0.65}}$$

$$= 16.3 \text{ lb/(ft}^2 \cdot \text{h)} \qquad [79.7 \text{ kg/(m}^2 \cdot \text{h)}]$$

From Eq. (11.17),

$$L_c = 16.3 \left(\frac{30}{100} \right) \times 0.8$$

$$= 3.91 \text{ lb/(ft}^2 \cdot \text{h)} \qquad [19.1 \text{ kg/(m}^2 \cdot \text{h)}]$$

$$\text{Sludge to be filtered} = 100 \text{ gal/min} \times 8.34 \text{ lb/gal} \times 0.06 \times 60 \text{ min/h} \times 24/16$$

$$= 4500 \text{ lb/h} \qquad (2041 \text{ kg/h})$$

$$\text{Filter area required} = \frac{4500}{3.91} = 1151 \text{ ft}^2 \qquad (107 \text{ m}^2)$$

11.12
PRESSURE FILTRATION

Pressure filtration is applicable to almost all water and wastewater sludges. The sludge is pumped between plates that are covered with a filter cloth. The liquid seeps through the cloth, leaving the solids behind between the plates. The medium may or may not be precoated. When the spaces between the plates are filled, the plates are separated and the solids removed. The pressure exerted on the cake during formation is limited to the pumping force and filter closing system design. Filters are designed at pressures ranging from 50 to 225 lb/in^2 (345 to 1550 kPa). As the final filtration pressure increases, a corresponding increase in dry cake solids is obtained. Most municipal sludges can be dewatered to produce 40 to 50 percent cake solids with 225-lb/in^2 (690-kPa) filters. Filtrate quality will vary from 10 mg/l suspended solids with precoat to 50 to 500 mg/l with unprecoated cloth, depending on the medium, type of solids, and type of conditioning. Conditioning chemicals are the same as used in vacuum filtration (lime, ferric chloride, or polymers). Materials such as ash have also been used. It should be noted that where 40 to 50 percent cakes are reached industrial applications usually employ a filter aid to increase the shear resistance as pressure is applied. Coal fines are used in Germany to increase the fuel value. In the U.S., 20 to 30 percent lime and ferric chloride is frequently used, which reduces the Btu content of the cake. If the cake is to be incinerated, filter aids with high Btu content should be considered. Poorly filterable sludges yielding highly compressible cakes can be conditioned with skeleton builders to produce a more porous and incompressible cake. For oily sludges both hydrated lime and fly ash have been successfully used.[19] A membrane filter press operation is shown in Fig. 11.22. A pressure filter is shown in Fig. 11.23. Example 11.7 illustrates a pressure filter design.

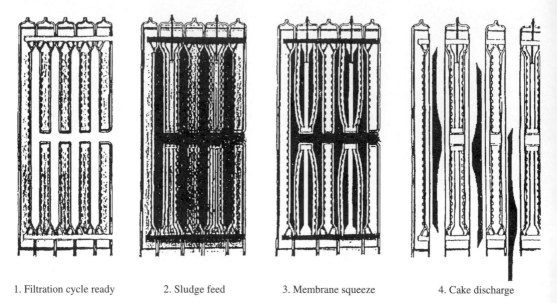

1. Filtration cycle ready   2. Sludge feed   3. Membrane squeeze   4. Cake discharge

FIGURE 11.22
Membrane filter press.

FIGURE 11.23
Plate and frame pressure filter.

EXAMPLE 11.7. Size a plate and frame pressure filter to dewater sludge using the following data:

> Average loading = 13,300 lb/d (6030 kg/d) dry TSS
> Maximum loading = 25,000 lb/d (11,340 kg/d) dry TSS
> Average sludge concentration = 3.0%
> Minimum sludge concentration = 2.0%

A series of pilot tests were conducted and the following conditions were found:

> Total cycle time = 3.5 h[†]
> Average cake solids = 40%
> Minimum cake solids = 30%
> Cake density = 70 lb/ft^3 (1120 kg/m^3)
> Conditioning = 100 lb FeCL$_3$/ton(50 kg/t) dry solids
> +200 lb lime/ton(100 kg/t) dry solids

Size dewatering equipment to treat the average sludge load in 1 shift/d and the maximum load in 2 shifts/d in a 7-day week.

Solution.

Calculate the volume of sludge to be treated.

$$\text{Average volume} = \frac{13,300 \text{ lb dry SS/d}}{0.03 \text{ lb dry SS/lb sludge} \times 8.34 \text{ lb/gal}}$$

$$= 53,000 \text{ gal/d} \quad (200 \text{ m}^3/\text{d})$$

$$\text{Maximum volume} = \frac{25,000}{0.02 \times 8.34} = 150,000 \text{ gal/d} \quad (570 \text{ m}^3/\text{d})$$

Calculate the dewatered volume.

$$\text{Average} = \frac{13,300 \text{ lb/d} + (300 \text{ lb/ton} \times 5 \times 10^{-4} \text{ ton/lb} \times 13,300 \text{ lb/d})}{0.4 \text{ lb TSS/lb cake} \times 70 \text{ lb/ft}^3}$$

$$= 545 \text{ ft}^3/\text{d} \quad (15 \text{ m}^3/\text{d})$$

$$\text{Maximum} = \frac{25,000 \text{ lb/d} + (300 \times 5 \times 10^{-4} \times 25,000)}{0.30 \times 70} = 1369 \text{ ft}^3/\text{d} \quad (39 \text{ m}^3/\text{d})$$

Calculate the number of filter cycles required per day knowing the cycle time of 3.5 h and designing for a single pressure filter.

$$\text{Average number of cycles per day} = \frac{1 \text{ shift/d} \times 8 \text{ h/shift}}{3.5 \text{ h/cycle}} \cong 2 \text{ cycles/d}$$

$$\text{Maximum number of cycles per day} = \frac{2 \times 8}{3.5} \simeq 4 \text{ cycles/d}$$

[†] Allows sufficient time for cloth washing and cake removal.

Calculate the volume of dewatered sludge or pressure filter volume required per cycle.

$$\text{Average filter volume per cycle} = \frac{545}{2} = 273 \text{ ft}^3 \ (8 \text{ m}^3/\text{d})$$

$$\text{Maximum filter volume per cycle} = \frac{1369}{4} = 342 \text{ ft}^3 \ (10 \text{ m}^3/\text{d})$$

Select the size for the filter press. The volume per chamber of the press is 3.0 ft³ (0.085 m³). The maximum filter volume per cycle is 342 ft³. Therefore, we need a press with a minimum of 114 (342/3) chambers. A standard filter press should then be selected which meets this criteria.

Hydraulic presses have also been applied to further dewater filter-cake paper mill sludges for incineration. Board mill sludge has been dewatered to 40 percent solids from 30 percent solids at a pressure of 300 lb/in² (2070 kPa) and a pressing time of 5 min.

11.13
BELT FILTER PRESS

In a belt filter press, shown in Fig. 11.24, chemically conditioned sludge is fed through two filter belts and is squeezed by force to drive water through these belts. Variations of this device are successfully used to dewater municipal and industrial sludges.

A belt filter press (shown schematically in Fig. 11.25) employs not only the concept of cake shear with simultaneous application of pressure but also low-pressure filtration and thickening by gravity drainage. An endless filter belt runs over a drive and guide roller at each end like a conveyor belt. The upper side of the filter belt is supported by several rollers. Above the filter bed a press belt runs in the same direction and at the same speed. The drive roller for this belt is coupled with the drive roller of the filter belt.

The press belt can be pressed on the filter belt by means of a pressure roller system whose rollers can be individually adjusted horizontally and vertically. The sludge to be dewatered is fed on the upper face of the filter belt and is continuously dewatered between the filter and press belts. Note how the supporting rollers of the filter belt and pressure rollers of the pressure belt are adjusted in such a way that the belts and the sludge between them describe an S-shaped curve. This configuration induces parallel displacement of the belts relative to each other due to the difference in radius, producing shear in the cake. After dewatering in the shear zone, the sludge is removed by a scraper. Available data indicate that over a range of 0.5 to 12 percent dissolved solids, a filter is relatively insensitive to concentration, but is very sensitive to rate of flux per unit area. On the average belt, washwater flow approximately equals the sludge application rate. A belt filter design is illustrated in Example 11.8.

FIGURE 11.24
Belt filter press.

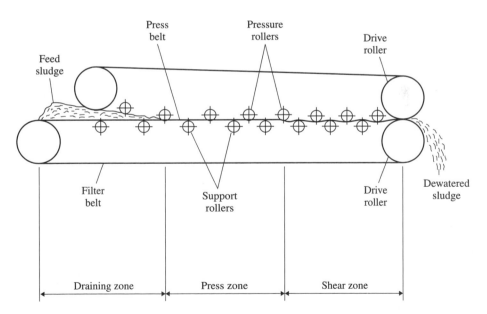

FIGURE 11.25
Schematic of a belt filter press.

EXAMPLE 11.8. Design a belt filter to dewater 86,600 gal/d (330 m³/d) of 2 percent thickened sludge. The sludge is produced from a pulp-and-paper wastewater treatment facility and comprises approximately 23 percent waste activated and 77 percent primary sludge. Pilot scale tests were conducted on a 0.5-m belt width pilot press with the values given in Table 11.5.

The following design criteria have been selected to fall in the midrange the operating specification of the full-scale belt press:

Cake total solids	30%
Solids capture	95%
Belt speed	10 ft/min (3 m/min)
Throughput	200 lb/(h · 0.5 m) (181 kg/h/m)
Polymer usage	5 lb/ton dry solids (2.5 kg/t)

Solution.

Design for two 8-h shifts with a total of 14 h of belt press operation 5 d/week as follows:

$$0.0866 \text{ million gal/d} \times 20{,}000 \text{ mg/l} \times 8.34 = 14{,}445 \text{ lb/d} \quad (6550 \text{ kg/d})$$

$$\frac{14{,}445 \text{ lb/d} \times 7 \text{ d}}{14 \text{ h/d} \times 5 \text{ d}} = 1445 \text{ lb/h} \quad (660 \text{ kg/d})$$

$$\text{Required belt width} = \frac{1445 \text{ lb/h}}{200 \text{ lb/(h} \cdot 0.5 \text{ m)}} = \frac{1445 \text{ lb/h}}{400 \text{ lb/(h} \cdot \text{m)}} = 3.6 \text{ m}$$

TABLE 11.5

	Run number				
	1	**2**	**3**	**4**	**5**
Inlet TS, %	1.76	1.76	2.34	2.34	2.34
Sludge flow, gal/min	15	28	15	20.1	28
Sludge throughput, lb/h	132	247	176	235	328
Cake TS, %	30.2	28.5	31.5	30.9	35.1
Solids capture, %	95.7	94.4	95.2	94.7	94.3
Polymer dosage, lb/ton dry solids	8.3	11.3	5.1	4.6	4.5
Belt speed, ft/min	5	10	5	10	20
Upper belt pressure		5 bar	(72.5 lb/in²)		
Lower belt pressure		5 bar	(72.5 lb/in²)		
Belt tensions		45 bar	(653 lb/in²)		

Note:
gal/min = 3.785 × 10⁻³ m³/min
lb/h = 0.4536 kg/h
lb/ton = 0.5 kg/t
ft/min = 0.3048 m/min

11.14
SCREW PRESS

A screw press (Fig. 11.26) may be employed to dewater thickened sludge or for further dewatering of conventionally dewatered sludge. The screw press employs tapered shafts to create a gradual forced volume reduction of the material to be dewatered. The moisture is removed through a perforated screen. The press is equipped with a shaft rotary steam joint allowing for the application of steam pressures up to 100 lb/in^2. The press used for further dewatering is designed to handle pre-dewatered cake with solids content in excess of 15 to 20 percent total solids.

11.15
SAND BED DRYING

For small industrial waste treatment plants, sludge can be dewatered on open or covered sand beds. Drying of the sludge occurs by percolation and evaporation. The proportion of the water removed by percolation may vary from 20 to 55 percent, depending on the initial solids content of the sludge and on the characteristics of the solids. The design and use of drying beds are affected by climatic conditions (rainfall and evaporation). Sludge drying beds usually consist of 4 to 9 in (10 to 23 cm) of sand over 8 to 18 in (20 to 46 cm) of graded gravel or stone. The sand has an effective size of 0.3 to 1.2 mm and a uniformity coefficient less than 5.0. Gravel is graded from 1/8 to 1 in (0.32 to 2.54 cm). The beds are provided with underdrains spaced from 9 to 20 ft apart (2.7 to 6.1 m). The underdrain piping may be vitrified clay laid with open joints having a minimum diameter of 4 in (10 cm) and a minimum slope of about 1 percent. The filtrate is returned to the treatment plant.

Wet sludge is usually applied to the drying beds at depths of 8 to 12 in (20 to 30 cm). Removal of the dried sludge in a "liftable state" varies with both individual judgment and final disposal means, but usually involves sludge of 30 to 50 percent solid.

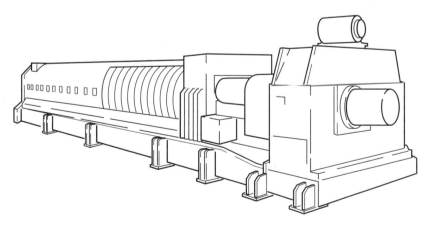

FIGURE 11.26
Screw press (*Courtesy of ANDRITZ*).

In many cases, the bed turnover can be substantially increased by the use of chemicals. Alum treatment can reduce the sludge drying time by 50 percent. The use of polymers can increase the rate of bed dewatering and also increase the depth of application. Bed yield has been reported to increase linearly with polymer dosage.

A rational method has been developed by Swanwick[6] based on the observed dewatering characteristics of a variety of sewage and industrial waste sludges. In this procedure, sludge after drainage (usually 18 to 24 h) is permitted to air dry to the desired consistency. The moisture difference (initial − final) is that which must be evaporated. Depending on the cumulative rainfall and evaporation for the geographical area in question, the time required for various times of the year for evaporation of this moisture is computed. The required bed area may then be determined. Bed yield as related to specific resistance is shown in Fig. 11.27.

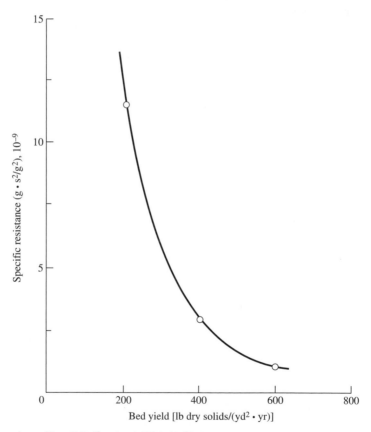

Note: lb/(yd^2 · yr) = 0.543 kg/(m^2/yr)

FIGURE 11.27
Relationship between the specific resistance of the sludge and drying bed yield as liftable sludge (*after Swanwick*).[6]

11.16
FACTORS AFFECTING DEWATERING PERFORMANCE

The sludge mixture (i.e., primary versus activated sludge) will affect both the chemical conditioning requirements and the final cake solids. Data from a pulp-and-paper mill are shown in Fig. 11.28.[7] Barber and Bullard[8] showed that a well-flocculated activated sludge produced a 15 to 16 percent solids cake on a belt filter press compared to 10 percent with an extracellular polysaccharide bulking sludge. In addition, the flocculant sludge used 57 percent less dewatering polymers. The effect of waste activated sludge and dewatering is shown in Fig. 11.29. Thickening and dewatering of wastewater sludges is shown in Table 11.3.

11.17
LAND DISPOSAL OF SLUDGES

Land disposal of wet sludges can be accomplished by lagooning or by application of liquid sludge to land by truck or spray system. Liquid sludge may also be carried by pipeline to a remote agricultural or lagoon site.

Lagooning is commonly employed for the disposal of inorganic industrial waste sludges. Organic sludges usually receive aerobic and anaerobic digestion prior to lagooning to eliminate odors and insects. Lagoons may be operated as substitutes for

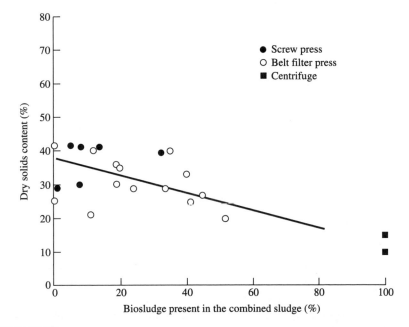

FIGURE 11.28
The results from dewatering of mixed primary and waste activated sludge from a pulp and paper mill with different methods.

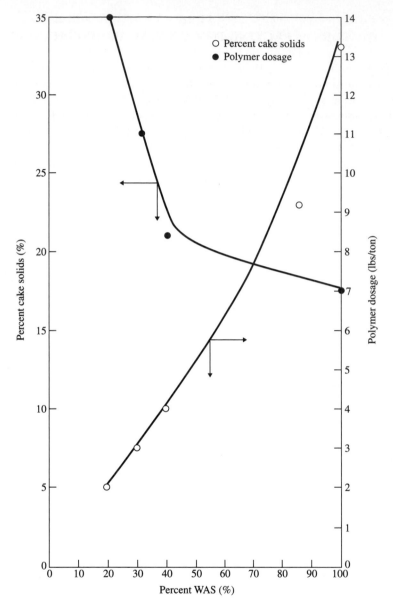

FIGURE 11.29
Screw press performance for waste sludge from a virgin pulp and recycle mill.

drying beds, with the sludge periodically removed and the lagoon refilled. In a permanent lagoon, supernatant liquor is removed. When it is filled with solids, the lagoon is abandoned and a new site selected.

In general, lagoons should be considered where large land areas are available and the sludge will not present a nuisance to the surrounding environment.

Benthal stabilization occurs in a lagoon as a result of anaerobic or a combination of aerobic and anaerobic mechanisms. Below the surface aerobic layer, anaerobic conditions prevail in which methane gas is evolved, as well as other products of anaerobic decomposition. Ammonia as well as some of the less-reduced products generated in the anaerobic layer (primarily organic acids) diffuse into the aerobic layer in which they are oxidized. Rich[15] has reported an average benthal stabilization rate of 80 g/(m$^2 \cdot$ d) of biodegradable solids at 20°C. Under these conditions, as much as 63 percent of total carbon stabilization can occur via methane fermentation. Assuming oxidation of all ammonia and BOD released to the water, the oxygen uptake would be 86 g O$_2$/(m$^2 \cdot$ d). The sludge should concentrate in the bottom of the lagoon to 2.5 to 3.0 percent solids.

Assuming continuous input to the lagoon and withdrawal once a year, the average annual stabilization rate is estimated as 68 g biomass/(m$^2 \cdot$ d). Example 11.9 illustrates a sludge lagoon design.

EXAMPLE 11.9. Design a lagoon to stabilize the sludge generated from an activated sludge plant treating 1.0 million gal/d (3785 m^3/d) of a wastewater with a BOD of 425 mg/l. The activated sludge plant operates at a sludge age of 45 d. The mean temperature is 20°C, a = 0.55 g, b = 0.1/d, t = 0.71 d, MLVSS = 3000 mg/l and 80 percent is volatile, S = 10 mg/l.

Solution.

X_d is calculated from Eq. (6.6.):

$$X_d = \frac{0.8}{1 + (0.1)(45)(0.2)} = 0.42$$

$$\Delta X_v = [0.55(425 - 10) - 0.1 \times 0.42 \times 3000 \times 0.71](8.34) \times 1.0$$

$$= 1158 \text{ lb VSS d} \quad (525 \text{ kg/d})$$

or

$$\frac{1158}{0.8} = 1448 \text{ lb SS/d} \quad (657 \text{ kg/d})$$

Assume 75 percent of the VSS will degrade in the lagoon; the area required is

$$A = \frac{R_a}{B_{av}}$$

in which R_a is the loading rate in lb/d of VSS to be degraded and B_{av} is the average annual stabilization rate [600 lb VSS/(acre $\cdot$ d) or 0.067 kg/(m$^2 \cdot$ d)].

The required area is

$$A = \frac{1158 \text{ lb/d} \times 0.75}{600 \text{ lb/(acre} \cdot \text{d)}}$$

$$= 1.45 \text{ acres or } 63{,}162 \text{ ft}^2 \quad (5868 \text{ m}^2)$$

The sludge depth accumulation can be calculated by first finding the residual sludge:

$$\left(\frac{1158}{0.8} - 1158 \times 0.75\right) \text{ lb residue SS/d} \times 365 \text{ d/yr}$$

$$= 211{,}500 \text{ lb residual sludge/yr} \quad (96{,}000 \text{ kg/yr})$$

At 3 percent solids, the volume of sludge is

$$\frac{211,500 \text{ lb/yr}}{0.03 \times 62.5 \text{ lb/ft}^3} = 112,800 \text{ ft}^3/\text{yr} \quad (31,950 \text{ mm}^3/\text{yr})$$

The sludge depth accumulation will then be

$$\frac{112,800 \text{ ft}^3/\text{yr}}{1.45 \text{ acre} \times 43,560 \text{ ft}^2/\text{acre}} = 1.8 \text{ ft/yr} \quad (0.55 \text{ m/yr})$$

If the oxygen requirements are 760 lb O_2/(acre · d) [0.086 kg/(m^2 · d)], the total O_2 requirement will be 1102 lb/d (760 × 1.45). Using a 10-ft (3.05-m) water column and an aerator transfer efficiency of 10 percent, the required airflow will be

$$\text{Required airflow} = \frac{1102 \text{ lb/d}}{1440 \text{ min/d} \times 0.1 \times 0.232 \text{ lb } O_2/\text{lb air} \times 0.0746 \text{ lb air/ft}^3}$$

$$= 442 \text{ standard ft}^3/\text{min} \quad (12.5 \text{ m}^3/\text{min})$$

If each diffuser operates at 4 standard ft³/min, the number of diffusers will be 110.

In several cases, biological sludges after aerobic or anaerobic digestion have been sprayed on local land sites from tank wagons or pumped through agricultural pipes. This employs multiple applications at low dosages from 100 dry tons/acre (22.4 kg/m^2) for average conditions to 300 tons/acre (67.3 kg/m^2) in areas of low rainfall.

Excess activated sludge has been disposed of in oxidation ponds in which algal activity maintains aerobic conditions in the overlaying liquids while the sludge undergoes anaerobic digestion. This procedure has been successfully employed for municipal activated sludge at Austin, Texas, and excess activated sludge from a petrochemical plant in Houston, Texas. Lagoon loading rates of 600 lb VSS/(acre · d) [0.0673 kg/(m^2 · d)] have been employed.

Many organic sludges can be incorporated into the soil without mechanical dewatering. Surface application can be accomplished by spreading from a truck or spraying. Sludge may also be injected into the soil 8 to 10 in (20 to 25 cm) below the surface by a mobile unit, as shown in Fig. 11.30. Injection offers the advantage of minimizing surface runoff and odor problems. An important consideration is the heavy metals content of the sludge. At a pH greater than 6.0, heavy metals will exchange for Ca^{2+}, Mg^{2+}, Na^+, and K^+. This natural ability to exchange heavy metals by the soil is called the cation exchange capacity (CEC) and is expressed in milliequivalents per hundred grams of dry soil. The amount of heavy metals from sludge is influenced by such factors as pH and aerobic or anaerobic conditions. The CEC of sandy soil may vary from 0 to 5 while clay soils will have a CEC between 15 and 20. The nutrient content of the sludge will support the growth of plants. The organic portion of the soil will also chelate heavy metals. In accordance with USEPA regulation 503.13, ceiling concentrations of metals for land disposal shall not exceed the concentrations shown in Table 11.6. USEPA annual pollutant loadings are given in Table 11.7.

FIGURE 11.30
Sludge injection vehicle.

TABLE 11.6
Cumulative pollutant loading rates

Pollutant	Cumulative pollutant loading rate, kg/ha
Arsenic	41
Cadmium	39
Copper	1500
Lead	300
Mercury	17
Nickel	420
Selenium	100
Zinc	2800

TABLE 11.7
Annual pollutant loading rates

Pollutant	Annual pollutant loading rate, kg/ha per 365-d period
Arsenic	2.0
Cadmium	1.9
Copper	75
Lead	15
Mercury	0.85
Nickel	21
Selenium	5.0
Zinc	140

Prior to incorporation, sludges should receive a minimum degree of stabiliza-tion. Chow[16] has recommended aerobic digestion of 15 d to reduce the volatile content to less than 55 percent.

Each crop has a nutrient requirement (N, P, K, etc.). The annual quantity of sludge that can be incorporated depends on the available nitrogen content of the sludge and the nitrogen uptake of the selected crop. Excess application of sludge can result in oxidation of ammonia to nitrate, which can go back into the ground-water. Since all the applied organic nitrogen is not available to the crops in the same year, there is a sequential removal of organic nitrogen. Normally, about 40 percent of the organic nitrogen applied in the first year is available for crop growth that year. Subsequently, 20, 10, 5, and, 2.5 percent of the organic nitrogen is avail-able for the second, third, fourth, and fifth years, respectively.

The Illinois EPA[17] recommends a minimum depth of earth cover to the annual water table of 10 ft (3 m) with a permeability rate of 2 to 20 in/h (5 to 51 cm/h). A maximum land slope of 8 percent is recommended. A minimum soil pH of 6.5 should be maintained. The design of a sludge land incorporation system is shown in Example 11.10.

EXAMPLE 11.10. Design a land incorporation system for an excess activated sludge. The sludge characteristics are

Quantity, gal/d	6560
Amount, lb/d	3500
NH_3^-N, mg/l	235
Org-N, mg/l	865
SS, mg/l	63,000
PO_4, mg/l	30

Note:
gal/d = 3.785 × 10^3 m³/d
lb/d = 0.4536 kg/d

Solution.

Metal analysis of sludge:

Metal	mg/kg (dry solids)
Al	700
Cd	3.0
Ca	105,000
Cr	400
Cu	60
Fe	6000
Pb	30
Ni	150
Zn	120
K	150

For example, consider cadmium. The total Cd in the sludge is

$$3 \; \frac{\text{mg Cd}}{\text{kg sludge}} \times 3500 \; \frac{\text{lb sludge}}{\text{d}} \times \frac{\text{kg}}{2.2 \text{ lb}} = 4773 \; \frac{\text{mg Cd}}{\text{d}}$$

For each acre (0.414 hectare), one can apply, without exceeding the allowable yearly loading rate of 1.9 kg/ha (Table 11.7),

$$1.9 \; \frac{\text{kg Cd}}{\text{ha}} \times 0.405 \; \frac{\text{ha}}{\text{acre}} \times 10^6 \; \frac{\text{mg}}{\text{kg}} = 769{,}500 \; \frac{\text{mg Cd}}{\text{d}}$$

The years to reach the maximum allowable loading of 39 kg/ha (Table 11.6) at the 1.9 kg/ha yearly rate would be

$$39 \text{ kg/ha} \times 1.9 \text{ kg/(ha} \cdot \text{yr)} = 20.5 \text{ yr}$$

But, for the 4773 mg Cd/d in the sludge, a somewhat longer time is available:

$$39 \text{ kg/ha} = \frac{39 \times 10^6 \text{ mg Cd}}{4773 \text{ mg Cd/d}} = 8170 \text{ d} = 22.3 \text{ yr}$$

The maximum annual loading in pounds per acre is

$$1.9 \; \frac{\text{kg}}{\text{ha} \cdot \text{yr}} \times \frac{0.414 \text{ ha}}{\text{acre}} \times 2.2 \; \frac{\text{lb}}{\text{kg}} = 1.73 \text{ lb/(acre} \cdot \text{yr)}$$

Agronomic loading:
The maximum allowable nitrogen loading for Bermuda grass is 0.0224 kg/(m^2 · yr). This equates to 200 lb/(acre · yr). For subsurface incorporation, NH$_3$ availability is 100 percent and org-N availability is 40 percent.
    The available N for the first year of application is

$$235 \text{ mg NH}_3{}^-\text{N/l} + 865 \text{ mg org-N/l} \times 0.4 = 581 \text{ mg N/l}$$

$$= 0.00486 \text{ lb/gal}$$

The sludge loading is therefore

$$\frac{200}{0.00486} = 41{,}152 \text{ gal/(acre} \cdot \text{yr)} \qquad [0.0385 \text{ m}^3/(\text{m}^2 \cdot \text{yr})]$$

The acres required is thus

$$\frac{6560 \times 365}{41{,}152} = 58 \text{ acres} \qquad (234{,}720 \text{ m}^2)$$

In subsequent years of application, additional organic nitrogen conversion should be considered.

Oily sludges have successfully been disposed of on land. Recent data published indicate the following:

1. The oil degradation rate is directly related to the percentage of oil in the soil.
2. Fertilization improved the degradation rate.
3. Aeration (tilling) frequencies vary (from 1 week to 2 months).
4. Between 380 and 400 m^3 (2000 and 2500 bbl) of oil per hectare should be degraded in an 8-month growing season.
5. Sludge farming is about one-fifth as expensive as incineration.

11.18
INCINERATION

The sludge cake must be disposed of. This can be accomplished by hauling the cake to a land disposal site or by incineration.

The variables to be considered in incineration are the moisture and volatile content of the sludge cake and the thermal value of the sludge. The moisture content is of primary significance because it dictates whether the combustion process will be self-supporting or whether supplementary fuel will be required. The thermal values of sludges may vary from 5000 to 10,000 Btu/lb (1.16×10^7 to 2.33×10^7 J/kg).

Incineration involves drying and combustion. Various types of incineration units are available to accomplish these reactions in single or combined units. In the incineration process, the sludge temperature is raised to 212°F (100°C), at which point moisture is evaporated from the sludge. The water vapor and air temperature are increased to the ignition point. Some excess air is required for complete combustion of the sludge. Self-sustaining combustion is often possible with dewatered waste sludges once the burning of auxiliary fuel raises the incinerator temperature to the ignition point. An autogenous sludge is one in which there is a favorable ratio of water to volatiles. Generally speaking, that ratio must be slightly better than 2 lb H_2O/lb volatiles for the sludge to approach autogenous conditions. The primary end products of combustion are carbon dioxide, sulfur dioxide, and ash.

Incineration can be accomplished in multiple-hearth furnaces in which the sludge passes vertically through a series of hearths. In the upper hearths, vaporization of moisture and cooling of exhaust gases occur. In the intermediate hearths, the volatile gases and solids are burned. The total fixed carbon is burned in the lower hearths. Temperatures range from 1000°F (538°C) at the top hearth to 600°F (316°C) at the bottom. The exhaust gases pass through a scrubber to remove fly ash and other volatile products. This is shown in Fig. 11.31.

In the fluidized bed, sludge particles are fed into a bed of sand fluidized by upward-moving air. A temperature of 1400 to 1500°F (760 to 815°C) is maintained in the bed, resulting in rapid drying and burning of the sludge. Ash is removed from the bed by the upward-flowing combustion gases.

PROBLEMS

11.1. An aerobic digester has a retention time of 10 days at 20°C and the following characteristics:

Reaction rate k_d = 0.155/deg
Influent SS = 10,500 mg/l
Volatile = 85%
Nondegradable VSS = 32%
X_d = 0.68
b = 0.1/d
Flow = 100,000 gal/d (379 m³/d)

(*a*) Calculate the composition of the digester effluent, TSS, VSS.
(*b*) Calculate the oxygen requirements including nitrification.

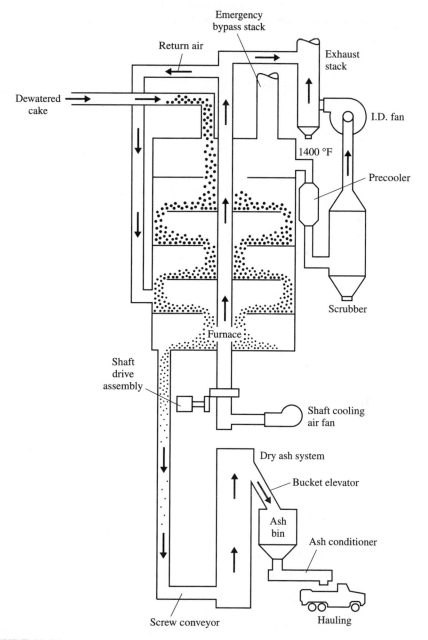

FIGURE 11.31
Multiple-hearth system.

11.2. A thickener is operating under the following conditions:

Solids flux = 20 lb/(ft² · d) [97.7 kg/(m² · d)]
Maximum underflow concentration = 45%
Area = 600 ft²(55.7 m²)

A change in plant operations modifies the characteristics of the sludge to a flow of 180,000 gal/d (681 m³/d) with a concentration of 1 percent. The thickening characteristics are those of Example 11.3. What will be the new maximum underflow concentration?

11.3. A refinery has a sludge consisting of waste lime from a neutralization process and several oily wastestreams. The composite waste has a solids concentration of 52.6 g/l and an average flow of 29,600 gal/d (112 m³/d).

Given: The design coefficients and design operating conditions as shown below:

		Design coefficients				Design operating conditions		
Process	μ, cP	$\dfrac{1-s}{2}$	m	n	R_o	Vacuum or pressure, lb/in² gage	Cake solids, %	Cycle time, min
Vacuum filter	1	0.087	0.548	−0.562	3.5	9.8	34	6
Filter press	1	0.299	0.306	−0.559	7.63	100	40	1200

Determine:
(a) The required vacuum filter area based on a 7 d/week operation with 8 h/week for precoating.
(b) The required filter press area based on a 7 d/week operation with 4 h of downtime for every cycle.

Use the relationship:

$$L = \frac{35.7P^{(1-s)/2}(c^m)(t)^n}{(\mu R_o)^{1/2}}$$

where L = filter loading, lb/(ft² · h)
P = vacuum pressure, lb/in² gage
c = solids deposited per unit volume of filtrate, g/ml
t = cycle time, min
μ = viscosity, cP
R_o = cake resistivity

Note 1: $c = \left[\dfrac{\text{sludge moisture }(\%)}{\text{sludge solids }(\%)} - \dfrac{\text{cake moisture }(\%)}{\text{cake solids }(\%)} \right]^{-1}$

Note 2: The above relationship can be used to determine the loading on a filter press where

$$P = \text{filter pressure, lb/in}^2 \text{ gage}$$

REFERENCES

1. Stien, R., C. E. Adams, and W. W. Eckenfelder: *Water Res.,* vol. 8, p. 213, 1974.
2. Andrews, J. F., and K. Kambhu: "Thermophilic Aerobic Digestion of Organic Solid Waste," Final Progress Report, Clemson University, Clemson, S.C., 1970.
3. Dick, R. I.: "Thickening," in *Process Design in Water Quality Engineering,* E. L. Thackston and W. W. Eckenfelder, eds., Jenkins, Austin, Texas, 1972.
4. Coackley, P., and B. R. S. Jones: *Sewage Ind. Waste,* vol. 28, pt. 8; p. 963, 1956.
5. Bernard, J., and A. J. Englande: "Centrifugation," in *Process Design in Water Quality Engineering,* E. L. Thackston and W. W. Eckenfelder, eds., Jenkins, Austin, Texas, 1972.
6. Swanwick, J. D.: *Advances in Water Pollution Research,* vol. II, p. 387, Pergamon Press, New York, 1963.
7. Saunamaki, R.: *Wat. Sci. Tech.,* vol. 35, no. 2-3, p. 235, 1997.
8. Barber, J. B. and Bullard, C. M.: *Proc. WEF Industrial Waste Conf.,* Nashville, 1998.
9. Eckenfelder, W. W.: *Principles of Water Quality Management,* CBI, Boston, Mass., 1980.
10. Bassett, P. J. et al.: *Proc. 33rd Ind. Waste Conf.,* Purdue University, 1978.
11. Dickey, R. O., and R. C. Ward: *Proc. 36th Ind. Waste Conf.,* Purdue University, 1978.
12. Podusks, R. A. et al.: *Proc. 35th Ind. Waste Conf.,* Purdue University, 1980.
13. Leonard, R. J., and J. W. Parrott: *Proc. 33rd Ind. Waste Conf,* Purdue University, 1978.
14. Miner, R. G.: *J. WPCF,* vol. 52, pt. 9(2), p. 389, 1980.
15. Rich, L.: *Water Res.,* vol. 16, pt. 9(1), p. 399, 1982.
16. Chow, V.: *Sludge Disposal on Land,* 3M Co., St. Paul, Minn., 1979.
17. Illinois EPA: "Design Criteria for Municipal Sludge Utilization on Agricultural Land," Tech. Policy, WPC-3, 1977.
18. Deeny, K. et al.: *Proc. 40th Purdue Industrial Waste Conf.,* Purdue University, 1985.
19. Zall, J. et al.: *J. WPCF,* vol. 59, no. 7, 1987.

12

MISCELLANEOUS TREATMENT PROCESSES

12.1
LAND TREATMENT

A wide variety of food-processing wastewaters, including meat, poultry, dairy, brewery, and winery wastewaters, have been applied successfully to the land. Disposal of industrial wastes by irrigation can be practiced in one of several ways, depending on the topography of the land, the nature of the soil, and the characteristics of the waste:

1. Distribution of waste through spray nozzles over relatively flat terrain.
2. Distribution of waste over sloping land which runs off to a natural watercourse.
3. Disposal through ridge and furrow irrigation channels.

Irrigation

Irrigation includes those systems where loading rates are about 2 to 4 in/week (5 to 10 cm/week) and a crop is grown. Methods of application include various sprinkler systems, ridge and furrow, and surface flooding. It is desirable to have spray application periods followed by rest periods in the ratio of approximately 1:4 or greater, for example 0.5 h spray:2 h rest.

Screened waste is pumped through laterals and sprayed through sprinklers located at appropriate intervals, as shown in Fig. 12.1. The waste percolates through the soil, and during this process the organics undergo biological degradation. The liquid is either stored in the soil layer or discharged to the groundwater. Most spray irrigation systems use a cover crop of grass or other vegetation to maintain porosity in the upper soil layers. The most popular cover crop is reed canary grass (*Phalaris arundinacea*). This grass develops an extensive root system, has a relatively large leaf area, and is tolerant to adverse conditions. There is a net waste

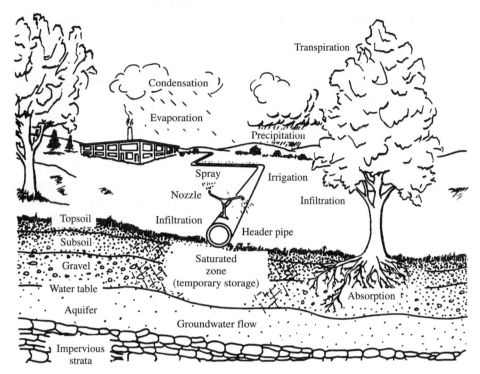

FIGURE 12.1
Spray irrigation system.

loss by evapotranspiration (evaporation to the atmosphere and absorption by the roots and leaves of plants). This may amount to as much as 10 percent of the waste flow.

Loamy well-drained soil is most suitable for irrigation systems; however, soil types from clays to sands are acceptable. A minimum depth to groundwater of 5 ft (1.5 m) is preferred to prevent saturation of the root zone. Underdrain systems have been used successfully to adapt to high groundwater or impervious subsoil conditions.

Water-tolerant perennial grasses have been used most commonly because they take up large quantities of nitrogen, are low in maintenance, and keep the soil infiltration rates as high as possible. Seasonal canning wastewaters are often used to irrigate corn or annual forages to coincide with the production of wastewater.

In some cases, wastes have been sprayed into woodland areas. Trees develop a high-porosity soil cover and yield high transpiration rates. A small elm tree may take up as much as 3000 gal/d (11.4 m³/d) under arid conditions.

The principal factors governing the capacity of a site to adsorb wastewater are:[1]

1. *Character of the soil.* A sandy type of soil will have a high filtration rate; clay will pass very little water.
2. *Stratification of the soil profile.* Some soils will exhibit clay lenses, which are a barrier to flow.

3. *Depth to groundwater.* The quantity of wastewater that can be sprayed on a given area will be proportional to the depth of soil through which it must travel to the groundwater. Sufficient soil depth must be available, however, to effect biological degradation of the organics.
4. *Initial moisture content.* The capacity of the soil to adsorb water is proportional to the initial moisture content of the soil.
5. *Terrain and ground cover.* A cover crop will increase the quantity of water that can be absorbed by a given area. A sloping site will increase the runoff.

In a spray irrigation field, water absorbed into the ground is held in capillary suspension until approximately 95 percent of saturation. Additional water will flow into the groundwater under a head equal to the distance from the ground surface to the groundwater table. The discharge of wastewater may be at a steady rate or at a short-term rate. The capacity to absorb at a steady rate is proportional to the overall coefficient of permeability for the soil between the ground surface and the groundwater table:[1]

$$Q = 328 \times 10^3 KS \tag{12.1}$$

where Q = gal/(min · acre)
K = overall coefficient of permeability, ft/min
S = degree of saturation of soil (near 1.0 for steady-rate application)

or

$$Q = 1.00KS \tag{12.1a}$$

for Q in m³/(min · m²) and K in m/min.

The coefficient of permeability K depends on the soil characteristics, as shown in Table 12.1. The overall coefficient will depend on the variation in soil characteristics for various depths below the surface and can be computed as

$$K = \frac{H}{H_1/K_1 + H_2/K_2 + \ldots + H_n/K_n} \tag{12.2}$$

where H is the total depth to the groundwater in feet, $H_1, H_2, \ldots, H_n$ are the thickness of layers of soil profile, and K_1, K_2, etc., are the average permeability coefficients for each layer. Soil borings are usually necessary to determine the overall coefficient K. Care should be taken to avoid missing clay lenses that may be pres-

TABLE 12.1
Variation of permeability with soil characteristics[1]

Description of fine component	K, ft/min
Trace fine sand (0.10%)	1.0–0.2
Trace silt (0–10%)	0.8–0.04
Coarse and fine silt (10–20%)	0.012–0.002
Fissured clay soils (50%) and organic soils (50%)	0.0008–0.0004
Dominating clay soils (up to 100%)	<0.0002

Note: ft/min = 0.3048 m/min

ent. The short-term rate is proportional to the capillarity of the soil and the initial moisture content. It is usually considerably higher than the steady rate.

Rapid Infiltration

Rapid infiltration systems are characterized by percolation of most of the applied wastewater through the soil and into the subsurface. The method is restricted to use with rapidly permeable soils such as sands and sand loams. This type of system is normally thought to be associated with recharge or spreading basins, although in food-processing applications, high-rate sprinkler systems have been used to provide distribution of the wastewater.

In rapid infiltration systems, plants play a relatively minor role in terms of treatment of the applied wastewater. Physical, chemical, and biological mechanisms operating within the soil are responsible for treatment. The more permeable the soil, the farther the wastewater must travel through the soil to receive treatment. In very sandy soils, this minimum distance is considered to be approximately 15 ft (5 m).

Overland Flow

Overland flow is a fixed film biological-treatment process. In overland flow, land treatment wastewater is applied at the upper reaches of the grass-covered slopes and allowed to flow over the vegetated surface to runoff collection ditches. The wastewater is treated by a thin film down the length of the slope. The process is best suited to slowly permeable soils but can also be used on moderately permeable soils that have relatively impermeable subsoils.

Wastewater is usually applied by sprinklers to the upper two-thirds of slopes that are 150 to 200 ft (46 to 61 m) in length. A runoff collection ditch or drain is provided at the bottom of each slope. Treatment is accomplished by bacteria on the soil surface and within the vegetative litter as the wastewater flows down the sloped, grass-covered surface to the runoff collection drains. Ideally, the slopes should have a grade of 2 to 4 percent to provide adequate treatment and prevent ponding or erosion. The system may be used on naturally sloped lands or it may be adapted to flat agricultural land by reshaping the surface to provide the necessary slopes.

The characteristics of land treatment systems are summarized in Table 12.2.

Waste Characteristics

In addition to soil conditions, there are several waste characteristics that require consideration in a spray irrigation system. Suspended solids should be removed from the waste, either by screening or by sedimentation, before it is sprayed. Solids will tend to clog the spray nozzles and may mat the soil surface, rendering it impermeable to further percolation. An excessively acid or alkaline pH will be harmful

TABLE 12.2
Comparative characteristics of land treatment systems[2]

Feature	Irrigation	Rapid infiltration	Overland flow
Hydraulic loading rate, cm/d	0.2–1.5	1.5–30	0.6–3.6
Land required, Ha[†]	24–150	1.2–24	10–60
Soil type	Loamy sand to clay	Sands	Clay to clay loam
Soil permeability	Moderately slow to moderately rapid	Rapid	Slow

† Field area in hectares not including buffer area, roads, or ditches for 3785 m³/d (1 million gal/d) flow.

to the cover crop. High salinity will impair the growth of a cover crop and in clay soils will cause sodium to replace calcium and magnesium by ion exchange. This will cause soil dispersion, and as a result drainage and aeration in the soil will be poor. A maximum salinity of 0.15 percent has been suggested to eliminate these problems.[3]

The soil is a highly efficient biological treatment process, and the performance of a system is usually governed by the hydraulic capacity of the soil as opposed to the organic loading rate. Oxygen exchange into soils depends on the air-filled pore spaces. In saturated soils, oxygen transfer will be similar to that in oxidation ponds. In well-drained soils, oxygen exchange of the surface is rapid because of mass flow. Below the first 4 in (10 cm), however, oxygen exchange is slow because it depends on diffusion.

Research by Jewell[4] showed that organic loading rates onto soil can exceed 16,000 lb/(acre · d) [1.79 kg/(m² · d)] on a COD basis without exceeding bacterial capabilities. Adamczyka[5] reported problems with irrigation systems loaded at 2000 to 5000 lb/(acre · d) [0.22 to 0.56 kg/m² · d)] of BOD but no problems at 500 lb/(acre · d) [0.056 kg/(m² · d). The problems of the very high BOD loadings can include damage to vegetation, odors, and leaching of undegraded organics into the subsurface. Loadings in the range of 535 lb/(acre · d) [0.060 kg/(m² · d)] are generally acceptable for irrigation and rapid infiltration systems. For overland flow the limiting loading rate has not been defined but is probably 134 to 180 lb/(acre · d) [0.017 to 0.020 kg/(m² · d)]. Nutrient addition in the form of nitrogen or phosphates may be necessary for nutrient-deficient wastes unless the fields are adequately fertilized.

Design of Irrigation Systems

A model has been developed[6] for irrigation that includes mass balances of water and nitrogen and provides estimates of nitrogen losses to the groundwater. The principal mechanism for nitrogen removal is crop growth.

The land application system has four components: pretreatment, wastewater transmission to the irrigation site, a lagoon for storage of the wastewater during time periods in which irrigation is infeasible, and the irrigation site. The principal variables are

$$Q_m = \text{wastewater flow to land application site } (L^3/T)$$

$$A = \text{irrrigated land area } (L^2)$$

$$r = \text{average wastewater application rate } (L/T)$$

If Q_m is in million gallons per day, A in acres, and r in inches per week,

$$\frac{Q_m}{A} = \frac{r}{258} \tag{12.3}$$

If Q_m is in cubic meters per day, A in square meters, and r in centimeters per week,

$$\frac{Q_m}{A} = \frac{r}{700}$$

If T is the number of weeks of the irrigation season and P and ET are irrigation season precipitation and evapotranspiration (in inches), respectively, the amount of water entering the groundwater below the irrigation site is $7Q_mT/A + 0.02715(P - \text{ET})$, in million gallons per acre. If P and ET are in centimeters,

$$\text{Flow to groundwater (m}^3) = 7Q_m \,(\text{m}^3/\text{d})\,\frac{T}{A\,(\text{m}^2)} + 0.01(P - \text{ET})$$

Similarly, if n is the nitrogen concentration of the pretreated wastewater (in milligrams per liter) and NC is the nitrogen removal by the growing crop (M/L^2), an estimate of nitrogen entering the groundwater from the irrigation site is

$$7(8.34)nQ_m\, T/A - \text{NC}$$

if NC is in pounds per acre, or

$$7 \times 10^{-3}nQ_m \,(\text{m}^3/\text{d})\,\frac{T}{A\,(m^2)} - \text{NC}$$

if NC is in kilograms per square meter.

If groundwater standards require the average nitrogen concentrations in seepage water to be less than the drinking water standard of 10 mg/l,

$$\frac{7nQ_mT/A - \text{NC}/8.34}{7Q_mT/A + 0.02715(P - \text{ET})} < 10 \tag{12.4}$$

which reduces to

$$\frac{Q_m}{A} < \frac{\text{NC}}{58.4T(n - 10)} + \frac{0.0388(P - \text{ET})}{T(n - 10)} \tag{12.4a}$$

In metric units

$$\frac{7nQ_mT/A - \text{NC}/ \times 10^3}{7Q_mT/A + 0.01(P - \text{ET})} < 10$$

which reduces to

$$\frac{Q_m}{A} < \frac{143\text{NC}}{T(n-10)} + \frac{(P-\text{ET})}{70T(n-10)}$$

If the nitrogen application in the wastewater is to be equal to or less than the crop requirement, NC, then

$$\frac{7(8.34)nQ_mT}{A} < \text{NC} \tag{12.5}$$

or

$$\frac{Q_m}{A} < \frac{\text{NC}}{58.4nT} \tag{12.5a}$$

or in metric units

$$7 \times 10^{-3}nQ_m\frac{T}{A} < \text{NC}$$

or

$$\frac{Q_m}{A} < \frac{143\text{NC}}{nT}$$

Equations (12.4) and (12.5) constrain the nitrogen loading rate at the land application site. The liquid loading rate will be constrained by the drainage capacity of the soil, $\bar{r}$ (in inches per week or centimeters per week):

$$r < \bar{r} \tag{12.6}$$

These equations will define the land area requirements for a specific wastewater. A design is shown in Example 12.1.

EXAMPLE 12.1. A food-processing wastewater is to be land-irrigated. Compute the area requirements for the following conditions:

1 million gal/d (3785 m^3/d)
500 mg/l BOD
25 mg/l N

Solution.

Regulations limit the application rate to 3 in/week (7.6 cm/week) and 500 lb BOD/(acre · d) [0.056 kg/(m^2 · d). For this maximum loading rate,

$$\frac{Q_m}{A} = \frac{r}{258\ (\text{acre} \cdot \text{in/week})/(\text{million gal/d})}$$

$$\frac{1}{A} = \frac{3}{258}$$

$$A = 86\ \text{acres} \qquad (35\ \text{ha} = 350{,}000\ \text{m}^2)$$

The BOD in terms of pounds per day is

$$\text{lb BOD/d} = \frac{500\ \text{mg BO}}{1} \times 1\ \text{million gal/d} \times 8.34\frac{\text{lb/million gal}}{\text{mg/l}}$$

$$= 4170\ \text{lb/d} \qquad (1893\ \text{kg/d})$$

For this BOD loading rate, a relatively small area is adequate:

$$A = \frac{4170 \text{ lb/d}}{500 \text{ lb BOD/(acre} \cdot \text{d)}} = 8.34 \text{ acres} \qquad (3.40 \text{ ha} = 34,000 \text{ m}^2)$$

However, the nitrogen for the crop requirement with reed canary grass is 200 lb/acre (0.022 kg/m^2). The spraying period for the cannery is 12 weeks. Therefore,

$$\text{lb N} = \frac{25 \text{ mg N}}{1} \times \text{million gal/d} \times 8.34 \ \frac{\text{lb/million gal}}{\text{mg/l}} \times 12 \text{ weeks} \times 7 \ \frac{\text{d}}{\text{week}}$$

$$= 17,512 \text{ lb N} \qquad (7951 \text{ kg})$$

$$A = \frac{17,514 \text{ lb N}}{200 \text{ lb N/acre}} = 88 \text{ acres} \qquad (35.9 \text{ ha} = 359,000 \text{ m}^2)$$

or

$$\frac{Q_m}{A} = \frac{NC}{58.4nT}$$

$$\frac{1}{A} = \frac{200}{58.4 \times 25 \times 12}$$

$$A = 88 \text{ acres} \qquad (35.9 \text{ ha} = 359,000 \text{ m}^2)$$

The area requirement is 88 acres, dictated by the crop requirement rather than the regulatory maximum loading or the BOD.

Performance of Land Application Systems

Wastes that have been successfully disposed of by spray irrigation include cannery,[7] pulp and paper,[8] dairy, and tannery wastes. Spent sulfite liquor[9,10] has been applied up to a rate of 116 lb solids/(d $\cdot$ yd^2) [63 kg/(d $\cdot$ m^2)]; this resulted in up to 95 percent removal of BOD through a 10-ft (3-m) soil layer. Boxboard[8] wastes have been sprayed on a blown silt loam with a gravel underlay and alfalfa cover crop at a short-term rate of 0.7 in/h (1.8 cm/h) and a total daily rate of 0.21 to 0.56 in/d (0.5 to 1.4 cm/d). This is equivalent to 0.2 to 1.0 acres per ton of production per day [0.89 to 4.40 m^2/(kg/d)]. Strawboard cook liquor including beater-washer water and machine water[8] has been sprayed twice a week for 6 h/d at a rate of 0.5 in/d (1.3 cm/d). A rate of 1.2 in/d (3 cm/d) resulted in ponding. Kraft mill wastes have been found to require 1.5 acres per ton of production (6.7 m^2/kg) or more.[8] Runoff of airport deicing fluid (ADF) consisting of propylene glycol plus additives were treated on soil.[11] Waste activated sludge amendments on the order of 955 mg/kg soil, in addition to lime amendments and tilling, enhanced the degradation rate. Diluting ADF solutions in excess of 20 percent by weight prior to soil application is necessary to prevent inhibition. The degradation followed first-order kinetics.

Total petroleum hydrocarbons were degraded in soil.[12] Soil aeration was accomplished by ploughing the soils monthly, and nutrients were incorporated into the soils by adding dry fertilizers. Complete destruction was achieved in 50 days. Typical data for spray irrigation systems are shown in Table 12.3. Results

TABLE 12.3
Spray irrigation of industrial wastes

Waste	Acres sprayed	Application rate, gal/ (min · acre)	Duration	Average loading lb BOD/ (acre · d)	lb SS/ (acre · d)	Reference
Tomatoes	5.63	178	7.5 h/d	413	364	1
	6.40	86	3.7 h/d	155	139	1
Corn	2.28	153.5	10 h/d	864	500	1
Asparagus and beans	0.90	282.0	5.6 h/d	22.5	356	1
Lima beans	6.65	65	16 h/d	65	46	1
Cherries	2.24	96.5	17 h/d	807	654	1
Paperboard	1.30	77	3 h/5 d			8
Hardboard	100	42	12 h/10 d			8
	300	24	18 h/10 d			
Strawboard	1.5	94	6 h/3 d			8
Kraft	70	98	8 h/week			8

Note:
gal/(min · acre) = 9.35×10^{-7} m³/(min · m²)
lb/(acre · d) = 1.12×10^{-4} kg/(m² · d)
acre = 4.05×10^3 m²

for food-processing wastewaters are shown in Table 12.4. Overland flow systems are summarized in Table 12.5.

12.2
DEEP-WELL DISPOSAL

Deep-well disposal involves the injection of liquid wastes into a porous subsurface stratum that contains noncommercial brines.[13] The wastewaters are stored in sealed subsurface strata isolated from groundwater or mineral resources. Disposal wells may vary in depth from a few hundred feet (< 100 in) to 15,000 ft (4570 in) with capacities ranging from less than 10 to more than 2000 gal/min (38 to 7550 l/min). Wastes disposed of in wells are usually highly concentrated, toxic, acidic, or radioactive, or wastes high in inorganic content which are difficult or excessively expensive to treat by some other process.

    The disposal system consists of the well and pretreatment equipment necessary to prepare the waste for suitable disposal into the well. A casing, generally of steel, is cemented in place to seal the disposal stratum from other strata penetrated during drilling of the well. An injection tube transports the waste to the disposal stratum as shown in Fig. 12.2.

    Oil or freshwater is used to fill the annular space between the injection tube and the casing and extends to but is sealed from the injection stratum. Leaks in the injection tube or drainage to the casing can be detected by monitoring the pressure of the fluid.

TABLE 12.4
Spray irrigation of food-processing wastewaters[5]

Design flow, million gal/d	Overall application rate, in/week	Spray field scheduling			Total spray area, acres	Number of spray fields	Crop cover	Soil	Pretreatment	Product
		Application rate, in/h	Use/rest ratio	Spray time, h						
0.2	2.7	0.25	1:3	0.33	23	4	Reed canary grass	Coarse and fine sediments	Activated sludge	Dairy cream, butter, cheese, powdered milk
0189	0.68	0.037	1:8	5.25	85.6	6	Reed canary grass	Low plastic clay	Lagoons	Beets, cabbage
1.8	5.5	0.25	1:5	20	61	5	Forested with heavy undergrowth	Genesee fine sandy loam	Screening and pH adjustment	Beets, corn, peas
1.0	3.15	0.30	1:7	12	80	4	Reed canary grass, and clover	Ontario loam and Cazenovia silt loam	Aerated lagoons	Corn, beans, peas, carrots, potatoes
0.64	3.53	0.21	1:28	6	73.5	7	Reed canary grass	Williamson and Wallington loam	pH adjustment and aerated lagoons	Beans, cherries, beets, apples
0.8	2.7	0.15	1:8	8	80	9	Reed canary grass	Schoharie silty loam	Lagoons	Sauerkraut, beans
0.26	3.36	0.2	1:7	12	20	4	Reed canary grass	Gravelly loam	Screening, aerated lagoons	Beans, apples
0.025	2.6	0.11	1:6	18	9	5	Reed canary grass	Chenango gravel	Screening	Cheddar cheese
0.1	1.0	0.1	1:17	12	27	9	Japanese millet	Fine, sandy loam and silt loam	Lagoons	Cherries, prunes, apples
0.047	2.02	0.12	1:9	12	6	6	Reed canary grass	Sandy silt	Screening pH adjustment lagoon	Cider, vinegar

Note:
gal/d = 3.79 × 10⁻³ m³/d
in/week = 2.54 cm/week
acre = 4.05 × 10³ m²

TABLE 12.5
Selected overland flow systems[2]

Name and location	Loading rate, cm/d	Slope length, m	Type of wastewater
Campbell Soup Co.,			
Chestertown, Maryland	1.0	53–76	Poultry
Hunt-Wesson Foods,			
Davis, California	2.2	51	Tomato
El Paso, Texas	2.5	62	Meat packing
Middlebury, Indiana	1.5	79	Poultry
Campbell Soup Co.,			
Napoleon, Ohio	1.5	53–61	Tomato
Paris, Texas	1.6–3.6	60–90	Vegetable soup
Sebastopol Co-op,			
Sebastopol, California	1.5	45	Apple
Nabisco, Inc.			
Woodbury, Georgia	0.9–11	80–84	Peanuts, pimientos
Frito-Lay Inc.,			
Wooster, Ohio	0.5–1.0	67	Vegetable

The system includes a basin to level fluctuations in flow, pretreatment equipment, and high-pressure pumps. Pretreatment requirements are determined by the characteristics of the wastewater, compatibility of the wastewater and the formation water, and the characteristics of the receiving stratum.

Pretreatment may include the removal of oils and floating material, suspended solids, biological growths, dissolved gases, precipitatable ions, acidity, or alkalinity. A typical system is shown in Fig. 12.3.

The best disposal areas include sedimentary rock in the unfractured state, including sandstones, limestones, and dolomites and unconsolidated sands. Fractured strata should be avoided because, if a vertical fissure exists, groundwater contamination may result.

The well-head pressure is related to the difference between the bottom-hole pressure and the reservoir pressure. Cores of the injection location are needed to evaluate the porosity and permeability of the stratum and any possible reactions between the wastewater and the stratum.

Although wastewaters should be generally free of suspended solids, some vugular formations will accept suspended solids without problems or an increase in injection pressure. In some cases, injection can be increased by well stimulation, which involves the injection of mineral acids to dissolve calcium carbonate and other acid-soluble particulates that tend to plug the stratum. Mechanical procedures involve scratching, swabbing, washing, and underreaming the well bore and shotting the uncased stratum with explosives or hydraulic fracturing.

The cost of a well system will be affected by the depth of the injection well, type of formation, geographic location, waste volume, required pretreatment, and injection pressure.

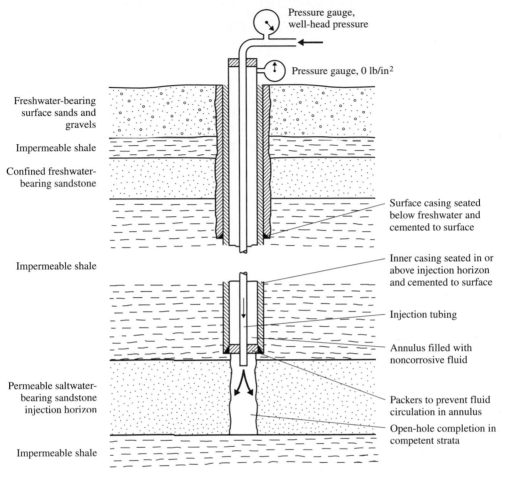

FIGURE 12.2
Schematic diagram of a waste injection well completed in a competent sandstone (*after Donaldson*[14]).

12.3
MEMBRANE PROCESSES

Membrane filtration includes a broad range of separation processes from filtration and ultrafiltration to reverse osmosis. Generally, those processes defined as filtration refer to systems in which discrete holes or pores exist in the filter medium, generally in the order of 10^2 to 10^4 nm or larger. The efficiency of this type of filtration depends entirely on the difference in size between the pore and the particle to be removed.

The various filtration processes relative to molecular size are shown in Fig. 12.4 and Table 12.6. Membrane separation technologies for wastewater treatment are shown in Table 12.7.

Reverse osmosis employs a semipermeable membrane and a pressure differential to drive freshwater to one side of the cell, concentrating salts on the input or

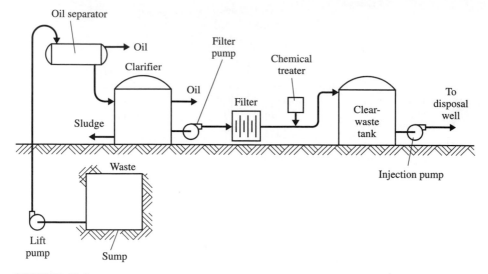

FIGURE 12.3
Typical subsurface waste-disposal system (*after Donaldson*[14]).

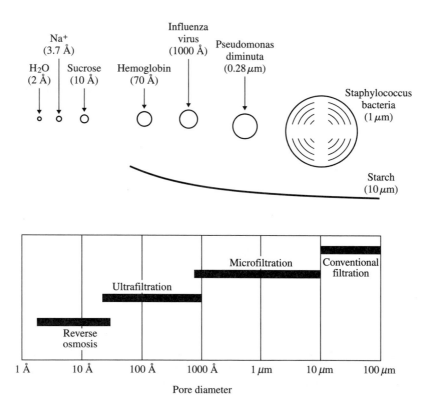

FIGURE 12.4
Membrane processes and pore sizes.

TABLE 12.6
Membrane processes

Material to be removed	Approximate size, nm	Process
Ion removal	1–20	Diffusion or reverse osmosis
Removal of organics in true solution	5–200	Diffusion
Removal of organics: subcolloidal—not in true molecular dispersion	200–10,000	Pore flow
Removal of colloidal and particulate matter	75,000	Pore flow

TABLE 12.7
Membrane separation technologies for wastewater treatment[15]

Feature	Micro-filtration	Ultra-filtration	Nano-filtration	Reverse osmosis	Per-vaporation
Suspended solids removal	Excellent	Impractical	Impractical	Impractical	N/A
Dissolved organic removal	N/A	Excellent[†]	Excellent[†]	Excellent[†]	Good[§]
VOC removal	N/A	Poor	Fair[†]	Fair–good[†]	Excellent
Dissolved inorganic removal	N/A	N/A	Good (function of salt species)	Very good (90–99% removal)	N/A
Osmotic pressure effects	None	Minor	Significant	High	None
Concentration capabilities	Up to 5% total solids	Up to 50% total organics	Up to 15%[‡]	Up to 15%[‡]	N/A
Permeate quality	Excellent	Excellent	Good	Excellent	Excellent
Energy requirements	1–3 bars	3–7 bars	5–10 bars	15–70 bars	<25% of distillation
Capital costs ($/GPD)	0.15–1.5	0.15–1.85	0.15–1.5	0.15–1.5	1.85–4.00
Operating cost ($/1000 liter feed rate)	0.15–1.10	0.15–0.80	0.20–0.80	0.25–0.80	0.80–1.30

All technologies are application-specific, and testing is required to develop precise data. Estimating specific capital and operating costs is difficult because some wastestreams may require special materials of construction or additional design considerations as a result of high potential.
[†] Function of molecular weight.
[‡] Function of osmotic pressure.
[§] Function of vapor pressure.
N/A—Not applicable.
GPD—Gallon per day.

rejection side of the cell. In this process, freshwater is literally squeezed out of the feedwater solution.

The reverse osmosis process can be described by considering the normal osmosis process. In osmosis, a salt solution is separated from a pure solvent into a solution of less concentration by a semipermeable membrane. The semipermeable membrane is permeable to the solvent and impermeable to the solute. Such an arrangement is shown in Fig. 12.5. The chemical potential of the pure solvent is greater than that of the solvent in solution, and therefore drives the system to equilibrium. If an imaginary piston applies an increasing pressure on the solution compartment, the solvent flow through the membrane will continue to decrease. When sufficient pressure has been applied to bring about thermodynamic equilibrium, the solvent flow will stop. The pressure developed in achieving equilibrium is the osmotic pressure of the solution, or the difference in the osmotic pressure between the two solutions if a less concentrated salt solution is used instead of pure solvent in the right chamber of the cell.

If a pressure in excess of the osmotic pressure is now applied to the more concentrated solution chamber, pure solvent is caused to flow from this chamber to the pure solvent side of the membrane, leaving a more concentrated solution behind. This phenomenon is the basis of the reverse osmosis process.

The criteria of membrane performance are the degree of impermeability (how well the membrane rejects the flow of the solute) and the degree of permeability (how easily the solvent is allowed to flow through the membrane). Cellulose acetate membranes provide an attractive combination of these criteria. Membrane technology uses cross-flow filtration in which the feed stream flows across (tangential to) the membrane surface, as shown in Fig. 12.6. Cross-flow filtration reduces fouling and concentration polarization. Membrane types are shown in Fig. 12.7. Each of these design configurations uses cellulose acetate membranes, except certain of the hollow fine-fiber systems that employ a nylon polymer. The more common membranes are described below:

1. *Tubular.* Manufactured from ceramics, carbon, or any number of porous plastics, these tubes have inside diameters ranging from 1/8 inch (3.2 mm) up to approximately 1 inch (2.54 cm). The membrane is typically coated on the inside of the tube, and the feed solution flows through the interior from one end to the other, with the "permeate" or "filtrate' passing through the wall to be collected on the outside of the tube.
2. *Hollow fiber.* Similar to the tubular elements in design, hollow fibers are generally much smaller in diameter and require rigid support such as is obtained from the "potting" of a bundle inside a cylinder. As with tubular elements, feed flow is usually down the core of the fiber.
3. *Spiral wound.* This device is constructed from an envelope of sheet membrane wound around a permeate tube that is perforated to allow collection of the permeate or filtrate.
4. *Plate and frame.* This device incorporates sheet membrane that is stretched over a frame to separate the layers and facilitate collection of the permeate.

With regard to membrane element configuration, the important physical characteristics of the various membrane element device configurations available today are listed in Table 12.8.

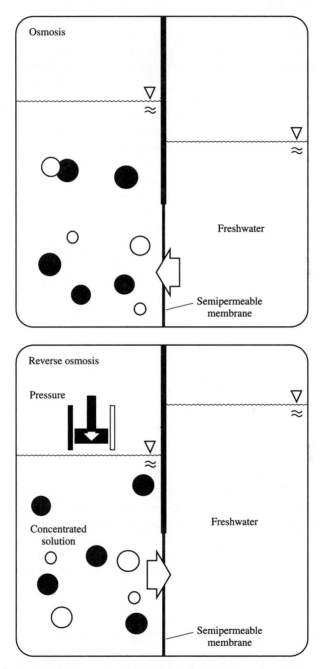

FIGURE 12.5
Osmosis and reverse osmosis.

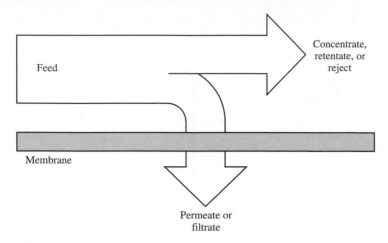

FIGURE 12.6
Cross-flow filtration: The one input feed stream is divided into two output streams—the concentrate and the permeate streams.

Because of the propensity of suspended or precipitated materials settling out on the membrane surface and plugging the membrane pores, turbulent flow conditions must be maintained (Reynolds numbers in excess of 2000). For high recovery systems, this usually requires recycling a significant percentage of the concentrate back to the feed side of the pump. The addition of this concentrate stream into the feed solution obviously increases the dissolved solids concentration, further increasing osmotic pressure.

In order to minimize membrane fouling, pretreatment is required for the removal of suspended matter, bacteria, and precipitatable ions. A typical reverse osmosis process schematic is shown in Fig. 12.8. The design and operating parameters for a reverse osmosis system are summarized from Agardy.[16]

Pressure

The water flux is a function of the pressure differential between the applied pressure and the osmotic pressure across the membrane. The higher the applied pressure, the greater the flux. However, the pressure capability of the membrane is limited, so the maximum pressure is generally taken to be 1000 lb/in^2 (6895 kPa or 68 atm) gauge. Operating experience dictates in the 400 to 600 lb/in^2 (2758 to 4137 kPa or 27 to 41 atm) gauge range, with 600 lb/in^2 (4137 kPa or 41 atm) gauge normally being the design pressure.

Temperature

The water flux increases with increasing feedwater temperature. A standard of 70°F (21°C) is generally assumed and temperatures of up to 85°F (29°C) are

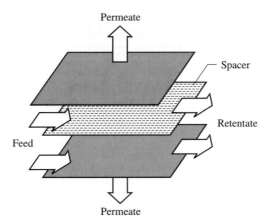

Plate-and-frame module

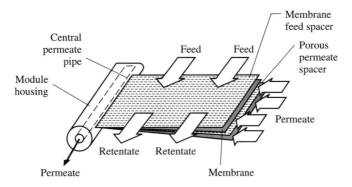

Spiral-wound module

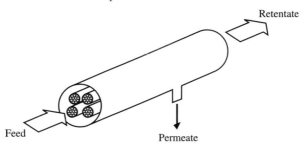

Tubular module

FIGURE 12.7
Membrane shapes.

TABLE 12.8
Common membrane characteristics

Element configuration	Packing density[†]	Suspended solids tolerance
Spiral wound	High	Fair
Tubular	Low	High
Plate and frame	Low	High
Hollow fine fiber	Highest	Poor

[†] Membrane area per unit volume of space required.

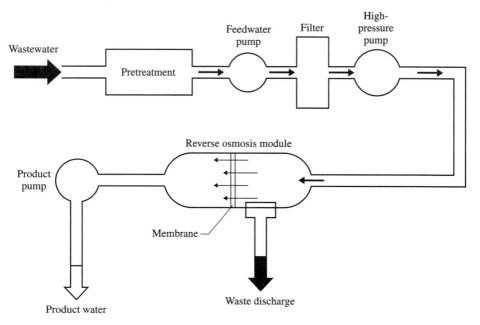

FIGURE 12.8
Basic reverse osmosis process schematic.

acceptable. However, temperatures in excess of 85°F (29°C) and up to 100°F (38°C) will accelerate membrane deterioration and cannot be tolerated for long operating periods.

Membrane Packing Density

This is an expression of the unit area of the membrane, which can be placed per unit volume of pressure vessel. The greater this factor the greater will be the overall flow through the system. Typical values range from 50 to 500 ft²/ft³ (160 to 1640 m²/m³) of pressure vessel.

Flux

Although the flux rates for hollow fibers are from 6.0×10^{-3} to 10.2×10^{-3} m³/ (d · m²) [0.15 to 0.25 gal/(d · ft²)] versus 6.1×10^{-1} to 10.2×10^{-1} m³/(d · m²) [15 to 25 gal/(d · ft²)] for sheet systems, the stacking densities for fibers are about 10 times greater, making the sizes of the two systems competitive. This flux tends to decrease with length of run and over a period of 1 or 2 years of operation might be reduced by 10 to 50 percent.

Recovery Factor

This consideration actually represents plant capacity and is generally in the range of 75 to 95 percent, with 80 percent being the practical maximum. At high recovery factors there is a greater concentration in the process water as well as in the brine. At higher concentrations salt precipitation on the membrane increases, causing a reduction in operation efficiency.

Salt Rejection

Salt rejection depends on the type and character of the selected membrane and the salt concentration gradient. Generally, rejection values of 85 to 99.5 percent are obtainable, with 95 percent being commonly used.

Membrane Life

Membrane life can be drastically shortened by undesired constituents in the feed water, such as phenols, bacteria, and fungi as well as high temperatures and high or low pH values. Generally, membranes will last up to 2 years with some loss in flux efficiency.

pH

Membranes consisting of cellulose acetate are subject to hydrolysis at high and low pH values. The optimum pH is approximately 4.7, with operating ranges between 4.5 and 5.5.

Turbidity

Reverse osmosis units can be used to remove turbidity from feedwaters. They operate best if little or no turbidity is applied to the membrane. Generally, it is considered that the turbidity should not exceed one Jackson turbidity unit (JTU) and that the feedwater should not contain particles larger than 25 μm.

Feedwater Stream Velocity

The hydraulics of reverse osmosis systems are such that velocities in the range of 0.04 to 2.5 ft/s (1.2 to 76.2 cm/s) are common. Plate and frame systems operate at the higher velocity, while the hollow fine-fiber units operate at the lower velocities. High velocities and turbulent flow are necessary to minimize concentration polarization at the membrane surface.

Power Utilization

Power requirements are generally associated with the system pumping capacity and operational pressures. Values range from 9 to 17 kW $\cdot$ h/thousand gal (2.4 to 4.5 kW $\cdot$ h/m^3), with the lower figure taking into account some power recovery from the brine stream.

Pretreatment

The present stage of development of membranes limits their direct application to feedwaters having a TDS not exceeding 10,000 mg/l. Further, the presence of scale-forming constituents, such as calcium carbonate, calcium sulfate, oxides and hydroxides of iron, manganese, and silicon, barium and strontium sulfates, zinc sulfide, and calcium phosphate, must be controlled by pretreatment or they will require subsequent removal from the membrane. These constituents can be controlled by pH adjustment, chemical removal, precipitation, inhibition, and filtration. Organic debris and bacteria can be controlled by filtration, carbon, pretreatment, and chlorination. Oil and grease must also be removed to prevent coating and fouling of the membranes.

Cleaning

Since, under continuous use, membranes will foul, provision must be made for mechanical and/or chemical cleaning. Methods include periodic depressurizations, high-velocity water flushing, flushing with air-water mixtures, backwashing, cleaning with enzyme detergents, ethylene diamine, tetraacetic acid, and sodium perborate. The control of pH during cleaning operations must be maintained to prevent membrane hydrolysis. Approximately 1 to 1.5 percent of the process water goes to waste as a part of the cleaning operation, with the cleaning cycle being every 24 to 48 hours.

A summary of operational parameters is shown in Table 12.9.

Applications

Reverse osmosis has been applied to the treatment of plating wastewaters for the removal of cadmium, copper, nickel, and chromium at pressures of 200 to 300 lb/in^2

TABLE 12.9
Summary of system operational parameters for reverse osmosis[13]

Parameter	Range	Typical
Pressure, lb/in² gauge	400–1000	600
Temperature,°F	60–100	70
Packing density, ft²/ft³	50–500	—
Flux, gal/(d · ft²)	10–80	12–35
Recovery factor, %	75–95	80
Rejection factor, %	85–99.5	95
Membrane life	—	2
pH	3–8	4.5–5.5
Turbidity, JTU	—	1
Feedwater velocity, ft/s	0.04–2.5	—
Power utilization, kW · h/thousand gal	9–17	—

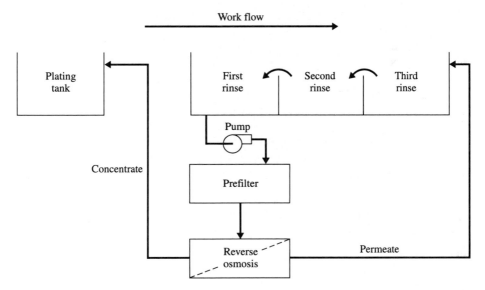

FIGURE 12.9
Treatment of plating wastewaters by reverse osmosis.

(1378 to 2067 kPa or 13.6 to 20.4 atm). The concentrated stream is returned to the plating bath and the treated water to the next-to-last rinse tank, as shown in Fig. 12.9.

Pulp mill effluents have been treated by reverse osmosis at a pressure of 600 lb/in² (4137 kPa or 41 atm). Wastestreams were concentrated up to 100,000 mg/l total solids. The flux was found to be a function of total solids level and varied from 2 to 15 gal/(d · ft²) [0.08 to 0.61 m³/(d · m²)].[17]

Oily wastes can be treated by ultrafiltration in which the permeate can be recycled as rinsewater and the concentrate can be hauled or incinerated, as shown in Fig. 12.10.

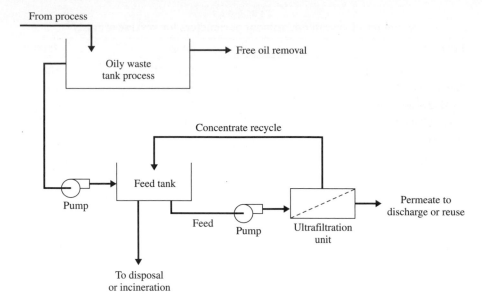

FIGURE 12.10
Treatment of oily wastewaters by ultrafiltration.

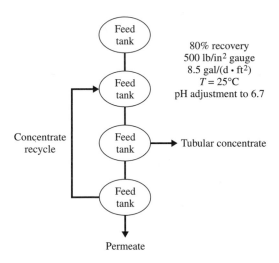

FIGURE 12.11
Two-stage membrane industrial landfill leachate treatment system.

Leachate from an industrial landfill was treated in a two-stage reverse osmosis system as shown in Fig. 12.11.[18] Performance of the treatment system is shown in Table 12.10. It was also found that improved operation resulted from biological pretreatment of the wastewater as shown in Fig. 12.12.[19] Membrane treatment of an organic chemicals wastewater is shown in Table 12.11.

An RO design is shown in Example 12.2.

TABLE 12.10

Summary of performance of organic chemical wastewater treatment by reverse osmosis

Parameter	Leachate	Permeate
pH	8.2	5.6
COD, mg/l	1948	7
BOD, mg/l	105	2
TKN, mg/l	612	9
Cl, mg/l	2504	33
Zinc, μg/l	630	440
Copper, μg/l	170	45
Lead, μg/l	100	15
Chromium, μg/l	170	60
Nickel, μg/l	150	40
Cadmium, μg/l	1.3	0.7
Arsenic, μg/l	12	4
Mercury, μg/l	0.5	—

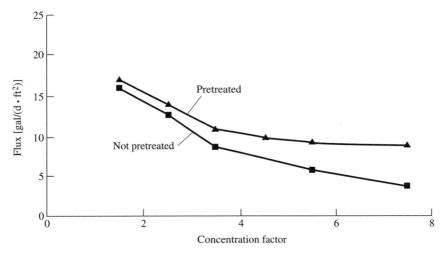

FIGURE 12.12

Effect of biological pretreatment on reverse osmosis performance.[19]

TABLE 12.11

Membrane filtration of a filtered industrial wastewater

Parameter	Unit	Influent	Effluent Nanofiltration	Reverse osmosis
Compound A	μg/l	140	ND(1)[†]	ND(1)
Compound B	μg/l	1.3	ND(1)	ND(1)
Compound C	μg/l	200	11	13
COD	mg/l	12,630	9000	4750
TDS	mg/l	32,950	13,550	1250
C. dubia, LC_{50}	%	1.3	8.8	8.8

[†] Nonobservable peak at the detection level indicated in parentheses.

EXAMPLE 12.2. The following information on a wastewater and reverse osmosis system is to be used for renovation. Design an RO system for this wastewater.

Plant design capacity, Q	10 million gal/d (37,850 m³/d)
Recovery factor, R	75%
Salt rejection factor, S	95%
Design pressure	600 lb/in² gauge (40.8 relative atm)
Feedwater temperature	80°F (27°C)
Total dissolved solids	600 mg/l

Solution.

System sizing. The wastewater (system feedwater) flow Q_w is

$$Q_w = \frac{Q}{R} = \frac{10 \text{ million gal/d}}{0.75} = 13.3 \text{ million gal/d} \quad (50,341 \text{ m}^3/\text{d})$$

The brine flow to be disposed of is then $13.3 - 10$, or 3.3 million gal/d (12,491 m³/d).

Assuming a flux of 20 gal/(d · ft²) [0.814 ml/(d · m²)], the membrane area needed, A, is

$$A = \frac{10 \times 10^6 \text{ gal/d}}{20 \text{ gal/(d·ft}^2)} = 500,000 \text{ ft}^2 \quad (46,450 \text{ m}^2)$$

Assuming a packing density of 250 ft²/ft³ (820 m²/m³), the total module volume, V, is

$$V = \frac{500,000 \text{ ft}^2}{250 \text{ ft}^2/\text{ft}^3} = 2000 \text{ ft}^3 \quad (56.6 \text{ m}^3)$$

Assuming 1 ft³ (0.0283 m³) of volume per module, 2000 modules would be required, and assuming 10 modules per pressure vessel, 200 pressure vessels would be required. The approximate total dissolved solids content of the product water, TDS_p, is

$$TDS_p = \frac{1}{R}(\text{wastewater TDS})(1 - S)$$

$$= \frac{1}{0.75}(600 \text{ mg/l})(1 - 0.95)$$

$$= 40 \text{ mg/l}$$

Power consumption. The power consumed by the pressurizing pump is

$$\text{Hydraulic hp} = \frac{(\text{wastewater flow in million gal})(\text{pressure in lb/in}^2 \text{ gauge})}{2.74}$$

$$= \frac{(13.3 \text{ million gal})(600 \text{ lb/in}^2 \text{ gauge})}{2.47}$$

$$= 3230 \text{ hp} \quad (2423 \text{ kW})$$

The brake horsepower which must be delivered by the pump motor is

$$\text{Brake hp} = \frac{\text{hydraulic hp}}{\text{pump efficiency}}$$

$$= \frac{3230 \text{ hp}}{0.75}$$

$$= 4000 \text{ hp} \quad (3000 \text{ kW})$$

If provision is desired to operate the system at higher pressures, i.e., up to 1000 lb/in gauge (68.05 relative atm), the calculated power can be corrected by multiplying by the ratio of the selected pressure to 600 lb/in^2 gauge (40.8 relative atm).

Land area required. The land area in acres required for a reverse osmosis system can be estimated by the formula

$$\text{Plant area} = 0.7 + 0.33Q$$

$$= 0.7 + 0.33(10)$$

$$= 4 \text{ acres} \quad (16{,}300 \text{ m}^2 = 1.6 \text{ ha})$$

Brine disposal. The approximate total dissolved solids content of the rejected brine, TDS_b, is

$$\text{TDS}_b = \frac{S}{(1 - R)}(\text{wastewater TDS})$$

$$= \frac{0.95}{1 - 0.75}(600 \text{ mg/l})$$

$$= 2280 \text{ mg/l}$$

The total dry weight of salt to be disposed of daily, W, is

$$W = 8.34 \,(2280 \text{ mg/l})(3.3 \text{ million gal/d}) = 62{,}600 \text{ lb/d} \,(28{,}400 \text{ kg/d})$$

12.4
PHOSPHORUS REMOVAL

Phosphorus can be removed from wastewater either chemically or biologically.

Chemical Phosphorus Removal

Chemical phosphorus removal involves precipitation with calcium, iron, or aluminum. Phosphorus is precipitated with calcium salts to low residuals, depending on the pH. The precipitate is a hydroxyapatite, $Ca_5OH(PO_4)_3$:

$$5Ca^{2+} + 7OH^- + 3H_2PO_4{}^- \rightarrow Ca_5OH(PO_4)_3 + 6H_2O$$

Between pH 9.0 and pH 10.5, precipitation of calcium carbonate competes with calcium phosphate. Unlike aluminum and iron, calcium phosphate solids nucleate and grow very slowly, especially at neutral pH. The addition of seed enhances the reaction, indicating the advantage of solids recycle. Residual soluble phosphorus with respect to pH is shown in Fig. 12.13. The calcium phosphate precipitate is finely divided so that the presence of $Mg(OH)_2$ floc aids in the removal of the calcium phosphate precipitate.

The lime requirements will be dictated by the hardness and the alkalinity, as shown in Fig. 12.14. At high pH levels, low soluble phosphorus levels are achieved but residual particulates may require postfiltration. Recarbonation for pH adjustment after precipitation may redissolve particulate phosphorus if incomplete removal exists prior to pH adjustment.

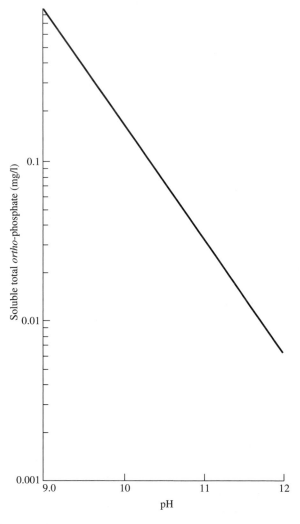

FIGURE 12.13
Ortho-phosphate versus pH for lime precipitation.[20]

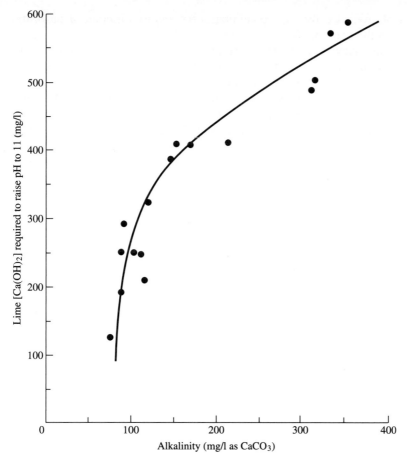

FIGURE 12.14
Lime required to raise the pH to 11 as a function of the wastewater alkalinity.[21]

Iron or aluminum can be employed as a direct precipitation of the metallic phosphate in the case of inorganic wastewater or simultaneous precipitation by addition of the coagulating chemicals at the end of the aeration step in the activated sludge process.

Theoretical phosphorus residuals from precipitation with aluminum are a function of pH and the Al/P ratio. The precipitate is usually a mixture of $Al(OH)_3$ and $AlPO_4$, although the $AlPO_4$ precipitation is favored over $Al(OH)_3$. The precipitate tends to be amorphous rather than crystalline.

In order to approach the predicted solubility, a dosage of 1.5 to 3.0 moles of aluminum per mole of phosphorus as P is required over a pH range of 6.0 to 6.5. If the water is alkaline, the pH should be lowered prior to alum addition to minimize $Al(OH)_3$ precipitation. Some turbidity may result from the alum addition.

When the alum is added to the activated sludge process, addition should be immediately prior to the final clarifier, in the case of completely mixed systems, or at the end of plug flow aeration basins. This is to avoid phosphorus precipitation in the biological process before microbial utilization and to minimize shear of the chemical flocs in the aeration basin.

In some cases, chemical requirements can be minimized by multiple-point addition: i.e., partial precipitation at the head end of the aeration basin with final precipitation at the end of the basin after microbial assimilation.

Iron can be added as $FeSO_4$ or as $FeCl_3$. The dosage is dependent on the dissolved oxygen level, the pH, biological catalysis, and the presence of sulfur and carbonates. Iron has been employed for phosphorus precipitation in biological treatment processes, but has the disadvantage of leaving some iron in the treated effluent. The iron dosage will range from 1.5 to 3.0 moles of iron (Fe^{3+}) per mole of phosphorus (as P). The optimum pH is 5.0, which is too low for conventional biological treatment. Precipitation at neutral pH values may produce a colloidal precipitate requiring a polymer to obtain a minimum total phosphorus residual.

The molar ratio of metallic ion to phosphorus increases as the effluent phosphorus concentration decreases, as shown in Fig. 12.15. These ratios are approximate because side reactions form by-products such as hydroxide and carbonates.

For simultaneous precipitation, the optimum pH is in the range of 7 to 8. For direct precipitation, the optimum pH is in the order of 6. For processes with a long precipitation time, such as simultaneous precipitation, the calcium concentration in the water has a beneficial effect with both Al and Fe. Alkalinity has a negative

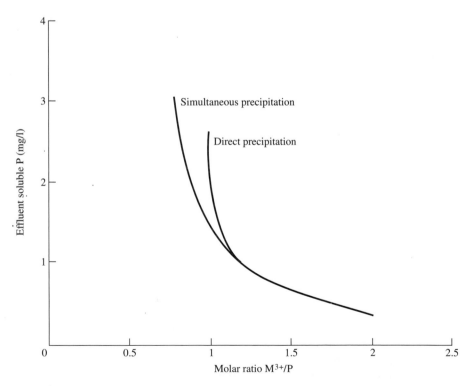

FIGURE 12.15
Phosphorus removal with aluminum and iron.

effect on direct precipitation, with short detention times, and the removal of alkalinity is essential for a low-P effluent. This is achieved to a certain extent automatically with the addition of Fe or Al since they behave as acids, although this condition will result in higher chemical consumption and increased sludge production. For low effluent phosphorus concentrations, effluent filtration may be required because of the high phosphorus content of the effluent suspended solids.

Biological Phosphorus Removal

Phosphorus, like N, is assimilated by bacteria and is removed from the water with the excess sludge. The degree of removal is related to the rate of sludge production, as discussed on p. 209. It has been found, however, that certain bacteria can remove more than the usual quantity of phosphorus as a result of the unusually high P content in the sludge mass.

The basic principle of biological P removal is shown in Fig. 12.16. The bacteria are exposed alternatively to anaerobic and aerobic conditions. Certain bacteria, notably *Acinetobacter,* possess the ability to absorb low-molecular-weight organics (e.g., fatty acids) under anaerobic conditions. The energy required for this is made available by the release of phosphorus bound as polyphosphates in volutin granules in the protoplasm of the bacteria. Under subsequent exposure to aerobic conditions, the organic matter is oxidized and energy is made available for growth and for the reaccumulation of phosphates into polyphosphates in the bacteria. The net effect is an excess content of phosphorus in the bacteria possessing this ability.

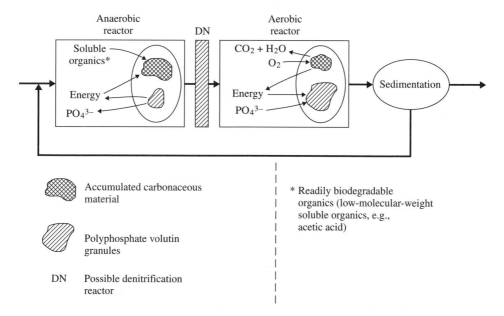

FIGURE 12.16
Biological release and uptake of phosphate, alternating between anaerobic and aerobic conditions.[22]

These bacteria have a competitive edge over other bacteria, since they can hoard the readily available organic matter for their own consumption. Under proper process conditions, they can flourish and dominate the population, resulting in an increased P content in the waste sludge.[23]

In addition to excess biological P uptake, chemical precipitation may also occur with calcium, magnesium, iron, and aluminum present in the wastewater. Recent data indicate that the physical-chemical removal increases with increasing biological P removal. One reason for the biologically mediated P precipitation may be the high phosphate concentration in the anaerobic reactor created by the biological P release.[24]

When nitrification and denitrification is not occurring, the influent soluble BOD_5/soluble P ratio should be greater than 10:1 in order to provide substrate for the bio-P organisms.[25] With nitrogen removal the influent COD/TKN must be greater than 12. The P content of the excess sludge is in the order of 5 to 7 percent. When the plant effluent has 10 to 30 mg/l suspended solids, the total content in the effluent will be 0.5 to 2.0 mg/l P. Under optimum conditions, the soluble P in the effluent can be reduced to 0.1 mg/l. A combination of bio-P removal with simultaneous precipitation with small dosages of Al or Fe will achieve low effluent P concentrations.[26]

The principle of a biological treatment plant with biological removal of phosphorus is the exposure of the organisms to alternating anaerobic and aerobic conditions. This can be implemented with or without nitrogen removal. The alternating exposure to anaerobic and to aerobic conditions can be arranged by recirculation of the biomass through anaerobic and aerobic stages, and an anoxic stage if nitrogen removal is also desired. Generalized plant flow sheets are shown in Fig. 12.17. Performance data for a tobacco wastewater treatment plant following the flow sheet shown in Fig. 12.17b is summarized in Table 12.12.

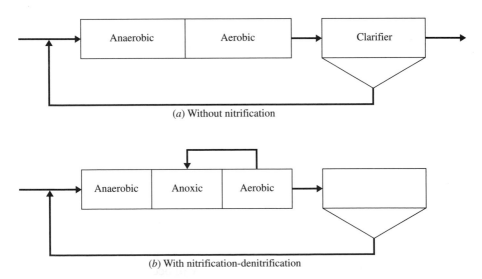

(a) Without nitrification

(b) With nitrification-denitrification

FIGURE 12.17
Biological phosphorus removal with and without nitrification-denitrification.

TABLE 12.12
Bio-P removal for a tobacco wastewater using the flow sheet in Fig. 12.17

	Phosphorus, mg/l	NH_3^-N, mg/l	COD, mg/l
Primary effluent	9.2	20	683
Anaerobic effluent	19.6	—	—
Aerobic effluent	1.4	3	99

12.5
GRANULAR MEDIA FILTRATION

Granular media filtration is employed for the removal of suspended solids as a pretreatment for low suspended solids wastewaters, following coagulation in physical-chemical treatment or as a tertiary treatment following a biological wastewater treatment process.

Suspended solids are removed on the surface of a filter by straining, and through the depth of a filter by both straining and adsorption. Adsorption is related to the zeta potential on the suspended solids and the filter media. Particles normally encountered in a wastewater vary in size and particle charge and some will pass the filter continuously. The efficiency of the filtration process is therefore a function of:

1. The concentration and characteristics of the solids in suspension
2. The characteristics of the filter medium and other filtration aids
3. The method of filter operation

Granular-medium filters may be either gravity or pressure. Gravity filters may be operated at a constant rate with influent flow control and flow splitting, or at a declining rate with four or more units fed through a common header. To achieve constant flow an artificial head loss (flow regulator) is used. As suspended solids are removed and the head loss increases, the artificial head loss is reduced so the total head loss remains constant. In a declining-rate filter design, the decrease in flow rate through one filter as the head loss increases raises the head and rate through the other filters. A maximum filtration rate of 6 gal/(min · ft²) [0.24 m³/(min · m²)] is used when one unit is out of service. The filter run terminates when the total head loss reaches the available driving force or when excess suspended solids or turbidity appear in the effluent.

Medium size is an important consideration in filter design. The sand size is chosen on the basis that it provides slightly better removal than is required. In dual-media filters, the coal size is selected to provide 75 to 90 percent suspended solids removal across 1.5 to 2.0 ft (0.46 to 0.6 m) of media. For example, if 90 percent suspended solids removal is desired across a filter bed, 68 to 80 percent should be removed through the coal layer and the remaining 10 to 25 percent through the sand layer. If the feed suspended solids particle size is larger than 5 percent of the granular medium particles, mechanical straining will occur.

A 25-μm particle will be mechanically strained by a 0.5-mm filter media. If the feed solids particles have a density of 2 to 3 times that of the suspending medium, then particles as small as 0.5 percent of the filter media particle size can be effectively removed by in-depth granular medium filtration.

Table 12.13 presents available media options. Two general types of monomedia are available, fine and coarse. The fine media are usually found in propriety-type filters such as the automatic backwash filter or the pulse-bed-type filter and rely on a straining mechanism for removal. Frequent backwashes (or pulsing) are required, especially during plant upset and/or high influent turbidity conditions. The coarse monomedia are usually much deeper than larger media requiring scour for cost-effective backwash. Operation of coarse medium filters is characterized by longer filter runs and the ability to respond to plant upset conditions. Dual media and multimedia have traditionally been used in potable water applications and their use has carried over into the tertiary filtration application wastewater treatment.

Filtration rate will affect the buildup of head loss and the effluent quality attainable. The optimum filtration rate is defined as the filtration rate that results in the maximum volume of filtrate per unit filter area while achieving an acceptable effluent quality.

Too high a filtration rate will permit solids to penetrate the coarse media and accumulate on the fine media. Too low a filtration rate is insufficient to achieve good solids penetration of the coarse media, resulting in head loss buildup at the top of the coarse media. Filtration rate will also influence effluent quality, depending on the nature of the particles to be removed.

The head loss through the filter is related to the solids loading, as shown in Fig. 12.18:

$$H = aS^n \tag{12.7}$$

where H = head loss, ft or m
           S = solids captured, lb/ft^2 or kg/m^2
      a, n = constants

For a given head loss, the filtration cycle depends on the influent suspended solids and the hydraulic flow rate. The type of coagulant used may also influence the head loss, as shown in Fig. 12.19.

TABLE 12.13
Media options

Type	Material	Size, mm	Depth, in
1. Monomedia			
(a) Fine	Sand	0.35–0.60	10–20
(b) Coarse	Anthracite coal	1.3–1.7	36–60
2. Dual	Sand,	0.45–0.6	10–12
	anthracite coal	1.0–1.1	20–30
3. Multimedia	Garnet,[†]	0.25–0.4	2–4
	sand,	0. 45–0.55	8–12
	anthracite coal	1.0–1.1	18–24

[†] Other types of material such as metal oxides are also used.
Note: in = 2.54 cm

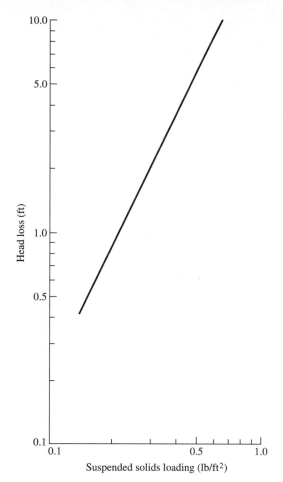

FIGURE 12.18
Head loss versus solids retention in a filter.

Improved suspended solids removals can be achieved by the addition of coagulants to the wastewater prior to filtration. The use of alum also results in the precipitation and removal of phosphorus through the filter. Flocculation is not needed since the filter serves as a flocculator. Effective mixing is required to disperse the chemicals and initiate the reaction. Since the suspended solids are removed by filtration rather than by sedimentation, 25 to 50 percent less chemicals are required in many cases. For most applications, a maximum of 100 mg/l suspended solids removal is used in order to avoid excessive backwash volumes.

Backwash systems used for the filtration of secondary effluent require an auxiliary scour system to effectively dislodge particulates. The two most frequently used backwash systems are: (1) water with auxiliary surface agitation and (2) water with air scour. Surface systems may be a fixed grid or rotary wash. The filter backwash rate is a function of medium size (both effective size and uniformity coefficient), medium type (specific gravity), and water temperature. Air scour rates are in the range of 1 to

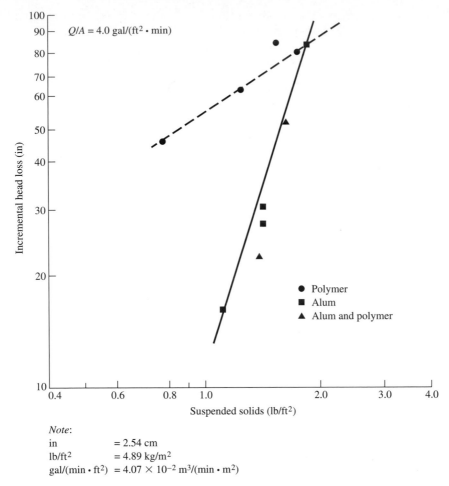

FIGURE 12.19
Head loss as related to solids deposit in a filter.

5 standard ft³/(min · ft²) [0.305 to 1.53 std m³/(min · m²)] with auxiliary surface wash rates in the range of 2 to 5 gal/(min · ft²) [0.081 to 0.204 m³/(min · m²)].

Granular medium filters can be fully automated by the use of pressure sensors that monitor the head loss across the filter. When the terminal head loss is achieved, the filter will automatically backwash. Turbidimeters serve as a secondary control such that the backwash sequence will be initiated when the turbidity reaches the allowable level in the effluent.

While considerable data are available for the design of filters treating domestic secondary effluents, industrial wastewaters require pilot plant studies to define the type of medium, filter flow rate, coagulant requirements, head loss relationships, and backwash requirements.

There are several types of filters available today. Three of the more common are the dual-media filter consisting of anthrafilt (coal) and sand, the Hydroclear filter, and the continuous backwash filter.[2]

A typical dual-media filter is shown in Fig. 12.20. The Hydroclear filter employs a single sand medium with an air mix for solids suspension and regeneration of the filter surface. Filter operation enables periodic regeneration of the medium surface without backwashing. Typical operating parameters are

Filtration rate	2–5 gal/(min · ft²) [0.081–0.204 m³/(min · m²)]
Media size	0.35–0.45 mm sand
Bed depth	10–12 in (25.4–30.4 cm)
Backwash rate	12 gal/(min · ft²) [0.5 m³/(min · m²)]
Air mix	0.25 standard ft³/(min · ft²) [0.076 std m³/(min m²)]
Terminal head loss	3.5 ft (1.07 m)
Backwash filtrate ratio	0.10

The Dynasand filter (DSF) is a continuous backwash, self-cleaning upflow deep-bed granular medium filter. The filter medium is cleaned continuously by recycling the sand internally through an airlift pipe and sand washer, as shown in Fig. 12.21. The regenerated sand is redistributed on top of the bed, allowing for a continuous uninterrupted flow of filtered water and reject water. Filtration performance is shown in Table 12.14. Filter design is shown in Example 12.3.

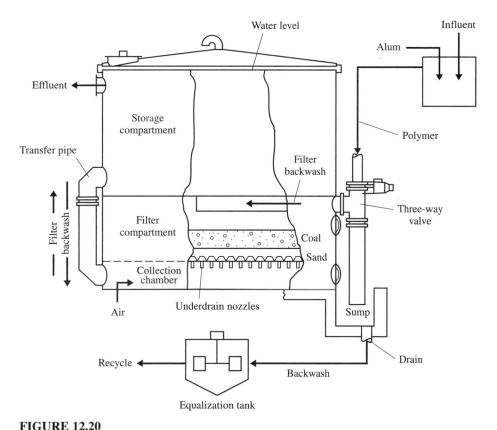

FIGURE 12.20
Typical automatic dual media filter.

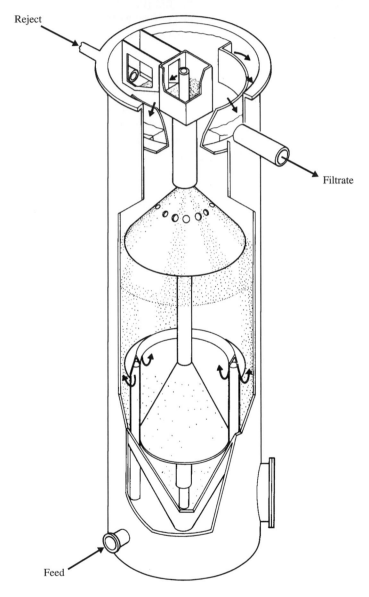

FIGURE 12.21
Dynasand filter (DSF) *(courtesy of Parkson Corporation).*

TABLE 12.14
Filtration performance

Filter type	Wastewater	Filter depth, ft	Hydraulic loading, gal/(min · ft²)	Percent removal SS	Percent removal BOD	Effluent, mg/l SS	Effluent, mg/l BOD
Gravity downflow	TF effluent	2–3	3	67	58	—	2.5
Pressure upflow	AS effluent	5	2.2	50	62	7.0	6.4
Dual media	AS effluent	2.5	5.0	74	88	4.6	2.5
Gravity downflow	AS effluent	1.0	5.3	62	78	5	4
Dynasand	Metal finishing	3.3	4–6	90	—	2–5	—
	AS effluent	3.3	3–10	75–90	—	5–10	—
	Oily waste-water	3.3	2–6	80–90[†]	—	5–10[†]	—
Hydroclear	Poultry	1	2–5	88	—	19	—
	Oil refinery	1	2–5	68	—	11	—
	Unbleached kraft	1	2–5	74	—	17	—

[†] Free oil.
Note:
ft = 0.305 m
gal/(min · ft²) = 4.07 × 10⁻² m³/(min · m²)

EXAMPLE 12.3. A 1 million gal/d secondary effluent is to be polished by a granular medium filter. If an 8 ft head loss is applied, calculate the filter run time to achieve 80 percent removal from an initial suspended solids of 70 mg/l. The hydraulic loading to the filter is 3.0 gal/(min · ft²).

Solution.

The required surface area is

$$A = \frac{1 \times 10^6}{3 \cdot 1440}$$

$$= 230 \text{ ft}^2$$

For a head loss of 8 ft, the suspended solids loading is 0.57 lb/ft² (see Fig. 12.18). The expected filter run is then

$$t = \frac{0.57 \cdot 230}{1 \cdot 0.8 \cdot 70 \cdot 8.34}$$

$$= 0.28 \text{ d or } 6.7 \text{ h}$$

12.6
MICROSCREEN

A microscreen is a rotary drum revolving on a horizontal axis covered with stainless steel fabric (Fig. 12.22). The water enters the open end of the drum and is filtered

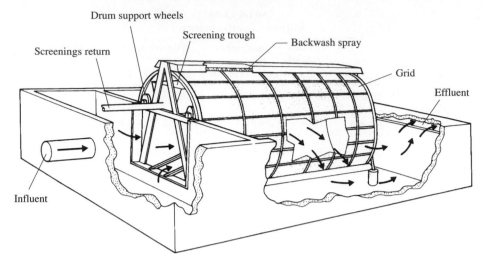

FIGURE 12.22
Microscreen *(courtesy of Envirex, Inc.).*

through the fabric, with solids being retained on the inside surface of the fabric. As the drum rotates, the solids are transported and continuously removed at the top of the drum by pumping effluent under pressure through a series of spray nozzles that extend the length of the drum. The head loss is less than 12 to 18 inches (30 to 46 cm) of water. The backwash water is 46 percent of the total throughput water. Peripheral drum speeds vary up to 100 ft/min (30.5 m/min) with hydraulic loadings of 2.5 to 10 gal/(d · ft^2) [0.1 to 0.4 m^3/(min · m^2)]. Periodic cleaning of the drum is required for slime control.

For filtration of secondary effluent a maximum solids loading of 0.88 lb/(ft^2 · d) [4.3 kg/(m^2 · d)] at a hydraulic loading of 6.6 gal/(min · ft^2) [0.27 m^3/(min · m^2)] has been reported. Lynam et al.[27] reported effluent suspended solids and BOD of 6 to 8 mg/l and 3.5 to 5 mg/l, respectively, with a 20-20 activated sludge effluent of a 23-m microscreen at 3.5 gal/(min · ft^2) [0.14 m^5/(min · m^2)]. For design purposes, the removal efficiencies treating a secondary effluent are given in Table 12.15.

The efficiency of the unit is suspended solids sensitive, as indicated by a decrease in the throughput rate from 60 to 13 gal/min (0.227 to 0.049 m^3/min) with an increase in influent suspended solids from 25 to 200 mg/l.[28]

TABLE 12.15
Removal efficiencies

Aperture screen, μm	Flow rate, gal/(min · ft^2) (submerged)	Percent removal	
		SS	BOD
35	10.0	50–60	40–50
23	6.7	70–80	60–70

Note: gal/(min · ft^2) = 4.07 × 10^{-2} m^3/(min · m^2)

REFERENCES

1. Eckenfelder, W. W., J. P. Lawler, and J. T. Walsch: "Study of Fruit and Vegetable Processing Waste Disposal Methods in the Eastern Region," U.S. Department of Agriculture, Final Report, 1958.

2. Crites, R. W.: *Proceedings of the Industrial Wastes Symposia,* Water Pollution Control Federation, 1982.

3. "Diagnosis and Improvement of Saline and Alkali Solids," U.S. Department of Agriculture, Handbook 60, 1954.

4. Jewell, W. J.: *Limitations of Land Treatment of Wastes in the Vegetable Processing Industries,* Cornell University, Ithaca, N.Y., 1978.

5. Adamczyka, A. F.: *Land as a Waste Management Alternative,* R. C. Loehr, ed., Ann Arbor Science, Ann Arbor, Mich., 1977.

6. Haith, D. A., and D. C. Chapman: "Land Application as a Best Practical Treatment Alternative," in *Land as a Waste Management Alternative,* R. C. Loehr, ed., Ann Arbor Science, Ann Arbor, Mich., 1977.

7. Crites, R.W. et al.: *Proc. 5th National Symposium on Food Processing Wastes,* 1974.

8. Gellman, I., and R. O. Blosser: *Proc. 14th Ind. Waste Conf.,* Purdue University, 1959.

9. Wisneiwski, T. F., A. J. Wiley, and B. F. Lueck: *TAPPI,* vol. 89, pt. 2, p. 65, 1956.

10. Billings, R. M.: *Proc 13th Ind. Waste Conf.,* Purdue University, 1958.

11. Wong, G., and R. Pfarrer: *Proc. 50th Purdue Industrial Waste Conf.,* Ann Arbor Press, Ann Arbor, Mich., 1995.

12. Bausmith, D. S., and R. D. Neufeld: *Proc. WEF.* vol. 3, Dallas, 1996.

13. Warner, D. L.: "Deep Well Injection of Liquid Waste," Environmental Health Series, U.S. Department of Health, Education, and Welfare, Cincinnati, April 1965.

14. Donaldson, E. C.: "Subsurface Disposal of Industrial Wastes in the United States," Bureau of Mines Information Circular 8212, U.S. Department of the Interior, Washington, D.C., 1964.

15. Cartwright, W. P.: *Chemical Engineering,* McGraw-Hill, September 1994.

16. Agardy, F. J.: *Membrane Processes, Process Design in Water Quality Engineering,* E. L. Tackston and W. W. Eckenfelder, eds., Jenkins Publishing Co., Austin, Texas, 1972.

17. Okey, R. W.: *Water Quality Improvement by Physical and Chemical Processes,* E. F. Gloyna and W. W. Eckenfelder, eds., University of Texas Press, Austin, Texas, 1970.

18. Logemann, F. P.: *Proc. 11th National Conf. Superfund '90,* Hazardous Materials Control Research Institute, 1990.

19. Kettern, J. T.: *Wat. Sci. Tech.* vol. 26, no. 1–2, p. 137, 1992.

20. Zoltek, J.: *J. WPCF,* vol. 48, p. 179, 1976.

21. Tshobanoglous, G.: *Proc. 12th Sanitary Engineering Conf.,* University of Illinois, 1970.

22. Arvin, E.: "Biological Removal of Phosphorus from Wastewater," *CRC Critical Rev. J.,* vol. 15, pp. 25–64, 1985.

23. Marais, G. V. R. et al.: *Water Sci. Technol.,* vol. 15, nos. 3–4, 1983.

24. Arvin, E.: *Water Sci. Technol.,* vol. 15, nos. 3–4, 1983.

25. Hong. S. N. et al.: *EPA Workshop on Biological Phosphorus Removal,* Annapolis, Md., June 1982.

26. Ekema, G. A. et al.: *Water Sci. Technol.,* vol. 15, nos. 3–4, 1983.

27. Lynam, B. et al.: *J. WPCF,* vol. 41, p. 247, 1969.

28. Convery, J. J.: *FWPCA Symposium on Nutrient Removal and Advanced Waste Treatment,* Tampa, Fla., 1968.

REFERENCES

BIBLIOGRAPHY

Adams, C. E., Jr., and W. W. Eckenfelder, Jr.: *Process Design Techniques for Industrial Waste Treatment,* Enviro Process, Nashville, Tenn., 1974.

Adams, C. E., D. L. Ford, and W. W. Eckenfelder, Jr.: *Development of Design and Operational Criteria for Wastewater Treatment,* Butterworth, Boston, 1981.

Azad, H. S. (ed.): *Industrial Wastewater Management Handbook,* McGraw-Hill, New York, 1976.

Benefield, L. D., C. W. Randall: *Biological Process Design for Wastewater Treatment,* Prentice-Hall, Englewood Cliffs, N.J., 1980.

Besselievre, E. B., and M. Schwartz: *Treatment of Industrial Waste,* McGraw-Hill, New York, 1976.

Busch, A. W.: *Aerobic Biological Treatment of Wastewater and Practices,* Oligodynamics Press, Houston, 1971.

Callely, A. G., C. F. Forster, and D. A. Stafford (eds.): *Treatment of Industrial Effluents,* Halstead Press, Somerset, N.J., 1976.

Eckenfelder, W. W.: *Water Pollution Control in the Pulp and Paper Industry,* Brown and Caldwell, Nashville, Tenn., 1999.

Eckenfelder, W. W., and P. Grau (eds.): *Activated Sludge Process Design and Control,* 2nd Ed., Technomic Publishing Co., Lancaster, Pa., 1998.

Eckenfelder, W. W., and J. Musterman: *Activated Sludge Treatment of Industrial Wastewaters,* Technomic Publishing Co., Lancaster, Pa., 1997.

Eckenfelder, W. W.: *Developing Industrial Water Pollution Control Programs,* Technomic Publishing Co., Lancaster, Pa., 1997.

Eckenfelder, W. W., *Principles of Water Quality Management,* CBI Publishing, Boston, 1980.

EPA Treatability Manual EPA-600/2-82-0012, 1982, vol. 1, *Treatability Data*; vol. 2, *Industrial Descriptions*; vol. 3, *Technologies for Control/Removal of Pollutants*; vol. 4, *Cost Estimating.*

Evans, F. L., III (ed.): *Ozone in Water and Wastewater Treatment,* Ann Arbor Science, Ann Arbor, Mich., 1972.

Ford, D. L. (ed.): *Toxicity Reduction—Evaluation and Control,* 2nd ed., Technomic Publishing Co., Lancaster, Pa., 1998.

Goransson, B. (ed.): *Industrial Wastewater and Waste,* vol. 2, Pergamon Press, Oxford, England, 1977.

Humenick, M. J.: *Water and Wastewater Treatment: Calculations for Chemical and Physical Processes,* Marcel Dekker, New York, 1977.

Industrial Wastes Tech. Conf. Proc., Water Env. Fed., Nashville, Tenn., 1998.

Industrial Wastes Tech. Conf. Proc., Water Env. Fed., New Orleans, La., 1997.

Jorgenson, S. E., and M. J. Gromiez: *Mathematical Models in Biological Wastewater Treatment,* Elsevier, Amsterdam, 1985.

Lankford, P. W., and W. W. Eckenfelder (eds.): *Toxicity Reduction in Industrial Effluents,* Van Nostrand Reinhold, New York, 1990.

Loehr, R. C.: *Pollution Control for Agriculture,* Academic Press, New York, 1977.

Middlebrooks, E. J., et al.: *Wastewater Stabilization Lagoon Design, Performance and Upgrading,* Macmillan, New York, 1982.

Nemerow, N. L.: *Industrial Water Pollution, Origins, Characteristics, and Treatment,* Addison-Wesley, Reading, Mass., 1978.

Orhon, D., and A. Nazik: *A Modeling of Activated Sludge Systems,* Technomic Publishing Co., Lancaster, Pa., 1994.

Patterson, J. W.: *Industrial Wastewater Treatment Technology,* 2nd ed., Butterworth, Boston, Mass., 1985.

Pitter, P., and J. Chudoba, *Biodegradability of Organic Substances in the Aquatic Environment,* CRC Press, Boca Raton, FL, 1990.

Proceedings of Annual Industrial Waste Conferences, Purdue University, West Lafayette, Ind.

Proceedings, A Consolidated Approach to Activated Sludge Process Design, Progress in Water Technology, vol. 7, no. 1, Pergamon Press, Oxford, England, 1975.

Proceedings of the Third National Conference on Complete Water Reuse, Cincinnati, Ohio, June 1976, American Institute of Chemical Engineers, New York, N.Y., 1976.

Proceedings of the First International Conference on Fixed Film Biological Processes, Kings Island, Ohio, 1982.

Proceedings of the Second International Conference on Fixed Biological Processes, Arlington, Va., 1984.

Rice, R., and J. A. Cotruvo: *Ozone/Chlorine Dioxide Oxidation Products of Organic Materials,* Ozone Press International, Cleveland, 1978.

Sanks, R. L., and T. Asano: *Land Treatment and Disposal of Municipal and Industrial Wastewater,* Ann Arbor Science, Ann Arbor, Mich., 1976.

Sarner, E.: *Plastic Packed Trickling Filters,* Ann Arbor Science, Ann Arbor, Mich., 1980.

Sawyer, C. N., and P. L. McCarty:*Chemistry for Environmental Engineering,* 3rd ed., McGraw-Hill, New York, 1978.

Schmidtke, N. W., and D. W. Smith: *Scaleup of Water and Wastewater Treatment Processes,* Butterworth, Boston, 1983.

Speece, R. E.: *Anaerobic Biotechnology for Industrial Wastewaters,* Archer Press, Nashville, Tenn., 1996.

Speece, R. E., and J. F. Maline, Jr. (eds.): *Application of Commercial Oxygen to Water and Wastewater Systems,* University of Texas Press, Austin, Texas,1973.

Stumm, W., and J. J. Morgan: *Aquatic Chemistry,* Wiley, New York, 1970.

Schroeder, E. D.: *Water and Wastewater Engineering Treatment,* McGraw-Hill, New York, 1977.

Vesilind, A. P.: *Treatment and Disposal of Wastewater Sludges,* Ann Arbor Science, Ann Arbor, Mich., 1974.

Wanner, J.: *Activated Sludge Bulking and Foaming Control,* Technomic Publishing Co., Lancaster, Pa., 1994.

Water—1974, I. Industrial Wastewater Treatment, AIChE Symposium Series, vol. 70, no. 144, American Institute of Chemical Engineers, New York, 1974.

Water—1976, II. Biological Wastewater Treatment, AIChE Symposium Series, vol. 73, no. 167, American Institute of Chemical Engineers, New York, 1977.

Verschurenen, K.: *Handbook on Environmental Data on Organic Chemicals,* Van Nostrand Reinhold, New York, 1977.

INDEX